AF746729

10 0349663 2

UNIVERSITY OF BRISTOL
UNIVERSITY LIBRARY
TYNDALL AVENUE
BRISTOL, BS8 1TJ

Date due for return (unless recalled earlier)

NOT FOR LOAN BEFORE 14 OCT 1987

RETURNED 21. JUN. 1991
18 JAN 1991

- 3 DEC 1987

28 JAN 1992

WITHDRAWN University of Bristol

10 FEB 19

07 FEB 1992

07 FEB 1992

RETURNED
24 APR 1989

RETURNED
12 MAY 1989

14. MAY
- 1 MAY 1992

08.APR.1998
10. JUN 98
30.SEP.2000
02.JAN.2001
17.JAN.2001
-1. JUN 2001
19. NOV 2001
21.FEB.2003
04.OCT.2004
11.FEB.2005
26.MAY.2005
12.OCT.2005
Used 12/13

The human uses of
flint and chert

The human uses of flint and chert

Proceedings of the fourth international flint symposium held at Brighton Polytechnic 10–15 April 1983

EDITED BY

G. DE G. SIEVEKING
British Museum, London

M. H. NEWCOMER
Institute of Archaeology, University of London

The right of the University of Cambridge to print and sell all manner of books was granted by Henry VIII in 1534. The University has printed and published continuously since 1584.

CAMBRIDGE UNIVERSITY PRESS

CAMBRIDGE

LONDON NEW YORK NEW ROCHELLE

MELBOURNE SYDNEY

Published by the Press Syndicate of the University of Cambridge
The Pitt Building, Trumpington Street, Cambridge CB2 1RP
32 East 57th Street, New York, NY 10022, USA
10 Stamford Road, Oakleigh, Melbourne 3166, Australia

© Cambridge University Press 1987

First published 1987

Printed in Great Britain at the University Press, Cambridge

British Library cataloguing in publication data
International Flint Symposium *(4th: 1983: Brighton Polytechnic)*
The human uses of flint and chert: proceedings of the fourth International Flint Symposium held at Brighton Polytechnic, 10–15 April 1983.
1. Flint 2. Man, Prehistoric——Tools
I. Title II. Sieveking, G. de G. III. Newcomer, M.H.
553.6′5 GV799.F/

Library of Congress cataloging in publication data
International Flint Symposium (4th: 1983: Brighton Polytechnic)
The human uses of flint and chert.

Includes index.
1. Flint—Europe—Congresses. 2. Chert—Europe—Congresses. 3. Micropaleontology—Europe—Congresses. 4. Stone implements—Europe—Congresses.
I. Sieveking, G. de G. (Gale de G.) II. Newcomer, M. H. III. Title.
QE391.Q2I56 1983a 930.1 85–17126

ISBN 0 521 26253 4

WV

WITHDRAWN
University of Bristol
UNIVERSITY OF BRISTOL LIBRARY

Contents

Contributors

K. Akoshima Department of Archaeology, Tohoku University, Sendai, Japan.

A. J. Ammerman Department of Anthropology, State University Binghampton, NY, USA.

L. H. Barfield Department of Archaeology, University of Birmingham, United Kingdom.

R. N. E. Barton, Quaternary Research Centre, University of Oxford, United Kingdom

C. A. Bergman, Institute of Archaeology, University of London, United Kingdom.

R. Bradley, Department of Archaeology, University of Edinburgh, United Kingdom.

J. Burton, Department of Anthropology, University of Papua New Guinea, Port Moresby, PNG.

D. Cahen, Institut Royal des Sciences Naturelles de Beligique, Brussels, Belgium.

S. N. Collcutt, Quaternary Research Centre, University of Oxford, United Kingdom.

C. H. Clayton, Exploration and production Division, B.P. Research Centre, Sunbury-on-Thames, Middlesex, United Kingdom.

J. Cook, British Museum, London, United Kingdom.

J. Dumont, Louis Berger Associates, East Orange, New Jersey, USA.

P. L. Gibbard, Sub-department of Quaternary Research, Botany School, University of Cambridge, United Kingdom.

R. Grace, Institute of Archaeology, University of London, United Kingdom.

I. D. Graham, Institute of Archaeology, University of London, United Kingdom.

D. R. Griffiths, Institute of Archaeology, University of London, United Kingdom.

P. Harding, Wessex Archaeological Trust, Salisbury, Wiltshire, United Kingdom.

D. L. Holmes, Department of Anthropology, Washington State University, Pullman, Washington, USA.

C. Karlin, Department de Préhistoire Musée de l'Homme, Paris, France.

K. W. Kintigh, Arizona State Museum, Tucson, Arizona, USA.

J. Lech, Institute of the History of Material Culture, Warsaw, Poland.

J. Lewin, Department of Geography, University of Wales, Aberystwyth, United Kingdom

R. Larick, Direction Regionale des Antiquites Prehistoriques, Ministere de la Culture et de la Communication, Bordeaux, France.

M. G. Macklin, Department of Geography, University of Newcastle Upon Tyne, United Kingdom.

A. E. Marks, Department of Anthropology, Southern Methodist University, Dallas, Texas, USA.

R. Mercer, Department of Archaeology, University of Edinburgh, United Kingdom.

G. Morris, Department of Archaeology, University of Cambridge, United Kingdom.

E. H. Moss, Institute of Archaeology, University of London, United Kingdom.

M. H. Newcomer, Institute of Archaeology, University of London, United Kingdom.

K. Ohnuma, Institute of Archaeology, University of London, United Kingdom.

G V. Robins, Institute of Archaeology, University of London, United Kingdom.

E. Rudebeck, Swedish State Excavation Service, Malmo, Sweden.

R. Schild, Institute of the History of Material Culture, Warsaw, Poland.

N. J. Seeley, Institute of Archaeology, University of London, United Kingdom.

A. Sherratt, Ashmolean Museum, University of Oxford, United Kingdom.

G. de G. Sieveking, British Museum, London, United Kingdom.

J. F. Simek, Department of Anthropology, University of Binghampton, NY, USA.

G. Smolla, Seminar Fur Vor-und-Fruhfgeschichte, University of Frankfurt, Germany.

P. Vaughan, Department of Anthropology, University of Pennsylvania, USA.

P. Volkman, Department of Anthropology, Southern Methodist University, Dallas, Texas, USA.

G. Weisgerber, German National Mining Museum, Bochum, Germany.

P. C. Woodman, Department of Archaeology, University College, Cork, Eire.

R. N. Zeitlin, Department of Anthropology, Brandeis University, USA.

Preface

One of the problems with any expanding field of scholarship is the information explosion: everyone seems to want to get in on the act. This band-waggon effect is especially noticeable at scientific meetings devoted to multi-disciplinary and applied studies. The purpose of such meetings seems to be to communicate and reinforce already achieved success, and to provide a rallying point for current practitioners, rather than to break new ground. A recent meeting of a multi-disciplinary archaeological group had almost 170 registered communications, nearly all applications of fairly standard techniques. This raises the problem in another form: everyone tends to say the same thing at such meetings. The proceedings that result from them, though often including excellent papers, have tended to be unfocussed, to fall apart at the edges. This is because the meeting itself has had no deliberate research aim: no educational purpose, or no built-in strategy for effecting it. An effective meeting requires clearly defined objectives and careful preliminary planning to ensure achievement. This is as true for a large meeting covering several different disciplines as it is for a limited size single-topic invitation-only discussion group.

It was with this in mind that the Brighton meeting of the Flint Symposium was limited to a small number of topical problems in the archaeological and scientific study of flint and chert: some to be discussed by a single well-placed contribution, others to be the subject of a seminar with parallel studies by current practitioners. Focus was also sharpened by commissioning review papers in the different fields.

The International Flint Symposium started in 1969 as a collaborative venture between archaeologists and geologists[1] and the balance between the earth and physical sciences and prehistoric archaeology has since been maintained; each undertaking independent studies as well as collaborative applications of science in the archaeological field. Rather than placing undue emphasis on the producer/user aspect of the science/archaeology relationship, we have tried to act on the advice of the physicist Martin Aitken that, 'the archaeologist should also learn about the aid that his discipline can give to other perfectly respectable branches of academic learning' (Aitken, 1982).

Such aims affected the choice of both scientific and archeological subjects at the Brighton Symposium. For example, it was intended to hold a seminar on microwear analysis of flint tools. To complement this a scientific seminar was organised on the SEM analysis of sand grain surface textures, which uses techniques that closely resemble those employed by microwear archaeologists. On the other hand, as a number of geological studies discussed the distribution of flint and chert, studies of the human dispersal of flint tools were commissioned from archaeologists working in the same geographic region. More generally, it was decided to limit the archaeological topics to techniques and results which lent themselves to the application of science and to those whose methods were those of the laboratory, where experiments were carried out, and measurements taken to validate a hypothesis.

The Proceedings that result from the Brighton Symposium are thus limited to eight subject areas: four scientific and four archaeological. The scientific subject areas are: (1) Cretaceous stratigraphy and its relation to the occurrence and natural distribution of flint and chert; (2) the Quaternary environments of the chalklands and residual flint deposits; (3) Geochemistry and the origin of flint; (4) SEM analysis of quartz and flint surface textures. Archaeological subject areas are: (1) spatial analysis and the dispersal and human transport of flint; (2) Flint mine technology and the exploitation of minor raw material sources; (3) Flint tool replication and the reconstruction of tool manufacturing sequences; (4) Flint tool microwear and surface texture analysis.

For reasons of size, it has been necessary to publish the Proceedings in two separate volumes: the present volume containing studies that are primarily of archaeological interest and a companion volume, *The Scientific Study of Flint and Chert*, which contains papers attributed to the first four subject areas. This splits the Symposium papers quite arbitrarily. Many participants have contributed, at least in part, to both volumes, as commissioned scientific review papers deal with aspects of particular archaeological significance and others describing scientific techniques occur under science, while their research applications are published here under archaeology. Such is the case with heat treatment (Griffiths *et al.*, this volume) flint microstructure (Bradley and Clayton, this volume) River gravels (Harding *et al.*, this volume) and also with the papers already mentioned dealing with flint dispersal and exchange in Eastern Europe. In these and other cases, the archaeological papers stand on their own feet and reference to the science volume is only necessary for those who wish to follow up some point of detail or science on which they are based.

[1] See preface to companion volume.

The subject areas covered in this volume are those in which research is active. Stimulated by the discovery of reliable scientific methods of identifying lithic sources, trade and exchange in their products has been a principal research area in America and the Near East for more than a decade. In Europe, flint mine research revived from the slumber that overtook it in the 1930s and has made considerable progress since the first Flint Symposium in 1969. (cf. Weisgerber *et al.*, 1980) Flint tool replication and refitting, and microwear research are subjects that have more recently achieved or regained popularity among specialists, and the application of mathematical models of spatial analysis to archaeological debris is another relatively recent development. Published monographs and collections of research papers exist in all subject areas. However, archaeological studies have further advanced since these appeared and we are able to include papers representing many different aspects of present research including a number of highly innovatory contributions. While this is primarily due to the Symposium's tactical role as a centre of inter-disciplinary discussion,[2] the proceedings have also been strengthened by additional papers, exploring avenues of research that seemed particularly promising at the Brighton meeting.

In spatial analysis, flint dispersal, and also in reconstruction of flint tool manufacturing sequences, our American colleagues have produced particularly forward looking studies. Larick's paper on the circulation of Solutrean points breaks new ground in a particularly simple and satisfying way, using classic geological stratigraphy to provide the first comprehensible model for a European Palaeolithic flint dispersal system. His use of the knapping qualities of different types of raw material and of the relationship between the distribution of raw material and that of living sites to build up a model of a small-scale economic system, should interest everyone studying the interaction of Pleistocene hunter-gather communities.

In the same spatial analysis field, both Zeitlin and Ammerman and co-workers give descriptions of newly developed techniques that are likely to be widely used in the future. The archaeological adaptation of the *k*-means analysis technique, earlier versions of which have been described elsewhere, should appeal to all field archaeologists, since it offers for the first time an opportunity to use field-data in a common-sense way to correct and amplify a distribution pattern formed by a mathematical spatial analysis model.

In the flint dispersal field also, Griffiths and Woodman publish a first application of a new technique for identifying chert sources and in addition there is an important summary by Barfield of the North Italian evidence, drawing attention to the different value placed by prehistoric communities upon the varying qualities of different lithic sources, to which attention is also drawn by Larick and by Burton. On the broader regional canvas, we are fortunate in being able to publish a further magisterial study by Sherratt of European Neolithic exchange systems, on this occasion confirmed and amplified by geological and archaeological studies by our East European colleagues (Lech; Schild, this volume; Takács-Biró, companion volume). Finally both Miller and Smolla (*inter-alia*) draw attention to limitations on the long-distance transport of lithic raw materials, in part based on calculations from original data of the load to be carried on the back of a single porter.

[2] Cf. Sieveking 1975.

One of the educational functions of the proceedings is to provide English language translations of work that would otherwise appear only as continental publications, unavailable to the reader. This is true both of the papers by Schild and Lech, though both are substantially unpublished, and also of the two valuable papers by Weisgerber, which are preliminary reports of field surveys that are intended to be published more fully in Germany. Weisgerber's paper on the Egyptian Chert Mines at Wadi El Sheikh is particularly important, since the prehistoric industrial remains at this site are so much better preserved than at any other mine complex so far recorded.

Both this site and the South Swedish flint mines discussed by Rudebeck are also interesting for the evidence they provide of the continuation of large-scale mining for chert well into the copper- and bronze-using period. Rudebeck, in particular, suggests that the same site was in use, with similar or identical mining techniques, by groups of different cultural origins over a period of more than a thousand years. While this would have occasioned no remark in the days of the Campignian hypothesis, it is surprising now that short-term radiocarbon dated flint mines, associated with not more than one cultural group, appear to be the rule in Western Europe (Smolla, this volume; Burleigh 1979).

Weisgerber in Egypt and Mercer in Dorset[3] both draw attention to the existence of different types of surface quarrying that are alternative mining strategies to the shaft and gallery mines, and it is good to learn from these authors that we are beginning to move away from the technological fallacy to the realisation that the approach adopted is controlled not only by geological considerations but also by labour supply.

In the field of experimental archaeology, the collected papers concentrate on the manufacture of flint tools and the reconstruction of manufacturing sequences and the other purposes for which stone artefacts, originally from the same blocks of raw material, may be re-assembled. A good summary of previous work on the re-fitting field is provided by Cahen; and two papers by Marks and Volkman and by Bergman and co-workers immediately provide valuable examples of additional uses to which the technique can be put, and which Cahan does not fully take into account.

Due to the bias in favour of techniques and technology in the present volume, the last-mentioned papers are perhaps the only ones we publish where classic Upper Palaeolithic flint typology is mentioned at all seriously, and between them they brilliantly expose its limitations; Marks and

[3] Surface features and trial pits described by Mercer are known at large scale flint mines in England, e.g. Grimes' Graves (Sieveking 1979), and in Eastern Europe.

Volkman, in particular, use the re-fitting technique as a small-scale site experiment to demonstrate the weakness of our present models of the development of Middle and Upper Palaeolithic flint technology, in a way that commands wholehearted attention.

Two further experimental archaeology papers, with which Harding's name is associated, emphasise the importance of the prehistorians measurement of activity-times. In the case of the manufacture of the stone axe, Harding brings to our attention how long is spent on the cosmetic treatment[4] that produces the final ground or polished surface (say 20–30 hours), as opposed to the short period of time (say 20–30 minutes) required to complete the first part of the shaping procedure that provides a fully formed flaked stone axe blade hardly less efficient than the final product. In this connection, attention may be drawn to the difference between the times Harding has recorded for the two operations and those put forward in some recently published Danish experiments, where times of between 1 hour 48 minutes and 2 hours 15 minutes were recorded for the preparation of six thin butted flint axes *up to the point where they were ready for polishing* (Hanson and Madson 1983, pp. 50–1: experiments III and IV).

Harding's times for producing flaked axes with a lenticular cross-section are supported by a long series of carefully controlled experiments (Newcomer and Sieveking 1980: average time 20 min). Though Danish thin-butted axes may be more difficult to make, can they really take four to six times as long to produce? It seems more likely that the contrast between British and Danish manufacturing times is due partly to the adoption of more deliberate production procedures, and the variation between the times recorded in the Danish experiments to the relative skills of different operators.[5]

An equally salutory time and motion study is provided by Harding and co-workers, who demonstrate, in a carefully controlled experiment, that flint hand axes may be transported considerable distances down river and change in appearance from a fresh to a considerably worn condition (in an environment fully comparable to that of British rivers during the Pleistocene), as a result of a handful of small-scale floods, each perhaps lasting not more than a few hours, taking place during a single annual season. The evidence of these experiments, which is borne out by similar flood records (Williams, companion volume, p. 163), would have greatly surprised those who less than a generation ago suggested, in classic accounts of hand axe deposition in river gravels, that a similar degree of abrasion to that recorded on the hand axes in this experiment, were the result of thousands of years of transport in a torrential Pleistocene river environment.

Two general approaches exist in microwear analysis: the study of surface abrasion and edge damage and the study of supposedly additive surface polishes.[6] In our microwear seminar Akoshima is the only contributor to devote a paper wholely to the recognised results of abrasion. *Microflaking Quantification* describes a developed laboratory technique for characterising and recording microflake scars, whose adoption should make possible comparison of data collected by different research workers. Such techniques are badly needed to describe systematically the large amount of potential data in this field (see Bull, companion volume).

The wear trace papers, in the main, are devoted to the discussion of data provided by microwear polishes. The recent development of polish studies is described by Cook and Dumont. We may summarise by saying that, since the appearance of experimental evidence suggesting that different types of microwear polish, such as wood polish, bone polish, hide polish and meat polish, can be recognised and the publication by Keeley of a number of papers and a monograph detailing his results in this field,[7] more than a few researchers have trained themselves in his techniques and papers and doctoral theses dealing with its application to archaeological sites have begun to appear in some numbers.

Several recent conferences have detailed these results. Keeley's procedure involves the use of high-power stereoscopic microscopy to recognise features thought to be characteristic of each type of polish (the 'high-power school', as opposed to the 'low-power school' of abrasion studies mentioned by Akoshima). The type of polish is identified by the experienced observer under the microscope, by comparison with typical specimens of polish of known origin. Polishes are poorly defined by conventional incident light photomicrography, so that no satisfactory objective record is obtained except in extreme (i.e. 'typical') cases. This has meant that each observer has had to rely on his own skill and experience to determine the boundary between polishes held to be characteristic of certain contact materials. The unsatisfactory nature of the recording procedure has attracted increasing attention, leading to the development of alternative methods of polish characterisation by interferometry (Cook and Dumont, 54; Dumont, Fig. 12.1, this volume) and image processing (Grace *et al.*, this volume).

The papers in this volume appear at a climacteric period in microwear polish studies, when the re-definition of this form of research is under way, though conventional polish identification is still undertaken. All our contributors show signs of this change in emphasis. The two microwear site reports include conventional polish identification with other types of evidence, but concentrate on an all-round

[4] Olaussen's suggestion (1982,62) that polishing strengthens the body and the blade of the axe against the risk of fracture is not substantiated. The well-known experiments suggesting that a polished flint axe is more efficient than a flaked axe as a tool for felling young trees, relate to the efficiency of the axe blade, and do not explain the presence of overall polish. In any case the experimental results are unconvincing (for bibliography, see Harding, this volume; Olaussen 1982; also Steensberg 1979); supposed differences in efficiency between the two tool types, as measured by comparative speed of tree felling, are likely to be due to experimental error.

[5] The production technique is relatively simple and production times for skilled operatives making thin-butted axes on a routine basis should not vary by more than 10 minutes.

[6] Though theoretically possible, additive polishes have not been demonstrated satisfactorily and certain well developed forms of polish, e.g. sickle gloss, appear to be due to abrasion (cf. Meeks *et al*, 1982. 336ff).

[7] Newcomer & Keeley 1977; Keeley 1980; see Cook and Dumont, this volume for additional bibliography.

examination of tool classes. Dumont discusses the number of use-episodes that can be identified on tools of different types and the interpretation of polish and abrasion traces in terms of different types of hafting and tool-use. Vaughan's Palaeolithic cave site report discusses the utilised zones shown by microwear traces on Palaeolithic scrapers, but also pays attention to the identification of (microwear) activity areas in the cave and to the percentage of the total assemblage in a single occupation level to show traces of use, an approach seldom applied since it was introduced by McBurney (1968).

The three remaining contributors in the microwear field discuss matters relating to the characterisation of contact polishes. Two independent contributions study variation of flint surface textures and its effect on polish formation. In the first paper, Bradley and Clayton demonstrate considerable variation in the microstructure of different Cretaceous chalk flint sources and show how this can affect the appearance of contact polishes. In the second paper, Holmes compares the surface microtopography of Brandon (Norfolk) flint with that of fine quality Egyptian chert and shows how polish formation varies with changes in surface texture. She also discusses post-depositional surface effects that resemble contact polishes. The authors conclude that variation in microstructure affects the rate of polish formation and the final appearance of some contact polishes: such poorly developed polishes may not exhibit features considered typical of contact materials.

Two final discussion papers (by members of the same research team) consider the development and application of an objective method of polish characterisation by image processing and compare results obtained by these means with those obtained by experienced specialists using Keeley's identification techniques. The first paper, by Grace, and Co-workers and delivered at the 1983 Flint Symposium, is a detailed description of the application of image processing to the characterisation of microwear polish textures, giving an explanation of the data provided by the new technique and some preliminary results illustrating its application. The second paper (Newcomer *et al*) is a report on a subsequent stage of the research, incorporating a modification of the technique itself and a summary of a series of detailed experiments on contact materials. A group of experienced microwear analysts each identify a series of 'unknown' contact polishes, exhibiting considerable operator variance in their results. Texture analysis of the same experimentally produced contact polishes shows, in this instance, that polishes produced by a single type of contact material such as wood or bone exhibit considerable variations, and in certain cases are indistinguishable from each other by this technique.

It is clear to anyone reading this paper that the identification of contact polishes is a great deal more difficult than has been assumed in the past, and that microwear analysts simply do not have sufficiently good control material at present to identify polishes on a routine basis. As a preliminary hypothesis we may accept, as do the authors of the paper, that there is likely to be a close resemblance between polishes produced by contact with material of similar consistency: 'soft' materials such as hide and meat; 'hard' materials such as wood and bone. There may still be architypical well-developed polishes, where more exact identifications are possible, as Keeley's research suggests, but this is something that will have to be demonstrated.

This conclusion will come as a disappointment to many who had pinned their hopes on the original technique. But it should not be concluded that the study of microwear polishes is necessarily a blind alley; merely that further research is needed to establish the value of the data it produces.

In fact, the difficulties that have been encountered in this field are not unique, but typical of such fields of study, where human skill is used for exact measurement and reference criteria are imperfect or lacking. The adoption of the SEM in the biological and earth sciences more than 15 years ago, opened up many new fields of investigation, where similar problems are still unsolved. (Bull, companion volume: introduction). As has been said by another investigator 'the discrimination that the human brain can achieve is not constant (on different occasions) . . . nor indeed is it constant for different people' (Whalley and Orford, companion volume: introduction). The solution plainly lies in some form of objective control of the data, in microwear studies as well as in other forms of microscopy.

Chalk flint and other forms of chert have been used by modern man and other hominids for millions of years; they are one of the most long-lived and widespread of all human raw materials for tool makers and the words 'flint tools' are often used more generally as a synonym for stone implements for many different rocks, as well as for those of other varieties of chert. In this book we discuss four unique aspects of these rocks related to their use by man: their characteristics as tool making materials; their location and extraction from natural deposits; their dispersal to their eventual owners; and their use and abandonment. Research in all these fields is continuing and the cover provided by this book is necessarily incomplete: the part has to stand for the whole.

A few years ago a materials scientist from Nasa came to my office with a request for a few kilos of Brandon Flint, to build into the nose-cone of a spacecraft for protective purposes. So far I have received no report on the results of this venture. But it reminds us that the scientific investigation of the properties of chert and the archaeological study of its use may still have some way to go.

Bibliography

Aitken, M. J. (1982). 'Archaeometry does not only serve archaeology'. In: *Future Directions in Archaeometry* (ed. J. S. Olin). Smithsonian Institution, Washington, DC.

Burleigh, R. (1979). 'Radiocarbon dates for flint mining': In *Proceedings of the Third International Symposium on Flint 24–27 May 1979*. *Staringia*, **5**, 45–6 Oldenzaal.

Hanson, P. V. & Madson, B. (1983). 'Flint axe manufacture in the Neolithic'. *J. Danish Archaeology*, **2**, 43–59.

Keeley, L. H. (1980). *Experimental Determination of Stone Tool Uses: a Microwear Analysis*. University of Chicago Press, Chicago.

Keeley, L. H. & Newcomer M. H., (1977). 'Microwear analysis of experimental flint tools: a test case'. *J. Arch. Sc.*, **4**, 29–62.

McBurney, C. B. M. (1968). 'The cave of Ali-Tappeh and the Epipalaeolithic in North Eastern Iran'. *Proc. Prehist. Soc.*, **34**, 385–413.

Meeks, N. D., Sieveking, G. de G., Tite, M. S. & Cook, J. (1982). 'Gloss and use wear traces on flint sickles and similar phenomena'. *J. Arch. Sc.*, **9**, 317–40.

Newcomer, M. H. & Sieveking, G. de G. (1980). 'Experimental flake scatter-patterns: a new interpretive technique'. *J. Field, Archaeology*, **7**, 345–52.

Olaussen, D. S., (1982). 'Lithic technological analysis of the thin-butted flint axe'. *Acta Archaeologica*, **53**, 1–87.

Sieveking, G. de G., (1975). 'The second Maastricht International Flint Symposium: results and conclusions'. In: *Proceedings of the Second Internaitonal Symposium on Flint 8–11 May 1975*. Maastricht (Ed. E. F. G. Engelen). *Staringia*, **3**, 86–7. Oldenzaal.

Sieveking, G. de G. (1979). 'Grimes Graves and European flint mining'. In: *Subterranean Britain* (ed. H. Crawford), pp. 1–43.

Steensberg, A., (1979). *Draved. An Experiment in Stone Age Agriculture*. National Museum, Copenhagen.

Weisgerber, G., Slotta, R. & Weiner, J. (eds.) (1980). *5000 Jahre Feuersteinbergbau*. Veroffenlichungen aus dem Bergbau-Museum, **22**, Bochum.

1

Refitting stone artefacts: why bother?

DANIEL CAHEN

The search for conjoinable prehistoric stone artefacts began over a century ago (Spurrell, 1880) but, except for Worthington Smith's remarkable work at Caddington (1894), this kind of research remained rather pointless. Later, reassembled artefacts helped to elucidate some technical questions, such as the significance of core rejuvenation flakes (Hamal-Nandrin & Servais, 1929) or microburins (Vignard, 1934). In recent years interest in refitting has grown, and it has become an integral part of the study of several prehistoric living sites (Cahen, Karlin, Keeley & Van Noten, 1980). This renewed interest reflects a new approach to stone industries, which instead of being viewed as 'fossils' are now seen as evidence of human behaviour. Such a change results from a new conception of prehistory, which stresses ethnological analysis rather than a preoccupation with chronology. The excavation and study of the open air Magdalenian site of Pincevent, which began in 1964, provides a good example of this change in emphasis (Leroi-Gourhan & Brézillon, 1966 and 1972).

The concept of refitting

The aim of conjoining artefacts is exactly the inverse of the prehistoric knapper, in that the intention is to reconstruct the original nodule from which flakes were struck. Conjoining has also been applied to other materials like pottery (Schietzel, 1975; Cahen & van Berg, 1979), bone (David, 1972; Poplin, 1976) or unworked hearth stones (Julien, 1972; Olive & Pigeot, 1982). When seen as a research strategy and not merely restoration, each refit establishes a spatial relationship between elements of the same block of material. The special nature of refitted stone artefacts results from the fact that their fragmentation is due to a technical, and thus intentional, process which establishes a strong relationship between the fragmentation itself and the dispersal of the fragments. Refitting of stone artefacts thus leads to the study of lithic technology, while the interaction of technology and spatial distribution of the refitted pieces yields more general conclusions relating to the economy, organisation of living areas, duration of occupation and reconstruction of activities and behaviour.

Some practical considerations

The search for conjoinable stone artefacts requires a basic knowledge of typology and technology, as it is necessary to assign the correct place to each piece within the three-dimensional puzzle. Indeed, random attempts to refit pieces would only result in endless frustration.

Particularities of the raw material such as texture, colour, veining, impurities and cortex can be useful guides in conjoining, and grouping pieces using these attributes can help in making the obvious joins. Problems can arise, however, since colour and texture can vary within the same nodule, and the patination and staining of artefacts can vary on the same site.

In addition to hints provided by the material itself, the archaeological context can be a useful guide to refitting, and artefacts from a given excavation unit or feature can be selected for conjoining. The progressive elaboration of refits can also help to suggest new directions for investigation or perhaps generate new hypotheses.

The efficiency and final results of a refitting project depend on the primary goal: if the goal is to examine the steps in a reduction sequence, this can be achieved without too much difficulty. If, however, the aim is the establishment of the pattern of spatial relationships or the internal chronology of a settlement then much more work will be required to reach a reasonable level of probability.

Refitting and lithic technology

If we agree that a study of lithic technology involves more than the simple statement that flakes are detached from a core, then the conjoining of artefacts constitutes a very powerful method of investigation, different from but complementary to flaking experiments. Conjoining shows how each individual artifact has been made and where each artefact fits in a sequence of operations, beginning with the search for and choice of raw material and ending with the abandonment of the utilised tool. Within a reduction sequence then, refitting allows one to distinguish between circumstantial factors, technical constraints and cultural choices, and to build up a technological pattern which may be peculiar to a site, a culture, or even an epoch.

The reassembling of the flint artefacts found in layer 4 from pit number 9 of the Linear Bandkeramik site of Place Saint-Lambert at Liège (Cahen, 1984) has demonstrated the existence of a very repetitive process of blade production. This material consists essentially of flaking waste and of the total of some 50 kg, nearly 80% has been refitted. This layer is not a primary context workshop but contains material dumped from such a workshop, after all the useable products had been selected.

The main characteristics of this reduction sequence are as follows:

- No discernible criterion for the choice of flint nodules: cylindrical, ovoid, conical and irregular shapes were selected, having only their rather large size in common.
- An intensive preparation with a hard hammer, yielding a pyramidal core with a single plain striking platform, achieved without regard for loss of volume or length (Fig. 1.2(*a*)).
- The face of the core (i.e. the surface where blades will be removed) may be prepared in three different manners:

1. Preparation of a crest by bifacial flaking, the length of which is usually shorter than the available length of the core. Indeed, most of the crested blades are shorter than the succeeding blades. The preparation of the crest always follows that of the flat striking platform (Fig. 1.1(*a*), (*b*)).
2. Preparation of a ridge by striking opposing flakes from the base and from the flat striking platform.
3. Preparation by cortical and semi-cortical flakes struck from the flat striking platform, creating a series of parallel ridges (Fig. 1.1(*e*)).

The two last methods are more common for cylindrical nodules. Minor reshaping of the core usually takes place immediately after the removal of the first blades. The aim of such a preparation is to create a core ready for the removal of a large series of rather regular blades. The first blades, including the crest, are usually considered preparatory waste and are thus discarded. Such a rigid conception contrasts with the more opportunistic blade technology of the Upper Palaeolithic. The removal of the blades is done with a soft hammer. The platform, which is invariably plain, and the lack of abrasion of the platform overhangs or previous blade negatives strongly suggest the use of a punch. Indeed, direct percussion would have been useless or even dangerous since the hammer would have hit the platform in several places (Fig. 1.1(*d*)).

The progress of the flaking is controlled by three methods:

1. The removal of core rejuvenation flakes, or 'tablets', modifies the angle between core platform and face and may eventually move the flaking orientation to another side of the core. In Linear Bandkeramik industries such core tablets are rather characteristic, being thick, tabular slices of the core, usually struck off in series (Fig. 1.2(*b*), (*c*), (*d*)).
2. The removal of axial flank flakes ('*flancs axiaux*'). Such flakes are characterised by the presence, on their dorsal aspect, of blade scars parallel to the flake percussion axis. Their aim is to restore the curved profile of the core face, when the face becomes too flat (Fig. 1.2(*a*)).
3. The removal of side-struck flank flakes ('*flancs lateraux*'), whose dorsal surfaces have scars left by blades detached perpendicular to the flake's percussion axis. They are intended to restore the keel of the core, its curve along a sagittal section. When struck off in series, they may prepare a unilateral crest (Fig. 1.1(*c*)).

These three methods may be utilised casually to bypass some flaws within the raw material or some flaking accident. The final products, the blades, are rather standardised both in size and shape. Their length lies between 8 and 12 cm. They are broad and robust and flat and blades with two dorsal ridges are preferred. The reduction sequence is stopped when the core no longer allows the removal of such blades.

Besides the reduction of flint nodules, blades were also produced by the flaking of cores made on flakes. Usually, such flakes result from the preparation of the striking platform of the nodules. The blank is oriented according to its greatest length and this side is shaped as a crest by unilateral retouch, while a platform is prepared by a concave truncation. Like a burin spall, the crested blade has an asymmetrical triangular cross-section while the next blades have a rectangular or trapezoidal cross-section (Figs. 1.3 and 1.4).

The reduction of cores made on flakes does not seem to have been very successful at the Place Saint-Lambert site. However, as there is an obvious lack of large flakes, one may venture the hypothesis that such flakes have been selected, before dumping, for further flaking. Several Belgian Linear Bandkeramik sites contain more cores made on flakes than on nodules and the latter are often reused as hammerstones. Indeed, in the material studied, almost half of the reconstructed nodules lacked their core (Fig. 1.1(*c*)). These observations suggest that the preparation of blades, directly from the nodule, was a specialised craft while the flaking of cores made on flakes was a much more generalised, perhaps 'domestic', operation.

In a rather similar perspective, continuing research conducted on the four Magdalenian open air sites in the Paris Basin (Pincevent (Karlin, 1972), Marsangy (Schmider, 1978; 1982), Etiolles (Pigeot, 1982) and Verberie (Audouze, Cahen, Keeley & Schmider, 1981)) has shown that these sites share the same basic technology of blade production. However, each site presents its own characteristics and further investigation and a closer comparison of these similarities and differences will perhaps elucidate which elements result from a shared cultural tradition and which are related to local constraints (availability, morphology, quality of the raw material) or to the peculiarities of each human group, or different activities.

Refitting and spatial analysis

The pattern of spatial relationships established by the conjoining of artefacts constitutes a basic frame for the study of the spatial organisation of a dwelling place. With few exceptions all reassembled items are contemporaneous as they result from the same activity. However, the interpretation of this pattern differs greatly according to whether the dispersion of artefacts is mainly vertical or horizontal.

1. *Vertical dispersion.* Recently Villa (1982) has analysed the implications of conjoining artefacts for the study of site formation processes using examples provided by the sites of

Fig. 1.1. Linear Bandkeramik site of Place Saint-Lambert, Liège, Belgium. Pit number 9, layer 4. (*a*) Reassembled nodule showing platform and crest preparation. (*b*) Reassembled nodule showing the exhausted core, platform and crest preparation. (*c*) Reassembled nodule, the core itself is lacking: side-struck flank flake. (*d*) Series of (broken) blades with their flat, unabraded platforms. (*e*) Reassembled nodule with no crest preparation at the first stage of the reduction sequence; presence of crest preparation after the removal of the first cortical blades.

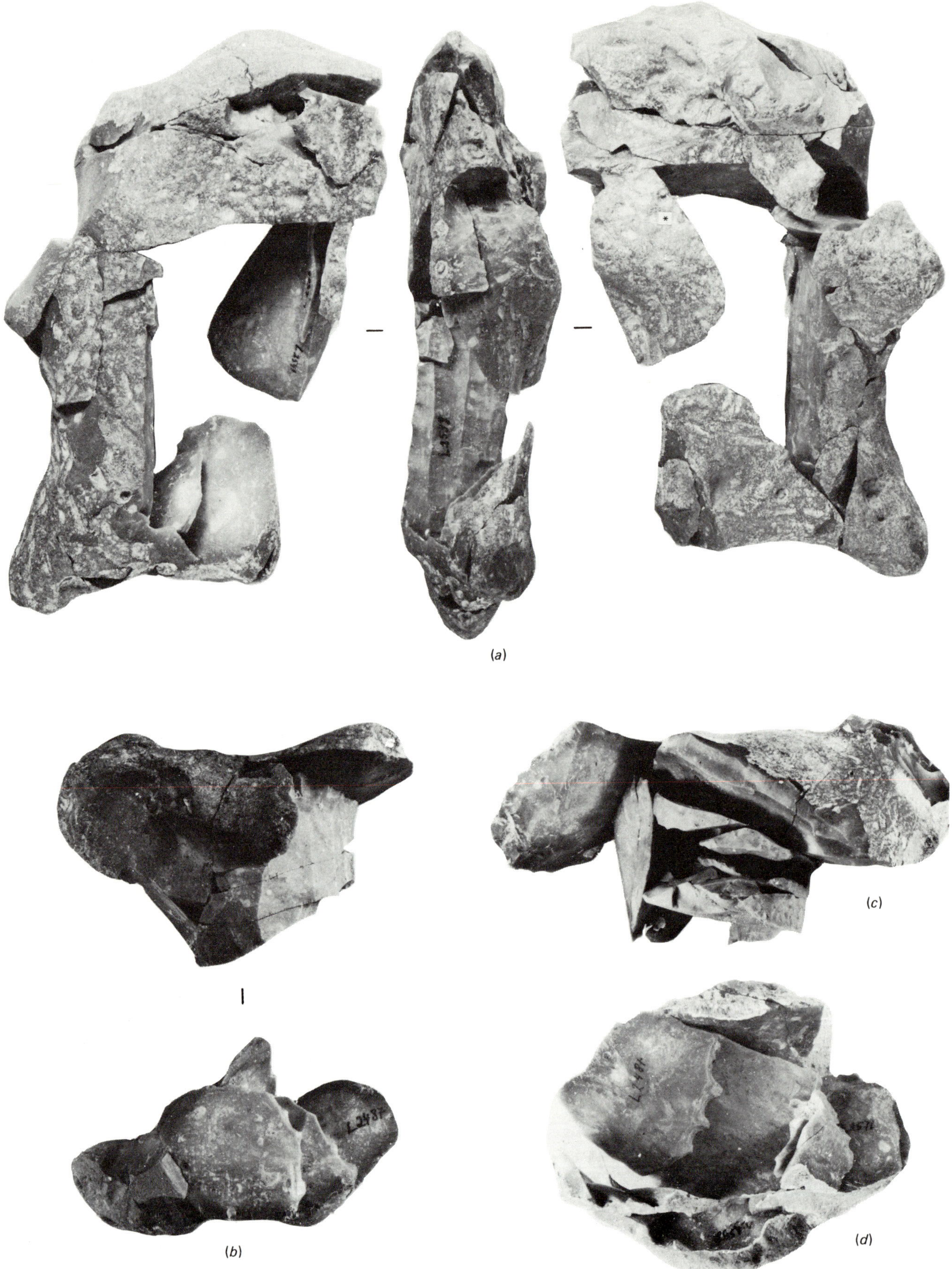

Fig. 1.2. Linear Bandkeramik site of Place Saint-Lambert, Liège, Belgium. Pit number 9, layer 4. (*a*) Reassembled nodule showing platform and base preparation and the exhausted core: axial flank flake. (*b*), (*c*) and (*d*) Series of core rejuvenation flakes.

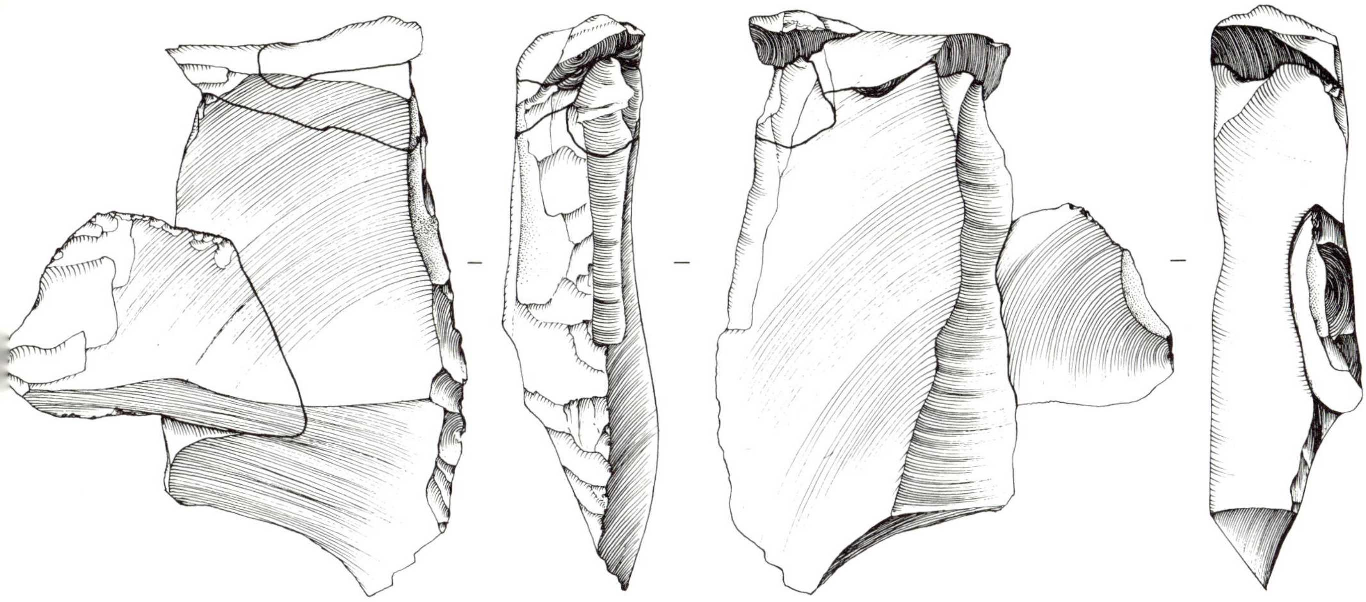

Fig. 1.3. Linear Bandkeramik site of Omal, Belgium. Pit number 3. Core on flake, with some preparatory flakes conjoined.

Gombe Point, Zaire (Cahen, 1976; Cahen & Moeyersons, 1977), Meer, Belgium (van Noten, 1978), Terra Amata, France (Villa, 1976) and Hortus Cave, France (Lumley, 1972; Bordes, 1975). In all these sites, conjoinable pieces were scattered at various depths, within a homogeneous sediment in the first two instances and within various stratigraphical units in the two latter cases. At Gombe and Meer, the vertical dispersion could not be explained by a reuse of older artefacts during a new occupation, as there was no difference in the physical state of the pieces, nor any correspondence between the sequence of removals and depth. In such cases, refitting suggests the existence of post-depositional disturbance not visible during the excavation. In stratified sites like Terra Amata and Hortus, the stratigraphic division is perhaps excessive and probably masks some severe disturbance. In such cases, some critical doubts concerning the homogeneity of the assemblage should be raised and the relevance of the ethnological reconstructions should be questioned.

A rather similar example is provided by the site of Hengistbury Head, England, where Barton & Bergman (1982) have reassembled numerous artefacts found at various depths, within a sandy homogeneous sediment. Moreover, they have conjoined artefacts which were previously attributed on typological grounds to separate occupations including a 'Mesolithic' burin and its 'Upper Palaeolithic' burin spall. The Linear Bandkeramic site of Place Saint-Lambert at Liege, discussed earlier, illustrates a slightly different case. The material found in layer 4 of pit number 9 has been repeatedly refitted with artefacts from higher and lower layers of the same feature. As layer 4 corresponds to the rubbish from a workshop, it is obviously impossible that a blade found in layer 8 (the lowest one) could have been utilised at this level before it was struck off at the level of layer 4, where the core, other blades and waste were found. In such a case one may conclude that the stratigraphic sequence in this pit is due to a succession of human activities and is not simply a straightforward accumulation of sediment due to natural processes.

2. *Horizontal distribution.* The search for conjoinable artefacts is most rewarding when these are found in undisturbed and well-excavated living floors. In such instances the gradual recognition and recording of the spatial patterning of the core reduction sequence provides a dynamic approach to the understanding of the organisation of the dwelling place. The limited number of sites that have been studied in this way prevents us from examining more than a few aspects of this type of investigation, but such work holds great promise for the future. Here we comment on results from the available studies.

Two of the four Magdalenian open air sites of the Paris Basin, Pincevent Habitation 1 (Leroi-Gourhan & Brézillon, 1966) and Verberie (Audouze *et al.*, 1981) have rather similar typology and organisation. Blades were apparently produced around the domestic hearth, which is surrounded by a concentration of flint artefacts including tools and waste. Some large accumulations of waste are found further away, while the cores have been removed (perhaps thrown) towards the periphery of the living site. The various different elements of the reduction sequence do not have the same distribution. While the waste, either in primary, manufactory, or in secondary, dump, context has a *collective* story (see Karlin & Newcomer, 1982), blades and tools seem independent and each artefact has an individual life-history (Cahen *et al.*, 1980). Their scatter must be interpreted with regard to factors such as function, hafting and to the location of activity and dumping areas, but seems unrelated to the place where the blanks were removed.

This kind of organisation contrasts with that observed at

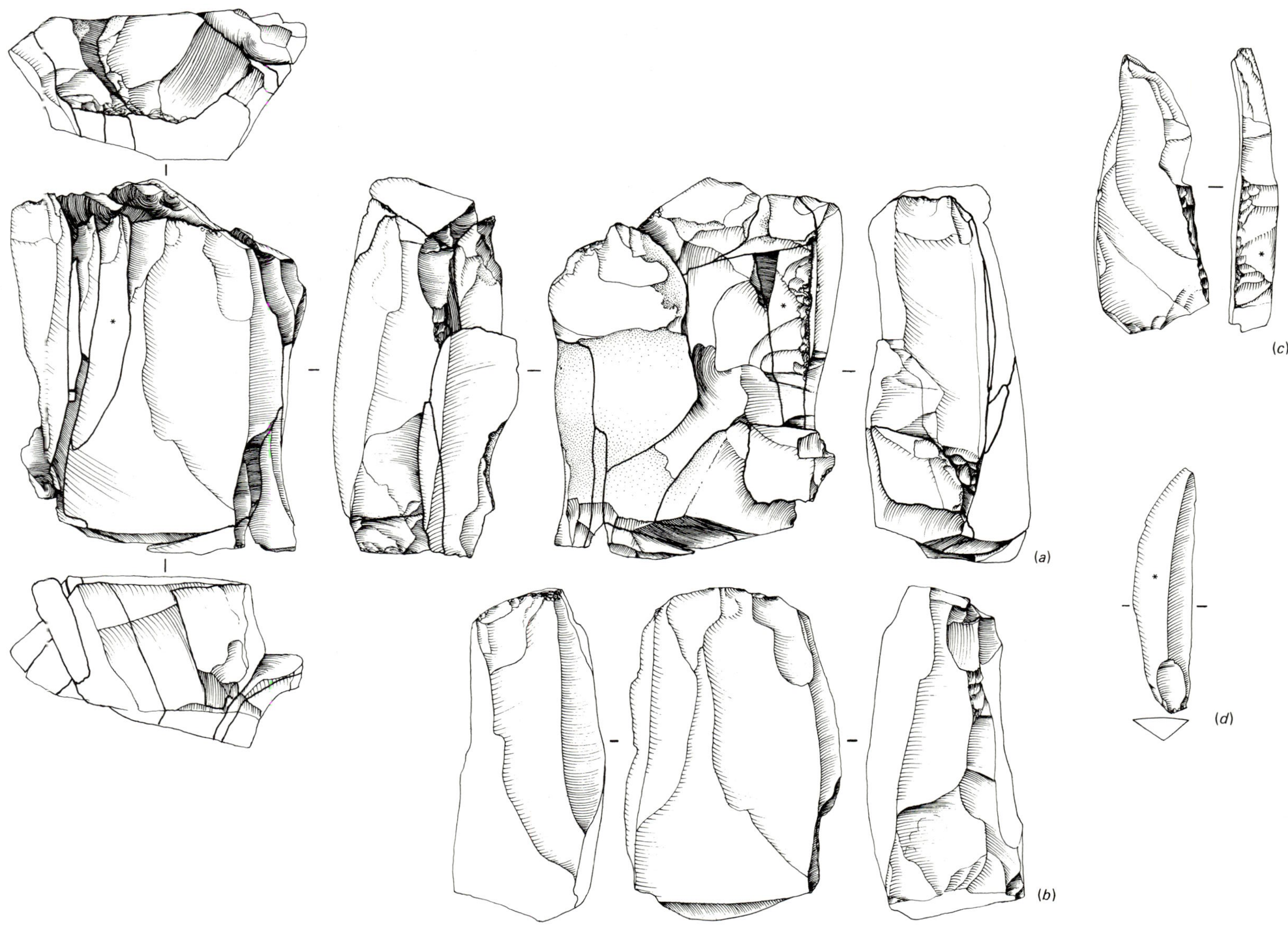

Fig. 1 4. Linear Bandkeramik site of Omal, Belgium. Pit number 3. (*a*) Reassembled core on flake. (*b*) The core alone. (*c*) and (*d*) Blades struck from this core.

the Tjongerian site of Meer, Belgium (Van Noten, 1978; Cahen, Keeley & Van Noten, 1979; Cahen & Keeley, 1980). There, it has repeatedly been observed that the reduction sequence was continuous from the flint nodule onward to the utilised and discarded tool. Tools were found with their manufactory and rejuvenation waste (retouch flakes, burin spalls) and with the preparatory waste of the core. Moreover, when several tools were made on successive blanks, they remain together and have been utilised for processing the same material.

At Pincevent and Verberie, a reserve of blanks is formed which can be utilised gradually according to future need, while at Meer tools are created as an immediate answer to a particular, existing, need. In the Magdalenian examples it was necessary to produce fairly standardised blanks to meet a large range of possibilities; in the Tjongerian case, the flaking and selection of blanks is oriented towards a particular goal. It is also not unreasonable to suppose that the Magdalenian tools were more systematically hafted than those of Meer (Cahen & Karlin, 1980).

Several factors may lead to the misinterpretation of spatial relationships. Such relationships are usually indicated on a plan by straight lines from artifact to artefact. Even if it is obvious that the line does not imply that the movement between two points was perfectly straight, it suggests at least a direct causal relationship. However, the fact that points A and B are linked by the conjoining of two or more artefacts found at these points does not imply a direct connection A–B or B–A in archaeological space. Indeed, such a relationship may be indirect, existing only with respect to a third point: C. At Verberie for instance, two opposite extremities (A and B) of the dwelling are repeatedly linked by the conjoining of tools made on successive blanks (Fig. 1.5). If direct, this relationship would imply that both areas are engaged in the same activity. However, as the tools each have an independent history, the spatial relationship between them exists only in relation to place C, where the blanks were struck off, probably somewhere near the hearth. The exact location of C remains unknown. At this site, most of the preparatory waste comes from another area, D, which appears to be an accumulation of dumped material. In this case the waste material has been removed from the place of production, C, to the dump area, D, while the tools were used independently in A and B (Fig. 1.5). It is also possible that the two blanks were kept for some time in a reserve, E, before being retouched. The problem is that if it is easy to point out areas like A, B and D, area C is more difficult to find while E, though likely, remains totally conjectural.

Technique+space=time

In an archaeological deposit, the fourth dimension of human activities, time, can be approached in two ways. When the thickness of the deposit exceeds that of one artefact, the microstratigraphy of such an accumulation yields an objective but relative chronology which, however, gives no idea of duration (Pigeot, 1982). On the other hand,

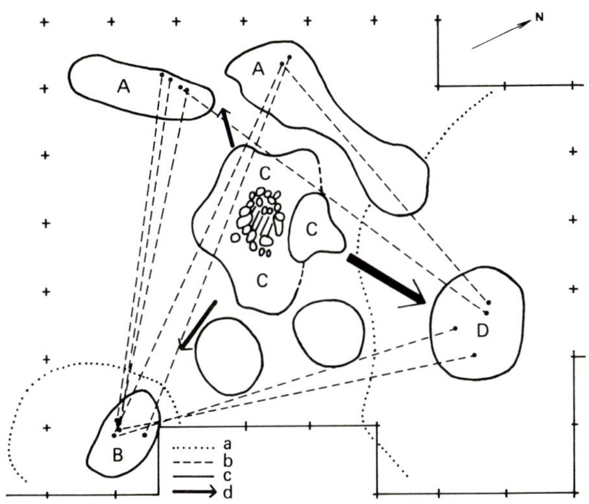

Fig. 1.5. Magdalenian open air dwelling of Verberie, Oise, France. A and B: location of tools. C: presumed location of blade production. D: accumulation of flint waste. (*a*) Limits of dumping areas (flint, bone, stone). (*b*) Conjoined artefacts (A–B or B–A: tool–tool or tool–blank; A–D or B–D: tool–blank-waste). (*c*) Limits of archaeological features. (*d*) Presumed relationships between some archaeological features (C–D: removal of waste; C–A and C–B: movement of tools).

the reduction sequence, when reconstructed, arranges the artefacts in a technical succession, whose duration may be estimated, for instance, by comparison with experimental flint knapping. Microstratigraphy is strictly limited in space and relates primarily to the process of accumulation. Refitting concerns a particular set of artefacts related to an activity and is thus equally limited. However, both approaches are complementary and may be combined, thanks to the conjoining of accumulated artefacts. It then becomes possible to build up a more general chronology for the whole occupation, which may sometimes then be extended to allow several occupations within the same site to be recognised (Pigeot, Taborin & Olive, 1976; Baffier, David, Gaucher, Julien, Karlin, Leroi-Gourhan & Orliac, 1982; Audouze & Cahen, 1982).

Such time-consuming research may be considered unrewarding or anecdotal since the occupation itself, as a whole, represents only a very short moment compared to the prehistoric time scale. From this point of view, all the effort involved illuminates at best only a very short part of the daily life of a few individuals. However, if it is necessary to consider the period of occupation as a time unit, the establishment of its internal chronology yields some important and general conclusions. For instance, if it is possible to recognise the activities performed at the beginning and at the end of the occupation, respectively, and if one can demonstrate which elements have been brought to the site and which have been removed, the correlation of these two sets of data allows us to extend the interpretation beyond the limits of the site itself. At Pincevent, for example, the exotic cream-coloured flint present on the site probably represents equipment carried by the Magdalenians when they arrived. The repetitive pattern of this particular flint from habitation to habitation, indicates that the hunters always came from the same area (Leroi-Gourhan & Brézillon, 1972). At Verberie, on the other hand, two flint nodules, among the most carefully and skilfully prepared, were probably knapped right at the end of the occupation. None of the blanks removed from these nodules have yet been found. This suggests that the blanks were taken away from the site by the departing Magdalenians (Fig. 1.6; Audouze *et al.*, 1981: Fig. 16 and 15). Such results make it possible to consider a prehistoric site not as a single, isolated, unit but as a stage within a cycle of occupations and activities. From such a point of view, palaeolithic nomadism is no longer a theoretical model but a tangible reality.

As for spatial organisation, the interpretation of the temporal structure may be hindered by several factors. As blanks and tools have an individual history, independent of the collective one of the waste, for the most part they cannot be fitted into a precise chronology. Indeed, in the technical sequence, a platform preparation flake is surely earlier than the blade removed from this platform. However, this technical relationship does not imply any priority between the spots where this flake and this blade have been found. Similarly, a blank produced at the very beginning of the occupation may have been retouched, utilised and discarded only at the very end of the stay.

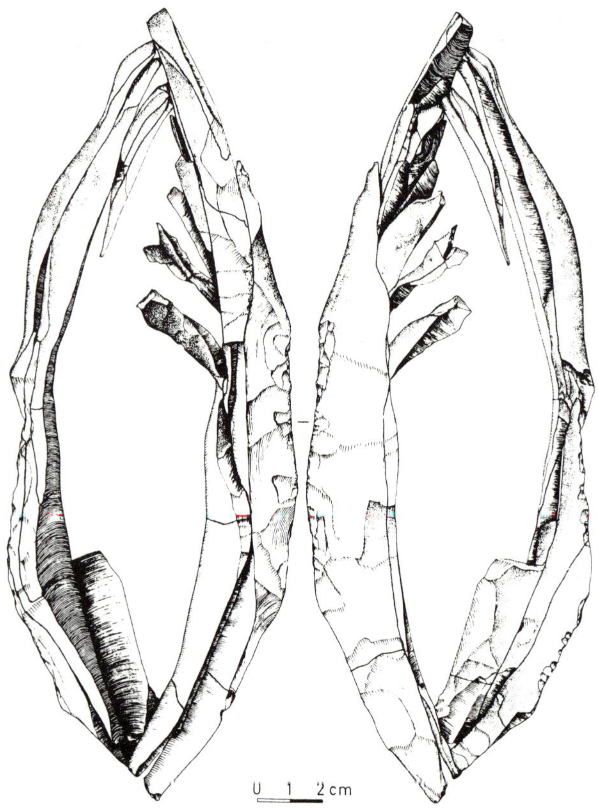

Fig. 1.6. Magdalenian open air dwelling of Verberie, Oise, France. Reassembled series of preparatory blades struck from opposite platforms. Neither core nor blanks have been found at the site.

The reconstruction of the reduction sequence does not imply that the sequence was continuous through time. Continuity or discontinuity of the sequence may have important implications for the interpretation not only of the chronology but also of the organisation of activities. A continuous sequence will usually be attributed to a single flint knapper. On the other hand, a discontinuity may perhaps indicate the participation of two or more craftsmen. Another question then arises: How can we demonstrate continuity or discontinuity within an archaeological deposit? Such a demonstration rests mostly on an assumption linking time and space: a spatial interruption within the reduction sequence (i.e. within the distribution of its elements) will be considered as a possible sign of discontinuity, especially if it corresponds to a technical break in the process. Thus, a change of location between the preparation of the core and the removal of the blades, or between two series of blades, indicates a discontinuity in the process, the length of which perhaps exceeds the time necessary to move from one location to another. But if this movement is not accompanied by a conceptual modification of the process, it remains rather difficult to assume that one or more flint knappers are concerned. C. Karlin provided a good example with core B. 22–159 from Pincevent Habitation I (Cahen *et al.*, 1980). This core was prepared and reduced near hearth II and afterwards modified and reprepared near hearth III. In such a case, the assumption that the change of location corresponds to a discontinuity of some length and even to a change of knapper appears reasonable. However, all these assumptions primarily relate to the waste material and not the blades or the tools.

Conclusion

The search for conjoinable artefacts is a time-consuming operation which immobilises an archaeological assemblage for rather long periods. Therefore, before entering upon such an analysis, it is necessary to evaluate carefully its purpose and value. The method is very helpful in providing a better knowledge of lithic technology and may also help us to understand the archaeological context. Furthermore, when circumstances are most favourable it may allow us to recognise an organisation through space and time within a prehistoric dwelling-place. Such results are not an automatic consequence of the method: even if one succeeds in refitting several artefacts found in a fluviatile deposit, this achievement does not turn this rather disturbed site into a primary context occupation. At best, it demonstrates that disturbances were limited in lateral extent.

Each method of study used in the analysis of prehistoric material offers certain possibilities and has its own limitations. Taken by itself, the refitting of prehistoric artefacts has little significance since the reconstruction of a block of flint has, taken alone, limited interest. However, there is no denying the feeling of closeness to prehistoric man, that is given by rejoining flakes that were struck apart thousands of years ago.

Acknowledgements

I am indebted to Mr M. Splingaer who made the illustrations for Figs. 1.1 and 1.2; to Mrs Y. Baele for Figs. 1.3, 1.4 and 1.6 and to Mrs S. Jansen for Fig. 1.5. I am particularly grateful to Dr F. Audouze for her help and discussion concerning Verberie and to Dr M. Otte for the Place Saint-Lambert site. Dr M. Newcomer helped with the translation.

Bibliography

Audouze, F., Cahen, D., Keeley, L. H. & Schmider, B. (1981). 'Le site magdalénien du Buisson Campin à Verberie (Oise).' *Gallia Préhistoire*, **24**, 99–143.

Audouze, F. & Cahen, D. (1982). 'Décryptage d'un habitat magdalénien de plein air.' In: *Les habitats du Paléolithique supérieur*, vol. 2, pp. 210–20. Roanne-Villerest.

Baffier, D., David, F., Gaucher, G., Julien, M., Karlin, C., Leroi-Gourhan, A. & Orliac, M. (1982). 'Les occupations magdaléniennes de Pincevent: problèmes de durée.' In: *Les habitats du Paléolithique supérieur*, vol. 2, pp. 243–71. Roanne-Villerest.

Barton, R. N. E. & Bergman, C. A. (1982). 'Hunters at Hengistbury: some evidence from experimental archaeology.' *World Archaeology*, **14**, 237–48.

Bordes, F. (1975). 'Sur la notion de sol d'habitat en préhistoire paléolithique.' *Bulletin de la Société Préhistorique Française*, **72**, 139–42.

Cahen, D. (1976). 'Nouvelles fouilles à la pointe de la Gombe (ex-pointe de Kalina), Kinshasa, Zaire.' *L'Anthropologie*, **80**, 573–602.

Cahen, D. (1984). 'Technologie du débitage laminaire.' In: *Les fouilles de la Place Saint Lambert à Liège, vol.* I*: la zone orientale*. Etude et Recherches Archéologiques de l'Université de Liège (in press).

Cahen, D. & Karlin, C. (1980). 'Les artisans de la préhistoire.' *La Recherche*, **11**, 1258–68.

Cahen, D., Karlin, C., Keeley, L. H. & Van Noten, F. (1980). 'Méthodes d'analyse technique, spatiale et fonctionnelle d'ensembles lithiques.' *Helinium*, XX, 209–59.

Cahen, D. & Keeley, L. H. (1980). 'Not less than two, not more than three.' *World Archaeology*, **12**, 166–80.

Cahen, D., Keeley, L. H. & van Noten, F. L. (1979). 'Stone tools, toolkits and human behaviour in Prehistory.' *Current Anthropology*, **20**, 661–83.

Cahen, D. & Moeyersons, J. (1977). 'Subsurface movements of stone artefacts and their implications for the prehistory of Central Africa.' *Nature*, **266**, 812–13.

Cahen, D. & van Berg, P. (1979). *Un habitat danubien à Blicquy, vol.* I. *Structures et industrie lithique. Archaeologia Belgica*, 221. Bruxelles: S.N.F.

David, F. (1972). 'Témoins osseux.' In: Leroi-Gourhan & Brézillon, *Fouilles de Pincevent*, pp. 295–320.

Hamal-Nandrin, J. & Servais, J. (1929). 'Contribution à l'étude de la taille du silex aux différentes époques de l'âge de la pierre. Le nucléus et ses différentes transformations.' *Bulletin de la Société préhistorique française*, **26**, 541–52.

Julien, M. (1972). 'Témoins relatifs au feu.' In: Leroi-Gourhan & Brézillon, *Fouilles de Pincevent*, pp. 279–94.

Karlin, C. (1972). 'Le débitage.' In: Leroi-Gourhan & Brézillon, *Fouilles de Pincevent*, pp. 263–77.

Karlin, C. & Newcomer, M. (1982). 'Interpeting flake scatters: an example from Pincevent.' *Studia Praehistorica Belgica*, **2**, 159–65.

Leroi-Gourhan, A. & Brézillon, M. (1966). 'L'habitation n° 1 de Pincevent près de Montereau (Seine-et-Marne).' *Gallia Préhistoire*, IX, 263–385.

Leroi-Gourhan, A. & Brézillon, M. (1972) *Fouilles de Pincevent. Essai d'analyse ethnographique d'un habitat Magdalénien (la section 36)*. Paris: CNRS.

Lumley, H., de (1972). *La grotte moustérienne de l'Hortus*. Etudes Quaternaires, I. Marseille: Université de Provence.

Olive, M. & Pigeot, N. (1982). 'Methodes d'analyse de deux foyers pierreux d'Etiolles: U5 et P15.' In *Les habitats du Paléolithique supérieur*, vol. 1, pp. 99–102. Roanne-Villerest.

Pigeot, N., Taborin, Y. & Olive, M. (1976). Problèmes de stratigraphie dans un site de plein air. *Cahiers du Centre de Recherches de l'Université de Paris* I, **5**, 5–27.

Pigeot, N. (1982). 'L'organisation spatio-temporelle d'un habitat, vue à travers l'étude du débitage (Etiolles).' In: *Les habitats du Paléolithique supérieur*, vol. 1, pp. 80–89. Roanne-Villerest.

Poplin, F. (1976). *Les grands vertébrés de Gönnersdorf. Fouille 1968*. Der Magdalenien-Fundplatz Gönnerdorf, 2. Wiesbaden: Fr. Steiner Verlag.

Schietzel, K. (1975). 'Haithabu. Ein Beitrag zur Entwicklung frühstädtischer Siedlungsformen Nordeuropa.' *Ausgrabungen in Deutschland*, **3**, Mittelalter II, Bonn; Rudolf Habelt.

Schmider, B. (1978). 'Données préliminaires sur les structures d'habitation du gisement magdalénien du Pré-des-Forges à Marsangy (Yonne). Fouilles 1974–1977.' *Cahiers du Centre de Recherches Préhistoriques*, **6**, 4–21.

Schmider, B. (1982). 'Analyse d'une aire de taille au Pré-des-Forges à Marsangy (Yonne).' In: *Les habitats du Paléolithique supérieur*, vol. 1, pp. 17–30. Roanne-Villerest.

Smith, W. G. (1894). *Man, the Primeval Savage*. London: Stanford.

Spurrell, F. C. J. (1880). 'On implements and chips from the floor of a Palaeolithic workshop.' *The Archaeological Journal*, **37**, 249–99.

Van Noten, F. (1978). *Les chasseurs de Meer*. Dissertationes Archaeologicae Gandenses, XVIII. Brugge: De Tempel.

Vignard, Ed. (1934). 'Triangles et trapèzes du Capsien en connexion avec leurs microburins.' *Bulletin de la Société préhistorique française*, **31**, 457–9.

Villa, P. (1976). 'Sols et niveaux d'habitat du Paléolithique inférieur en Europe et au Proche-Orient.' *Quaternaria*, XIX, 107–34.

Villa, P. (1982). 'Conjoinable pieces and site formation processes.' *American Antiquity*, **47**, 277–90.

2

Technological variability and change seen through core reconstruction

ANTHONY E. MARKS AND PHILLIP VOLKMAN

Introduction

Recently, the conjoining of lithic artifacts has become a valuable, if rather specialized, tool of prehistorians. Its two most effective uses, so far, have been to document otherwise undetectable vertical artifact displacement (Cahen & Moeyerson, 1977; Villa, 1982) and, in a more complex mode, to use the spatial distributions of reconstructed core elements and other conjoins to assist in various site interpretations, including evidence for multiple versus single occupations, intrasite interactions, the presence of activity areas, and more (Cahen, Keeley & van Noten, 1979; Hietala & Marks, 1981).

There is, however, another potential use for conjoining artefacts which has not been extensively exploited; that is, in the study of lithic technology and, in particular, in the study of core reduction strategies. To date, technological studies of stone flaking mainly have utilized either statistically significant attribute coassociations (e.g. Ferring, 1979; Fish, 1979; Jelinek, 1981; Wilmsen, 1970), or have used analogies supplied through experimental archaeology. The better of these latter studies (e.g. Bordes & Crabtree, 1969; Semenov, 1964; Tixier *et al.*, 1980) have been carried out by scholars who themselves have achieved a competency in flint knapping which, at the least, approaches the once lost knapping skills of the European Upper Paleolithic.

Yet, while experimental procedures which replicate the morphology of prehistoric artefacts often permit significant insights into the mechanisms and modes of production which are responsible for specific attribute states (e.g. Newcomer, 1975) they can do no more than to demonstrate ways in which a given artifact form *might* have been produced. Only through reconstructions can one tell how an artifact was *actually* produced. This is particularly important when assumptions concerning method of production are based upon artifact morphology, as for Levallois pieces, where the definition rests upon a method of production (Bordes, 1961), while its recognition, in each case, rests upon a set of morphological attributes.

Although recognized that the Levallois method may be achieved through a number of specific reduction strategies (Bordes, 1980; Tixier *et al.*, 1980), it is assumed that each results in the same 'predetermined' morphology. Thus, in practice, an artifact is considered to have been produced by the Levallois method if its morphology is consistent with artifacts experimentally produced using the Levallois method.

In as much as this type of observation is essentially classificatory, having so far little analytic impact on the comparison of assemblages (Bordes, 1980, 1981), it might not be very important whether or not 'Levallois morphology' had or had not been achieved by a method based on conscious predetermination through prior core shaping, provided that the morphological criteria were consistently applied. However, in the Levant, the apparent cultural continuity from Middle Paleolithic to Upper Paleolithic does make this distinction significant. The ability to recognize the actual method of production from final artifact form is central to views of how the change took place from a Levallois technology to a 'punch' blade technology.

Although for the past 30 years it has been held that there was cultural continuity from the Levantine Middle Paleolithic through to the Levantine Upper Paleolithic (Garrod, 1951), there has been no basic agreement on the nature of this transition, particularly as it concerns stone technology.

Prior to 1951, Neuville attempted to synthesize Levantine prehistory and to put it into relative chronological and developmental frameworks (Neuville, 1934). Although he tended to view most of the Paleolithic as being developmental, he did not believe that the Levantine Mousterian developed into the Levantine Upper Paleolithic, in spite of the fact that he saw significant 'Mousterian' traits in the early Upper Paleolithic (Neuville, 1934, 1951). Rather, he believed that the beginning of the Upper Paleolithic corresponded with the arrival in the Levant of still somewhat primitive but essentially modern men (Neuville, 1934, p. 245). Perhaps his unwillingness to see a developmental transition between the Middle and Upper Paleolithic lay in his belief that the Upper Paleolithic entered the Levant well prior to the demise of the local Mousterian (Neuville, 1934, p. 249), thereby precluding development of one out of the other.

Although, at first, Garrod did not share Neuville's view that the early Upper Paleolithic (Neuville's Phase I) consisted of both traditional Mousterian and Upper Paleolithic elements, preferring to believe that such 'assemblages' were the result of geological mixing (Garrod & Bate, 1937), she did, along with Neuville, seem to believe blade technology was, in some way, correlated with modern man

(Garrod, 1951, p. 129). By 1951, however, Garrod had changed her position as a result of new data from Lebanon which had been published in the late 1940s (Haller, 1946; Ewing, 1947). She reanalysed her own materials from el Wad, in Palestine, and decided that Neuville, more or less, had been correct, after all, in his description of the earliest Levantine Upper Paleolithic (Phase I) (Garrod, 1951). More importantly, Garrod went a significant step further than Neuville in proclaiming that Neuville's Phases I and II, in somewhat different form than he described them, were 'truly intermediate' between the Levantine Mousterian of the Middle Paleolithic and the Aurignacian of Upper Paleolithic Phase III and were, as well, actually *transitional* between them (Garrod, 1951, p. 129). In doing this, Garrod defined a transitional industry, the Emiran, which, like Neuville's Phase I, had both Middle Paleolithic and Upper Paleolithic elements. The Emiran, according to Garrod, was made up of 'a flake element indistinguishable from the Upper Levalloiso–Mousterian; the new features are the Emireh point, the appearance in increasing numbers of true blades and blade tools and typical narrow blade cores' (1951, p. 128). In addition, typical Middle Paleolithic tools (sidescrapers and points) were made on flakes, often Levallois, produced by Middle Paleolithic methods, while Upper Paleolithic tools (endscrapers, burins, and backed blades) were made on true blades produced from narrow blade cores.

The transitional nature of the Emiran lay essentially in the presence of both Levallois method and 'true' blade technology in the same industry. Yet, Garrod did not see any transition between these two basically different methods of core reduction (Garrod, 1951, p. 129):

> . . . it is interesting to note how, within the culture itself, we find an absence of forms transitional from one to the other. The striking of true blades (probably by the use of a punch) appears rather as a new discovery than as an evolution from the older flaking technique . . I would suggest . . that here is a case of invention of a new technique within an industry of old tradition.

The concept and definition of the Emiran stood for many years, although various of its details were attacked or questioned by other scholars. Perhaps the most important and telling of these came in 1969 from Bar-Yosef & Vandermeersch (1972) who presented convincing arguments that all the sites containing the Emiran, except et-Tabban in the Judean Desert, had been affected by erosion from internal spring activity, mixing together late Mousterian and Upper Paleolithic deposits. They also reiterated various arguments (Stekelis, 1954, 1961; S. Binford, 1966) against using the Emireh point as a type fossil. Although they reported nothing particularly new, their timing was right, especially in light of new studies from Lebanon.

In spite of their impact on Garrod, the preliminary published reports on Ksar Akil (Ewing, 1947) and Abu Halka (Haller, 1946) failed to generate wide attention. This, and the absence of other significant publications on these sites, left them something of a mystery. However, in 1969, John Waechter, who then controlled the Ksar Akil materials, showed them to a number of archaeologists attending a conference at the Institute for Archaeology, University of London. There was a general feeling that, while the Emiran of Palestine might not have historic reality, the as yet unpublished transitional materials from Ksar Akil probably did. Although the Ksar Akil transitional assemblages were studied by Azoury (1971), it was the publication of apparently comparable materials from Antelias Cave (Copeland, 1970) which provided a new model for the technological transition from the Middle to the Upper Paleolithic.

This Lebanese transition, as described by Copeland was as follows (Copeland, 1970, p. 137):

> What has now appeared is the definite entity distinguishable in the Central Levant (Abu Halka IV *f* and *e*, joined now by both Ksar Akil and Antelias) each with a developing sequence of levels of one industry (sic); this does not have equal or even comparable numbers of Middle and Upper Paleolithic tools, but what it does have is Upper Paleolithic tool-types made on Middle Paleolithic blanks. More specifically . . . /it/ . . . is represented as a sudden emphasis on tools (i.e., burins and endscrapers) which up till then had played a minor role in the underlying assemblages, accompanied by the virtual disappearance of those types . . which till then had dominated the Mousterian tool kit. At the same time, the previous older flint working techniques were still in use.

In short, the typology had changed but not the technology. Yet, the technology was not simply Levallois: it was 'evolved' Levallois, which was described as being basically unidirectional during core reduction (Copeland, 1970, p. 105). Thus, in essence, the transitional nature of this industry lay in the combination of Middle and Upper Paleolithic elements on the same artifact, the blank being Middle Paleolithic and the tool Upper Paleolithic (Copeland, 1970, p. 177). Because of a stratigraphic hiatus, it is not clear how this 'evolved' Levallois technology led into true blade technology, but it certainly appeared to emphasize blank elongation in that virtually all Levallois pieces from Antelias were either elongated points or blades (Copeland, 1970). Upper Paleolithic 'punch' blades were very rare but did increase in number through time (Copeland, 1970), as was also the case at Ksar Akil (Copeland, 1975).

The concept of different tempos of change for typology and technology was new to Levantine prehistory and the idea of an 'evolved' Levallois reduction technique provided a framework into which could be fit increasing blank elongation.

In spite of gaps in the data, what more could one want in the way of a transition? Yet, how did it come about and what exactly was this 'evolved' Levallois technique which produced Levallois blades and elongated points from unidirectionally flaked cores? From where did such a reduction strategy come?

The answers to such questions were quite beyond us until the discovery of the site of Boker Tachtit in the Central Negev, Israel* (Marks, 1977). Boker Tachtit consisted of four main occupation floors which, combined, spanned the

* Work in the Central Negev has been founded, since 1969, by eight National Science Foundation grants to the senior author.

transition from terminal Mousterian to early Upper Paleolithic. Since each floor represented a very brief period of time but had on it moderate accumulations of stone artifacts (Hietala & Marks, 1981), each was well suited for detailed technological studies, including the reconstruction of cores and the conjoining of other artifacts. Through these reconstructions and other more limited artifact conjoins (Marks, 1983*b*), it was possible to model not only the general patterns of core reduction through the transitional period but also to determine the range of technological variation during each occupation. In addition, these reconstructions made it possible to correlate actual reduction strategies with their resulting products and to document to what extent certain final artifact morphologies could be tied to specific reduction techniques. The results of these studies were often unexpected and have direct bearing on not only the technological transition from the Middle to the Upper Paleolithic in the Levant, but also, on a more general level, on the assumptions that the same artifact morphology must be the result of the same method of production.

Since the assemblages from Boker Tachtit have already seen some preliminary publication (Marks, 1977, 1981; Hietala & Marks, 1981) and much more detailed publication has just appeared (Marks, 1983*a*, 1983*b*; Marks & Volkman, 1983; Volkman & Kaufman, 1983), this paper will only deal with those aspects of the Boker Tachtit assemblages which relate directly to core reduction techniques and the types of blanks produced by them. Even this limited treatment should provide the data needed to understand the nature and the complexity of the technological transition which led out of the Middle Paleolithic and into the Upper Paleolithic.

First, however, a few, brief notes about the site and its data. Boker Tachtit is located within a terrace of the Nahal Zin, just north of Avdat in the Central Negev, Israel (Fig. 2.1). The site contains four major stratified occupation floors within fine sediments of a spring-pool margin. Thus, the occupations were each submerged by fluctuations in the spring activity (Goldberg, in Marks, 1983*b*). The total duration of occupation is unknown but no one occupation was more than transient and the whole sequence can be dated to between 45 000 BC for the earliest occupation and the latest to no less than 38 000 BC (Marks, 1981).

Artifact samples from the four major floors differed but the two lowest (Levels 1 and 2) and the uppermost (Level 4) all provided samples which permitted extensive core reconstructions. For Level 1, there was complete or significant reconstruction of 24 blocks of raw material, involving no fewer than 19 cores, 23 tools, and 474 pieces of debitage. The Level 2 sample was larger, with 88 blocks of raw material reconstructed, involving 61 cores, 75 tools, and 2071 pieces of debitage. The Level 3 sample was small, because of post-occupational destruction of most of the floor, and only 9 reconstructions could be accomplished, including 9 cores and 97 pieces of debitage. The Level 4 reconstructions totalled 35, with 31 cores, 30 tools, and 631 pieces of debitage included. Thus, the descriptions of the core reduction techniques which follow are based upon

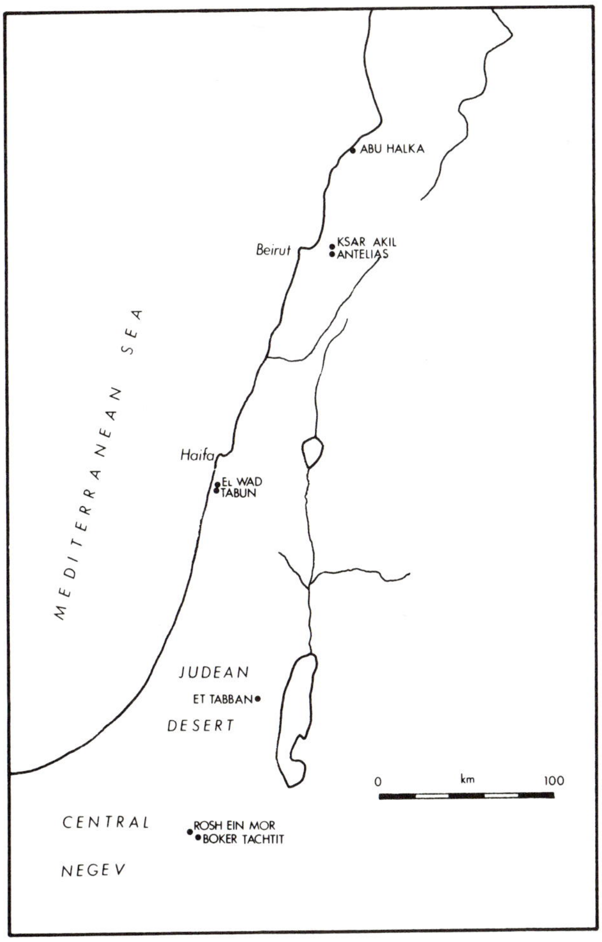

Fig. 2.1. Map of the Levant, showing sites mentioned in the text.

large samples and this study has avoided the main weakness of experimental work; there are sufficient numbers of reconstructed cores from each floor so that idiosyncratic variation is not taken to represent significant patterning. Therefore, although there are some examples of core reduction strategies which occur rarely or even as isolated phenomena (see Volkman, in Marks, 1983*b*), only the main patterns will be discussed here.

Level 1.

The dominant core reduction strategy consists of a specialized, opposed platform Levallois method, which normally produced at least one Levallois point per core, although on some successful point production was not always achieved, while on others more than one point was produced. This strategy involved a series of sequenced steps, summarized as follows:

1. The removal of most cortex from a nodule, during which process a *lame à crête* ridge was set up along the core's maximum length (Fig. 2.2).
2. The formation of two platforms, one at each end of the *lame à crête* ridge, by the removal of several small platform preparation flakes. Occasionally, this may involve the formation of small *lame à crête* ridges and their removal during the initial platform formation.

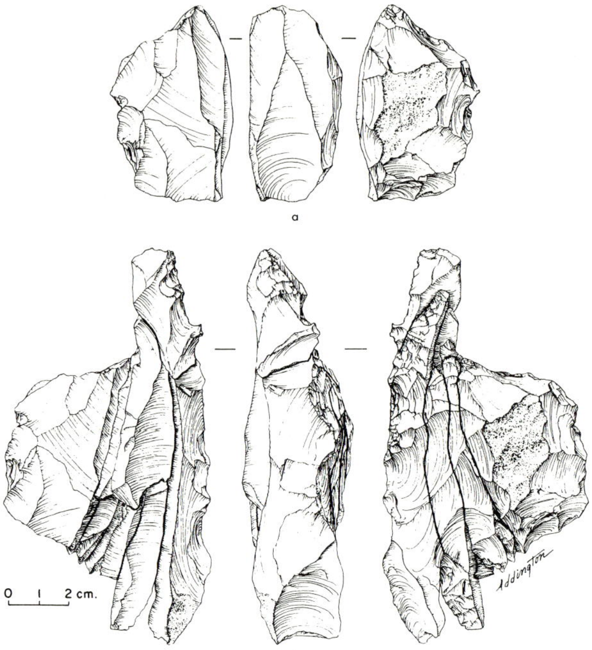

Fig. 2.2. (*a*) abandoned Levallois point core from Level 1. (*b*) same core reconstructed partly to show *lame à crête*. Note the extent to which the core size has been reduced by platform faceting.

3. The second major phase of core reduction begins with the removal of the main *lame à crête*, which is struck from one of the prepared platforms.
4. The striking from the opposed platform of a short series of blades (from one to five), using the *lame à crête* scar as the surface for the initial removals (additional platform faceting may or may not be done, depending on the specific situation after each blade removal).
5. The striking from the opposite platform of another short series of blades (from one to five), using both the new blade scars and what is left of the *lame à crête* scar as the surface for initial removals (in this case, platform rejuvenation usually takes place to correct for the negative bulb, resulting from the *lame à crête* removal and additional platform modification may take place, as needed).
6. The last removal of this blade series, after more platform faceting, is usually an elongated Levallois point, the shape and dorsal scar pattern of which have been predetermined by the junctures of negative scars left by the previous two series of blade removals. The number of blades removed in each series may well relate to the number it takes, in each case, to set up the appropriate scar configuration. The actual sequence of blade removals may vary at the very end (Fig. 2.3) without altering significantly either the final point morphology or the *arête* pattern.

After the point is struck, the core is either abandoned or the last two major steps are repeated, again ending with a Levallois point removal. In several reconstructions, two or more elongated Levallois points were struck, after each had been prepared by a series of blade removals (Fig. 2.4). That these cores are finally abandoned after the removal of a Levallois point clearly indicates that it was the desired end product.

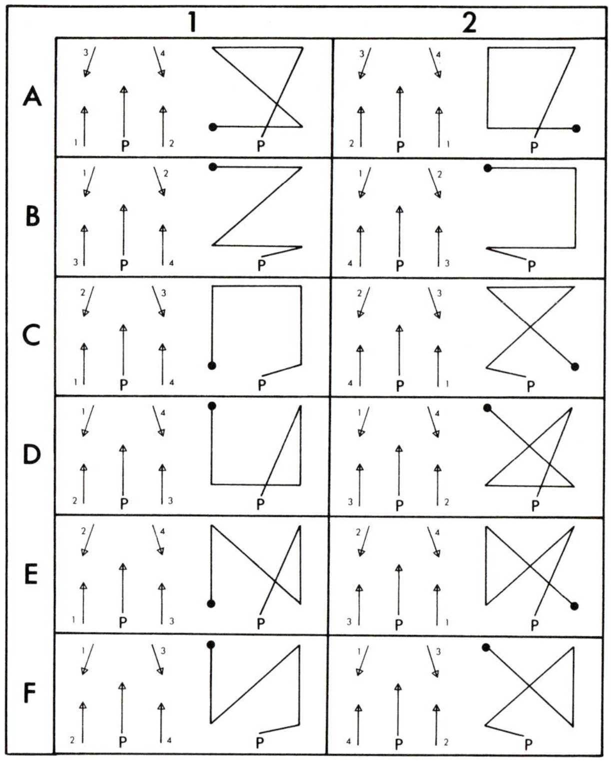

Fig. 2.3. Schematic chart of different specific blade removal sequences used in the final predetermination of a Levallois point.

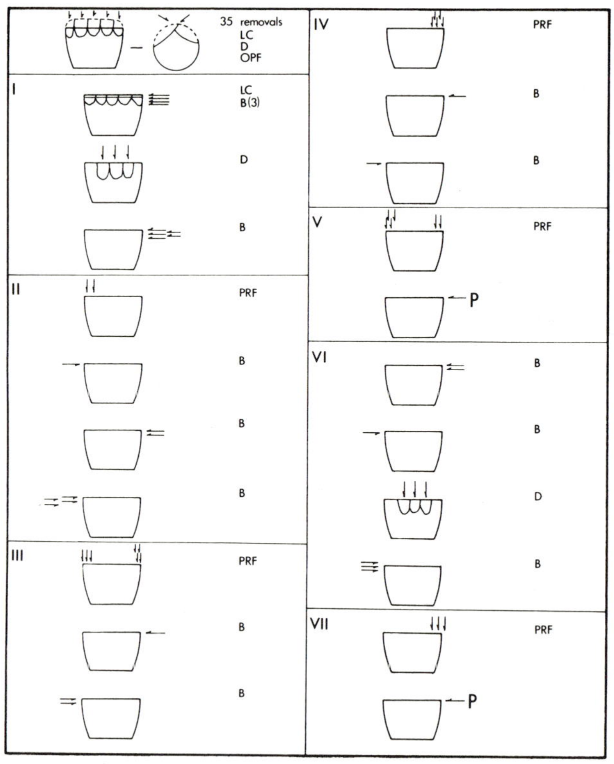

Fig. 2.4. Schematic representation of step-by-step reduction which resulted in the production of two Levallois points from Core H, Level 1, Boker Tachtit.

The resulting points have a very characteristic scar pattern, not so much in its morphology but in the opposed platform nature of the scars. The outer two scars always originate from the distal extremity, while the central scar or scars originate from the proximal end. Thus, the scar pattern is always bidirectional (Fig. 2.5).

While emphasis has been placed upon the Levallois points, the vast majority of the blanks removed are blades. Given the reduction strategy and the tendency for both platform faceting and bidirectional flaking, the majority fall morphologically into Levallois blades. However, the reconstructions make clear that they were not the end-products but were by-products of point manufacture. Therefore, they can hardly be considered Levallois.

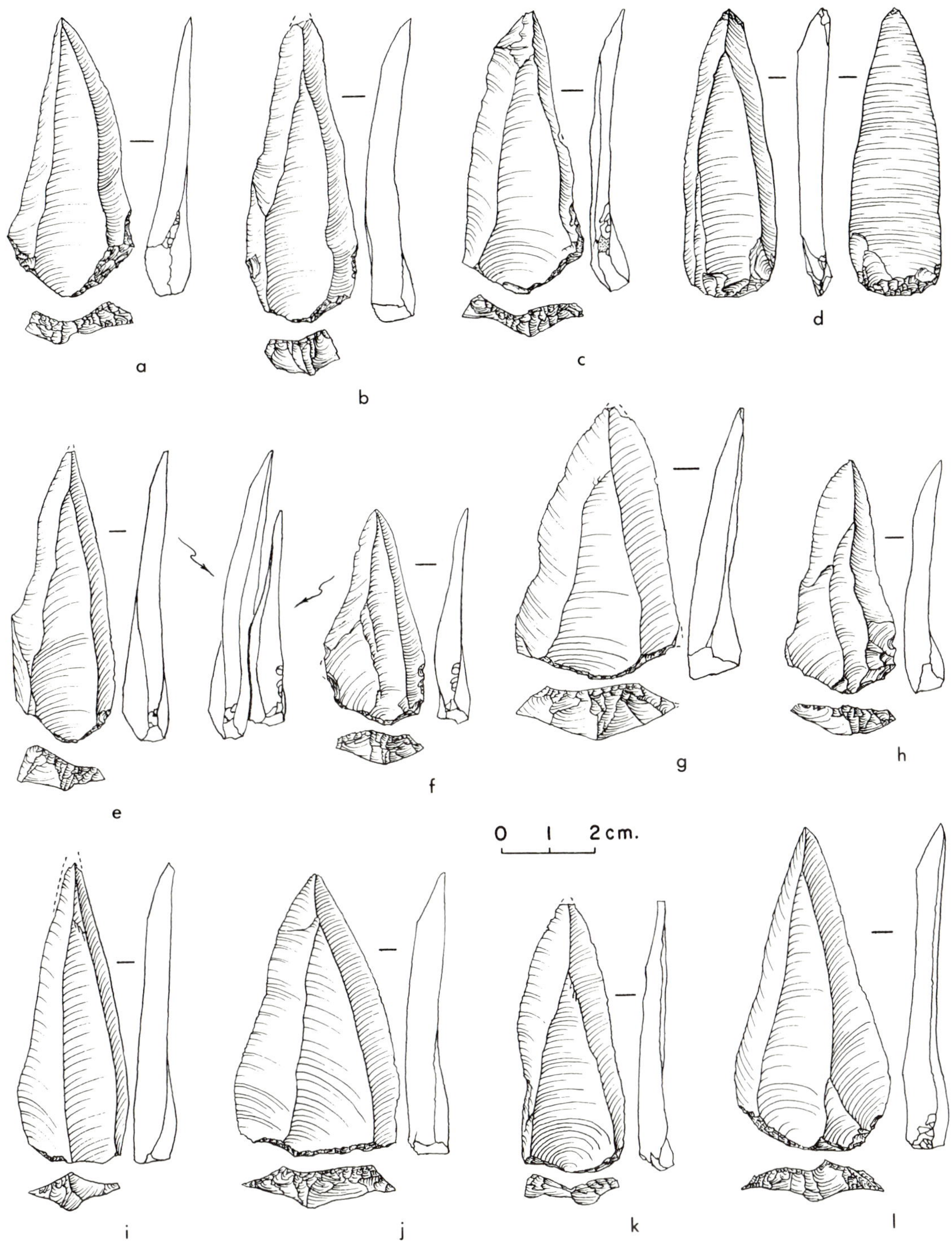

Fig. 2.5. (*a–c*) Levallois points from Level 1. (*a*) and (*c*) have blunted right edges adjacent to the butt. (*d*) Emireh point from Level 2. (*e–f*) conjoining Levallois points from an opposed platform point core. Note that (*e*) has two scars originating at the proximal end, while (*f*) is typical. (*g*) point from Level 2. (*h*) typical Levallois point from Level 3. (*i*) non-Levallois point with all scars originating from the proximal end. (*j–l*) non-Levallois points from Level 4 (1 retouched). (*k*) was struck from the core shown in Fig. 2.7.

Level 2.

The dominant core reduction strategies are the opposed platform Levallois point type and a simple opposed platform blade type which was present in very low frequency in Level 1. It is in Level 2 that a shift is evident from the Levallois point as a final product to the blade as a main product. This does not mean the abandonment of the Levallois point production strategy but, merely, its modification so that core reduction continues after point removal. Almost half of the reconstructed cores from which Levallois points were struck saw additional reduction such that the exhausted core shows no indication of point production. In addition, there is a tendency both to produce more blades during the setting up of the Levallois point configuration and, as well, to remove longer series of blades from individual platforms during this process (Fig. 2.6).

There is no question about the Levallois nature of the point production. However, the technology now shows a flexibility even within this Levallois method in that the *arête* pattern is not formed solely by the intersection of two scars from the distal end and one or more from the proximal. There are now some examples where only a single scar of the pattern originates from the distal core extremity (Fig. 2.5). In addition, Level 2 sees the development of single platform reduction strategies and the introduction of strategies which utilize two or more unopposed platforms.

A single platform reduction strategy appears here which becomes dominant in Level 4, the uppermost at Boker Tachtit. This initially calls for the selection of a thin, tabular nodule. A platform is produced at one end by the removal of one or more flakes, morphologically similar to core tablets. No *lame à crête* ridge is produced and all decortification takes place from this single platform, as blades are peeled from one thin edge of the nodule. Platform rejuvenation may take place either by the removal of true core tablets or by the removal of small flakes which facet the striking platform.

A variation which is seen first in Level 2 is the selection of large flakes to use as cores. This is like selecting a thin nodule but there is no cortex and the dorsal and ventral surfaces are very regular and usually parallel. To prepare one lateral, thin edge, from which the blades are to be struck, a unifacial *lame à crête* is made, along with platforms at each end. Blades are then removed as described for all the opposed platform cores.

Perhaps the most striking innovation in Level 2 was the selection of large blocks of raw material for reduction. At times, this appears to have been done so as to increase the size of blanks produced; opposed platform rejuvenation quickly reduces the effective length of any core (Fig. 2.2). However, much of the time these blocks merely were split into two or more pieces with each being treated as a separate core.

While the Levallois method is still present in Level 2, the other strategies are not Levallois as they fail to predetermine the specific shape of any removed blank. The techniques utilized are those also utilized in some Levallois technology (faceted platforms, opposed platform flaking, numerous refacetings of the platforms, etc.); however, these are more directly related to the mechanical requirements of hard hammer flaking than they are to any one method of core reduction. It is clear, however, that Level 2 exhibits reduction strategies formed by combinations of old and new approaches to flint knapping. In this sense, it is truly transitional and quite heterogeneous.

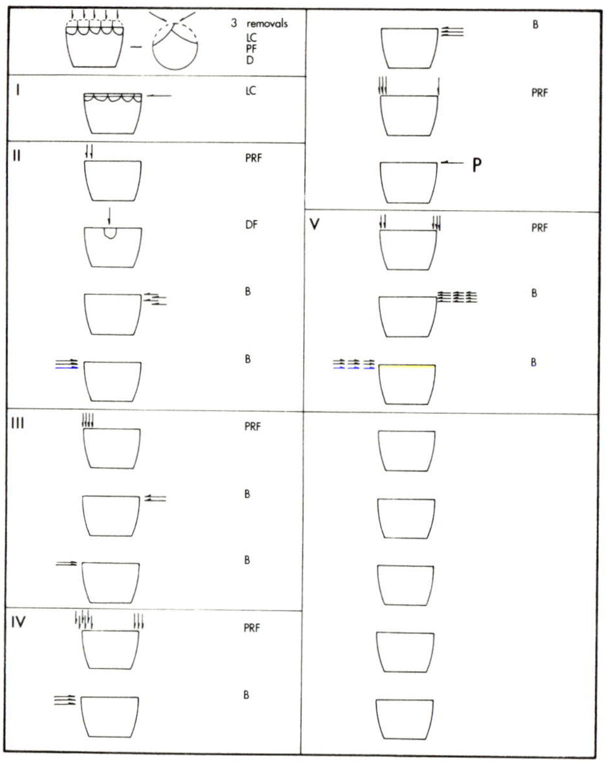

Fig. 2.6. Schematic representation of step by step reduction of a Level 2 point core.

Level 3.

Because of the small area in primary context, only six cores could be extensively reconstructed, but their patterns are consistent with those of Level 2. In spite of the few reconstructions, of over 30 recovered cores most indicate opposed platform flaking. Yet, from the dorsal scar patterns on blades, it is apparent that there was an increase in their production from either true single platform cores or from longer blade series struck from one platform of opposed platform cores. In Level 1, 39% of the blades have unidirectional scar patterns. This increases to 43% in Level 2 and to 50% in Level 3. By Level 4, 76% of the blades exhibit unidirectional patterns, always from the proximal end.

Level 4.

The dominant reduction strategy is a single platform technique, seen on 43 of 52 abandoned cores. Other, poorly represented, strategies include opposed platform, two platforms unopposed, and multiple platform.

Even within this rather uniform reduction strategy, there are some variations which are sufficiently reoccurring that they cannot be considered idiosyncratic. Three basic varieties can be seen, each associated with a different width of utilized raw material. When the nodule is thin, often tabular, the platform was produced by a single blow which removed a core tablet-like flake from one extremity. Blades were then simply peeled back from the original cortex covered edge of the nodule. Control of the platform was done by either the removal of a second flake which took off the whole top of the core or, more often, through the limited use of fine faceting. Thus, many of the blades produced have faceted platforms. In this variety, it is always a thin edge of the nodule which was chosen as the main flaking surface. This reduction strategy first appears in Level 2, although it is much more prevalent here. In addition to tabular nodules, large flakes were also often treated in this manner.

The second variety of this single platform reduction strategy came into play when a nodule of medium width was chosen. Because of its size, it normally has no thin edge, being either pointed oval or squatly rectangular in cross-section. In all cases, however, there is a natural crest along its long axis. The mass is not so large that initial platform formation needed to be modified. Rather, the platform was set up by the removal of a single flake, but because of the shape of the raw material, there was a marked tendency to remove blades from a fairly large portion of the core's circumference (i.e., one-half to two-thirds). These blade removals were interrupted when the platform was no longer suitable, and the platform was rejuvenated by the removal of true core tablets.

Because of the larger area forming the flaking surface, occasionally pieces were removed which morphologically parallel the elongated Levallois points of Levels 1 through 3. However, from the reconstruction it is clear that they are not the result of conscious predetermination, and cores were never abandoned after the removal of such a point. Unlike the true Levallois points, there was no return to a patterned sequence of removals after a point has been produced and, if another point happened to come off the same core later, it was by chance. In addition, there are a significant number of triangular blanks with a Y-*arête* pattern which are well off axis. The points are, therefore, only part of a continuum.

A final variety occurred when a large block of raw material was used. This strategy is like the other two, except in the method of platform production and rejuvenation. Initially, the platform was set up by faceting through the removal of several large flakes. Blades were then peeled off around a large part of the core circumference. When the platform needed rejuvenation, it was done piece-meal with additional small scale faceting which, each time, affected only a small part of the platform.

These cores are often pyramidal in shape, with large platforms and lateral edges converging toward the distal end (Fig. 2.7). The wide platform caused by normal hard hammer percussion on those blades struck from such a core often resulted in pieces which were triangular in shape. In a large series of removals, a certain number tended to have symmetric triangular or leaf shaped morphology. These often were morphologically identical to Levallois points (Fig. 2.5). Like those already described for Level 4, these cannot be considered Levallois because their forms were not consciously predetermined by prior core shaping.

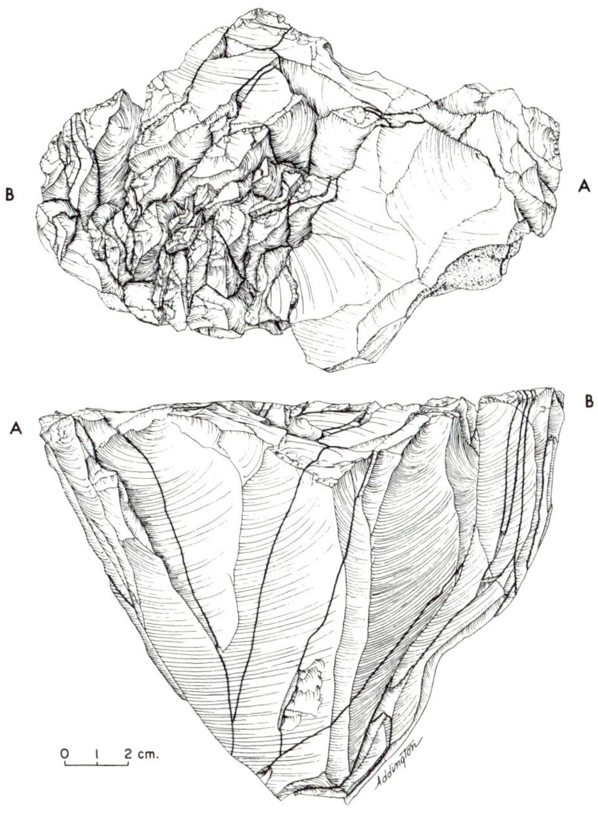

Fig. 2.7. Level 4, single platform blade core which produced seven points with Levallois morphology (see Fig. 2.5(*k*)).

Discussion

The core reduction strategies at Boker Tachtit indicate a shift from a Levallois point core technology, through a phase of considerable heterogeneity and flexibility in reduction strategies, finally to a generally standardized non-Levallois blade core technology. This latter might best be defined as a hard hammer, single platform, blade core strategy which does not utilize complex initial core preparation. This non-Levallois method, however, does produce both some morphologically typical Levallois points and what traditionally would have been considered Levallois blades (i.e., they tend to have faceted platforms and multiple dorsal scars). It is obvious that the pattern of blanks described for Boker Tachtit, Level 4, well matches that at Antelias and Ksar Akil. Even the minor right edge blunting of the points is the same.* There is no doubt in our

* Although little mention has been made of typology, both Ksar Akil and Boker Tachtit have virtually only retouched Upper Paleolithic tool types. However, the chamfered piece is probably missing at Boker Tachtit (see Marks, 1983*a* and *b* for details).

minds, after examining the Ksar Akil material, that the reduction strategy which produced most of the blanks in Levels XXIII through XXII was the same as that used at Boker Tachtit, Level 4 (Marks, 1983*a*).

The Boker Tachtit reconstructions make it necessary to confront the problem of Levallois piece recognition which, for blades at least, has been of some concern for the past several years. In addition, these reconstructions throw the whole question of Levallois points together with the blades.

As early as 1975, Jelinek suggested that true blades from Tabun, Unit IX (Garrod's level D), might be by-products of Levallois point production (Jelinek, 1975, p. 310). The same was suggested by Crew (1976), while studying the Early Levantine Mousterian site of Rosh Ein Mor. Their spatial coassociation there was confirmed at a later date (Hietala & Stevens, 1977), if not their reduction sequence association. In fact, after some unsuccessful attempts to differentiate between Levallois blades and normal blades based upon platform characteristics (Jelinek, 1975, p. 304), Jelinek finally essentially discarded the dichotomy, placing all blades with parallel edges and parallel dorsal scar patterns into a category of 'prismatic blades' (Jelinek, 1981). A somewhat less decisive resolution to the problem of Levallois blade recognition was proposed by Copeland (1981), who suggested that three groups should be recognized; Levallois, non-Levallois, and 'intermediate' (Copeland, 1981, p. 10). While Copeland was referring to all Levallois elements, she clearly has most practical and theoretical concerns with Levallois blades (Copeland, 1981, pp. 7–8).

The very clear evidence from all levels of Boker Tachtit is that blades with morphological characteristics as those described for Levallois blades (Tixier *et al.*, 1980), in fact, were mainly either preparatory removals for Levallois point shaping or, in Level 4, were nothing more than blades struck with a hard hammer technique from large, faceted, single platform cores. It is suggested, therefore, that the whole concept of the Levallois blade be abandoned in the Levant. The single exception we would make would be those where the dorsal scar patterns are formed by centripedal flaking (i.e., essentially an elongated Levallois flake) (Bordes, 1980: Figs. 1–3). Even in the Early Levantine Mousterian, well removed from the transitional levels at Ksar Akil, the characteristics referred to as 'evolved' Levallois by Copeland pertain. In a sample of 581 blades identified as Levallois by Crew at Rosh Ein Mor and used for attribute studies, a full 84% had unidirectional scar patterns originating from the proximal end and, with almost all of these, the blade edges were either somewhat converging or were parallel. Of the remaining some 15%, 3.8% have bidirectional scar patterns, and only 1.3% could be considered 'elongated Levallois flakes'. The rest are blades with scars indicating change of flaking orientation but with the major scars originating from the proximal extremity. Thus, even in this Early Levantine Mousterian, it is likely that the technique of core reduction was fully as 'evolved' as in the Lebanese transitional assemblages. Simply, it serves no purpose to call these blanks Levallois. As Copeland (1981, p. 7) has so clearly noted, the definition of Levallois blades is quite inconsistent with the definition of the Levallois method (Bordes, 1961, p. 14). Given the proportionately large number of these blades in Early Levantine Mousterian context and their relative paucity in Phase 2 Levantine Mousterian, as well as the traditional habit of lumping them with Levallois flakes, it is likely that their removal from the Levallois category might well make a Levallois Index more meaningful.

The problem of Levallois points is different, since it has been shown that often a point is *the* desired end-product of a core reduction sequence. Unlike classic Levallois flakes, however, there are a number of different reduction strategies which will purposefully produce them (Bordes, 1961, 1980; Tixier *et al.*, 1980; Guichard & Guichard, 1965). The question is whether it is possible to distinguish among these strategies and distinguish between them, as representing Levallois method, and the comparable points such as those from Boker Tachtit, Level 4, on the basis of final point morphology.

Because of the special shape and *arête* pattern of Levallois points, all reduction strategies which produce them purposefully, or otherwise, must be either unidirectional or bidirectional in their penultimate stage. It is simple, based on direction of flake scar, to determine which of the two is involved and, with some experience, the actual order of preparatory removals can be determined (Crew, 1975). Beyond that, we doubt that the point morphology can provide a sure guide to reduction method. On the other hand, there are other data in any assemblage which should give clues. For instance, it seems clear that the recognition of Levallois points, *per se*, is more convincing if struck Levallois point cores are associated. Large numbers of such cores speak eloquently for the presence of a true Levallois point method, even if the points themselves are missing. The reverse, unfortunately, is not necessarily true, since Levallois point cores may be further reduced until unrecognizable, as often happened in Boker Tachtit, Level 2. This is a particular problem when the points indicate a unidirectional reduction strategy. From the Boker Tachtit experience, we would suggest that such points, unassociated with struck Levallois point cores, but associated with large numbers of blades with unidirectional scar patterns, are probably not Levallois in any real sense. It is much more likely that they form part of a blade producing complex which includes a hard hammer mode of detachment and, at least, some pyramidal shaped core flaking surfaces.

In retrospect, we are tempted to suggest that this was probably the dominant method of point and blade production at the Early Levantine Mousterian site of Rosh Ein Mor, in spite of the original attribution of these elements to a Levallois method (Marks & Crew, 1972; Crew, 1976). In short, we are suggesting that our understanding of the Early Levantine Mousterian has been clouded by uncritical acceptance of the idea that artifact morphology is a sure guide to method of production. In the Levant, at least, this needs careful re-evaluation. Perhaps it is time to dispense with assumptions about broad technological *methods* and attempt to reconstruct specific reduction strategies.

Detailed descriptions of such strategies are likely to provide a sounder base of understanding technological developments than will over used and vaguely defined terms such as Levallois is today.

Bibliography

Azoury, I. (1971). A Technological and Typological Analysis of the Transitional and Early Upper Paleolithic Levels of Ksar Akil and Abu Halka. University of London: PhD dissertation.

Bar-Yosef, O. & Vandermeersch, B. (1972). 'The stratigraphical and cultural problems of the passage from Middle to Upper Paleolithic in Palestinian caves.' In: *The Origin of Homo Sapiens*, edited by F. Bordes. Paris: UNESCO, pp. 221–6.

Binford, S. R. (1966). 'Me'arat Shovakh: Mugharet-esh-Shebbabiq.' *Israel Exploration Journal*, **16**, 18–32, 96–103.

Bordes, F. (1961). *Typologie du Paléolithique Ancien et Moyen.* Publications de l'Institut de Préhistoire de l'Université de Bordeaux, Mémoire No. 1. Bordeaux: Imprimeries Delmas.

Bordes, F. (1980). 'Le débitage Levallois et ses variants.' *Bulletin de la Société Préhistorique Française*, **77**, 45–9.

Bordes, F. (1981). 'Vingt-cinq ans aprés: le complexe mousterien revisité. *Bulletin de la Société Préhistorique Française*, **78**, 77–87.

Bordes, F. & Crabtree, D. (1969). 'The Corbiac blade technique and other experiments.' *Tebiwa*, **12**, 1–21.

Cahen, D. & Moeyersons, J. (1977). 'Subsurface Movements of Stone Artifacts and their Implications for the Prehistory of Central Africa.' *Nature*, **266**, 812–15.

Cahen, D., Keeley, L. H. & Van Noten, F. L. (1979). 'Stone tools, tool kits and human behavior in prehistory.' *Current Anthropology* **20**, 661–83.

Copeland, L. (1970). 'The early Upper Paleolithic flint material from Levels VII–V, Antelias Cave, Lebanon. *Berytus*, **19**, 99–143.

Copeland, L. (1975). 'The Middle and Upper Paleolithic of Lebanon and Syria in the light of recent research.' In: *Problems in Prehistory: North Africa and the Levant*, edited by F. Wendorf & A. E. Marks. Dallas: Southern Methodist University Press, pp. 317–50.

Copeland, L. (1981). 'Levallois or Non-Levallois? Reflections on some technical features of Hummal, Level IA (El-Koum, Syria). Comision V.' *Terminología de la Prehistoria del Cercano Oriente.* X Congreso, Union Internacional de Ciencias Prehistoricas y Protohistoricas, pp. 1–19. Preprint.

Crew, H. L. (1975). 'An evaluation of the relationship between the Mousterian complexes of the eastern Mediterreanean: a technological perspective.' In: *Problems in Prehistory: North Africa and the Levant*, edited by F. Wendorf & A. E. Marks. Dallas: Southern Methodist University Press, pp. 427–38.

Crew, H. L. (1976). The Mousterian site of Rosh Ein Mor. In: *Prehistory and Paleoenvironments in the Central Negev, Israel. Vol. 1. The Avdat Aqev Area, Part 1*, edited by A. E. Marks. Dallas: Southern Methodist University Press, pp. 75–112.

Ewing, J. F. (1947). 'Preliminary note on the excavation at the Paleolithic site of Ksar Akil, Republic of Lebanon. *Antiquity*, **21**, 186–96.

Ferring, R. C. (1979). Technological variability and change in the Late Paleolithic of the Negev. Unpublished Ph.D. dissertation, Southern Methodist University, Dallas.

Fish, P. R. (1979). *The Interpretive Potential of Mousterian Debitage.* Arizona State University. Anthropological Research Papers No. 16.

Garrod, D. A. E. (1951). 'A transitional industry from the base of the Upper Paleolithic in Palestine and Syria.' *Journal of the Royal Anthropological Institute*, **81**, 121–9.

Garrod, D. A. E. & Bate, D. M. A. (1937). *The Stone Age of Mount Carmel*, Vol. 1. Oxford: Clarendon Press.

Guichard, J. & Guichard, G. (965). 'The Early and Middle Paleolithic of Nubia: a preliminary report.' In: *Contributions to the Prehistory of Nubia*, edited by F. Wendorf. Dallas: Southern Methodist University Press, pp. 57–116.

Haller, J. (1946). 'Notes de préhistoire phénicienne: l'Abri d'Abou Halka (Tripoli).' *Bulletin du Musée de Beyrouth*, **6**, 1–20.

Hietala, H. J. & Marks, A. E. (1981). 'Changes in spatial organization at the Middle to Upper Paleolithic transitional site of Boker Tachtit, Central Negev, Israel.' In: *Préhistoire de Levant*, edited by J. Cauvin & P. Sanlaville. Paris: Editions du Centre National de la Recherche Scientifique, pp. 305–18.

Hietala, H. J. & Stevens, D. E. (1977). 'Spatial analysis: multiple procedures in pattern recognition studies.' *American Antiquity*, **42**, 539–59.

Jelinek, A. J. (1975). 'A preliminary report on some Lower and Middle Paleolithic industries from the Tabun Cave, Mount Carmel (Israel). In: *Problems in Prehistory: North Africa and the Levant*, edited by F. Wendorf & A. E. Marks. Dallas: Southern Methodist University Press, pp. 297–315.

Jelinek, A. J. (1981). 'The Middle Paleolithic in the southern Levant from the perspective of the Tabun Cave.' In: *Préhistoire du Levant*, edited by J. Cauvin & P. Sanlaville. Paris: Editions du Centre National de la Recherche Scientifique, pp. 265–85.

Marks, A. E. (1977). 'The Upper Paleolithic sites of Boker Tachtit and Boker: a preliminary report.' In: *Prehistory and Paleoenvironments in the Central Negev, Israel. Vol. 2. The Avdat/Aqev Area, Part 2 and the Har Harif*, edited by A. E. Marks. Dallas: Department of Anthropology, Southern Methodist University, pp. 61–80.

Marks, A. E. (1981). 'The Upper Paleolithic of the Negev.' In: *Préhistoire du Levant*, edited by J. Cauvin & P. Sanlaville. Paris: Editions du Centre National de la Recherche Scientifique, pp. 343–52.

Marks, A. E. (1983*a*). 'The Middle to Upper Paleolithic transition in the Levant.' In: *Advances in World Archaeology*, edited by F. Wendorf & A. Close. Vol. 2. New York: Academic Press, pp. 51–98.

Marks, A. E. (1983*b*). (editor) *Prehistory and Paleoenvironments in the Central Negev, Israel, Vol. III. The Avdat/Aqev Area, Part 3.* Dallas: Department of Anthropology, Southern Methodist University.

Marks, A. E. & Crew, H. L. (1972). 'Rosh Ein Mor, an open-air Mousterian site in the Central Negev, Israel.' *Current Anthropology*, **13**, 591–3.

Marks, A. E. & Volkman, P. W. (1983). 'Changing core reduction strategies: a technological shift from the Middle to the Upper Paleolithic in the southern Levant.' In: *The Mousterian Legacy: Human Biocultural Change in the Upper Pleistocene*, edited by E. Trinkaus. British Archaeological Reports, International Series, pp. 13–33.

Neuville, R. (1934). 'Le Préhistorique de Palestine.' *Revue Biblique*, **43**, 237–59.

Neuville, R. (1951). *Le Paléolithique et le Mésolithique du Désert de Judée*, Archives de l'Institut de Paléontologie Humaine, Mémoire 24.

Newcomer, M. H. (1975). 'Punch technique and Upper Paleolithic blades.' In: *Lithic Technology: Making and Using Stone Tools*, edited by E. Swanson. The Hague: Mouton Publishers, pp. 97–102.

Semenov, S. A. (1964). *Prehistoric Technology*, translated by M. W. Thompson. London: Cory, Adams & Mackay.

Stekelis, M. (1954). 'Nouvelles fouilles dans la Grotte de Kébarah.' *Cronica del* IV *Congreso Internacional de Ciencias Prehistoricas y Protohistoricas, Madrid*, pp. 385–89.

Stekelis, M. (1961). 'Iraq-el-Baroud, nouvelle grotte préhistorique au Mont Carmel.' *Bulletin of the Research Council of Israel*, **10G**, 302–20.

Tixier, J., Inizan, M. & Roche, H. (1980). *Préhistoire de la pierre taillée; Vol. 1, Terminologie et technologie.* Valbonne: Centre de Recherche et d'Etudes Préhistoriques.

Villa, Paola (1982). 'Conjoinable Pieces and Site Formation Processes.' *American Antiquity*, **47**, 276–90.

Volkman, P. W. & Kaufman, D. (1983). Reassessment of the Emireh Point as a possible type fossil for the technological shift from the Middle to the Upper Paleolithic in the Levant.' In: *The Mousterian Legacy: Human Biocultural Change in the Upper Pleistocene*, edited by E. Trinkaus. British Archaeological Reports International Series, pp. 35–51.

Wilmsen, E. N. (1970). *Lithic Analysis and Cultural Inference: A Paleo-Indian Case.* Anthropological Papers, No. 16. Tucson: The University of Arizona Press.

3

Intentional breakage in a late Upper Palaeolithic assemblage from southern England

C. A. BERGMAN, R. N. E. BARTON, S. N. COLLCUTT AND G. MORRIS

This paper explores the use of intentional breakage in a late upper palaeolithic assemblage from Hengistbury Head (Dorset), dated to 12 500±1150 BP (Barton & Huxtable, in press). The deliberate breaking of flaked stone artefacts was extensively practised in prehistoric times as evidenced by finds covering a wide geographic and temporal range.

A flake or blade can be treated in a variety of ways after being detached from the core. It may be regarded as waste; on the other hand it can be modified by burin blow technique, retouch or intentional breakage (Bordes *fracture volontaire*, 1953). At Hengistbury Head an intentional break is a percussion-induced fracture which produces two substantial pieces. The resulting break surfaces are usually transverse to the long axis of the blank and have no pronounced concavity. The blow is delivered well away from the lateral edges of the piece and clearly differs from microburin technique.

Depending on the needs of the knapper a blank may be broken at any interval along its length. Some of our examples have only their proximal or distal ends removed while others are segmented into as many as six pieces. Preliminary microwear analysis suggests some of these segments were used without being retouched.

History of research

In 1953 Bordes described a series of Mousterian flakes as being intentionally broken. The presence of percussion-induced features such as cones lent support to his idea that these pieces were broken deliberately. Bordes noted that the blow was usually delivered on the ventral surface; that some examples had opposed cones showed they were broken on an anvil.

In the following years several authors discussed material similar to that described by Bordes (see, for example, Pradel, 1957, 1959). Delarue & Vignard (1958) examined three upper palaeolithic sites and demonstrated that intentional breakage was used in tool manufacture and repair. Eloy (1954) introduced an entirely different type of break. He described breaks on the tips of Mousterian points which differed from Bordes' examples in the position of the blow and the resulting features. This type of break has recently been studied by Benito del Rey (1982) at the Mousterian site of Las Grajas, Spain.

Mace began the first systematic investigation of the upper palaeolithic site on Hengistbury Head in 1957. Her published report (1959) discussed the use of intentional breakage in burin manufacture and the modification of unretouched flakes and blades. The belief that these pieces were deliberately broken was supported by the fact that several breaks refitted.

Semenov (1964, pp. 63–4) described broken pieces from the upper palaeolithic site of Kostienki I, USSR. Due to the uniform shapes and sizes of the broken blades, he felt that they were segmented with the use of a flint intermediary or 'punch'. As the resulting pieces were small, he suggested that these artefacts had been hafted.

Bosinski & Hahn (1972) presented material from the Magdalenian site of Andernach, West Germany, together with some experimental results. They noted lips, conchoidal fracture marks and 'negative bulbs' on pieces thought to be intentionally broken.

Experimental studies have recently been the focus of work by Owen (1982) and Roche & Tixier (1982). Using a variety of methods to snap flakes and blades, Owen shows that 'impact scars' are found only on pieces which have been struck. She also notes that percussion-induced fractures have more varied break surfaces than other types of snaps. Although Roche & Tixier (1982) concentrate on knapping accidents, they describe aspects of break morphology and spontaneous effects which also appear in our archaeological sample and experimental work.

The technique has been recognised at the rock-shelter of Ksar 'Akil, Lebanon (Newcomer, 1972, p. 106; Tixier & Inizan, 1981) and at the neolithic settlement site of Shams ed-Din Tannira, Syria (Azoury & Bergman, 1980), where it was used in burin manufacture. The later upper palaeolithic sites of Meer, Belgium (van Noten, 1978; Cahen *et al.*, 1980) and Olbrachcice, Poland (Burdukiewicz, 1981) have refitted examples of burins on intentional breaks. The technique was used during the British Iron Age to make lathe bits (Calkin, 1953); it survived until recently as part of the process of gunflint manufacture (see Mason, 1978, on gunflints).

The site

Hengistbury Head is situated in southern England, some five miles east of Bournemouth in Dorset (NGR SZ 175905) (Fig. 3.1). Today, Hengistbury forms a major promontory

Fig. 3.1. The location of Hengistbury Head (with kind permission of Elek Books Ltd, London).

which projects into the Channel and partly flanks the western entrance to the Solent. On its northern side lies the protected harbour of Christchurch into which flow the major Wessex rivers of the Stour and Avon.

The upper palaeolithic site lies on the exposed southern side of the headland in a shallow depression which runs out to the cliff-edge at 11 m OD (Fig. 3.2). Most of the artefacts occur near the cliff-edge in windblown sand deposits about a metre in thickness. The archaeological horizon comprises a well-defined scatter of flints with a few pieces of red ochre and several flat fragments of sandstone (Barton & Bergman, 1983). The artefacts are concentrated in a band about 20 cm thick with the number of finds diminishing with increasing depth. Extensive refitting has shown they belong to one upper palaeolithic assemblage (Barton & Bergman, 1982).

Excavations at the site have taken place twice before – in 1957 by A. Mace (1959), and by J. Campbell in 1968–9 (Campbell, 1977). During this time more than 200 m^2 (136 m^2 by J.C.) were examined, producing over 6500 artefacts of which 361 were retouched tools. The new excavations began in 1981 and are continuing. So far they have resulted in the recovery of 117 retouched tools and more than 3000 artefacts from an area of 43 m^2 near the cliff-edge.

The local flint on the headland derives mainly from cryoturbated Pleistocene gravels and occurs in the form of small cobbles marked with surface chattering. The refitting of cores at the site has shown that several were made on blocks weighing more than two kilogrammes; these blocks are clearly distinguishable from the local flint by their size and well-developed cortex. Today, the nearest known source of this material is a Cretaceous chalk escarpment lying about 12 km south-east of the site on the Isle of Wight. Other chalk outcrops exist inland of the site further to the west.

Blades and bladelets are well represented in the assemblage. Their butts are either plain or faceted; many are small and narrow, indicative of the blow falling close to the edge of the core's platform. Bulbs of percussion are generally diffuse and there is frequently a tiny lip between the butt and the ventral surface; these features suggest a soft hammer flaking mode (Newcomer, 1975; Ohnuma & Bergman, 1982). Many of the blades show signs of platform abrasion; however, it has been noted that blades with large faceted butts generally have no signs of this technique to remove overhang.

The refitting of blades and flakes to cores has helped explain particular aspects of the flaking technology. For example, the technique of platform faceting was employed only when necessary to modify the platform angle (see Karlin, 1972). This is well illustrated by the fact that blades with plain butts and faceted butts have been refitted to the same end of an opposed platform core. The presence of core tablets shows that other methods were also used for rejuvenating platforms.

From observations on reassembled cores, the flaking method seems to follow a consistent pattern. After preparing a striking platform at one end of the nodule, the knapper crested the block's long axis. The removal of the crested piece was followed by the detachment of long blades, some with cortical surfaces. Core reduction was accelerated by faceting of the platform and in one example over 63 flakes and blades were produced from the core before it was abandoned.

A detailed analysis of the retouched tool assemblage will not be presented here as the site is still being excavated. The present collection from the three excavations consists of over 470 retouched tools. Whilst most tools are made on blades and bladelets many scrapers and burins are made on flakes.

Relatively high numbers of *end-scrapers* have been recovered. Out of approximately 100 examples from the site well over half are made on flakes. They are nearly always made on thick blanks which are sometimes cortical. Some of the tools have breaks that might have occurred during resharpening (see Fig. 3.6 (1)).

Only a few *composite tools* have been found including a burin on a truncation combined with an end-scraper (Mace, 1959, p. 44, fig. 6).

There is only one typical *piercer*, from Campbell's excavations, made at the distal end of a retouched blade.

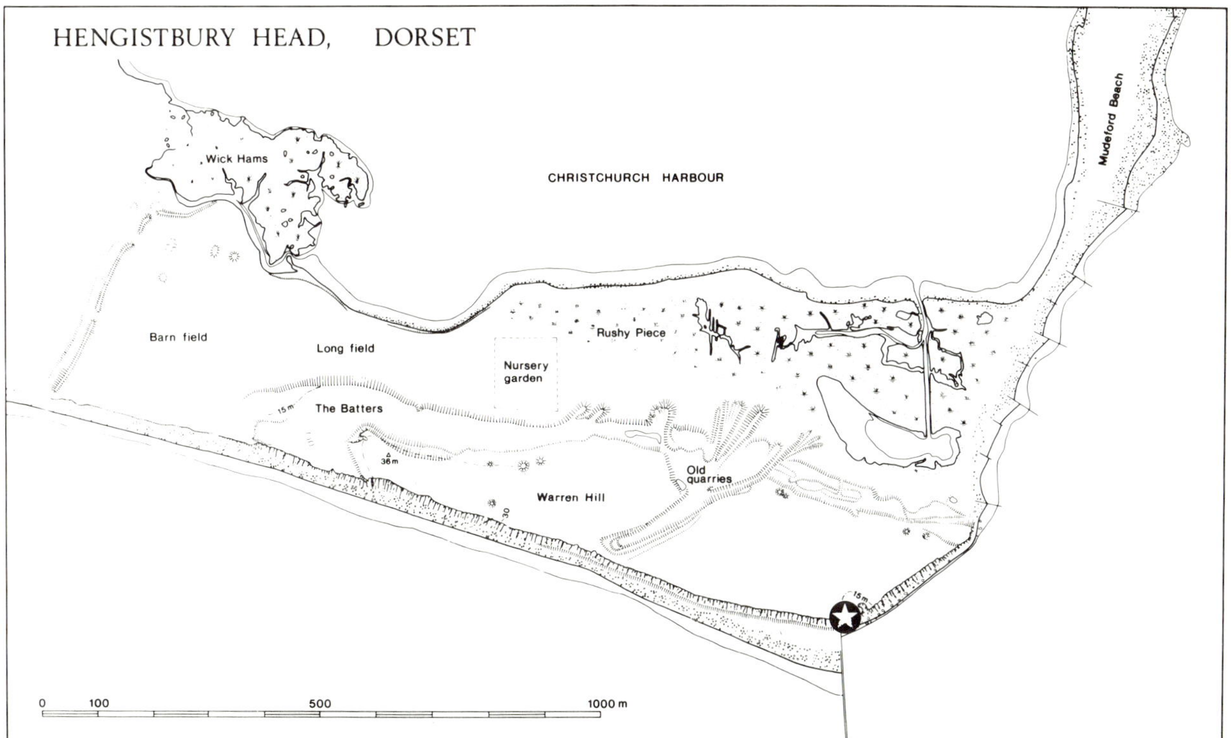

Fig. 3.2. Map of Hengistbury Head showing the excavation indicated by a star.

The few examples of *becs* occur on the distal ends of thick blanks. The scarcity of these tools compares well with some late glacial assemblages from north Germany (Schwabedissen, 1954).

More than 90 *burins* come from the site, with those on truncations and breaks providing the most common types. The majority were recovered from the Mace excavations; an important concentration occurs in the western part of the site. Localised associations of tools such as these may hint at the presence of specialised activity areas.

The *shouldered point* of Hamburgian type is rare, with only five examples, one of which is a surface find. A few thick tanged pieces also exist, but none so far that resemble the Bromme-type points already reported from the headland (Mace, 1959).

Backed blades and bladelets with semi-abrupt and abrupt retouch down one edge are the most common forms in the tool assemblage. Approximately two-thirds of over 200 backed pieces are burnt.

The intentional breaks

The present combined assemblage from the three excavations contains 105 artefacts which have one or more breaks considered to be intentional. These represent approximately 10% of the more substantial flakes and blades. There are 144 separate break surfaces.

At Hengistbury, the blade core flaking sequence produces different types of débitage, ranging from small 'chips' to flakes and blades. Some pieces are then carefully chosen to make specific tools; for example, the kind of blank selected to be broken would never be used to make a shouldered point.

A comparison with a random sample of flakes and blades from the site, shows that deliberate breaks tend to be wider and thicker (Fig. 3.3). Thin blanks easily shatter on impact and are not suitable for making tools by this technique. Cortex occurs on 45% of all deliberate breaks; one-half of these only have 20% or less of their dorsal surfaces covered. The products of core preparation and rejuvenation are seldom used as blanks. The sample includes only one fragment of a core tablet and three crested blade fragments.

An examination of the intentionally broken pieces from Hengistbury produced a list of eight major features which occur on or near the break surface. The following descriptions of these attributes relate to the intentional breaks and are not meant to cover more general aspects of knapping.

1 *Point and cone of percussion.* These are the most widely recognised characteristics of intentional breakage and result from impact. As Bordes observed in 1953, the use of an anvil sometimes produces opposed cones. This effect occurs when the force of the blow passes through the piece and rebounds off the anvil. The point and cone of percussion on a deliberate break are identical to those found on the ventral surface of a flake (Fig. 3.4 (1): coded '*a*' on Fig. 3.5).
2 *Negative cone of percussion.* Clearly, if a cone forms on one break surface it will leave a shallow negative scar on the opposite surface (Fig. 3.4 (2): also coded 'a' on Fig. 3.5).
3 *Wedge-shaped fracture lines.* When a break is initiated fine cracks may develop on either surface of a blank and spread out towards the lateral edges. In some cases they run right through the piece and a small triangular wedge (Bosinski & Hahn, 1972) of flint is detached (Fig. 3.8 (1), (5): coded 'b' on Fig. 3.5).
4 *Incipient cones.* When a blow is delivered with insufficient force to produce a break, an incipient cone (Breuil, 1932) may form. The cone usually occurs on the surface that receives the blow; however, it can develop when force is directed up from an anvil (Fig. 3.7 (5): coded 'c' on Fig. 3.5).
5 *Crushing on dorsal ridges.* Ridges represent the high points of a flake or blade's dorsal topography and are easily subjected to crushing from the hammer or anvil (Fig. 3.7) (5): coded 'd' on Fig. 3.5).
6 *Spontaneous effects.* Spontaneous effects develop when a piece is breaking and are entirely accidental (see Newcomer, 1976; Tixier, Inizan & Roche, 1980, p. 64). They can imitate the effects of a burin blow or deliberate retouch and make the identification of retouched tools difficult (Fig. 3.6 (2), (4), (6), (7)).
7 *Lips.* For the purpose of this paper, a lip is a tongue-shaped protrusion which occurs at the intersection of a break and the dorsal or ventral surface. Many pieces have short lips; we recorded the presence of a lip if it projected at least 5 mm beyond the surface of the break. These lips differ from those illustrated by Bordes (1970, Fig. 3.7 (6)–(12)) in that they do not have a stepped appearance (Fig. 3.4 (3), (4): coded 'f ' on Fig. 3.5).
8 *Conchoidal fracture marks.* Conchoidal fracture marks are the result of a passage of a curved shock wave through the piece. The rings are concentric around the point of initiation of the fracture which may sometimes, but not always, be a point of impact (Fig. 3.4 (4): coded 'f' on Fig. 3.5).

In order to understand the formation of the above attributes and their relationship to each other we have produced over 300 experimental break surfaces. The various experimental stages (blank and break manufacture, attribute recognition and study) were carried out by several different people, sometimes under blind test conditions. The knappers used different flint types, hammers and anvils, and varied the point of anvil contact. The power of the blow differed as a function of the skill of the knapper and resistance of the flint to fracture. Given the size of the upper palaeolithic blanks, an anvil and direct percussion were always used during the experiments. Other methods (e.g. snapping in the fingers, or use of a punch) proved impossible or difficult to control.

Our attribute list contains several features which we feel result from immediate contact with a hammer or anvil: points and cones of percussion, incipient cones and dorsal ridge crushing. These features can be called 'contact

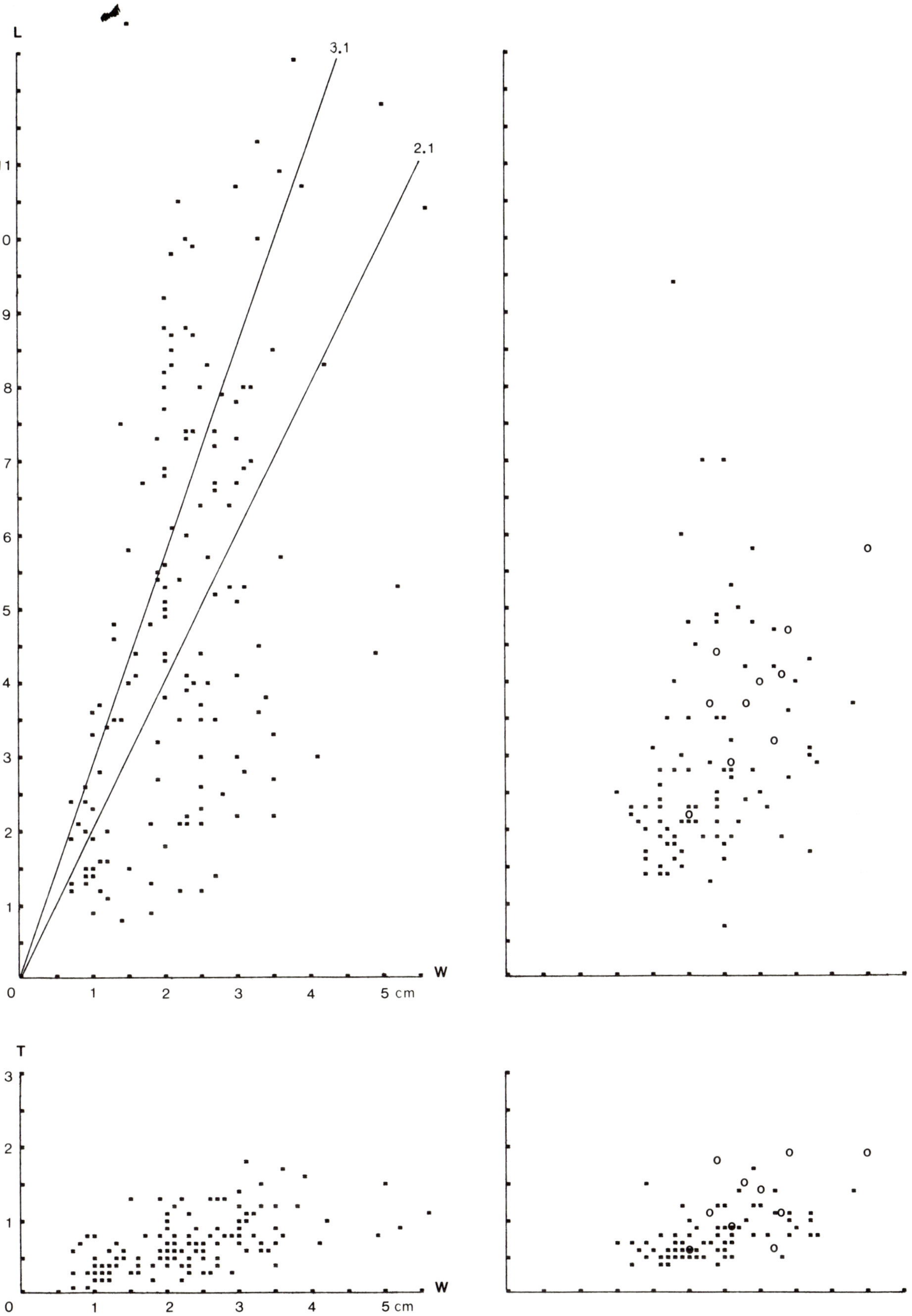

Fig. 3.3. Dimensions of flakes and blades, and intentional breaks. Left side: Length/Width and Thickness/Width scatter diagrams of a random sample of 150 flakes and blades. Right side: the same dimensions for 104 intentionally broken pieces. The burins on intentional breaks are shown as open circles.

Fig. 3.4. (1) Point and cone of percussion on an intentional break. (2) Negative cone of percussion on an intentional break. (3) and (4) Lips on intentional breaks. (5) Experimental break with a small flake detached spontaneously.

features' and are diagnostic of intentional breakage. Another group of attributes which includes wedge-shaped fracture lines, lips and conchoidal fracture marks is not necessarily indicative of intentional fracture. These features have also been noted on knapping accidents.

Theoretically, there are many thousand combinations involving our eight attributes and their location on a piece. In the experimental and archaeological data we observed nearly two hundred. In our classification scheme (Fig. 3.5) we have been able to reduce the number of classes considerably, without losing touch with the real observations. Spontaneous effects have been ignored in the scheme because of the difficulty in recognising such features in an archaeological context with absolute certainty.

It is a logical premise that one type of point and cone of percussion, positive or negative, cannot occur without the production of the corresponding type on the refitting surface. For this reason they are considered to be of equal significance and a single symbol is used in the classification (Fig. 3.5).

In the archaeological and experimental data the location and orientation of wedge-shaped fracture lines, lips and conchoidal fracture marks form two constant patterns, one the inverse of the other. These patterns may be seen in the two entries in the left-hand box of Fig. 3.5; they are referred to as the dorsal and ventral 'WLC (wedge-shaped fracture, lip, conchoidal fracture) system', respectively. Each attribute in the system may be considered to be of equal significance to the others. In Fig. 3.5 the classes are shown with all three attributes present; this should be read as a polythetic grouping, which indicates that any combination of one, two or three attributes was observed on a piece. In our experiments, the orientation of the WLC system is not dependent on where the blow falls.

We believe that these two patterns, involving wedge-shaped fracture lines, lips and conchoidal fracture marks, are the result of a single mechanical process called flexure. When a flint is flexed or 'bent' one surface is subjected to tensile stress, the other to compressive stress. The break is initiated at the surface under tensile stress and a curved fracture front spreads through the piece leaving conchoidal fracture marks. The wedge-shaped fracture lines are more or less successful breaks reacting to the same stress system. As the curved front approaches the far side, which is under compressive stress, the fracture place is increasingly deflected from its original straight path and terminates in a lip. In intentional breakage flexure is produced by a direct blow or by a rebound shock from the anvil just as readily as by snapping the blank in the fingers or during a knapping accident (see Hayden, 1979, for papers on fracture mechanics).

Approximately 6% of the breaks in both data batches

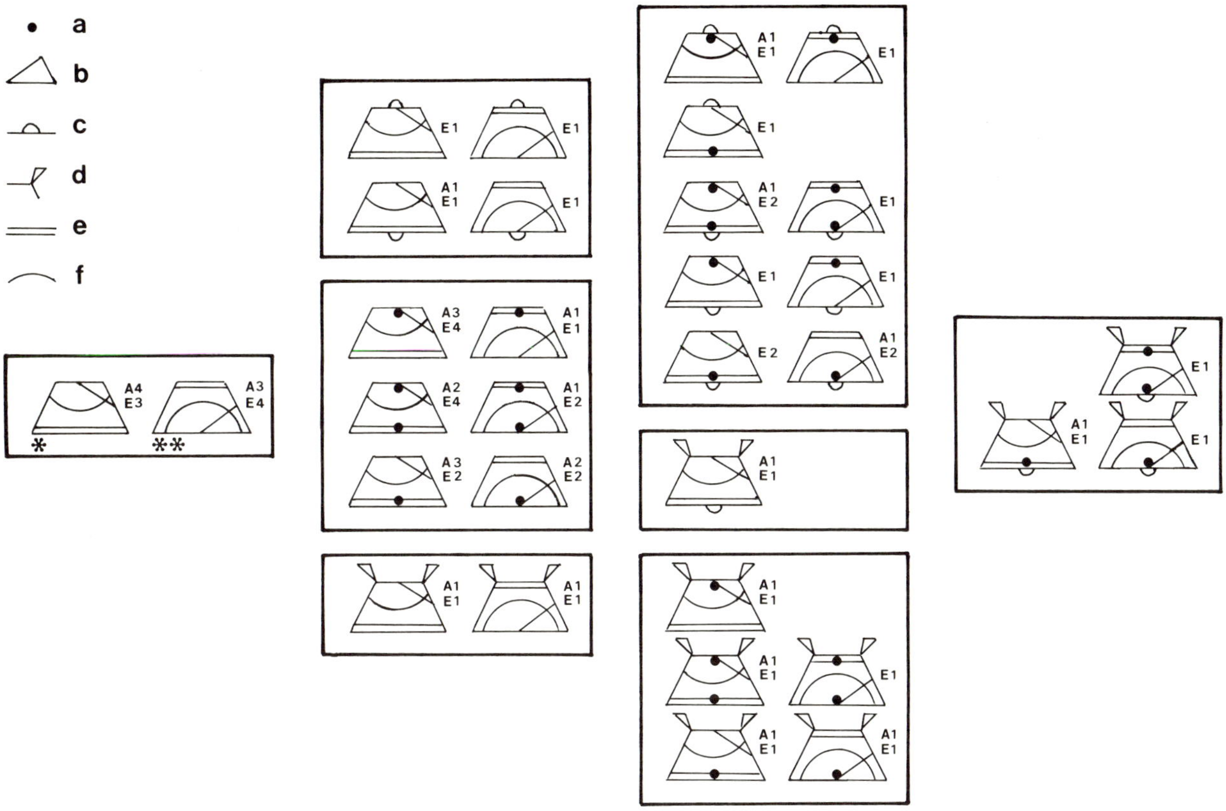

Fig. 3.5. Classification scheme of intentional breaks. Attributes: a – point and cone of percussion; b – wedge-shaped fracture line; c – incipient cone; d – dorsal ridge crushing; e – lip; f – conchoidal fracture marks. *dorsal WLC system; **ventral WLC system. Data: A – archaeological; E – experimental; 1 – rare; 2 – moderate; 3 – frequent; 4 – very frequent (frequency levels based upon actual discontinuities in the data). A total – 134; E total – 261. Pieces shown ventral side down.

have only contact features and as such could not be integrated into our classification scheme (Fig. 3.5). These pieces probably have conchoidal fracture marks which have been obscured by a coarse grain, burning or post-depositional modification of the flint. The presence of contact features, however, leaves no doubt that the archaeological examples are intentionally broken.

Given the number of unquantifiable variables, the match between the experimental data, consisting of actual intentional breaks, and the archaeological data, consisting of supposed intentional breaks, is remarkably good. All eight attributes occur in both samples and it is unlikely that any major variable in the system has been overlooked. This evidence, of course, in no way precludes the possibility that a small number of accidental breaks has been included in the archaeological sample.

A variety of spontaneous effects was noted during our experiments. In an archaeological context, they may make the identification of retouched tools difficult or impossible. A relatively common type of spontaneous retouch on our experimental pieces consists of a number of shallow scars that may appear in two different locations: (1) the retouch occurs on the break surface (Fig. 3.6 (7)) and (2) the retouch starts on the break surface and travels onto either dorsal or ventral surface (Fig. 3.6 (6): Rigaud, 1966, t.II, p. 3). The first type of retouch occurs as the edge of the break comes in contact with the hammer or anvil. The second seems to result when the piece bends as it breaks and presses off a series of tiny flakes; significantly this retouch has been noted on accidencal breaks. In both cases the retouch may be regular and continuous.

As a blank is breaking a small sliver of flint may be detached (see Bergman & Newcomer, 1983); the result imitates the effect of a burin blow. The spall is quite narrow but may travel a considerable distance down the edge (compare Fig. 3.6 (4) with Fig. 3.6 (5)).

The lateral edges which are thin and weak may become notched or denticulated due to rocking on the anvil or slipping of the hammer. The notches are Clactonian (see Bordes, 1961, p. 43) and occur on either dorsal or ventral surface, sometimes well away from the break (compare Fig. 3.6 (2) with Fig. 3.6 (3)).

Major damage can happen spontaneously on the break surface itself. Prominent lips may have large notches which are 'retouched' rather than Clactonian (Fig. 3.7 (4)). Sometimes a small flake is spontaneously detached and its scar will appear on one break surface only (Fig. 3.4 (5)). Such a detachment may result in a segment having different parts of its break surface showing conchoidal fracture marks originating from opposite directions.

During manufacture of intentional breaks, accidental ones sometimes occur. If there is anvil contact beneath the accidental break it will be identical to a deliberate one; in this case contact features may occur on one side only. With no anvil contact some combination of the WLC system attributes will result. We have also observed the formation of the WLC system without contact features on an end-scraper which broke accidentally during resharpening (Fig. 3.6 (1)). Accidental breaks occur at any distance from the point of percussion and may be accompanied by some of the spontaneous effects noted above.

In summary, there is no firm correlation between the attribute associations and the surface, dorsal or ventral, on which the blow is delivered. Any attribute may occur on either side; however, incipient cones are much more common on the surface struck by the hammer.

Secondly, there is a relationship between break attributes and the hammer type, hard (quartzite) or soft (antler). A hard hammer produces many more incipient cones than a soft hammer; sometimes several form after a single blow. Larger 'spontaneous flakes' are more commonly produced with a hard hammer; they are also produced quite easily by a hard (quartzite or limestone) anvil and a soft hammer. A hard hammer is more likely to produce contact features on both surfaces of the blank. These observations agree with those of Ohnuma & Bergman (1983) who note the 'diffuse' nature of ventral features on flakes detached by a soft hammer.

Thirdly, when contact features originate from both surfaces, we believe it indicates the use of an anvil. This situation arises in approximately 15% of the archaeological data and 35% of the experimental breaks; the latter group was entirely made with a hammer and anvil. A factor contributing to this difference could be the effect of different types of hammers.

Finally, we have never observed contact features on unsupported accidental breaks or on knapping accidents occurring during blank manufacture. We have, however, observed some of the other attributes (the WLC system and some spontaneous effects) on these accidental breaks (see Roche & Tixier, 1982). Nevertheless these attributes have been noted with no accompanying contact features on approximately 25% of our deliberately broken experimental pieces.

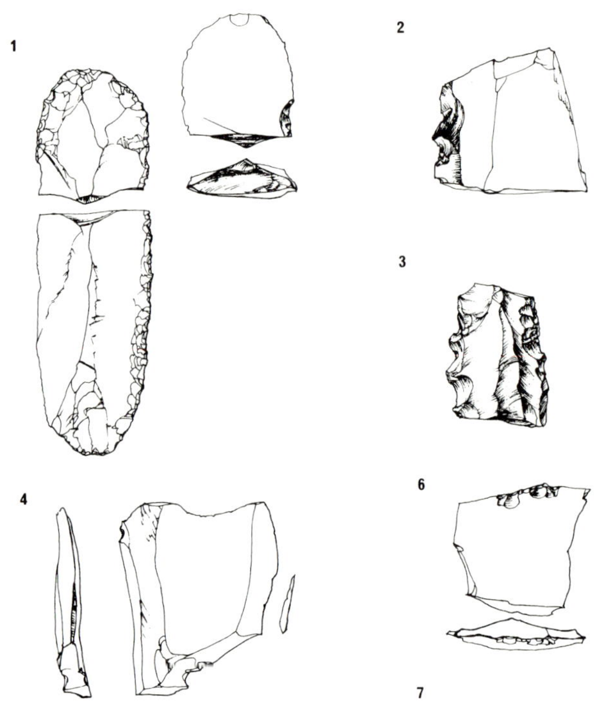

Fig. 3.6. (1) End-scraper accidentally broken while being resharpened. (2) Experimental break with notches formed spontaneously. (3) Archaeological example, intentionally broken with possible accidental edge notches. (4) Experimental piece showing spontaneous burin-like break. (5) Archaeological example of burin-like break. (6) and (7) Two experimental breaks showing spontaneous retouch with different orientations.

Refitting

Refitting has already been used to solve problems regarding the site's stratigraphy (Barton & Bergman, 1982). It has also proved valuable in the recognition of intentionally broken flakes and blades. A certain percentage of archaeological breaks have features which are not exclusively the result of deliberate breakage; as stated above accidental breaks may produce identical patterns. The grey area between intention and accident can be substantially narrowed by refitting. For example, one burin has a break associated with wedge-shaped fracture lines and conchoidal fracture marks. Both of these features have been noted on knapping accidents, when a blade bends and breaks as it is detached from a core. This particular piece, however, refits to a mesial segment with a break that has these features and a cone of percussion as well; the coarse grain of the flint masked the negative cone on the piece with the burin (Fig. 3.8 (1)).

Typological, technological and behavioural information can also be gained by refitting. A refitted burin on a break shows how one palaeolithic knapper created razor-sharp edges on a cortical flake (Fig. 3.7 (1)). After breaking the distal end of the blank, two spalls were removed which travelled up and onto the dorsal surface along the length of both lateral edges. Technologically, the tool is similar to a burin even though its morphology is quite different; the results resemble a tranchet blow more than a burin blow.

One multiple truncation burin is made on a segment which is only 24 mm long (Fig. 3.8 (3)). A refitted spall terminates at the break and clearly shows that the tool was made after the flake or blade was broken up.

Twice we have been able to refit four segments to form major parts of the original blades. One of these has at least two pieces missing and is made up of unretouched mesial segments (Fig. 3.7 (3)). The other has one fragment which was retouched after being broken off (Fig. 3.7 (2)); microwear analysis shows that this particular piece has polish formed by use. It is entirely possible that these segments were used as tools in order to economise on raw material.

Tool manufacture and repair

When a flake or blade is intentionally broken both waste products and blanks suitable for tool manufacture are produced. Some segments may be selected for use as tools without extra modification. In addition, an existing retouched tool may be broken as part of the process of repair.

Most of our sample consists of broken pieces which have not been further modified by retouch or by burin blow technique; these form over half (63) of our sample of 105

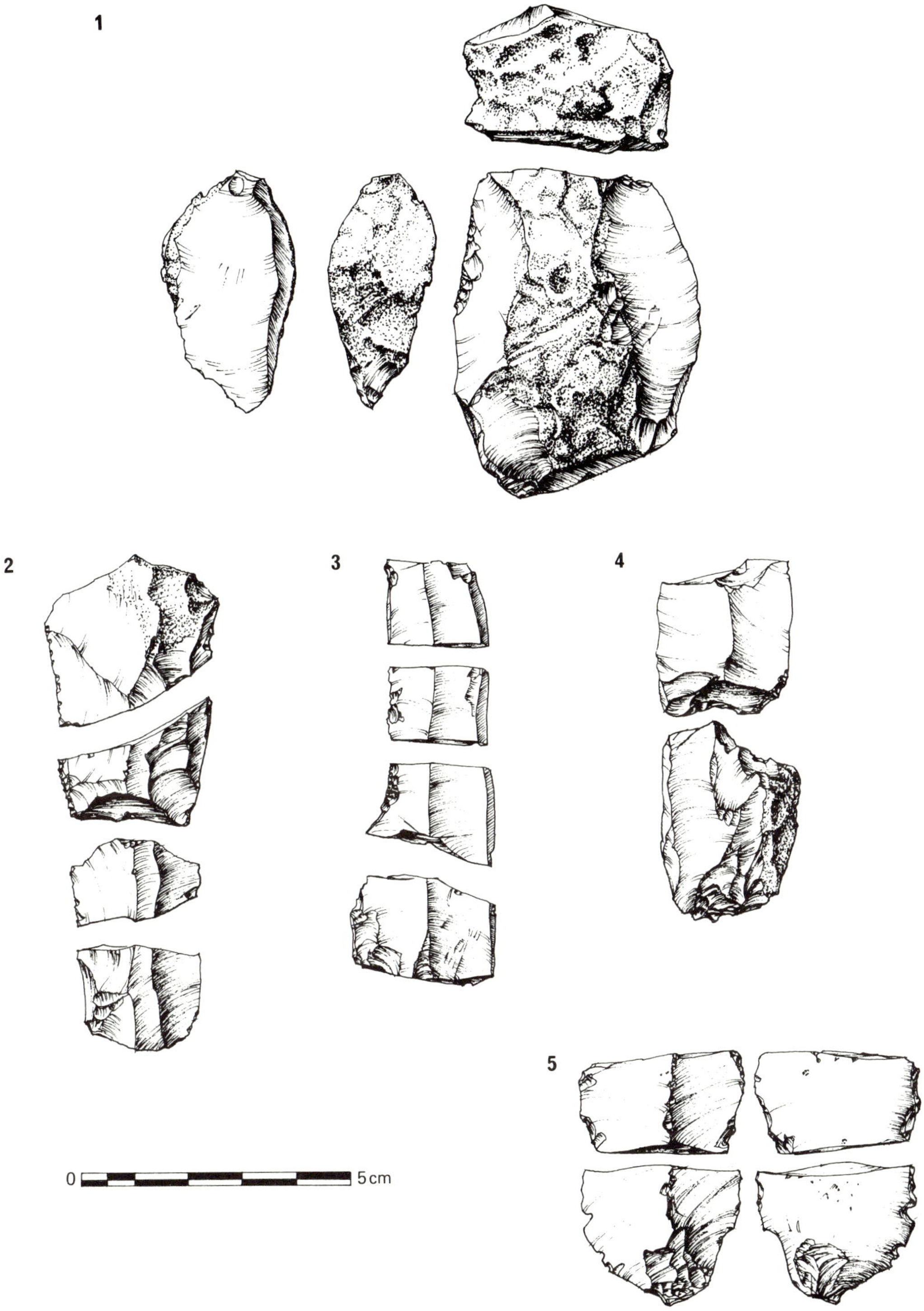

Fig. 3.7. (1) A typical burin on an intentional break with refitting spall. (2) Refits of four intentionally broken pieces forming part of the original blade. (3) Similar refitted blade; first segment and third from top show burin-like breaks along edges. (4) Refitting intentional breaks with spontaneous notch on lip. (5) Refitted intentional breaks with incipient cones (ventral) and dorsal ridge crushing.

UNIVERSITY OF BRISTOL LIBRARY

artefacts. Some of these are certainly the by-products of tool manufacture, as with the burins on intentional breaks. However, the preliminary microwear study shows that this was not always the case: some unretouched segments have polish due to use.

Ten burins are made on intentional breaks (Fig. 3.8 (2), (4), (5)); the break surface is used as a platform on which the blow is struck to detach a spall. The spalls tend to be much wider and longer than those which formed spontaneously in our experiments. These tools have a mean length of 39 mm and width of 34 mm; the standard deviations for these measurements are 10 mm and 7 mm respectively. They tend to be fairly thick, averaging about 13 mm with a standard deviation of 5 mm. The scatter diagrams show that these pieces are generally larger than most intentional breaks (Fig. 3.3). One of the tools is made on a broken core tablet, while the rest are on fragments of flakes and blades. Seven examples are mesial segments, two are distal fragments and the remaining one is made on a blank with the distal end removed. Two of the burins are multiple; one has its spalls detached from opposed break surfaces (Fig. 3.8 (5)), while the other has spalls removed from the same break (Fig. 3.7 (1)). In the same category of retouched tools made on intentionally broken blanks is the multiple truncation burin described above.

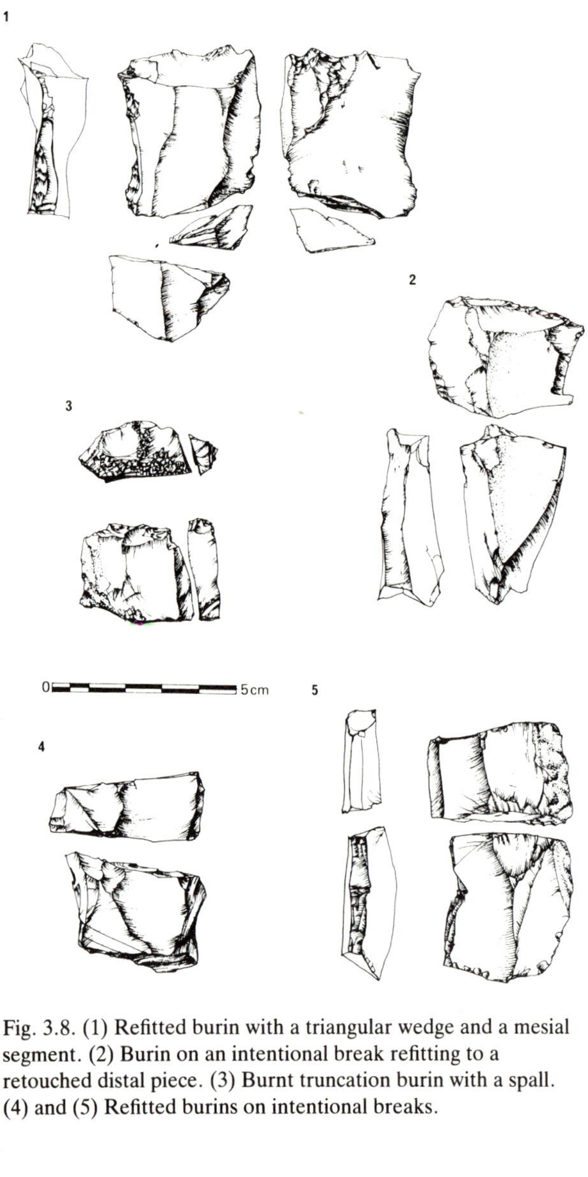

Fig. 3.8. (1) Refitted burin with a triangular wedge and a mesial segment. (2) Burin on an intentional break refitting to a retouched distal piece. (3) Burnt truncation burin with a spall. (4) and (5) Refitted burins on intentional breaks.

Sometimes, a flake or blade is broken up and a segment is then retouched. The retouch on these pieces is usually fine or semi-abrupt and occurs along all or part of a lateral edge. Retouch is also found at the intersection between the beak and the dorsal or ventral surface. Similar retouch occurs spontaneously on some of our experimental pieces and is difficult to separate from that produced by utilisation or deliberate modification.

Tools are occasionally broken, presumably for resharpening or repair. Two burins have their lateral facets 'truncated' by the effects of a blow. Besides creating a new burin edge, a platform is made from which additional spalls may be detached.

A few end-scrapers occur on intentionally broken pieces; four of these are rather short with lengths less than 25 mm. One tool exhibits rounding on the scraper edge similar to that found on some end-scrapers from the Magdalenian site of Andernach in Germany (see Bosinski & Hahn, 1972, Tafel 62, 63, 64). Due to their size, it is highly likely that these scrapers have been broken off larger blanks in order to repair their edges. The other two examples are large enough to have been made on blanks which had first been broken; however, it is impossible to tell whether any of these pieces had been retouched before or after the break. A few burins also present the same problem. Aside from these tool-types, a number of retouched pieces have lines of retouch interrupted by a subsequent deliberate break.

Microwear

Eight intentionally broken artefacts were chosen for examination under the microscope. The purposes of this limited study were two-fold: (1) to test the suitability of these pieces for microwear analysis and (2) to gain some insight into the function of the breaks.

All of the artefacts were examined under low magnifications ranging from 4× to 20×. This is useful in locating areas of edge-damage. After the initial assessment two of the artefacts were subjected to a detailed study under higher magnifications (50×, 100× and 200×).

The two intentional breaks chosen for this detailed study are mesial segments. One (J33, 21) is retouched and refits to a group of three pieces (Fig. 3.7 (2)). The microscopic polish which has affinities to that produced by working bone, occurs on the lateral edges and break surfaces. The presence of polish at these points definitely shows that the piece was used after breakage. The other example (C 1840) refits to a burin (Fig. 3.8 (1)). This tool has polish at the corners formed by the intersection of the lateral edges and the break surfaces; the polish could come from woodworking. Experiments have shown that both the break surface and its corners present stronger working edges than the relatively thin and weak lateral edges.

Discussion

We suggest that contact features are indicative of intent as we have never observed their formation in knapping accidents. The fact that about 45% of the archaeological breaks lack these features is not surprising when compared with our experiments. Approximately 25% of the experimental breaks have no contact features; all of these were broken by percussion.

A grey area exists between intentional and accidental breaks (this was also noted by Owen, 1982); individual examples may be difficult to separate. Refitting of over 30% of our sample has allowed us to extend inferences to a large group of individual breaks. As in the example noted previously, we have refitted breaks without contact features to others which have them.

The exact methods by which blanks were broken at Hengistbury Head are unclear. Our experiments have shown that the subject is extremely complex. The morphology of the breaks seems to indicate the use of two types of hammers. The comparatively low frequency of contact features, especially incipient cones, hints at the use of a soft hammer. On the other hand, breaks associated with clusters of incipient cones are most likely to be produced by a hard hammer. Direct evidence for flaking mode is absent from the site. We have recovered no hammerstones; a few refitted fragments of sandstone may be the remains of an anvil.

In our archaeological sample about 15% of the pieces have contact features originating from both the dorsal and ventral surface. This provides good evidence for the use of an anvil. Our experimental results show that the actual number of pieces broken on an anvil is probably higher than this figure indicates. Only 35% of the experimental breaks have opposed contact features even though they were all made on an anvil. As stated earlier, variability in hammer or anvil hardness may account for the difference in these figures.

Finally, the position of incipient cones on the artefacts points to a tendency for the blow to be delivered ventrally. The ventral side certainly presents a flat surface suitable for striking. However, our experimental data have shown that this problem is more complex than suggested by previous authors.

In terms of their horizontal distribution these artefacts cluster in a relatively small area on the western side of the site (see Fig. 3.9). Only a small number were found in the 136 m^2 excavated by Campbell. Most intentional breaks come from Mace's area; we are unable to position the refitted examples due to the fact that no exact co-ordinates were taken during her excavations.

The number of intentionally broken pieces from Hengist-

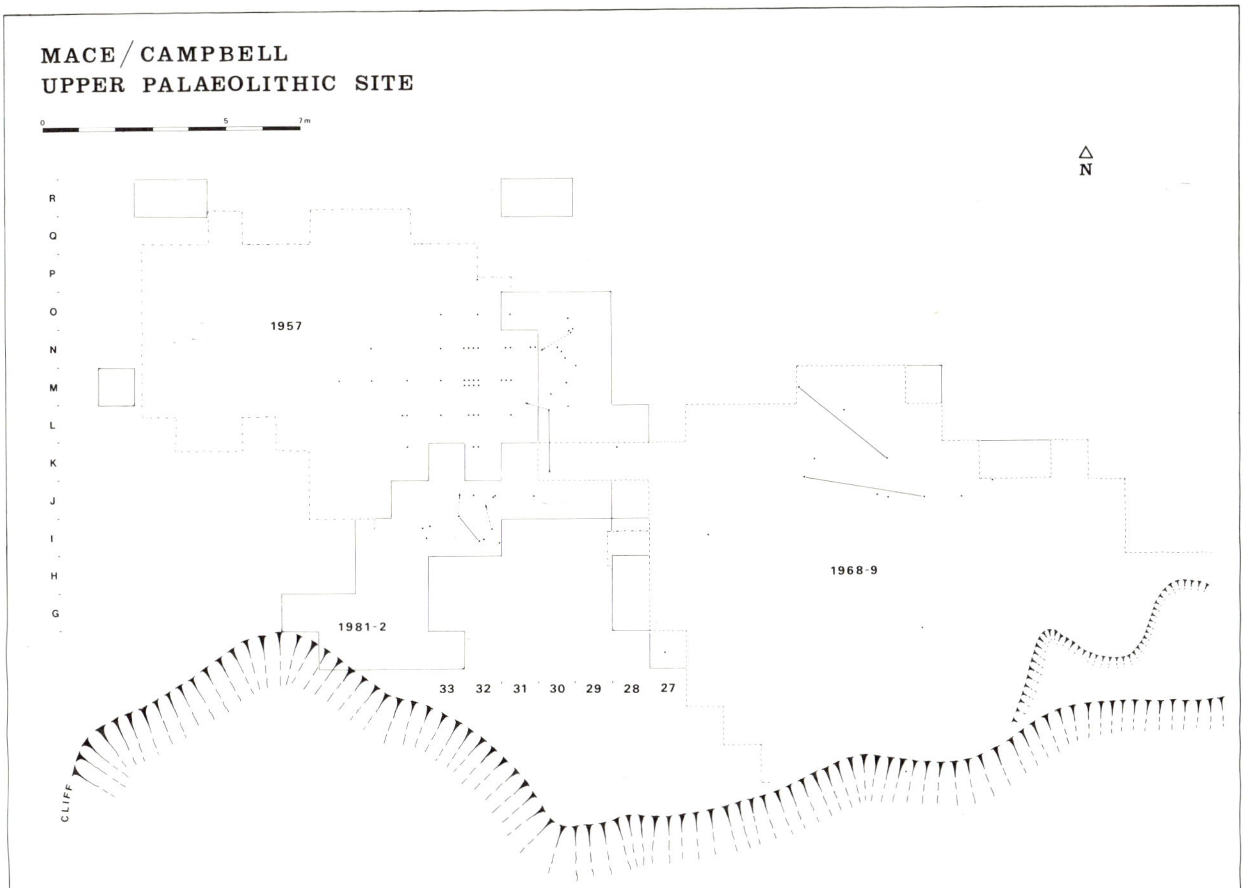

Fig. 3.9. Horizontal distribution of intentional breaks from the three excavations, with refits indicated where possible.

bury Head make it clear that the technique is an important element in the assemblage. Intentional breakage presents a good method of maximising raw material and it is interesting to note that the large flint nodules knapped at the site seem to have been imported from some distance. It is easy to imagine that the prospect of a long walk carrying nodules weighing two kilogrammes or more inspired these people to economise.

Acknowledgements

We would like to thank Mark Newcomer for reading an earlier draft of this paper and for offering useful comments. We also wish to thank Penny Robinson for her help at the British Museum. Christine Wilson drew the stone tools and Spence Ingerson typed the manuscript. Thanks are also due to Nick Pollard who helped with the production of the diagrams.

Bibliography

Azoury, I. & Bergman, C. A. (1980). 'The Halafian Lithic Assemblage of Shams ed-Din Tannira'. *Berytus*, XXVIII, 127–243.

Barton, R. N. E. & Bergman, C. A. (1982). 'Hunters at Hengistbury: some evidence from experimental archaeology.' *World Archaeology*, **14** (2), 237–48.

Barton, N. & Bergman, C. (1983). 'The hunters of Hengistbury.' *The Illustrated London News*. February, p. 42.

Barton, R. N. E. & Huxtable, J. (1983). 'New dates for Upper Palaeolithic and Mesolithic occupations at Hengistbury Head, Dorset.' *Antiquity*, tome 220: pp. 133–5.

Benito del Rey, L. (1982). 'Outils fractures intentionnellement dans le Moustérien de la Grotte de las Grajas, à Archidona (Malaga, Espagne).' *Bulletin de la Société Préhistorique Française*, **79/8**, 231–9.

Bergman, C. A. & Newcomer, M. H. (1983). 'Flint arrowhead breakage: examples from Ksar Akil, Lebanon.' *Journal of Field Archaeology*, **10**, 238–43.

Bordes, F. (1953). 'Notules de typologie paléolithique: I. Outils moustériens à fracture volontaire.' *Bulletin de la Société Préhistorique Française*, L, 224–6.

Bordes, F. (1961). *Typologie du Paléolithique: ancien et moyen* Bordeaux: Delmas.

Bordes, F. (1970). 'Observations typologiques et techniques sur le Périgordien supérieur de Corbiac (Dordogne).' *Bulletin de la Société Préhistorique Française*, **67** (4), 105–13.

Bosinski, G. & Hahn, J. (1972). 'Der Magdalénien-Fundplatz Andernach (Martinsberg). Beiträge zum Paläolithikum in Rheinland. *Rheinische Ausgrabungen*, **11**, 81–257.

Breuil, H. (1932). 'Les industries à éclats du paléolithiques ancien. 1. Le Clactonien.' *Préhistoire*, **1**, 125–90.

Burdukiewicz, J. M. (1981). 'The flint technology of the Hamburgian Culture (Olbrachcice, S.W. Poland).' *Third International Symposium on Flint* 24–27 May 1979. Maastricht, pp. 67–70.

Cahen, D., Karlin, C., Keeley, L. H. & van Noten, F. (1980). 'Mèthodes d'Analyse Technique, Spatiale et Fonctionnelle d'Ensembles Lithiques.' *Helinium*, XX, 209–59.

Calkin, J. B. (1953). ' "Kimmeridge Coal-Money". The Romano-British Shale Armlet Industry.' *Proceedings of the Dorset Natural History and Archaeological Society*, **75**, 45–71.

Campbell, J. B. (1977). *The Upper Palaeolithic of Britain: A study of Man and Nature in the Late Ice Age*. 2 Vols. Clarendon Press, Oxford.

Delarue, R. & Vignard, E. (1958). 'Intention et fractures moustériennes sectionnant des racloirs.' *Bulletin de la Société Préhistorique Française*, LV, 370–4.

Eloy, L. (1954). 'Fractures intentionelles observées sur des outils moustériens.' *Bulletin de la Société Préhistorique Française*, LI, 29–31.

Hayden, B. (ed.) (1979). *Lithic Use-Wear Analysis*. Academic Press, New York.

Karlin, C. (1972). 'Le Débitage.' *In:* Leroi-Gourhan A. et Brézillon M. (eds.) 'Fouilles de Pincevent. Essai d'analyse ethnographique d'un habitat magdalénian.' (La section 36), Vol. 1 VIIe supplément a *Gallia Préhistoire*, pp. 263–77.

Mace, A. (1959). 'The excavation of a Late Palaeolithic open-site on Hengistbury Head, Christchurch, Hants.' *Proceedings of the Prehistoric Society*, XXV, 233–59.

Mason, H. J. (1978). *Flint – The Versatile Stone*. Providence Press, Cambridgeshire.

Newcomer, M. H. (1972). An analysis of a series of burins from Ksar Akil (Lebanon). Unpublished Doctoral Thesis, University of London.

Newcomer, M. H. (1975). ' "Punch Technique" and Upper Palaeolithic blades.' *In:* Swanson, E. H. (ed.) *Lithic Technology: Making and Using Stone Tools*, pp. 97–102. Mouton Publishers, The Hague.

Newcomer, M. H. (1976). 'Spontaneous retouch.' Second International Symposium on Flint, *Staringia*, **3**, Maastricht, Nederlandse Geologische Vereniging, pp. 62–4.

Ohnuma, K. & Bergman, C. A. (1982). 'Experimental studies in the determination of flaking mode.' *Bulletin of the Institute of Archaeology*, **19**, 161–70.

Owen, L. (1982). 'Analysis of experimental breaks on flint blades and flakes. *Studia Praehistorica Belgica* NFWO Contact Groep (Prehistorie-Préhistoire) Groupe Contact FNRS, pp. 77–87.

Pradel, L. (1957). 'Intention et fractures moustériennes.' *Bulletin de la Société Préhistorique Française*, LIV, 382–6.

Pradel, L. (1959). 'A propos des fractures intentionelles au moustérien.' *Bulletin de la Société Préhistorique Française*, LVI, 31–2.

Rigaud, J. Ph. (1966). Étude typologique détaillée des burins de gisement périgordien de Corbiac (Comm. de Lembras), et comparaison avec quelques séries du Périgordien supérieur et final, et du Protomagdalenian. Tome I et II. Diplome d'Études Supérieurs. Université de Bordeaux.

Roche, H. & Tixier, J. (1982). 'Les Accidents de Taille.' *Studia Praehistorica Belgica* NFWO Contact Groep (Prehistorie–Préhistoire) Groupe Contact FNRS, pp. 65–76.

Schwabedissen, H. (1954). *Die Federmesser-Gruppen des nordwest-europäischen Flachlandes.* Karl Wachholtz Verlag, Neumunster.

Semenov, S. A. (1964). *Prehistoric Technology*. Translated by M. W. Thompson, Cory, Adams & Mackay, London.

Tixier, J., Inizan, M.-L. & Roche, H. (1980). *Préhistoire de la Pierre Taillée. 1. 'Terminologie et technologie.'* CREP, Valbonne.

Tixier, J. & Inizan, M.-L. (1981). 'Ksar 'Aqil. Stratigraphie et Ensembles Lithiques dans le Paléolithique Supérieur Fouilles 1971–1975.' *In:* Cauvin, J. & Sanlaville, P. (eds.), *Préhistoire du Levant*. Colloques Internationaux du CNRS. 598: 353–72.

van Noten, F. (1978). *Les Chasseurs de Meer*. Dissertationes Archaeologicae Gandenses, 18. Bruges: De Tempel.

4

Flint chips from Pincevent

M. H. NEWCOMER AND C. KARLIN

The tiny fragments of flint which are inevitably detached when making stone tools are often an embarrassment to the conscientious excavator of a prehistoric site: how should these chips be recorded and what useful information could be gained from studying them? Earlier publications have examined the technology of retouch chips (Frison, 1968) and papers by van Noten (1978) and Fladmark (1982) have discussed flint dust and chips as constituents of archaeological sediments which can help locate areas of past flaking activity. Our paper, on the other hand, focusses on chips as individual artefacts, and looks at their forms and distribution at a single site, the late Magdalenian open-air hunting camps on the banks of the Seine at Pincevent (Leroi-Gourhan & Brézillon, 1966, 1972).

Our first task must be to attempt a definition of a chip: any formal definition must take into account the size and type of raw material being worked, but at Pincevent where smallish blocks of fine-grained flint are the rule, a chip can be defined as any piece of unretouched flint whose broadest surface is less than 1 cm^2 (Karlin, 1972, p. 263). It is easiest to study and make sense of chips which are simply tiny flakes with the normal features of conchoidal fracture like butts and bulbs of percussion, but the term should also cover fragments of larger flakes or even shatter fragments with no features of intentional fracture. Chips are a special category of waste flake in that we can always be certain that they are by-products and not wanted for themselves: the knapper's attention is always on the scars left or edge shape created by removing chips and not on the chip itself – indeed, a whole series of chips may be detached in a single gesture as when making a scraper by direct percussion. This point is perhaps worth underlining since with larger pieces it is often hard to distinguish 'waste' (literally discards) from 'blanks' (potentially useful products) and metrical studies of assemblages of 'waste flakes' lump together flakes intended for, say, scraper manufacture with flakes from, say, thinning axes.

Chip form and technology

Flint chips are by far the most common form of artifact at Pincevent but, paradoxically, the least studied. As with our previous work at the site (Karlin & Newcomer, 1982), we approached the study of chips from two directions, archaeological and experimental. It is obvious that tools like end-scrapers and borers are made by detaching retouch chips and equally obvious from examining reconstructed blade cores that there are gaps between the proximal ends of refitted blades which should be filled by debitage chips. To see what form these retouch and debitage chips might take, we made replicas of the Pincevent blade cores and tools in the local flint and saved all the chips. Comparing experimental and palaeolithic chips, we were able to confirm a distinction between retouch and debitage chips and could isolate two different types of chip associated with blade production (Fig. 4.1).

Debitage chips

Blade debitage at Pincevent was normally done by creating a guiding ridge or crest by bifacial flaking on the front (and often back) of the core, and making a striking platform at an acute angle to this crest (Karlin, 1972). Preparation and repair of the striking platform yields the two types of debitage chips: the first results from making minor changes in the angle and shape of the platform surface by facetting. Facetting was used both to isolate a point or 'spur' (*éperon*) on which to strike and to make other small adjustments between core tablet removals. Chips from facetting the platform surface are typically short and fan-shaped, with the distal end formed by a hinge or step fracture (Fig. 4.2). When hinged, the distal end often has a thin lip of flint extending beyond the hinge at the junction of the dorsal surface and the hinge. Similar chips may be produced in retouching, especially when making the 'scaled' retouch scars of Aurignacian or Quina retouch (Lenoir, 1973), but at Pincevent this kind of retouch is not used and the core platform surface chip is the most diagnostic chip type and a sure indicator of blade debitage.

A second type of debitage chip is made when removing overhangs on the front of the core between blade removals. Each time a blade is detached, its bulb leaves a concavity at the junction of the striking platform and the core front. Reducing this concavity – and strengthening the platform edge – is done by retouching away the overhang, sometimes simply by rubbing the platform with a hammer-stone (Bordes, 1967, Fig. 5). When this retouching is done over a ridge on the core front, the chips run a short way down the core, guided by the ridge, and typically have small butts, feathered edges, parallel or converging lateral edges and straightish profiles (Fig. 4.3). If earlier abrasion of the platform edge has been extensive, some of this may be carried away on the proximal end of the core front chip.

Most other chips made during debitage, for example

those from operations like cresting, do not seem to have special shapes, though they are more likely to have some cortex on their dorsal surface. Very occasionally one finds the thin chip which forms the bulbar scar on a larger flake or blade, and this is also a definite indicator of debitage.

Retouch chips

Special forms of retouch chips like the burin spall, microburin and Krukowski microburin are well known and do not need further discussion here: we are concerned rather with the chips produced when making the retouched parts of tools like end-scrapers, truncations, piercers or borers and backed bladelets, the common tool forms at Pincevent. Tools from all the habitations at Pincevent are usually quite thin and there is very little stepped and scaled retouch and no carinated tools, so the retouch chips are generally much smaller and thinner than debitage chips. All forms of tools at Pincevent yield chips which are short, lack cortex, have curved profiles and normally have feathered edges (Fig. 4.4). In our opinion it is not possible to assign specific retouch chips to a given tool class (except when the chip carries away a recognisable part of the tool), but in a blind test conducted at Pincevent in July of 1978, sets of experimentally made debitage chips and retouch chips were shown, unlabelled, to one of us (C.K.), who consistently identified the chips from retouching end-scrapers and backed bladelets as distinct from debitage chips. In the context of a North American buffalo kill and butchering site, Frison (1968) has identified five kinds of retouch flakes

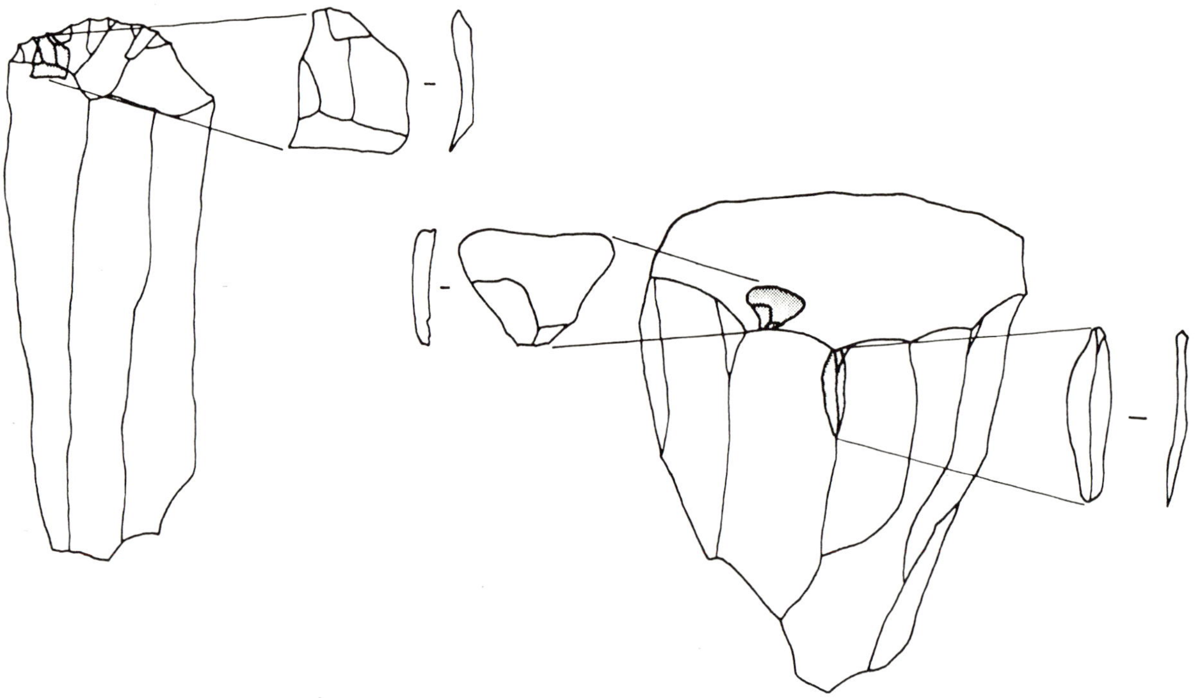

Fig. 4.1. Schematic drawings showing the origin of a retouch chip from an end-scraper (left), a core platform surface chip and core front chip (right).

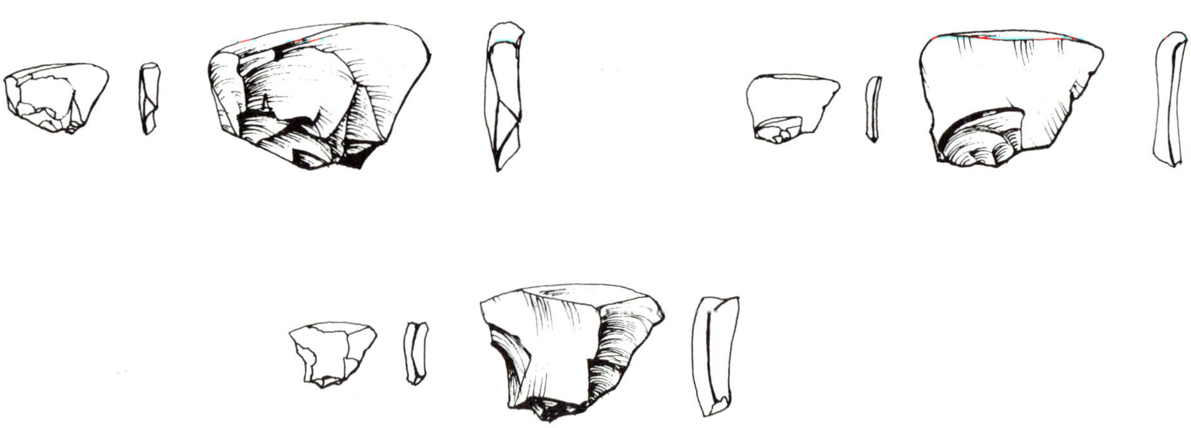

Fig. 4.2. Core platform surface chips from Pincevent: the unshaded outline is life size.

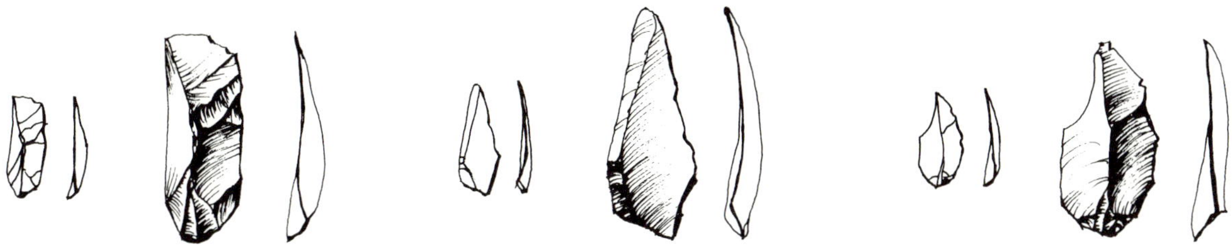

Fig. 4.3. Core front chips from Pincevent: unshaded outlines are life size.

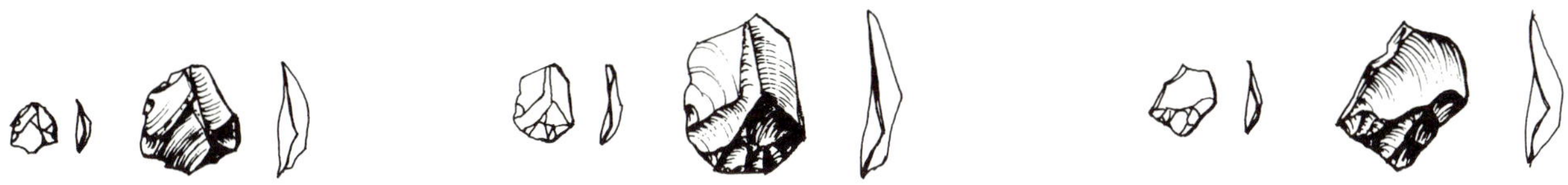

Fig. 4.4. Retouch chips from Pincevent: unshaded outlines are life size.

and chips from re-edging bifaces and scrapers, some of which resemble those from Pincevent. We have found it more difficult, however, to distinguish sharpening chips from manufacturing chips: the problem is the same as that with burins, where a 'sharpening' burin blow may be given on a totally unused burin tip, simply to alter a badly angled or poorly made – but not dull – working end. Chips from sharpening scrapers may have undercuts at the proximal end on the dorsal surface which are the traces of a macroscopically worn edge, but a similar undercut can be made by poorly executed retouch or by the abrupt retouch used to make truncations and becs.

Chip context

Using the classification of chips we have proposed, what can chips tell us about the occupations and activities at Pincevent? To answer this question we must first look at the methods used to recover and record chips during the excavation. In our experience, many excavators of prehistoric sites recognise chips as artefacts but do not know how to get information from them and consequently discard them, with or without recording their presence. At Pincevent, the systematic recovery of tiny chips is made possible by the fact that the stone industry is almost exclusively in flint, and the sediments which cover the living floors are very fine river silts which allow one to pick out chips as small as 1 mm across relatively easily. The recording system used at Pincevent classes chips with ochre particles, charcoal flecks and tiny fragments of bone; these *témoins fugaces* (Leroi-Gourhan & Brézillon, 1972, p. 323) move upward and downward from the living floors through the action of earthworms and roots and are removed as they are exposed rather than being left in place with the larger pieces of flint, bone and the like. Chip scatters are planned at 1:5 and as it is considered impractical to number and bag each chip separately, all the chips from a 20 cm square area within the metre square excavation unit are bagged together. When digging in areas rich in small objects, especially around hearths or *vidanges*, the sediment is water sieved through a 2 mm mesh.

With the chips found and recorded, how can they help our interpretations? One challenge of minimally disturbed sites like Pincevent is how to explain why artefacts are found where they are: have they been dropped, thrown, kicked aside or perhaps carried to the place we find them? As chips are by-products, too small to throw and not worked for themselves, they can be safer indicators of where past activities took place than larger artefacts which were probably routinely moved from the places where they were actually used. We believe that chips would only be moved from the place they were made if scooped up inadvertently with larger pieces of flint (as was the case with chips in the debitage heap in square 078 of section 26 – see Karlin & Newcomer, 1982), or if incorporated in the ashes of a fire which were subsequently scraped up and dumped.

If most of the chips we find have not moved significant distances, we should be able to use them to tell us where debitage took place and where tools were made, especially since our experiments have shown that when flaking close to the ground, as was the rule at Pincevent, chips do not fly very far (see Newcomer & Sieveking, 1980). Concentrations of retouch chips have so far eluded us, and this may be due to a real lack of special areas for tool making, as the dispersed distributions of burin spalls imply (compare Leroi-Gourhan & Brézillon, 1966, Fig. 65, and 1972, Fig. 71), or to the fact that the most common form of retouched tool, the backed bladelet, produces the tiniest and most difficult to recover retouch chips.

On the other hand, concentrations of debitage chips are very common and are most often associated with the other waste from blade making like core tablets, crested and broken blades and discarded cores. When such chips are found without the larger waste pieces, as was the case in Habitation 1 (Leroi-Gourhan and Brézillon, 1966: 339) the best interpretation is that the larger pieces were swept up to clear a domestic area and that only the hard to pick up chips remain as evidence of blade making here. If debitage chips *have* been moved along with other debitage products, as might happen easily when flaking was done over a skin or basket, the distribution of chips within the scatter can help to determine whether the material was moved or not: with *in situ* scatters, the chips are spread horizontally across the heap but concentrated at the bottom, and there will also be a fair quantity of the flint dust which invariably accompanies knapping. When debitage has been moved however, the dust is rarer and the chips may be concentrated in one part of the heap and may be found at the bottom or mingled throughout.

Conclusion

In concluding these notes on chips from Pincevent, two points need to be underlined: the first is that the proposed classification of two types of debitage chips and one type of retouch chip is meant for Pincevent and other similar blade industries, and that different raw material and other kinds of debitage and retouched tools will produce their own types of chip.

The second point concerns chip numbers: the proportion of chips found at Pincevent that can be assigned to the three categories we have defined is relatively small, perhaps 30–40%, but adding criteria like size – especially relative thickness – it seems clear that the vast majority of chips at Pincevent derive from debitage. Analysis of chips from other upper palaeolithic sites like Le Placard (Charente) where several thousand Solutrean shouldered points were found (Smith, 1966, p. 259), might, however, show quite a different distribution.

We hope these notes on chips from Pincevent will encourage some discussion of this hitherto neglected class of flint artifact, and would suggest that the humble chip is sometimes a better clue to prehistoric behaviour than the largest flake or even retouched tool.

Bibliography

Bordes, F. (1967). 'Considérations sur la typologie et les techniques dans le Paléolithique.' *Quartär*, **18**, 25–55.

Fladmark, K. R. (1982). 'Microdebitage analysis: initial considerations.' *Journal of Archaeological Science*, **9** (2), 205–20.

Frison, G. (1968). 'A functional analysis of certain chipped stone tools.' *American Antiquity*, **33**, 149–55.

Karlin, C. (1972). 'Le Débitage.' *In:* Leroi-Gourhan, A. & M. Brézillon, *Fouilles de Pincevent. Essai d'analyse ethnographique d'un habitat magdalénien. (La Section 36).* Editions du CNRS, Paris, 263–77.

Karlin, C. & M. Newcomer. (1982). 'Interpreting flake scatters: an example from Pincevent. *Studia Praehistorica Belgica*, **2**, 159–65.

Lenoir, M. (1973). 'Obtention expérimentale de la retouche de type Quina.' *Bulletin de la Société Préhistorique Française*, **70**, 10–11.

Leroi-Gourhan, A. & M. Brézillon. (1966). 'L'habitation magdalénienne no. 1 de Pincevent pres Montereau (Seine-et-Marne).' *Gallia Préhistoire*, **9** (2), 263–385.

Leroi-Gourhan, A. & M. Brézillon. (1972). *Fouilles de Pincevent. Essai d'analyse ethnographique d'un habitat magdalénien. (La Section 36).* Editions du CNRS, Paris.

Newcomer, M. H. & G. de G. Sieveking. (1980). 'Experimental flake scatter-patterns: a new interpretive technique.' *Journal of Field Archaeology*, **7** (3), 345–52.

Smith, P. E. L. (1966). *Le Solutréen en France.* Bordeaux: Delmas.

van Noten, F. (1978). *Les Chasseurs de Meer.* De Tempel: Brugge.

5

An experiment to produce a ground flint axe

P. HARDING

Introduction

This paper describes an experiment to reproduce, by grinding, flint axes similar to the type commonly found in the Neolithic period. Grinding the blade of a flaked axe strengthens it by removing irregularities present along the edge. By similarly grinding the body of the axe, it is possible to achieve an improved fit within the haft which reduces the possibility of the axe twisting and snapping. Such a development was of great importance at a time of forest clearance.

The efficiency of ground flint axes has been demonstrated in numerous experiments, principally by Lepic (Evans, 1897, p. 162), Jacob-Friesen (Nietsch, 1939), Jorgensen (1953), Iversen (1956), Saraydar & Shimida (1971), Carneiro (1979) and Harding & Young (1979). These experiments have not described axe manufacture, as genuine pieces were used, although Harding & Young described subsequent resharpening with a carborundum.

Experiments to manufacture ground axes have, however, been conducted by several authors with widely differing results. Dr F. A. Forel (Evans, 1897, p. 36) took four hours to flake and grind a stone axe in 1875 and Skavlem (Pond, 1930) pecked and ground an axe in a similar time in the 1920s. Sehested (1884), by comparison, using a granite grindstone, took 36 hours to produce a ground flint axe. More recently, in a series of experiments to both manufacture and use flint and stone axes similar to the type common in the Scanian Neolithic, Olausson (1983) achieved results similar to those of the earlier workers. In a series of experiments, supplemented by data by Andersen (1980–81), she showed that square sectioned axes could be produced from finely knapped blanks in approximately four hours. She also confirmed Dickson's (1981) observations that it was possible to grind stone axes more quickly than those made of flint.

My own experience includes the manufacture of four flint axes, three of which will be quoted here.

Aims

The occurrence of sandstones on the Marlborough Downs in Wiltshire, which were used in prehistoric times for grinding, stimulated the author to test their effectiveness in the production of ground flint axes. On the basis of the results obtained in the production of the first flint axe using such a stone, a second experiment was designed:

i to examine the effects on grinding efficiency by adding a variety of grinding media;
ii to compare any differences of surface finish which resulted from using the different grinding components;
iii to test methods of sharpening the blade.

The experiment was then repeated using a block of sandstone from Cheshire to note any differences in efficiency between the two types of sandstone. A general comparison was also made with results of similar experiments to grind axes made of stone from the Great Langdale axe factory in the Lake District.

Axes

Two of the axes were made of flint from Brandon in Suffolk, while the third was made of fresh flint quarried from Salisbury Plain in Wiltshire. They were produced using the technique described by numerous writers (Warren, 1921; Newcomer, 1971; Bradley & Sampson, 1978; Newcomer & Sieveking, 1980) for the production of large bifacial tools. This involves (i) rough work with a hard stone hammer, (ii) thinning and shaping, and (iii) finishing using a red deer antler. All stages were carried out by direct percussion and striking platforms were prepared by 'turning the edge' to strengthen them before percussion. The roughouts of axes 2 and 3 were broadly similar in size, weight and overall quality. Such roughouts can normally be produced by an experienced knapper in less than 30 minutes.

The first axe was fully ground on moist sandstone (sarsen), while the second axe discussed (Fig. 5.1), which forms the major part of this paper, was ground in a similar fashion with the addition of the grinding media. This technique was repeated for the third axe. Grinding was generally, but not exclusively, perpendicular to the blade of the axe, and each face of each axe was ground alternately.

Grindstones

Grindstones are known from all parts of Europe and the British Isles, either on natural rock surfaces, or as portable grinding stones (Dechelette, 1908, p. 523ff). A wide range of rock types was utilized but the most suitable is a quartz-rich sandstone, soft enough to allow a regeneration of abrasive grains as the stone is eroded. These abrasive grains also prevent friction between the axe and the grindstone surface below (Semenov, 1964, p. 69).

The stone selected for the first two experiments was sarsen, a durable sandstone bound by silica cement (Bowen and Smith, 1977, p. 187). It occurs relatively commonly throughout Wessex but is most densely concentrated on the downs of North Wiltshire. It is best known for its use in the construction of large ritual monuments, notably Stonehenge and Avebury. However, there is evidence for its use as a grindstone in prehistoric times. Several areas of sarsen worn smooth by grinding were found on stone 18 of the West Kennett Long Barrow (Piggott, 1962, p. 19 and pl. xiii). Less securely dated examples include Stone 37 (Stone 19b of Smith, 1965, p. 223) of the West Kennett Avenue, and a broken sarsen on Overton Down, three kilometres east of Avebury (Lacaille, 1966). They have characteristically shallow dished areas, often with deeper 'V' or 'U' sectioned grooves, which have been interpreted as the

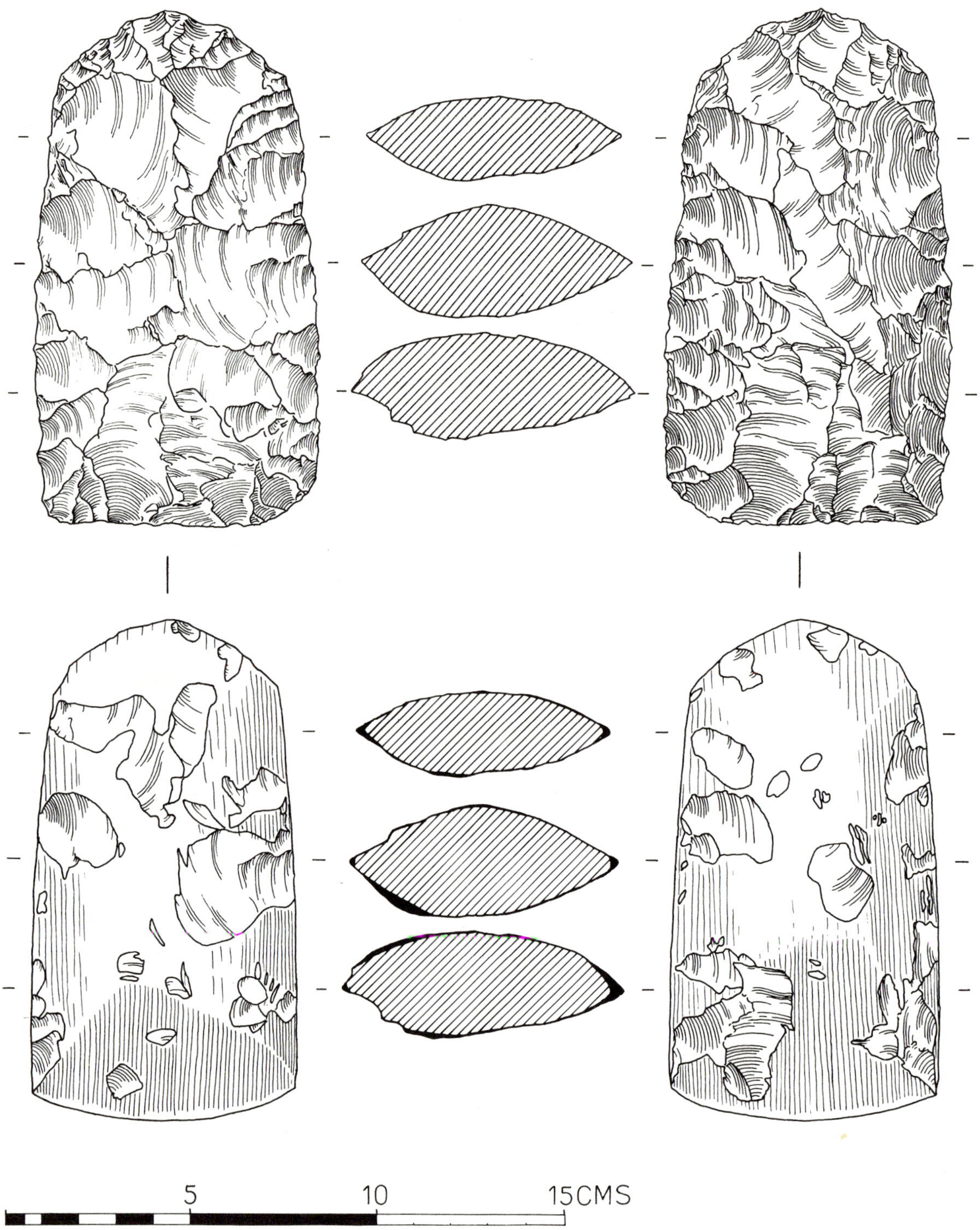

Fig. 5.1. Experimental axe 2: unground (above), ground (below). Solid shading on the lower profiles indicates flint removed by grinding.

result of grinding the edges and the blades of axes (Dechelette, 1908, p. 524).

The sarsen used to grind the second axe had already been used for approximately 12 hours, to grind two Great Langdale stone axes. It had developed both the distinctive dished area and deep 'V' sectioned groove.

The grindstone used in the final experiment was a quartz-rich ferruginous sandstone with mica, mainly muscovite, from Cheshire. It is distinctly less durable than sarsen.

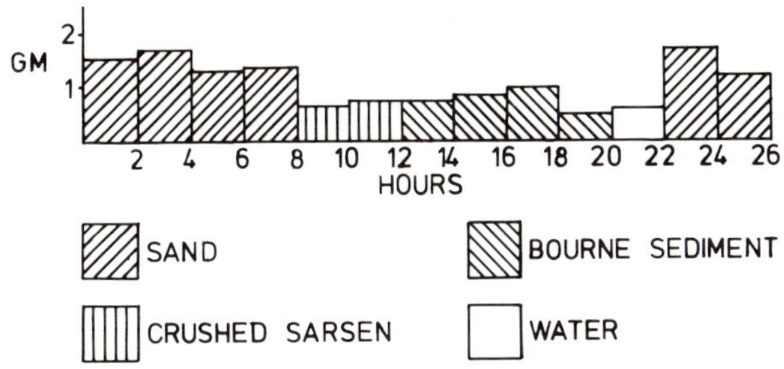

Fig. 5.2. Experiment 2: diagram illustrating weight loss during the first 26 hours grinding with different media.

Grinding media

Grinding media were used because of the durable nature of the sarsen. Ideally they act as a mobile abrasive and prevent friction between the grinding surfaces. The experience of grinding axe 1 suggested that it might be more efficient to add such a medium. This has also been observed by Callahan when grinding stone axes (Olausson, 1983, p. 28).

Three media were chosen which would have been available to prehistoric man:

- i Sharp sand: This was beach sand from Hengistbury Head, Dorset.
- ii Crushed loosely cemented sarsen: Some sarsen is loosely cemented and, especially when burnt, becomes sugary in texture; this can be reduced to powder by pounding. It is however, unsuitable if used alone, as it crumbles too easily.
- iii Sediment from a Dorset stream: This consists of coarse sharp sand, with larger pebbles which must be removed, and also includes chalk. Such sediment can be found in any chalk stream.

Water was used in all experiments, both as a lubricant, and to carry away waste material. In the second and third experiments it also helped to bind the medium together in the form of a paste. Although it is not certain that water was used in prehistoric times, two of the known grindstones, those in the West Kennett Long Barrow and the West Kennett Avenue, may have been dragged to their present locations from the course of the River Kennett, less than a kilometre away.

The grindstones were placed in a plastic bowl with water and enough grinding medium to make a paste. The paste was changed at two-hourly intervals during the first twenty-six hours of the second experiment, to offset any loss of abrasion to the grains, but was retained throughout the third experiment.

Weight loss

The first axe was ground for a total of 20 hours, but unfortunately figures of weight loss were not recorded. The second axe was ground for 40 hours, during which time it was reduced from 371.25 gm to 352.5 gm (5%). The third axe, however, was reduced from 363.6 gm to 354.5 gm (2%) in only eight hours, with 0.9 gm removed in an additional two hours to modify the blade. Although most of the surface of this axe was ground, proportionally more unground areas remain than on axe 2. The weight loss of both these axes was recorded at two-hourly intervals.

Figure 5.2 shows the rate of flint loss during the first 26 hours grinding in experiment 2 with each medium. The addition of sand to the grindstone is confirmed as the most efficient medium. The average loss of 1.5 gm for two hours makes it twice as effective as the other media, which show no clear distinction in efficiency when used with the same grindstone. The conclusion is substantiated by an average weight loss of 2.3 gm in each two hours grinding with Cheshire sandstone and sand. It is almost certain that the less durable nature of the Cheshire sandstone is responsible for a large part of this dramatic difference in weight loss. Its increased abrasiveness may render the addition of sand unnecessary, although this was not tested, nor were the alternative media used.

Crushed sarsen proved to be too fine-grained to be effective as an abrasive. The process was made worse by the addition of the time taken to pound the sarsen. Stream sediment, conversely, was too coarse and contained constituents which lacked abrasiveness. It was noticable that all three media became more silty and less effective the longer they were used.

Water by itself was totally unsuitable as it was pushed out from between the two grinding surfaces and, in effect, the axe was ground on dry sarsen. The abrasive surfaces became clogged with abraded particles and friction increased associated with the production of heat. More physical effort was therefore necessary. One of the largest weight removals was achieved by grinding both surfaces of the axe simultaneously in the 'V' sectioned groove. When used in conjunction with sand and water, 1.7 gm of flint were removed in two hours. This method was also used, with notable success, for the third axe, when 2.7 gm were removed during a two hour spell.

Surface texture

The surface finish produced by each grinding method was also considered, so that direct comparisons could be made with Neolithic axes.

Two broad distinctions resulted which were determined by the inclusion of the grinding medium. Both axes ground on wet sarsen developed a visually smooth, very glassy surface, with directional striations. The formation of this surface probably resulted from friction and production of heat. The inclusion of a grinding medium, in the later experiments, resulted in a far less glassy surface, with slightly deeper striations. There was no clear variation in the finish produced by each medium. It is probable that a similar finish might be expected from the softer sandstone used without the added medium. These observations have been confirmed microscopically (Fig. 5.3).

Most Neolithic axes conform to the type of finish produced by the inclusion of some form of grinding medium. However, some axes have been found where the blade is ground to a glassy polish (M. Pitts, pers. comm.). This suggests that the grinding medium may either have been omitted during the final stages of production or subsequent re-sharpening, or may be indicative of the durability of the grindstone.

Where a grinding medium is used, it is important that it should contain some coarse texture, as experimentation with mud produced similar results to pure water.

The blade

This usually received the greatest attention on ground axes. Even where the body of the axe was poorly ground, the blade was often highly finished.

Grinding on the blade of the second axe began after eight hours and was concentrated here after 28–30 hours work. No similar observations were made for axes 1 and 3. The biggest problem on all the axes was that the edge tended to chip on the upper surface as the lower was being ground. Various techniques were tried to solve this:

i prolonged use of the dished area;
ii use of the 'V' sectioned groove;
iii use of a hand whetstone.

The first option proved to be the least efficient. It required considerable patience to continue grinding long enough to remove all the irregularities on the edge and also produced most edge damage, particularly to the corners of the blade. This occurred if the profile of the grindstone did not match that of the blade; usually the grindstone hollow was too deep. Chipping continued until the two were nearly equal. Damage also resulted from the downward pressure on the axe during grinding.

The use of the groove proved to be potentially the most efficient method of the three. It has, however, only been successfully tested on an experimental Great Langdale axe. The method is particularly effective before the groove becomes too deep or rounded at the base. It therefore has a limited life for this function; nevertheless it might be possible to associate the shallow, acute sectioned grooves of Stone 37 of the West Kennett Avenue with blade sharpening, and the deeper 'U' sectioned grooves of the Overton Down sarsen with basic edge grinding (Fig. 5.4). The former all lie approximately parallel to natural fissures in

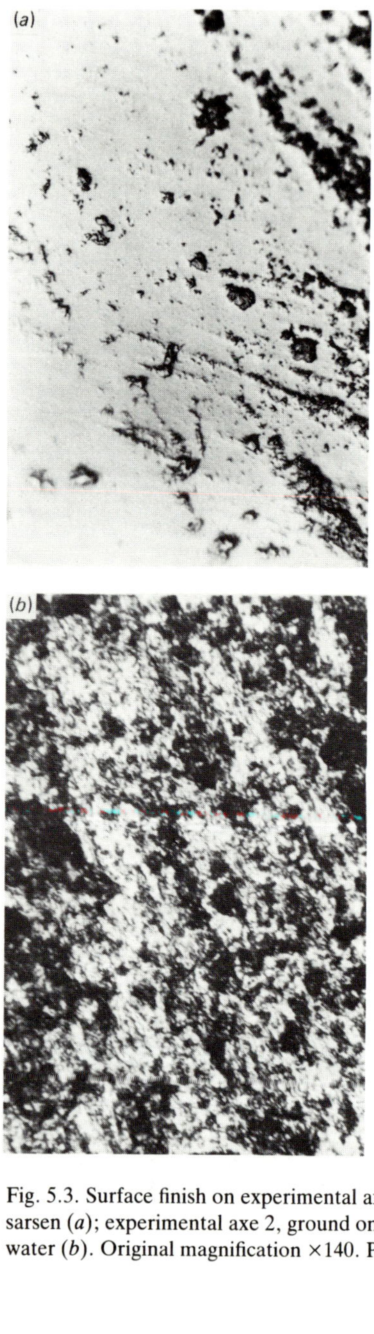

Fig. 5.3. Surface finish on experimental axe 1, ground on moist sarsen (*a*); experimental axe 2, ground on sarsen with sand and water (*b*). Original magnification ×140. Photos: E. H. Moss.

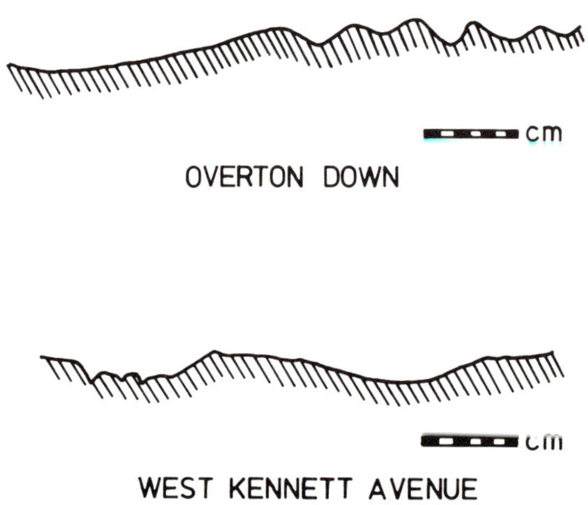

Fig. 5.4. Profiles through two sarsen grinding stones, showing dished grinding hollows and 'U' and 'V' sectioned grooves (from Lacaille, 1966).

the rock, which would have provided an initial seating for the blade, before the groove was fully established.

The use of a small hand-held sandstone rubber was found to be more time consuming than the first two methods but produced no edge damage. This method might be most suitable for resharpening the blade to remove chips which occur during use.

The second and third axes produced in these experiments were sharpened on the sarsen grindstone using both the grinding hollow and the groove. The latter had become too rounded at the base to be used by itself, but by first using it to blunt and consequently strengthen the blade edge damage was reduced and the blade could then be sharpened in the grinding hollow without further chipping. The fact that some Neolithic axes have been found with a distinctly blunted blade (M. Pitts, pers. comm.) may suggest that sharpening was sometimes considered to be unnecessary, the emphasis in use being on a strong edge.

Great Langdale axes

Similar experiments have also been undertaken to manufacture axes made of stone from the Great Langdale axe factory in the Lake District. This is a long-term project and will be published separately (Harding & Moffatt, in preparation); however broad comparisons relating to grinding time and weight loss may be of interest.

The stone from the Great Langdale axe factory (Group VI) is an epidotized tuff (ash). It flakes with a conchoidal fracture and, when ground, often produces axes of fine quality.

Four axes have been ground on sarsen with sand and water. The average grinding time has been calculated to be nearly seven hours per axe, in contrast to the broad figure of 26–30 hours for a flint axe ground by the same technique. The reduction in time is not reflected by a similar reduction in the quantity of stone removed. The stone axes, on average, lost 15% of their weight; (a rate of approximately 7.8 gm per hour), compared to the 5% weight loss (approximately 0.75 gm per hour) for the flint axe.

This is in broad accordance with the observations of Dickson (1981, pp. 106–7) that stone axes can be ground two or three times faster than those made of flint.

Discussion

This paper began with the aim of examining several aspects of the manufacture of ground flint axes. The use of weight loss as a method of showing grinding efficiency was successful. It has indicated that in axe manufacture grinding times are subject to much broader variations than flaking times. This may be explained by

- i The quality and size of the roughout. Grinding times can be reduced if the contour of the roughout is improved by removing pronounced flint ridges by soft hammer flaking or pressure flaking.
- ii The quality of the finished axe. Prehistoric axes show considerable variation in their degree of finish. Some merely involve the removal of flake ridges while others show few or no remnant flake scars.
- iii The skill of the workman. This will vary both with experience and physical strength.
- iv The abrasiveness of the grindstone. The choice of grindstones, in prehistoric times, was often governed by what was most readily available. The fact that most flint axe factory sites show no evidence for on-site grinding may be explained, not only by the time factor, but also because most flint mines do not have a supply of suitable grindstones. The sheer bulk of the material makes it more practical to transport finished roughouts than to import grinding benches. The fact that the most efficient experimental grinding stone was the less durable rock makes it possible that traces of some grinding benches may have been lost as a result of subsequent natural erosion.

Definite conclusions about the surface finish which results from using different grindstones cannot be reached, but it does seem possible that certain types of surface finish denote characteristics of the grindstone used, particularly durability.

It is important that other grindstone types should be tested experimentally and details of suitability and surface finish recorded.

Finally it has been possible to propose methods of sharpening the blade. It has been found that the blade can be ground more effectively if it is strengthened by blunting before the edge is applied. This reduces the tendancy for the edge to chip while it is being sharpened.

This work hopefully throws some light on the practical aspects of ground flint and stone axe production, and their relationship to the grindstone. Continued research, combined with patience, will help to add more to our complete understanding of these artefacts.

Acknowledgements

I wish to thank Dr Emily Moss for her practical help and comments in the microscopic examination, Chris Gingell and Mike Pitts for their help and advice throughout, and Mrs Margaret Couling who typed the manuscript.

Bibliography

Andersen, I. (1980–81). *Tildannelse og slibning af firesidet flintø kse.* Privately printed. Denmark.

Bowen, H. C. & Smith, I. F. (1977). 'Sarsen stones in Wessex: The Society's first investigations in the evolution of the landscape project.' *Ant. J.*, **57**, 185–96.

Bradley, B. & Sampson, C. G. (1978). 'Artifacts from the Cottages Site.' In: *Paleoecology and Archaeology of an Acheulian Site at Caddington, England.* C. G. Sampson (ed.), Dallas, pp. 83–137.

Carneiro, R. L. (1979). 'Tree felling with a stone axe: An experiment carried out among the Yanamano Indians of southern Venezuela.' In: *Ethnoarchaeology*. C. Kramer (ed.), New York, Columbia University Press.

Dechelette, J. (1908). *Manuel d'Archeologie Préhistorique celtique et Gallo-Romaine*. I. Paris: Picard.

Dickson, F. P. (1981). *Australian Stone Hatchets. A Study in Design and Dynamics*. Sydney.

Evans, J. (1897). *The Ancient Stone Implements of Great Britain*. London, 2nd edition.

Harding, A. & Young, R. (1979). 'Reconstruction of the hafting methods and function of stone implements.' *Stone axe studies*. CBA Research Report No. 23. London, pp. 102–5.

Iversen, J. (1956). 'Forest clearance in the Stone Age.' *Scientific American*, **194**, 36–41.

Jørgensen, S. (1953). 'Forest clearance with stone axes.' *Fra Nationalmuseets Arbejdsmark*, pp. 36–43, 109–10.

Lacaille, A. D. (1966). 'Polissoirs decouverts recemment en Angleterre.' *Revista da Faculdade de Letras de Lisboa*. Ser. 3, **10**, 61–75.

Newcomer, M. H. (1971). 'Some quantitative experiments in handaxe manufacture.' *World Archaeology*, **3**(1), 85–94.

Newcomer, M. H. & Sieveking, G. de G. (1980). 'Experimental flake scatter-patterns: a new interpretative technique.' *JFA*, **7**, 345–52.

Nietsch, H. (1939). Wald und Siedlung im vorgeschichtliche Mitteleuropa, *Mannus Bibliothek*, No. 64. Leipzig.

Olausson, D. S. (1983). 'Flint and groundstone axes in the Scanian Neolithic.' *Scripta Minora*, 1982–1983: 2. GWK Gleerup Lund, pp. 22–66.

Piggott, S. (1962). *The West Kennett Long Barrow Excavations 1955–56*. HMSO, London.

Pond, A. W. (1930). *Primitive methods of working stone. Based on experiments of Halvor L. Skavlem*. Wisconsin, Logan Museum Bulletin, Beloit College, Vol. 2, No. 1.

Saraydar, S. & Shimida, I. (1971). 'A quantitative comparison between a stone axe and a steel axe.' *American Antiquity*, **36**, 216–17.

Sehested, N. F. B. (1884). 'Praktiske forsøg.' *Archaeologiske Undersøgelser 1878–1881*. Copenhagen.

Semenov, S. A. (1964). *Prehistoric Technology*. London.

Smith, I. F. (1965). *Windmill Hill and Avebury. Excavations by Alexander Keiller, 1925–1939*. Oxford University Press.

Warren, S. H. (1921). 'Excavations at the stone-axe factory of Graig-Lwyd, Pennaenmawr. *JRAI*, **51**, 165–99.

6

Experimental investigation of the heat treatment of flint

D. R. GRIFFITHS, C. A. BERGMAN, C. J. CLAYTON, K. OHNUMA, G.V. ROBINS AND N. J. SEELEY

Introduction

It is now widely accepted that siliceous rock may on occasion have been deliberately heated in antiquity to improve its flaking quality. Ethnographic evidence, though sometimes suspect, suggests that such deliberate heating of siliceous rocks may have been practised in many parts of the world including western Oceania, northern Australia, southern Africa, Bengal and many parts of North America (Olausson & Larsson, 1982; Hester, 1972). The archaeological discovery of apparently deliberate heat treatment has been most common in North America where it has been widely recognised, for example in Arizona, Arkansas, California, Florida, Idaho, Illinois, Kansas, Kentucky, New York, Ohio and Texas (Epstein, 1979). Heated artefacts have been reported from the Solutrean at Laugerie Haute, including an unstratified fragment of a laurel leaf preform retouched after heating (Bordes, 1967 and 1969). From the same site Collins (1973) has published six stratified artefacts made out of heated chert, but one of these is a laurel leaf the heat treatment of which he describes as being far from certain. A recent examination of a large number of Solutrean artefacts has not shown any systematic evidence of deliberate heat treatment (R. Larick, pers. comm.). Examples of heated artefacts from the Old World upper palaeolithic are extremely rare, and where they do occur they do not appear to be of any particular type. For example, the upper palaeolithic level 11 at Ksar Akil (Lebanon) yielded a core and end scraper made of heated chert. In contrast to this, the use of heated flint in North America seems associated with the production of particular types of artefacts such as bifacial points. The present evidence thus fails to demonstrate conclusively the deliberate use of heat treatment in the Old World Palaeolithic. The evidence for its use in the Old World Neolithic is considerably stronger (Inizian *et al.*, 1975).

Opinions as to the methods used to effect the heat treatment of siliceous stone are largely based on analogy with ethnographic evidence. The methods described vary considerably. Some ethnographic reports refer to the burial of stone beneath a fire and often to the long duration of heating. For example, the Plateau Shoshoni Indians of North America were said to practise heat treatment by placing selected pieces of stone in damp earth covered with a brisk fire for 24 hours, and the Reece River Shoshoni placed flint intended for flaking under fire ashes for a period of five nights (Hester, 1972).

Modern attempts at heat treatment have generally incorporated one or both of these features of burial and a long duration of heating. Mandeville & Flenniken (1974), using a probable aboriginal technique, heat treated chert in a pit filled with successive layers of burning coals, sand, chert preforms, more sand, more burning coals and a final capping of sand. The pit was left until cool before removing the chert. A similar technique was used more recently by Ray (1982).

Although reports of artefacts manufactured from heated chert are now quite numerous, there does not seem to be much evidence in the archaeological record of features resembling heat treatment pits. The investigations reported here were made in an attempt to discover whether simpler and quicker methods of heat treatment could have been practised using primitive techniques. The effects of various temperatures and durations were determined by heating flint in a laboratory oven followed by pressure flaking. The heating conditions attainable in small fires were then investigated using thermocouples placed in and around them. Flint preforms were heated in various ways using the fires and subsequently flaked to determine the success of the heat treatments. In total, about 160 preforms were heated and pressure flaked. The structural changes underlying the changes in working quality were investigated using optical and electron microscopy and studying weight loss on heating.

Laboratory heating experiments

Two types of flint were used in these experiments. One was the visually homogeneous black flint from Taflin's Quarry, Brandon, Norfolk, England. The other, from South Mimms chalk quarry near Potters Bar, Hertfordshire, England, was a mottled grey flint containing lighter, apparently coarser grained, regions.

Six preforms of various shapes and sizes of South Mimms flint were placed in a thermostatically controlled electric oven at 100 °C. The thermostat was then set to the desired heating temperature, which was maintained for 24 hours. The preforms were buried in sand to reduce the likelihood of fracture due to thermal shock. The preforms were allowed to cool to 150 °C in the closed oven, and were then removed. This process was used for peak temperatures of 150, 200, 250, 300, 350 and 400 °C, using a different set of six preforms at each temperature. All the South Mimms preforms heated to 400 °C suffered extensive internal fracture and were unsuitable for flaking.

It is widely accepted that suitable heating of chert can facilitate flaking by soft hammer percussion or pressure. In these experiments the heated preforms were pressure flaked and changes in their working quality assessed subjectively. In successfully treated flint, longer and more controlled flakes can be detached with less force. One cannot be certain of primitive man's assessment of pressure flaking quality. The authors suggest, however, that a subjective assessment of this quality by experienced modern flakers may be as valid a parallel to primitive man's assessment as a parameter somewhat arbitrarily derived from laboratory tests of mechanical properties. Accordingly, two of the authors experienced in pressure flaking (CAB and KO) assessed the quality of the preforms on a scale of 1 to 5, 5 being the best quality. Untreated flint, which was available for comparison, scored 2 by definition. The flakers desired, and were allowed, to give half integral scores. While making their assessment the flakers did not know what, if any, heat treatment a particular preform had been given, nor were they aware of the score given by the other flaker. A copper rod hafted in wood was used as the pressure flaking tool throughout these experiments.

The results of one of the flakers for the preforms heated for 24 hours were plotted as flaking quality versus temperature in Fig. 6.1. The other flaker's results were very similar, except that the absolute scores differed slightly. The error bars indicate the average score for the six preforms heated at a particular temperature plus and minus one standard deviation ($\bar{y} \pm 1y\sigma_{n-1}$). There is a gradual improvement in pressure flaking quality up to 250 °C. As is suggested by the error bars, the flakers considered some of the 300 and 350 °C preforms better than any of the 250 °C samples, while others they considered no better or worse than the untreated flint by virtue of being too brittle. Although it was not known at the time of the experiment, Inizian *et al.* (1976) had published work that concurred with the present finding that increasing the temperature of heating gave a gradual amelioration of flaking quality until excessive brittleness rendered the chert worse than its unheated counterpart.

Preforms that are too brittle persistently crush at the point of applied force before a flake is detached. As preforms became more brittle with increasing temperature of previous heating, so more abrasion at the point of applied pressure is required before flaking. The reason for this may be that the compressive strength of the flint decreases. The force of the pressure flaking tool then needs to be spread over a large area to prevent crushing. Abrasion may round the point of applied pressure and so spread the load. It may also reduce the tensile strength by forming sharp ended cracks on the surface. The flakers gave scores for the amount of abrasion required and for the force required to detach a flake. The former increased monotonically and the latter decreased monotonically with increasing temperature. Heating flint to higher temperatures often results in a mass of internal fractures (visible with the naked eye) that cause the flint to crumble into angular fragments

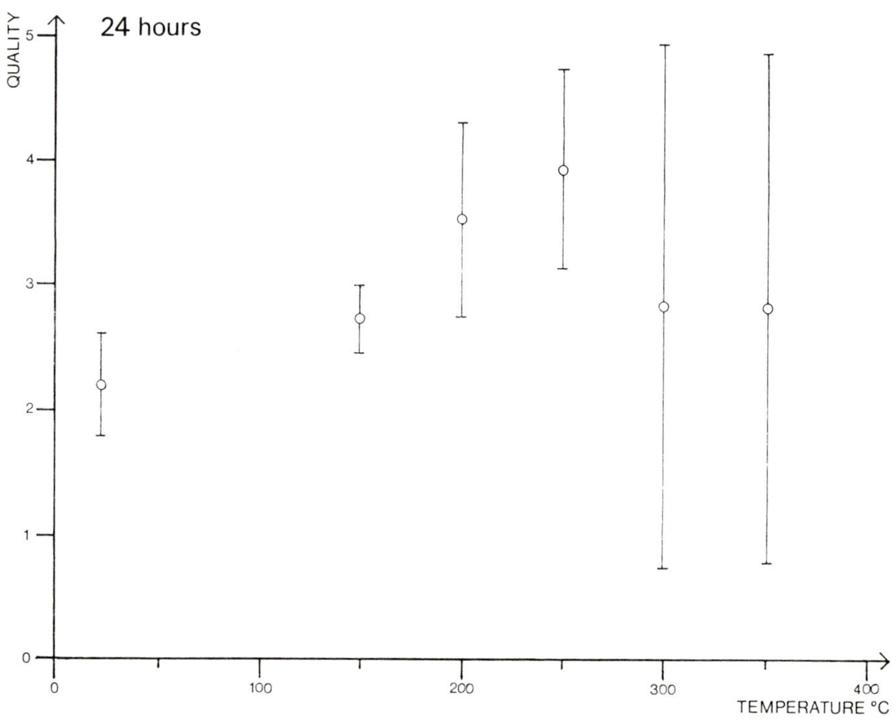

Fig. 6.1. Pressure flaking quality of South Mimms flint preforms held at the indicated temperatures for 24 hours.

when force is applied. The tendency to develop these internal fractures varies with provenance. Exposure to high temperatures may also cause flint to explode.

Both the flakers felt that the inhomogeneity of the South Mimms flint made assessment of the flaking quality difficult because the different regions responded to the heating to different extents. In order to overcome the difficulty and to investigate the effect of shorter heating periods the above laboratory heating and flaking experiments were repeated using Brandon flint placed in a preheated oven for one hour followed by cooling in the closed oven as before. The Brandon preforms were stood rather than buried in sand to make it more likely that the preforms would reach the temperature of the oven. In spite of the more abrupt heating only one of the six Brandon preforms heated to 400 °C suffered internal fracture. The results of pressure flaking are shown in Fig. 6.2. (As in Fig. 6.1 the error bars are $\bar{y} \pm 1y\sigma_{n-1}$ with six samples heated to each temperature). The results are similar to those of the 24 hour heating experiment except that the error bars at the lower temperatures are smaller and the optimum heat treatment was achieved at the slightly higher temperature of 300 °C. The former is probably due to the homogeneity of the flint, while the latter may be due to the shorter heating time and/or the difference in raw material.

To further test the efficacy of very short heating times in improving the pressure flaking quality of flint, South Mimms flint preforms were stood in a sand bath and placed in preheated oven for half an hour and then immediately removed. The preforms heated up to and including 350 °C were apparently not adversely affected by the abruptness of heating and cooling; the 400 °C preform again suffered internal fracture. As only a shallow surface layer of the preforms was tested by pressure flaking it is reasonable to assume that the region tested had indeed reached the temperature of the oven. The results of the pressure flaking (Fig. 6.3) show that improvement in the flaking quality of the flint has occurred to a significant and useful extent in this short heating time. (Only one preform was heated to each temperature.) The improvement in flaking quality at a given temperature is less for the shorter duration of heating, but a similar optimum improvement is attainable by employing a higher temperature. For example, the improvement in quality after half an hour at 350 °C is similar to that after 24 hours at 250 °C. The results of Figs. 6.1–3 are summarised in Fig. 6.4. (The half hour and 24 hour plots are directly comparable, both being South Mimms flint flaked by the same flaker during the same flaking session. The one hour plot is of Brandon flint flaked on another occasion.) The temperature required to reach optimum pressure flaking quality is likely to vary with raw material, and may also vary according to the preference of different flakers.

Campfire heating experiments

Having investigated the effects of temperature and duration of heating on flint pressure flaking quality and established the efficacy of short heating periods, the conditions attainable in small fires were investigated. The fires were built of fallen wood of mixed deciduous species that was dried indoors before use. The fires were started with small twigs and bark tinder. Wood of progressively greater diameter (up to 5 cm) was gradually added. All the wood

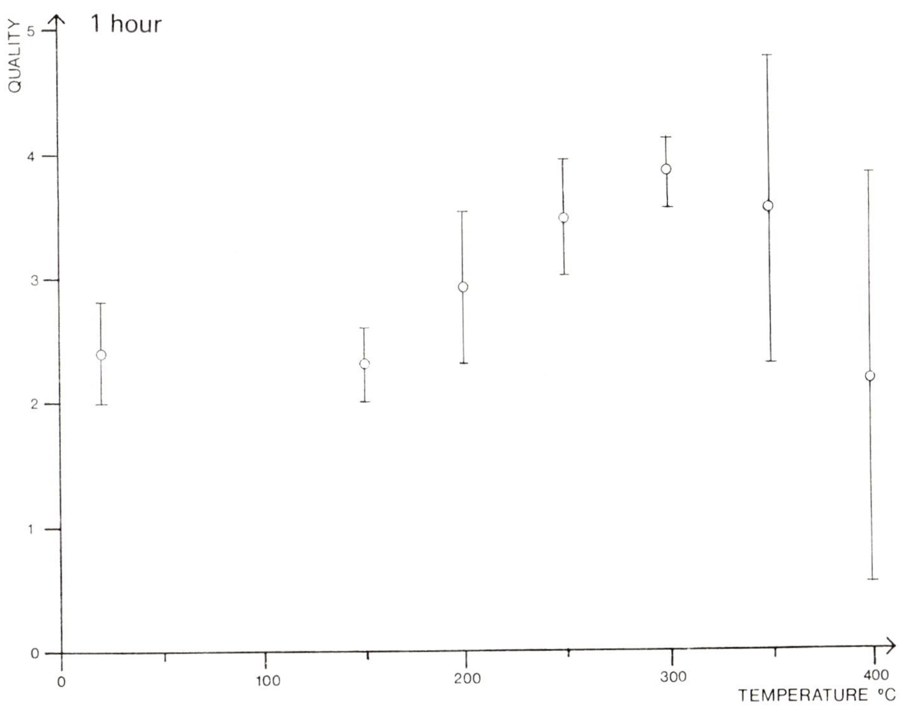

Fig. 6.2. Pressure flaking quality of Brandon flint preforms held at the indicated temperatures for one hour.

was added within the first 30–45 minutes. The fires were 50–80 cm in diameter, flamed for approximately 60–75 minutes, and glowed for some time longer. Chromel–alumel thermocouples (connected to a digital millivolt-meter) and flint preforms were placed in various positions in and around the fires. The fires were built when the ambient temperatures were 10–20 °C. The first fires were built on a patch of earth roughly swept clear so forming a shallow depression. It was noted when such a fire was built on a windy day that the ash bed was considerably disturbed

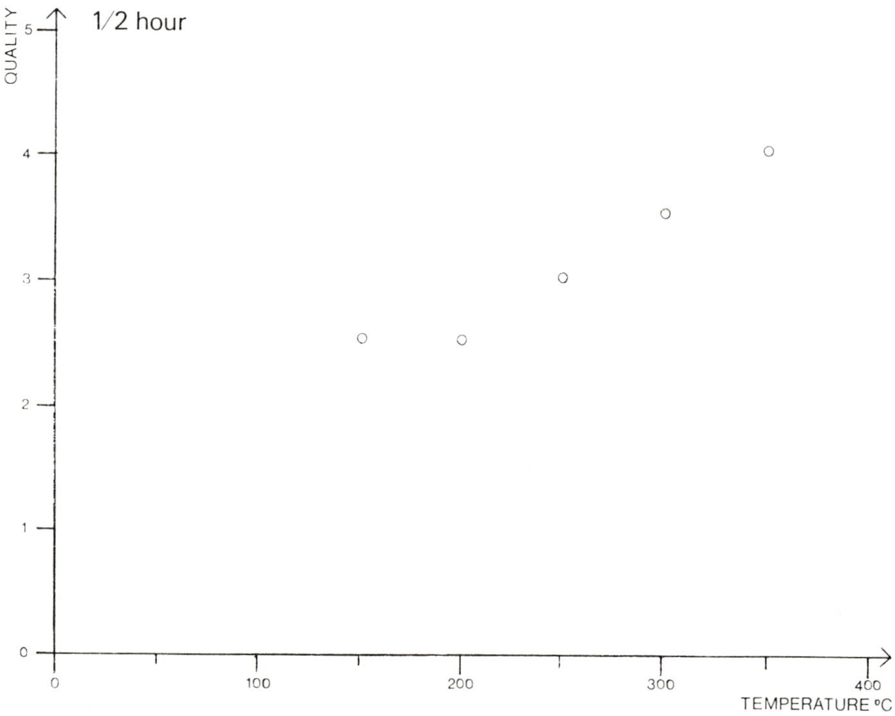

Fig. 6.3. Pressure flaking quality of South Mimms flint preforms held at the indicated temperatures for half an hour.

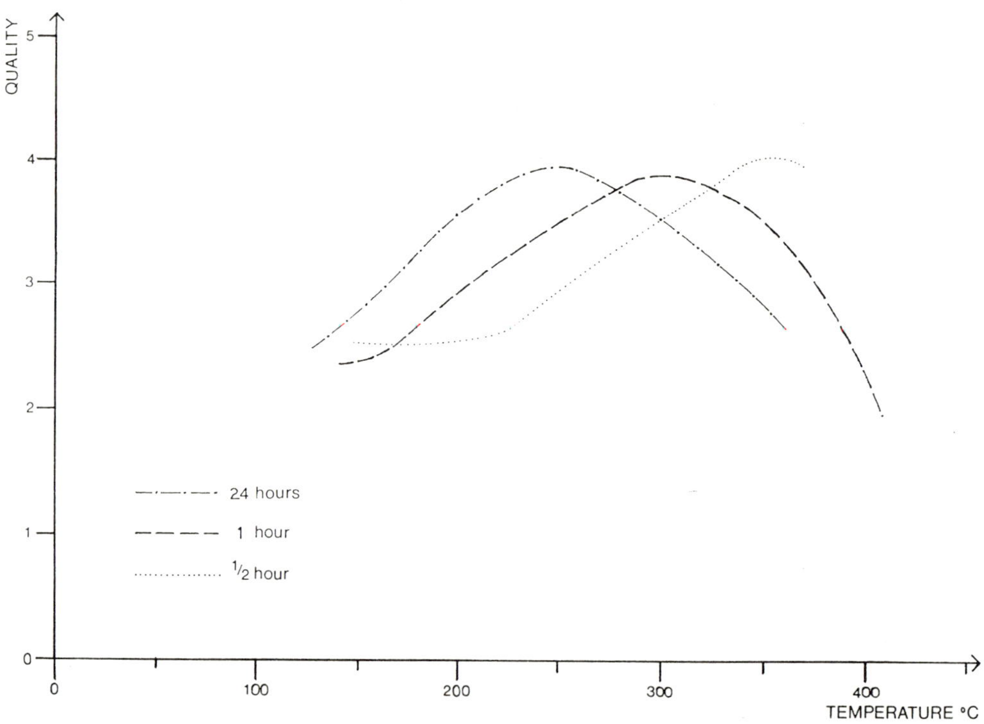

Fig. 6.4. Average pressure flaking quality of flint preforms held at the indicated temperatures for 24 hours, one hour and half an hour.

and blown into incandescence. In an attempt to reduce the effect of the wind and in response to the observation that prehistoric fires were often surrounded by hearthstones, a similar fire was built on another windy day with the addition of a fairly well fitting surrounding wall of broken bricks and stones some 20 cm high. Thermocouples were placed on the ground beneath the fires. Once the fires had become established, the ground temperature half a radius from the centre was higher than that at the centre in the fires without hearthstones, whereas in the fire with hearthstones the centre was generally hotter. The temperature difference between the two locations was generally 100–200 °C, and may have been due to differences in oxygen supply. This factor may also account for the ground temperatures under a fire built on a still day (generally ≤400 °C with peaks up to 540 °C) being some 200 °C lower than those of fires built on windy days (generally ≤600 °C with peaks up to 700 °C). Temperatures at the peripheries of the fires varied considerably as added wood acted first as a heat shield and then as a heat source. The generality of these remarks on temperature must remain in question pending further experimentation.

The ground surface under a fire during its early stages is too hot for successful heat treatment. All the preforms placed on the ground prior to lighting the fire suffered internal fracture. Direct contact with flames and incandescent wood also caused internal fracture. Attempts to place preforms on the ground under the fire once it had cooled a little would probably result in the temperature rising as oxygen was allowed in, as it does when a fire is poked. Little success was achieved with preforms placed near but not in the fires. They were generally insufficiently heated, although one of the preforms placed on a hearthstone showed a marginal improvement. Another preform at the periphery of a fire was over-heated by embers collapsing on it.

In an attempt to take advantage of the suitable ground level temperatures later in the development of the fires, but protect the preforms from the excessive early temperatures, two preforms were encased in coats of damp clay about 1 cm thick and placed on the ground in the centre of a fire before lighting it. It was a windy day and both preforms were internally fractured. Under the centre of the same fire two preforms had been buried in dry sand at depths of 1.1 cm and 2.3 cm. Both were very successfully heat treated. On a still day, however, the temperature at 3 cm depth in damp earth beneath the centre of a fire only reached around 70 °C, and a fire of much greater duration would have been necessary for successful heat treatment of flint buried in these conditions. It would thus appear that the successful heat treatment of artefacts buried or encased in soil using fires of brief duration is critically dependent on a number of variables, and would require considerable skill.

The most successful method of heat treatment discovered was simply to place the flint preforms (South Mimms or Brandon) on to the cooling bed of ash where it had recently ceased glowing but still had glowing embers beneath. (The preforms were dropped on from the bare hand rather than carefully placed.) Putting a preform on an area of ash immediately after it stopped glowing sometimes caused over-heating, but putting preforms on during the 15 minutes or so after an area of ash ceased glowing generally gave a good result. Although the time factor is fairly critical, the optimum time is clearly apparent by eye. Irrespective of other conditions, the colour of incandescence from the ash is indicative of its temperature, so this method has the advantage that it can be practised on fires of any size or duration.

It may be that the layer of ash beneath the preform collapses, so arresting combustion by excluding oxygen and forming an insulating layer through which heat from the embers below gradually passes. A thermocouple attached to the underside of a preform placed on an ash bed rose to 230 °C over ten minutes. The subsequent pressure flaking yielded a score of 3.5 for quality, which was in accord with the results of the laboratory heating experiments. The temperature in the midst of the ash below another preform which scored 3.5 was 320 °C.

An attempt to use this method on a windy day resulted in a higher incidence of thermal fracture due to the ash bed being disturbed and blown into incandescence. Nevertheless 70% of the preforms showed improvement due to heating (scored 2.5 or over for pressure flaking quality). The use of hearthstones on a similarly windy day prevented the ash bed from being much disturbed and gave better results. The proportion of preforms scoring 2.5 or over increased from 70% to 80%. (Untreated samples included in the batch as a control without the knowledge of the flaker all scored 2.) The proportion showing considerable improvement (scoring 3.5 or over for pressure flaking quality) increased from 30% to 50%. There was also a lower incidence of over-heating in the fire with hearthstones: 10% as opposed to 30%. This is particularly important because an insufficiently heated preform can be used as it is or heated again, whereas an over-heated piece is irreversibly damaged.

It was noted that preforms heated on the fires sometimes showed more improvement in one region than another, even when the flint itself was homogeneous. This heat treatment can still be useful in spite of being uneven. That primitive man used such material is particularly apparent in the Bald Eagle jasper of Pennsylvania where the change in working quality and a distinct colour change appear to arise from a common cause (Schindler *et al.*, 1982).

Structural changes caused by heating

The structural changes occurring during heating are conveniently monitored by observation of the associated weight loss. Figure 6.5 shows the percentage weight loss after heating the South Mimms flint preforms for 24 hours as described earlier. The measurements were made as soon as the preforms had cooled to room temperature. A curve has been drawn through the points representing the average percentage weight loss at each temperature, and the following arguments assume that the curve represents a genuine trend. In view of the variation between preforms and the

UNIVERSITY
OF BRISTOL
LIBRARY

consequent size of the error bars, the inferences should be taken as tentative until further data are available. The error bars represent the average percentage weight loss of the six preforms heated to each temperature plus and minus one standard deviation. There was no apparent correlation with preform size or surface area to volume ratio, but the large range of readings may have been caused by the inhomogeneity of the South Mimms preforms. There is a minimum in the weight loss curve around 300 °C. The weight loss is due to water loss (Micheelsen, 1966) and, all things being equal, this would be expected to be the same or greater after heating to 300 °C than after heating to 150 °C. The shape of the curve (Fig. 6.5) thus implies that a structural change is occurring between 150 and 300 °C that is obstructing water loss from the flint.

Water is present in the silica structure of flint as hydroxyl groups on grain and subgrain boundaries and along lattice defects, and as molecular water in microscopic and submicroscopic fluid inclusions and adsorbed on surfaces (Micheelsen, 1966). This distinction is clearly shown in the thermogravimetric curve (Fig. 6.6) where loosely bound molecular water is lost below 100 °C but the tightly bound

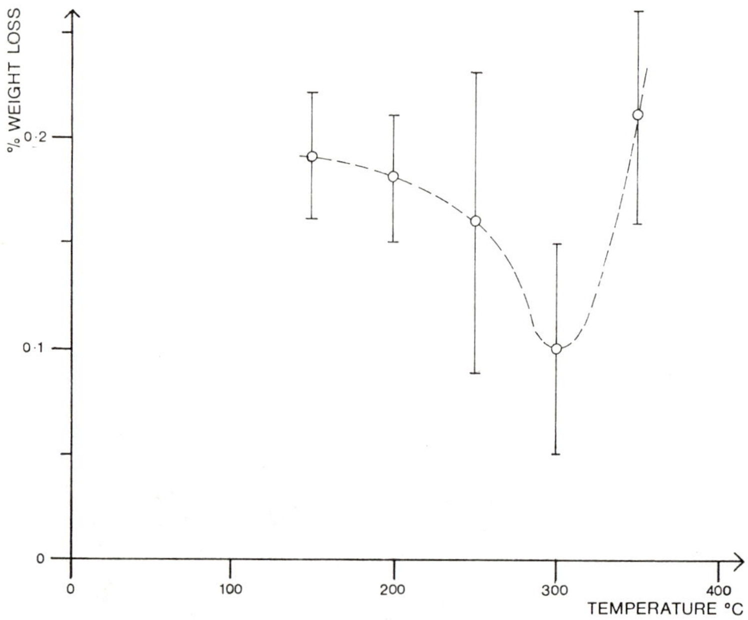

Fig. 6.5. Percentage weight loss from South Mimms flint preforms after heating at the indicated temperatures for 24 hours.

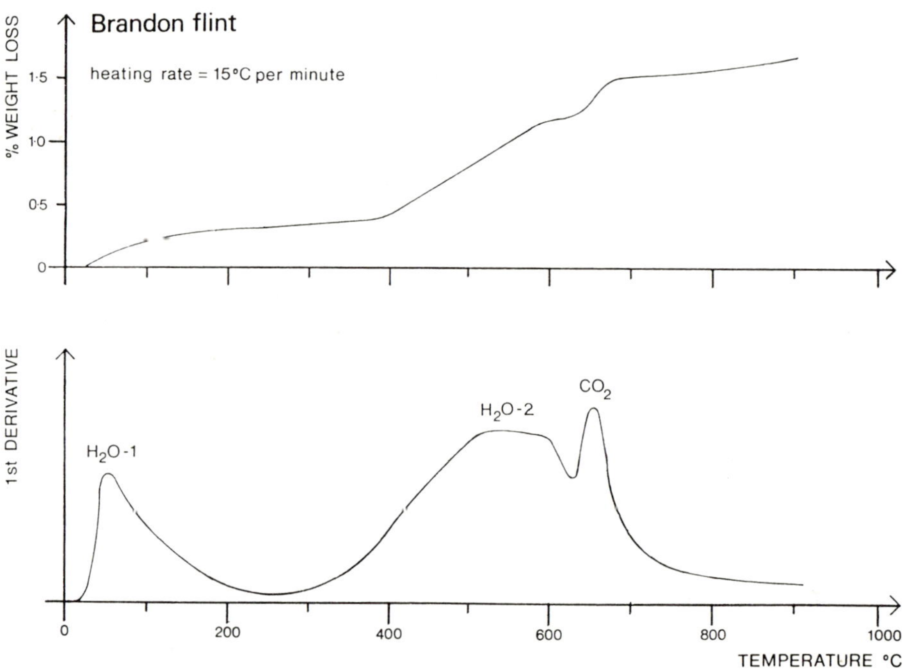

Fig. 6.6. Thermogravimetric trace and its first derivative from heating powdered Brandon flint.

Fig. 6.7. (A) General microstructure of Brandon black flint showing framework of globular lepispheres with interstitial chalcedony. Specimen etched for 20 seconds in 60% HF. Backscattered electron image. (B) Etched specimen of unheated Brandon flint showing detail of lepisphere and silicified microfossil (arrowed) framework and interstitial chalcedony (C). Note partial coalescence of lepispheres to form a more solid network. (C) Unetched fracture surface of low temperature heated (150 °C) Brandon flint. Note how fracture still passes through the chalcedony, leaving domed structures on surface (arrowed) reflecting the presence of lepispheres. (D) Fresh fracture surface of Brandon flint after heating to 200 °C. Vague traces of round lepispheric structures are just visible but fracture is able to pass through lepispheres as easily as the, now hardened, interstitial chalcedony. (E) as (D) but specimen heated to 300 °C. Distinction between lepispheres and chalcedony is now invisible as fracture propagates across lepispheres as easily as through chalcedony. (F) Etched specimen of flint shown in (E). Note the highly granular nature of the chalcedony visible in the pocket at centre of picture. This is caused by etching-out of sub-microscopic fluid inclusions formed during the heat treatment. This response to etching is different from that of the unaltered flint (B).

Fig. 6.8. (A) Detail of fracture surface of unheated South Mimms flint. Note the clear distinction between spherical lepispheres and interstitial chalcedony through which fracture preferentially propagates. (B) Same flint type as (A) after heating to 250 °C. Fracture now cuts across lepispheres to reveal coarser grained fabric inside (arrowed). (C) Detail of platy interstitial chalcedony in unheated South Mimms flint. (D) Early stages of natural (i.e. geological) recrystallisation in Brandon flint. The granular nature of the chalcedony is produced by etching out of the neoformed sub-microscopic fluid inclusions. More advanced stages of recrystallisation, such as that induced by heating the flint, result in larger and more numerous fluid inclusions which allow easier etching (Fig. 6.7(F)) even though the silica itself has a higher structural order and is more resistant to chemical attack.

structural water is not released until above 300 °C. A further weight loss between 600 and 700 °C represents the release of carbon dioxide caused by thermal decomposition of calcite in the flint (relic chalk left from the replacement process of flint formation: Clayton, 1986).

During natural recrystallisation, the structurally disordered silica in flint undergoes a transition to a more ordered structure in which water along sub-grain boundaries and lattice defects migrates into sub-microscopic fluid inclusions as defects in the structure are annealed (Clayton, 1986). It seems reasonable to assume a similar reaction during heat treatment. At temperatures up to 100 °C loosely bound molecular water and physically absorbed water on the surfaces of hydroxyl groups is lost. At higher temperatures annealing of the structure occurs and an increasing portion of this water becomes trapped in microfluid inclusions, as reflected in the falling weight loss curve up to 300 °C (Fig. 6.5). The presence of fluid inclusions eventually causes a waxy sheen or lustre on fractured surfaces, since inclusions tend to partially reflect light rather than simply refract it as does normal flint. This redistribution effect is not seen in the thermogravimetric trace (Fig. 6.6) because the fine grain size of the powdered sample results in total loss of any water released. Above 300 °C major structural recrystallisation occurs with structural water migrating into fluid inclusions and being lost. The migration into fluid inclusions is illustrated by the appearance of a broad nuclear magnetic resonance (NMR) signal attributed to liquid water in Brandon flint heated at 380 °C for one hour, but not after heating at 280 °C or lower temperatures (Symons, 1986). The loss of structural water shows as the H_2O–2 peak in the thermogravimetric trace (Fig. 6.6). The increasing susceptibility to crushing rather than simple fracture observed as flint was heated to higher temperatures may be due to microfracturing of the flint grains by hydraulic pressure. It may also be influenced by rates of cooling (Purdy & Brooks, 1971).

The significance of this recrystallisation in the flaking of heated flint is best understood by considering the larger scale structure of the material. Typical black flint is composed of a framework of spherical quartz aggregates called 'lepispheres' (5–20 μm diameter) and lesser amounts of silicified skeletal fragments, in an interstitial cement of structurally disordered microcrystalline chalcedony (Clayton, 1982 and 1983; Bradley & Clayton, 1986). In raw flint the interstitial chalcedony has a greater structural disorder and higher water content than the lepispheres, and is more prone to fracture. Consequently, fractures in this flint (such as those induced by pressure flaking) preferentially propagate through the chalcedony and around the lepispheres, leaving hemispherical projections or depressions in the fracture surface (Fig. 6.7). The direction of propagation of the fracture is thus influenced by the distribution and packing density of the lepispheres, features inherited from the sedimentary fabric of the chalk in which the flint grew (Clayton, 1983). In heated flint the interstitial chalcedony becomes annealed as structural water migrates to fluid inclusions. In this flint the chalcedony is structurally comparable to the lepispheres and fractures will propagate through the lepispheres as easily as through the chalcedony (Fig. 6.8). Pressure flaking of this material is therefore less affected by the original flint fabric and more easily controlled. The heated flint is more homogeneous and glass-like in its fracture properties.

Conclusion

Thermogravimetric and microscopic studies have indicated that recrystallisation of the microchalcedonic matrix of the flint is probably responsible for the change in flaking quality caused by heating. The heating experiments suggest that with a little experience the desired structural changes can be readily achieved on the cooling ash of an ordinary fire. Provided sufficient material is burnt to give a bed of ash the method can probably be used on fires of any size and duration. Its use could probably not be detected in the archaeological record, since the remains would be similar to those of an ordinary fire. Where artefacts do suggest the use of deliberate heat treatment there is thus no necessity to postulate the existence of as yet undiscovered heat treatment pits, unless their use is suggested by other evidence. While such pits may have been used in some instances, the method proposed in this paper has the appeal of simplicity.

Acknowledgements

DRG gratefully acknowledges the financial support of SERC.

Bibliography

Bordes, F. (1967). 'Considérations sur la Typoligie et les techniques dans le Paléolithique.' *Quartär*, **18**, 25–55.

Bordes, F. (1969). 'Traitement thermique du silex au Solutreen.' *Comptes Rendus des séances mensuelles de la Société Préhistorique Française*, **66**, 197.

Bradley, R. & Clayton, C. J. (1986). 'The influence of flint microstructure on the formation of microwear polishes.' In, G. de G. Sieveking & N. H. Newcomer (eds.) *The Human uses of flint and chert*, pp. 81–9, Cambridge University Press.

Clayton, C. J. (1982). 'Growth history and microstructure of flint.' *International Assoc. of Sedimentologists 3rd European Meeting*, Copenhagen, pp. 105–7.

Clayton, C. J. (1983). Geochemistry of chert formation in Upper Cretaceous flints. Unpublished PhD thesis, University of London.

Collins, M. B. (1973). 'Observations on the thermal treatment of chert in the Solutrean of Laugerie Haute, France.' *Proc. Prehist. Soc.*, **39**, 461–6.

Epstein, J. F. (1979). 'Flint technology and the heating of stone.' In: Schmandt-Besserat (ed.), *Early technology: Invited lectures on the Middle East at the University of Texas at Austin*, **3**, 27–38.

Hester, T. R. (1972). 'Ethnographic evidence for the thermal alteration of siliceous stone.' *TEBIWA* (J. of Idaho St. Univ. Museum), **15**, 63–5.

Inizian, M.-L., Roche, H. & Tixier, J. (1976). 'Avantages d'un traitement thermique pour la taille des roches siliceuses.' *Quaternaria*, **19**, 1–18.

Mandeville, M. D. & Flenniken, J. J. (1974). 'A comparison of the flaking qualities of Nehawka chert before and after thermal pretreatment.' *Plains Anthropologist*, **19**, 146–8.

Micheelsen, H. (1966). 'The structure of dark flint from Stevns, Denmark.' *Medd. fra Dansk Geol. Forening*, **16**, 286–368.

Olausson, D. S. & Larsson, L. (1982). 'Testing for the presence of thermal pretreatment of flint in the mesolithic and neolithic of Sweden.' *Journal of Archaeological Science*, **9**, 275–86.

Purdy, B. A. & Brooks, H. K. (1971). 'Thermal alteration of silica minerals: an archaeological approach.' *Science*, **173**, 322–5.

Ray, J. H. (1982). 'The effects of heat treatments on cherts from the Truman Reservoir.' *Missouri Archaeologist*, **43**, 67–85.

Schindler, D. L., Hatch, J. W., Hay, C. A. & Bradt, R. C. (1982). 'Aboriginal thermal alteration of a central Pennsylvania jasper: analytical and behavioral implications.' *American Antiquity*, **47**, 526–44.

Symons, M. C. R. (1986). 'Application of magnetic resonance methods to the study of cherts.' In, G. de G. Sieveking & M. B. Hart (eds.), *The scientific study of flint and chert*, pp. 249–58, Cambridge University Press.

7

The development and application of microwear analysis since 1964

JILL COOK AND JOHN DUMONT

Introduction

Over the last 100 years, many studies have set out to determine the purposes for which lithic assemblages or particular types of stone artefacts were made and used. These varied approaches to lithic use–wear analysis are summarised in Table 7.1. As far as artefacts made of flint and other siliceous raw materials are concerned, the most effective of these methods to date, and consequently the subject of this paper, is microwear analysis.

Microwear analysis involves the use of optical and/or scanning electron microscopy to observe and record the presence, character and distribution of modifications to the microtopography of the artefact surface caused by its use on one or more materials. Although microscopy had been used to examine technological, natural and use–wear traces on experimental and archaeological artefacts by workers such as Barnes (1939), Curwen (1930), and more consistently by Sonnenfeld (1962), it was the publication in Russian (1957) and subsequently in English (1964) of Semenov's book *Prehistoric Technology* that proved to be a turning point for microwear studies. Semenov's achievement was the recognition of the significance and potential of use–wear traces (striations, polishes, and pitting) for inferring the uses of archaeological artefacts. This transformed the purpose of microscopic investigation from a technique for accumulating incidental descriptive data on each object into a method of studying artefact use. By making focused observations of the nature and disposition of wear traces on an artefact surface, he showed that it was possible to ascertain how a tool had been used and then applied this information to support or eliminate competing hypotheses about the task it had fulfilled.

Table 7.1. *Approaches to lithic use-wear analysis*

Approach	Examples
Experimental work leading to inferences about archaeological artefacts	Spurrell, 1892; Curwen, 1930; Jones, 1980
Experimental research and observation of macroscopic wear traces and edge characteristics	Kantman, 1970*a*, 1970*b*; Odell, 1975; Tringham *et al.*, 1974
Experimental research and observation of microwear traces at medium range magnifications using incident light microscopy	Semenov, 1964; Keeley, 1980
Experimental research and observation using scanning electron microscopy	Anderson, 1980, 1981; Dies, 1981; Meeks *et al.*, 1982
Observation of use–wear traces without comparative experimental research	Lenoir, 1971; Rosenfeld, 1971
Interpretation of function by contextual inference	Frison, e.g. 1974
Observation of ethnographic artefacts of known use	Gould *et al.*, 1971; Kamminga, 1977
Detection and analysis of worked materials adhering to artefact surfaces	Anderson, 1980, 1981; Briuer, 1976

Semenov bases the recognition of traces resulting from use–wear and how they relate to the manner of tool use on observations made over several years on experimental, ethnographic and modern tools of known purpose and manner of use. He does not include details of this research, concentrating instead on detailed microwear analysis of archaeological objects. Thus, rather than seeking exact replication, the experimental work is used as a guide to interpreting archaeological evidence. In his introduction Semenov justifies this by saying that 'the experimental approach cannot serve as an independent method of study of the function of tools' (1964, p. 1) although 'it would be a serious error . . . to reject entirely the part of experiment' (*ibid.*, p. 2). This frequently forgotten truth is reiterated below and in the historical context, it is notable that studies which have produced the most satisfactory results from archaeological material (Anderson, 1980, 1981; Cahen & Keeley, 1980; Dumont, this volume, 1983 in press; Keeley, 1980; Meeks *et al.*, 1982, *inter alia*) have adopted the same cautious attitude to the purpose of experimental data and have not in fact given detailed, standardised descriptions of large numbers of 'controlled' tool use experiments. Indeed, it is very important that the collection of experimental data does not become an end in itself but should be related to archaeological research where it should be applied with appropriate caution (see, for example, Moss & Newcomer, 1982). However the absence of the experimental framework from *Prehistoric Technology* made many archaeologists sceptical about microwear analysis as they found it difficult to accept such definite interpretations presented purely on a basis of experience (ironically a criticism also levelled at typological methods of analysis). Indeed, this deficit and the lack of information about the sampling of assemblages and the success rate of identifying tool use have remained the major criticisms of Semenov's work.

Although a few workers, notably McBurney (1968), Rosenfeld (1971) and Wylie (1975), achieved some success by adapting Semenov's methods to their own studies, others such as Odell (1975) were unable to observe or interpret similar wear traces, probably because they failed to counteract the translucency of the materials examined and therefore could not observe the microtopography of the surface (Keeley, 1974). As a result studies of macroscopic damage (e.g. Hester *et al.*, 1973; Kantman, 1971; Odell, 1980; Tringham *et al.*, 1974) and edge characteristics such as thickness and angle (e.g. Kay, 1977; Keller, 1966; Wilmsen, 1968) proliferated. These papers tend to deal with variables that might influence the manner of tool use and either do not document archaeological applications or provide results from archaeological material that are of uncertain consequence. As far as the purpose of the tools is concerned, many of the variables investigated in these papers have been shown *not* to be use-specific (Anderson, 1981; Keeley, 1980, below) even in such general terms as scraping or cutting.

Although Semenov recognises both striations and polishes as significant use–wear traces, he makes no attempt to describe polish characteristics or to distinguish between different polishes, probably because his method of metallising surfaces to counteract translucency obscured such differences. Instead this breakthrough was to come from Keeley (1980) who discovered that different materials caused polishes with varying characteristics to develop on experimental tools and that these distinctions could also be seen on archaeological artefacts. Keeley's observations have subsequently been repeated by other workers (Anderson, 1981; Dumont, 1982; Frame, this volume; Gendel & Pirnay, 1982; Moss & Newcomer, 1982, *inter alia*). These workers have used incident light microscopes to examine uncoated specimens at magnifications of between ×150 and ×325. Workers using scanning electron microscopes (Gramsch, 1979; Kamminga, 1977; Pant, in de Lumley *et al.*, 1979) have not observed differences between polishes for technical reasons that will be discussed below.

Accepting that different materials do produce distinct polishes, although occasionally the distinctions may be difficult to recognise with complete certainty, the main problem with this technique has been the descriptions of polishes which to date have been given qualitatively in terms of brightness, texture, disposition and occasionally colour. These descriptions have attracted some criticism and demands have been made for more standardised, objective, less worker-dependent criteria. Such demands appear very reasonable and have been responded to in a variety of ways, none of which entirely solve the problem. Keeley (1980) shows that light reflection measured in microamperes through the microscope is both impractical and inadequate to the task of measuring polish brightness, whilst Semenov's & Shchelinski's (1971) luminous section method for measuring microroughness requires at least 15 measurements to be taken before a significant mean result can be achieved. The preliminary results of this work (*ibid.*, 1971) show that this metrical method does separate different degrees of microunevenness satisfactorily but the authors do not comment on its practical application to archaeological material which may not be feasible on a large scale. Dumont (1982, and below) has shown that interferometry provides a practical method of quantitatively assessing polish morphology, whilst Grace *et al.* (this volume) offer other possibilities. However, these techniques should be regarded as providing additional information about the altered surface rather than as simple alternatives to qualitative description, for as such they are not very meaningful. In fact, it is probably naive to imagine that quantitative assessment will ever satisfactorily replace qualitative description based on observation, photomicrography and the judgements of experienced analysts. That the residues of eclectic human activities will not always respond to standardised, quantitative assessment is certainly not a new problem for archaeologists, especially those dealing with stone tools, and using both empirical and metrical data is a familiar method of evaluation. It is therefore essential that novice microwear analysts should gain experience by studying reference collections where they are available, or by repeating experiments (*contra* Schiffer, 1979, p. 16), and that developments should equally be sought on the empirical (i.e. observation) side of microwear analysis. In this respect Meeks *et al.* (1982) have shown that high magnification (×500 to ×10 000) scanning electron microscopy can be used successfully to characterise and differentiate the microfeatures of various optical polishes and further research is in progress on micropolishes, whilst Anderson (1980, 1981) has identified organic residues in association with polishes, opening a completely new avenue of investigation.

Anderson's work also encompasses a topical area of microwear research: that of polish formation. It is too soon to evaluate the extent to which understanding the mechanism of polish formation will improve our ability to discriminate polishes produced on flint surfaces by contact with other materials. To date, the papers on this subject have been mainly theoretical (e.g. Del Bene, 1979; Dumont, 1982) and those that have been directed at specific uses have concentrated on the mechanism of so-called 'sickle gloss' and vegetable polish formation (e.g. Diamond, 1979; Kamminga, 1979; Witthoft, 1967), although Meeks *et al.* (1982) have also investigated natural polishes, and Anderson (1980, 1981) has extended her theory of polish build up to encompass bone and animal tissue. These papers propose three possible mechanisms of vegetable polish formation: abrasion (Diamond, 1979); friction and chemical interaction between the flint surface and plant stems (Meeks *et al.*, 1982); and the formation of amorphous silica gel under increased temperature, pressure, abrasion and friction causing plant cell elements to be trapped on the surface, so that the polish constitutes a surface build up of non-organic plant material (Anderson, 1980, 1981). Meeks *et al.* (1982) have indicated that in sickle gloss formation, abrasion is not the cause of polish formation but occurs concurrently with it. The same paper also illustrates cross-sections taken through polished sickles that show no indication of the surface build up suggested by Anderson (*op. cit.*) and suggests other interpretations for micrographs which apparently show this effect. The ball is clearly in the air, especially as some sickles, both archaeological and experimental, show no development of the distinct optical polish! In the light of Mansur's (1982) acceptance of Anderson's theory of polish build up and the consequent, rather unfortunate, use of terms such as 'additive' and 'filled-in' striations (*ibid.*, p. 222), it is perhaps necessary to add that it is vital for microwear analysts to regard these theories at least temporarily as hypotheses rather than definitive universally applicable explanations. Equally, features such as the 'striation comprised of polish' illustrated by Del Bene (1979, Fig. 2) do require investigation and explanation. Further research is clearly essential and it is to be hoped that this research will remain relevant to the archaeological problems that gave birth to it.

Theory and methods

There has been a tendency in recent years to regard microwear analysis as capable of becoming a method of directly answering increasingly complex questions relating to increasingly higher levels of intra- and inter-site interpretations of both behavioural and economic importance. The capability of microwear analysis to achieve this ideal level of significance is directly tied to the quantity, quality, comparability and comprehensiveness of the methods employed in research programmes designed to study experimental and archaeological material. This section of the paper reviews aspects of experimental and microscopic methods which are critical to the interpretation of microwear traces.

Semenov's caution, quoted above, regarding the role of experiments in the elucidation of archaeological tool function, remains an important guideline to microwear analysts and to the archaeologists/prehistorians who attempt to make use of the newly derived functional information. Experimentation, or more specifically the production of experimental microwear traces, has and will continue to be a necessary activity for microwear analysts. Although the importance of experimentation as both a teaching aid and as a guide for the interpretation of archaeological wear traces should not be underestimated, it would be wrong to consider it as an exercise leading to the 'replication' of archaeological traces to enable direct determinative comparisons without having first demonstrated that direct comparisons are possible.

Experimental wear traces have been found to be broadly comparable to archaeological traces (Anderson, 1981; Gendal & Pirnay, 1982; Keeley, 1980; Moss & Newcomer, 1982, *inter alia*) and, obviously, it is this relationship that has allowed microwear analysis to become a significant part of prehistoric research. Although there is no difficulty with this general level of comparability, workers such as Hayden & Kamminga (1979) and Schiffer (1979) are demanding an increase in experimental work with the aim of producing traces that are not only broadly similar to archaeological traces but 'replicate' them in order to provide specific and detailed causative explanations to the archaeological traces themselves. The making of direct one-to-one correlations necessitates the assumption that all the variables surrounding the formation of use–wear traces can be or are known and that their relative significance and inter-relationships can be explained. It is supposed that this state of affairs can be achieved by recognising and controlling variables such as pressure, contact duration, relative motion and worked surface condition experimentally and then assuming that these relationships operated at the same level of significance in the formation of particular archaeological traces. Although a 'very particular cause' (Moss & Newcomer, 1982, p. 291) can be empirically determined to have influenced the morphology of an experimental trace through rigorous experimental techniques, the experimentally determined 'cause' can be no more than suggestive of the archaeological 'cause' (see Meeks *et al.*, 1982, for example) and should not be used as an explicit explanation of the observed archaeological 'effect'. The temptation to equate known cause and known effect at the experimental level to the unknown cause and observed effect on artefacts is very strong, but in the absence of additional, independent information regarding the nature of the archaeological 'cause', any such correlation is potentially misleading. Functional information independent of experimentation is beginning to become available through the use of scanning electron microscopy (Anderson, 1980; Meeks *et al.*, 1982) and interferometry (Dumont, 1982), and may well be inferred from sites having exceptional qualities of organic preservation where the normally perishable results of tool function may be seen.

Wear traces such as edge damage (including edge rounding), polishes, and linear features (striations and linear arrangements of polish) occur as a direct result of tool contact against other materials under appropriate conditions of pressure, duration of contact, and relative motion of the contact surfaces (see Dumont, 1982, for full discussion). These conditions arise most commonly during tool use but can also be fulfilled on the surface of a site before their incorporation into the deposits, as well as during deposition. Furthermore, less than careful excavation and post-excavation handling and storage can produce yet more 'microwear traces' on the tools' surface (Gero, 1978). Thus the juxtaposed collection of microwear traces visible on archaeological specimens are not necessarily derived strictly from tool use and it is necessary to distinguish those traces likely to be functionally related from those that are spurious. This exercise is more easily accomplished with microwear polishes than with edge damage or striations.

The discrimination of different episodes of tool use and the identification of non-use related edge damage and striations is difficult. Changes in the orientation of striations on a tool's surface can indicate that the manner of tool use changed (assuming that they resulted from use), but unless or until striation morphology can be correlated to the type of causative worked material, any question regarding a change in contact material must be left unanswered (contra Mansur, 1982). Edge damage suffers from a similar problem. The sequence of edge damage production can only be determined if the scars are superimposed or, are linked together by a surface alteration of the tool, such as patination, staining or polish formation, but even this information cannot always demonstrate contemporaneity or the amount of time that has elapsed between the development of individual scars. Edge damage and striations form the entire body of the use–wear evidence available to researchers utilising the low magnification technique of microwear analysis. Although these traces are not as use-specific as polish and difficulties occur in the resolution of use-related traces from the general 'noise' of non-functional traces, the technique is of value in the microwear analysis of non-siliceous artefact raw materials that may not allow the preservation of polishes upon their surfaces or enable the formation of sufficiently variable and, hence, distinct polishes.

Unlike edge damage, microwear polishes tend not to vary according to the manner of tool use, but, rather, their morphology is highly correlated experimentally to the type of causative material worked (Keeley, 1980). Further, whereas edge damage has a restricted distribution along an edge where it can be easily modified or removed by subsequent events, polishes may form within a much greater area defined by the entire zone of contact with the worked material. Consequently, there is more chance of resolving different types of polish, distinguishing different episodes of tool use and differentiating non-use related traces. Archaeologically significant polishes are unlikely to be formed in any circumstances other than through use or hafting, though it is conceivable that under the extraordinary conditions of a densely packed midden deposit they could occur. The types of polish that can form on a tool's surface while within a deposit or during excavation or handling (storage) are generally sufficiently distinctive to allow identification and, hence, dismissal from the analysis. The non-use related polishes are of three broad types corresponding to mechanical abrasion, chemical alteration of the surface (Meeks *et al.*, 1982) and the addition of mineral deposits (staining, desert varnish) to the surface that can superficially resemble true microwear polish (*ibid.*, 1982).

It is not unlikely, though perhaps not common (see Dumont, 1983, for an example), that a tool that has been used on one material may undergo secondary use on a different material before finally becoming incorporated within the site deposits. The microwear traces relating to each period of use, if present, would then be juxtaposed and the correlation of these traces, especially edge damage and striations, to any particular period of use would be extremely difficult, if not impossible. The presence of dissimilar polishes produced by different worked materials could be an indication that more than one identifiable use of the tool has occurred, but one must bear in mind that one 'use' may well produce two types of polish if two types of contact material are concurrently worked: e.g. cutting meat on a wooden board.

Although through the study of polishes there is less difficulty in determining multiple use of an artefact on different worked materials and isolating the polishes that are of non-use origin, there is still the question of recognising and distinguishing different manners of tool use of a single artefact on the same material. The examination of the distribution and orientation of edge damage and striations may be suggestive but, as before, great difficulties surround their interpretation. The form of the differential distribution of polish along both aspects of a working edge can provide valuable information regarding the orientation of the tool during use, and the presence of linear arrangements of polish can help in the determination of the direction of such use. The observed final distribution of a single type of polish at or near the edge of a tool can result from either a single mode or multiple manners of tool use. The recognition and identification of multiple manners of use relies on careful analysis of the overall pattern of polish and careful argument from the available information.

A more tenuous assumption that must be made in order to make such correlations meaningful is that the range of polishes produced experimentally covers the entire range of polishes that could have been produced by the original tool users. The doubt that such a set of polishes has been acquired means that dogmatic determination of archaeological traces based on a limited range of possible experimental correlates should be guarded against. The technique of interferometry (Dumont, 1982) may well go some distance in reducing this doubt by providing a means of assessing particular characteristics of the worked materials directly from the morphology of the polishes, independent of analogy to experimental traces. In addition to providing this new set of information, the technique of interferometry enables the quantification of some aspects of polish morphology, such as the depth of striations and pits, the curvature of individual areas of polish, and the depth to which the polish extends within the stone microtopography, which have up to now lent themselves only to qualitative description.

The low and medium magnification techniques so far discussed are carried out by use of incident light microscopy which introduces its own practical problems and limitations. The problems associated with this research tool are well known to any microscopist and consist of: decreasing depth of field with increasing magnification, glare, poor contrast, and an occasional difficulty in surface recognition when slightly translucent materials are studied. Unfortunately the severity of these problems increases with increasing magnification and limits the relative usefulness of incident light microscopy to no more than 500 diameters at best. At greater magnifications, the range of elevation change of a flint microtopography exceeds that of the objective's depth of field and results in an incomprehensible if not incoherent image being delivered to the observer. Magnifications of this magnitude also produce intensified back-scattering of light that manifests itself as glare and results in increasingly poor image contrast. Although glare can be reduced by the insertion of a monochromatic or polarising filter, a reduction in the aperture or diaphram adjustment and/or by reducing the brightness of the light source, the image contrast also decreases due to the diminished image brightness. As a response to these practical difficulties and a desire to achieve higher magnifications in order to characterise the microtopography of different polishes, several researchers, notably Anderson (1980, 1981) and Meeks *et al.* (1982), have employed scanning electron microscopes (SEM) in their investigations.

The studies cited have shown that polishes and other use–wear traces, particularly striations, can be usefully studied in the SEM. Within the range ×100 to ×325 normally employed to observe polishes with incident light microscopes, the SEM is less effective, for although polished areas are visible, they all look the same at medium range magnification. This is because the light reflective qualities of polishes which permit their discrimination with the incident light microscope that forms an image by absorption of light in the specimen, are not detected by the SEM which uses the scattering of electrons rather than light to produce its image (compare Fig. 7.1(*a*) and (*b*), and Fig. 7.2(*a*) and (*b*)). However, above ×500 and particularly between ×2000 and ×10 000, the higher resolution and better depth of field of the SEM clearly separates individual microfeatures, provides excellent photomicrographs and allows confident assessment and interpretation of wear traces, as well as providing a further range of criteria for polish discrimination (see, for example, Meeks *et al.*, 1982). Anderson (1980, 1981) has also made use of microprobe facilities to determine the nature of materials adhering to artefact surfaces. The limitations on the size of objects that may be put into the SEM are similar to those effecting incident light microscopy of archaeological material but can be successfully overcome by taking moulds and making casts of possibly utilised areas. A minor drawback is the necessity of coating specimens, but coatings can be removed and are unnecessary for observations at less than ×1500 if an environmental chamber with a backscatter detector are available (Figs. 7.1(*a*) and 7.2(*b*)).

Of the methods discussed, medium range magnification work using incident light microscopes is both practical and successful. Interferometry and high magnification scanning electron microscopy add further refinements which provide new information about these traces and can be used to test and augment the descriptions and interpretations suggested by medium range work.

The relevance of microwear evidence

Improving methods of observation and quantification in microwear studies enhances the reliability and comparability of the initial descriptive information. The relevance of this information to archaeology and prehistory must then be established by reference to the questions posed about it. As usual these propositions are concerned with improving our knowledge of material culture and developing our understanding of economic and behavioural activities at the cultural as well as the intra- and inter-site level. The suitability of microwear data for supporting or refuting such propositions rests, as usual, on the extent to which valid generalisations may be made from the evidence. Two factors must be considered when using microwear information: sample size and the difficulty of extrapolating from specific tool use to function, then to economic and behavioural activities and finally to socio-cultural processes.

The time-consuming nature of microwear research virtually always means that assemblages have to be sampled. Sampling poses problems for many branches of archaeological research but the difficulties are increased for microwear analysts because of the constraints imposed by the condition of the artefacts, patinated and/or stained pieces as well as material damaged during post-deposition and the post-excavation period being unsuitable for inclusion, and because only a relatively small proportion of the material in suitable condition may show wear traces. Many workers fight shy of exact statements about the size and nature of their samples but Keeley's (1980) figures for the Middle Pleistocene sites considered in his study highlight the problem. Out of the 864 underived artefacts recovered from the 1969–70 excavations at the Golf Course Site, Clacton, Essex (Singer *et al.*, 1973), 248 (29%) were considered suitable for microscopic examination of which only 46 (5.3% of the total) showed any use traces. Similarly, of 267 pieces considered from the Lower Loam at Swanscombe, Kent, only 66 (25%) were suitable for examination and only 4 (1.5% of the total) showed use traces. To put it another way, out of over 38 000 bifaces listed in Roe's (1968) *Gazetteer of British Lower and Middle Palaeolithic Sites*, probably less than 1% are in a condition for microscopic examination and even if they all showed wear traces, most lack the contextual, spatial and environmental data required to support broader generalisations about tool function and specialisation within a particular typological category just on a simple intra-site level. Clearly, the availability of this non-material evidence is critically small and needs to be treated accordingly.

Equally, the nature of microwear evidence imposes some constraints on how it is applied to broader problems of prehistory. What the microwear analyst records from each artefact are the traces left by single historical events, each with unique characteristics which must be interpreted by reference to experimental material. This fact and the small size of the majority of samples prompts the question of whether microwear analysis can achieve an appropriate level of generality to be of use in developing our understanding of material culture and in promoting economic and behavioural studies, or whether it simply provides fascinating glimpses of some precise moments in the remote past.

To regard microwear as 'sheer phenomenology' is

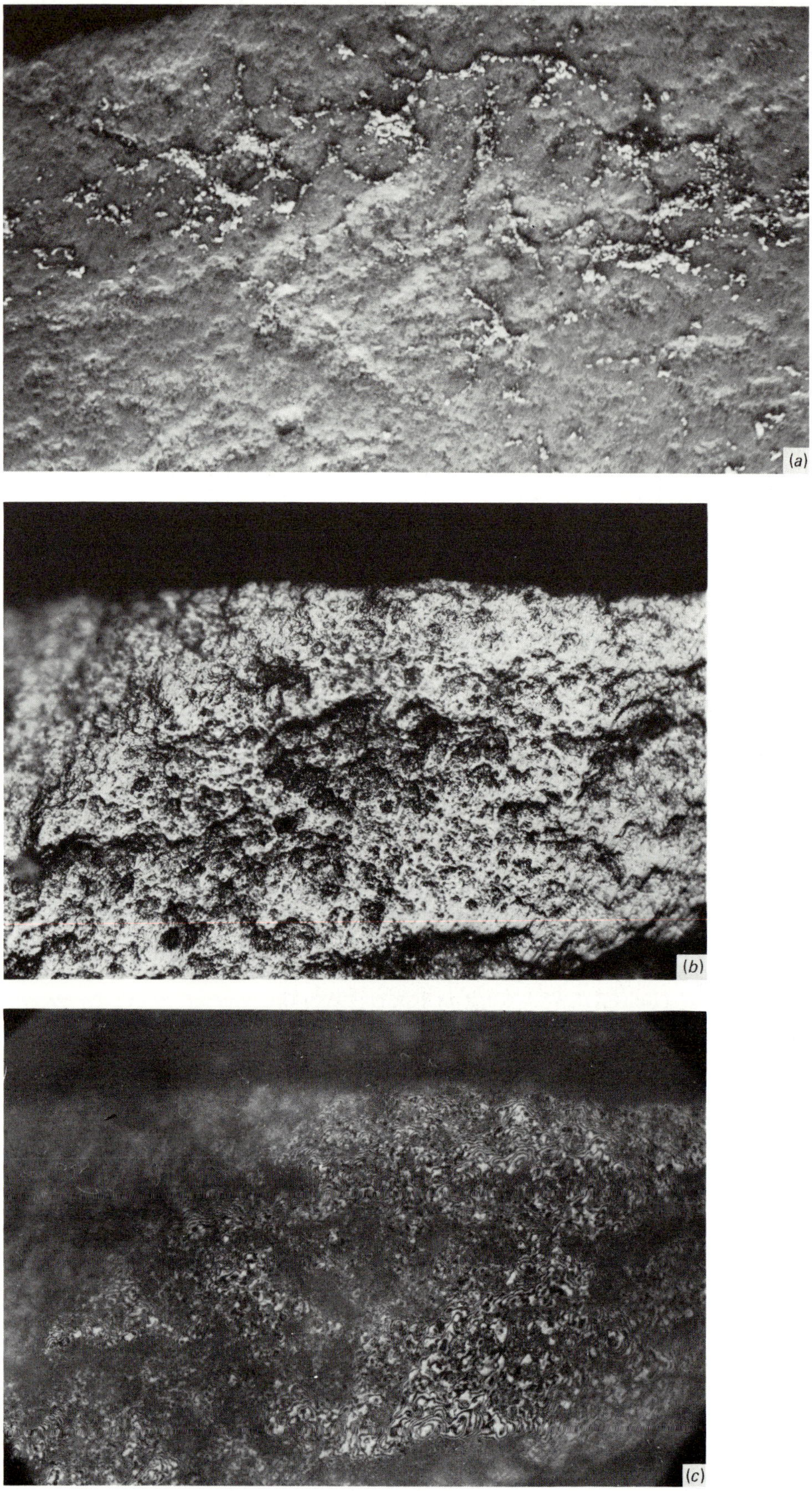

Fig. 7.1. Photomicrographs of the same area of wood polish on a utilised Mesolithic blade. Each provides information not found in the other. (*a*) SEM micrograph showing polish surface including an as yet unidentified superficial deposit taken on an ISI 60A with an environmental chamber. Specimen uncoated. (Original magnification (OM) ×250; dimensions (D) of the photographed area: 0.46×0.32 mm.) (*b*) Oblique incident light micrograph (OM: ×200; D: 0.56×0.37 mm). (*c*) Interference pattern on (*b*).

Fig. 7.2. Photomicrographs of the same area of antler polish on a truncated and backed Mesolithic blade. (*a*) SEM micrograph (data as in Fig. 7.1(*a*)). (*b*) Oblique incident light micrograph (data as in Fig. 7.1(*b*)).

unnecessarily pessimistic, since the authors of many of the studies cited above have judiciously transformed data from individual specimens to make valid generalisations about artefact function, the relationship of tool morphology to use, hafting, tool specialisation and craft processes on the intra-site level at which contextual, spatial, environmental and 'economic' data may provide corroborative evidence. In these cases the non-material evidence of manner of tool use and purpose provided by microwear analysis converts mere supposition about what might have been done at a site into more informed prediction of activities, adding detail to our knowledge of material culture and some criteria for accepting or refuting hypotheses about the character and purpose of a site.

Beyond this, microwear evidence can contribute to theories about tool function, site function and economic activity at the inter-assemblage level and to predictions about cultural evolution concerning technological and economic trends, increasing or decreasing specialisation or diversity and curation of tools. However, the extent to which we accept the possibilities provided by such predictions will depend on the degree to which we can justify such a partial record as representative of the whole range of past events, activities and systems.

Conclusions

Microwear analysis provides valuable non-material evidence about the past that is relevant at the specific artefact/assemblage level and to theory building in economic and behavioural studies. Since 1964, the quality of microwear evidence has improved with greater experience in the use of incident light microscopy and the introduction of quantitative and high magnification techniques to distinguish and characterise microwear traces more precisely, although the nature of the material may mean that the standardisation demanded by some critics will never be achieved. The limitation of small samples and the problem of extrapolating from wear trace to behaviour necessitates cautious handling of microwear data. However, even taking these inherent difficulties into account, microwear studies can be used to contribute to hypotheses of varying value, according to whether they are independently testable, beyond the intra-site level. In this light, microwear analysis can provide more than just a fascinating appendix to lithic analysis.

Bibliography

Anderson, P. C. (1980). 'A testimony of prehistoric tasks: diagnostic residues on stone tool working edges.' *World Archaeology*, **12** (2), 181–94.

Anderson, P. C. (1981). Contribution méthodologique à l'analyse des microtraces d'utilisation sur les outils préhistoriques. *Thèse presentée à l'université de Bordeaux 1.*

Barnes, A. S. (1939). 'De la manière dont la nature limite le travail humaine dans l'éclatement de silex.' *Bull. Soc. préhist. franç.*, **1**, 1–16.

Briuer, F. L. (1976). 'New clues to stone tool function: plant and animal residues.' *American Antiquity*, **41** (4), 474–84.

Cahen, D. & Keeley, L. H. (1980). 'Not less than two, not more than three.' *World Archaeology*, **12** (2), 166–80.

Curwen, E. C. (1930). 'Prehistoric flint sickles.' *Antiquity*, **4**, 179–86.

Del Bene, T. A. (1979). 'Once upon a striation: current models of striation and polish formation.' In: *Lithic Use-Wear Analysis* (ed., B. Hayden), pp. 167–77. Academic Press, New York.

Diamond, G. (1979). 'The nature of so-called polished surfaces on stone artefacts.' In: *Lithic Use-Wear Analysis* (ed., B. Hayden), pp. 159–66. Academic Press, New York.

Dies, K. F. (1981). 'Artefaktbruchflächen, ihre Morphologie und Sekundären Änderungen.' *Quartär*, **31/32**, 95–106.

Dumont, J. (1982). 'The quantification of microwear traces: a new use for interferometry.' *World Archaeology*, **14** (2), 206–17.

Dumont, J. (1983). 'An interim report of the Star Carr microwear study.' *Oxford Journal of Archaeology*, **2** (2), 127–45.

Dumont, J. (in press) 'A preliminary report on the Mount Sandel microwear study.' In: *Excavations at Mount Sandel* (ed., P. C. Woodman), Northern Ireland Archaeological Monographs no. 2. HMSO, Belfast.

Frison, G. C. (1974). *The Casper Site*. Academic Press, New York.

Gendel, P. A. & Pirnay, L. (1982). 'Microwear analysis of experimental stone tools: further test results.' *Studia Praehistorica Belgica*, **2**, 251–65.

Gero, J. M. (1978). 'Summary of experiments to duplicate post excavational damage to tool edges.' *Lithic Technology*, **7**, 34.

Gould, R. A., Koster, D. A. & Sontz, A. H. (1971). 'The lithic assemblages of the Western Desert Aborigines of Australia.' *American Antiquity*, **36**, 149–69.

Gramsch, B. (1979). 'Gebrauchspuren an Silexartefakten von Bilzingsleben (Erste Mitteilung).' *Ethnogr. -Archäeol. Z.*, **20**, 704–7.

Hayden, B. & Kamminga, J. (1979). 'An introduction to use-wear: the first CLUW.' In: *Lithic Use–Wear Analysis* (ed., B. Hayden), pp. 1–13. Academic Press, New York.

Hester, T. R., Gilbow, D. & Albee, A. D. (1973). 'A functional analysis of "Clear Fork" artefacts from the Rio Grande Plain, Texas.' *American Antiquity*, **28**, 90–6.

Jones, P. R. (1980). 'Experimental butchery with modern stone tools and its relevance to Palaeolithic archaeology.' *World Archaeology*, **12** (2), 153–65.

Kamminga, J. (1977). 'A functional study of use-polished eloueras.' In: *Stone Tools as Cultural Markers* (ed., R. V. S. Wright), pp. 205–12. Australian Institute of Aboriginal Studies, Canberra.

Kamminga, J. (1979). 'The nature of use-polish and abrasive smoothing on stone tools. In: *Lithic Use–Wear Analysis* (ed., B. Hayden), pp. 143–57. Academic Press, New York.

Kantman, S. (1970*a*). 'Esquisse d'un procédé analytique pour l'étude macrographique des "encoches".' *Quaternaria*, **13**, 269–80.

Kantman, S. (1970*b*). ' "Raclettes moustēriennes": une étude experimentale sur la distinction de retouches intentionelle et les modifications du tranchant par utilisation.' *Quaternaria*, **13**, 295–304.

Kantman, S. (1971). 'Essai sur le problème de la retouche d'utilisation dans l'étude du matériau lithique. Premiers résultats.' *Bull. Soc. préhist. franç.*, **68**, 200–4.

Kay, M. (1977). Wear analysis and unifacial scraping tool morphology: implications for studying individual use.' In: *The Individual in Prehistory – studies in variability in style in prehistoric technologies* (eds., J. N. Hill & J. Gunn), pp. 231–45. Academic Press, New York.

Keeley, L. H. (1974). 'Technique and methodology in microwear studies: a critical review.' *World Archaeology*, **5** (3), 323–36.

Keeley, L. H. (1980). *Experimental Determination of Stone Tool Uses.* Chicago University Press, Chicago.

Keller, C. M. (1966). 'Development of edge damage patterns on stone tools.' *Man*, **1** (4), 501–11.

Lenoir, M. (1971). 'Traces d'utilisation observèes sur un nucleus à lamelles.' *Bull. Soc. préhist. franç.*, **68**, 69–70.

Lumley, H. de (1979). 'L'homme de Tautavel.' *Dossiers de l'Archéologie*, **36**, 4–11.

Mansur, M. E. (1982). 'Microwear analysis of natural and use striations: new clues to the mechanism of striation formation.' *Studia Praehistorica Belgica*, **2**, 213–33.

McBurney, C. B. M. (1968). 'The cave of Ali-Tappeh and the Epi-Palaeolithic in northeastern Iran.' *Proc. Prehist. Soc.*, **34**, 385–413.

Meeks, N. D., Sieveking, G. de G., Tite, M. S. & Cook, J. (1982). 'Gloss and use-wear traces on flint sickles and similar phenomena.' *J. Arch. Sci.*, **9**, 317–40.

Moss, E. H. & Newcomer, M. H. (1982). 'Reconstruction of tool use at Pincevent: microwear and experiments.' *Studia Praehistorica Belgica*, **2**, 289–321.

Odell, G. H. (1975). 'Microwear in perspective: a sympathetic response to L. H. Keeley.' *World Archaeology*, **7**, 226–40.

Odell, G. H. (1980). 'Towards a more behavioural approach to archaeological lithic concentrations.' *American Antiquity*, **45** (3), 404–31.

Roe, D. A. (1968). *A Gazetteer of British Lower and Middle Palaeolithic Sites*. Council for British Archaeology research report no. 8, London.

Rosenfeld, A. (1971). 'The examination of use-marks on some Magdalenian end scrapers.' *British Museum Quarterly*, **35**, 176–82.

Schiffer, M. B. (1979). 'The place of lithic use-wear studies in behavioural archaeology.' In: *Lithic Use–Wear Analysis* (ed., B. Hayden), pp. 15–25. Academic Press, New York.

Semenov, S. A. (1964). *Prehistoric Technology*. Adams and Dart, Bath.

Semenov, S. A. & Shchelinski, V. E. (1971). 'Micrometrical study of wear traces on Palaeolithic stone tools.' *Soviet Archaeology*, **1**, 19–30.

Singer, R., Wymer, J., Gladfelter, B. G. & Wolff, R. G. (1973). 'Excavation of the Golf Course, Clacton-on-Sea, Essex.' *Proc. Prehist. Soc.*, **39**, 6–74.

Sonnenfeld, J. (1962). 'Interpreting the function of primitive implements.' *American Antiquity*, **28**, 56–65.

Spurrell, F. C. J. (1892). 'Notes on early sickles.' *Arch. J.*, **49**, 53–69.

Tringham, R., Cooper, G., Odell, G., Voytek, B. & Whitman, A. (1974). 'Experimentation in the formation of edge damage: a new approach.' *J. of Field Archaeology*, **1**, 171–96.

Wilmsen, E. N. (1968). 'Functional analysis of flaked stone artefacts.' *American Antiquity*, **33**, 156–61.

Witthoft, J. (1967). 'Glazed polish on flint tools.' *American Antiquity*, **32**, 383–8.

Wylie, H. G. (1975). 'Tool microwear and functional types from Hogup Cave, Utah.' *Tebiwa*, **17** (2), 1–31.

8

Preliminary investigation into the quantification of wear traces on flint tools

R. GRACE, I. D. GRAHAM AND M. H. NEWCOMER

Archaeologists can agree that an understanding of the function of prehistoric flint tools would be valuable in all aspects of prehistoric studies, whether for comparisons over time and space, or for reconstructions of activities at a single site. While there is still no general agreement as to how tool function can best be studied, it seems that the most useful results derive from microscopic study of wear traces on the edges and surfaces of the ancient tools themselves. Such studies typically involve the making of replicas of prehistoric tools, the experimental use of these tools in simulated prehistoric tasks, the examination of the wear on the experimental tools and, finally, the comparison of the experimentally produced wear with that observed on actual prehistoric tools and the inference of prehistoric tasks.

In the past five years or so, the most interesting work in the 'micro-wear' field has been undertaken by proponents of the high magnification approach as pioneered by Lawrence Keeley (1980). This approach involves the identification of a series of 'micro-wear polishes' on flint tools, each polish resulting from contact with a different material, such as bone, wood, antler, meat, hide or vegetable material. The advantage of this method is its ability to make specific inferences about prehistoric tool use, even in the total absence of any of the worked material on site (Cahen, Keeley & van Noten, 1979).

However, the current state of the art of micro-wear analysis has two major disadvantages. The first is the production of comparable results by different analysts. This problem derives from the subjective nature of polish descriptions. Such descriptive terms as 'greasy' and with 'innumerable tiny pits' (Keeley, 1980, p. 43), are often found in the literature. These terms are not specific enough to allow strict comparisons of results produced by different analysts. A 'greasy' polish can be bright, moderately bright or very bright (Keeley, 1980; Brink, 1978), and with 'innumerable tiny pits', without any mathematical quantification is a relative term depending on such features as the perception of the analyst, the range of magnification and so on. The relative brightness of a polish is also dependent on such things as the type of microscope in use, any filters that may be employed, and differences in lighting conditions. The Olympus Vanox, for example, as used at the Institute of Archaeology, has two separate adjustments affecting the illumination of the image through the eyepiece. The lamp has a variable dimmer control, and the light path through the microscope is also adjustable to allow different levels of light to fall on the subject.

Therefore even if two analysts are using the same type of microscope the viewing conditions have to be the same if the relative brightness of a polish is to be directly compared. The same polish may appear very bright under one set of conditions, moderately bright or even dull under another.

The relative nature of polish descriptions not only makes comparative studies difficult, but also exacerbates the second problem, as it extends the process of acquiring the ability to recognise and confidently identify polishes, the necessary 'apprenticeship' of the micro-wear analyst. Anyone beginning a micro-wear analysis has first to carry out experiments to produce polishes of known origin which he can visually recognise and use as models. In fact, descriptions of micro-polish only become intelligible when the apprentice has learned his trade. Initially the descriptions are more confusing than helpful and only after looking at a variety of polishes do terms such as looking like 'melted snow' (Keeley, 1980, p. 56), and features that have 'comet tails' (Vaughan, 1981, p. 132), have any meaning. The only way of communicating objective information on polishes at the moment is through photography. Apart from the technical difficulties of taking good quality photographs, such as the narrow depth of field that decreases as the magnification increases, without any standardisation of magnification and viewing conditions photographs produced by one analyst are often irrelevant to another. Photographs taken at up to 50 magnifications contain little useful information for an analyst working at 200-plus magnifications. The poor quality of reproductions of polish photographs in many journals makes them inadequate for conveying sufficient information.

This means that rather than being able to build on work already carried out, every micro-wear analyst has to start from the beginning. Even when the apprentice becomes a master of his craft, the accuracy of the functional interpretations is subject to individual ability and experience. A micro-wear analysis is a repeatable experiment, but, given the limited resources and the amount of time required it is unlikely that any results will be checked by independently repeating the original analysis.

The quantification of micro-wear polishes would enable them to be classified by mathematical criteria, rather than by individual expertise, and any micro-wear analysis so classified would be comparable with any other. The intention is to carry out this quantification with the use of image processing techniques. Extensive research has been done on image processing to facilitate the identification and classification of images. These techniques have been applied to such diverse problems as the classification of cloud formations, lunar landscape features and the improvement of X-ray photographs in medical diagnosis (Darling & Joseph, 1968; Hall *et al.*, 1971). In all these applications the problem is the same as that of classifying micro-polishes, the unit of study is an image made up of tone and texture. In this context texture is represented by the distribution of tone in the image. In a two-dimensional image we visually perceive texture as the spatial relationship of different tones. Where there is little tonal variation in an image the texture is perceived as smooth; where tonal variation between discrete features is marked, a coarser texture is indicated.

The first stage in using image processing to classify micro-polishes is to convert the image into a format suitable for computer input. This is carried out by a scanning digitiser. The negative is placed on a rotary drum and a narrow light beam is passed through to a photo-cell which measures the photographic density of each resolution area, resulting in an array of grey level scores. From these arrays a number of statistics can be calculated that differentiate between use–wear polishes. The image of an unused flint surface (Fig. 8.1) can be characterised by its grey level histogram, the histogram being of the frequencies of the grey levels in the digital array. The histograms of two unused flint surfaces (Fig. 8.2(*a*), (*b*)), demonstrate the consistency of the histograms in representing similar surface textures. These can be compared with the histograms of wood polish (Fig. 8.3; Fig. 8.4(*a*)–(*c*)) which are significantly different from those representing the unused flint surface. This difference is also present when comparison is made with histograms of other polish images (Fig. 8.5(*a*)–(*c*)). These histograms demonstrate that not only the difference between polished and unused flint can be characterised in this way, but also the difference between polishes produced by working different materials.

One way in which the distribution of dark and light tones characterises textural features is in the abruptness of tonal transitions. For example, two textural features defined by Vaughan (1981, p. 133), and used as diagnostic features for polish identification are pit depressions, which he describes as holes with sharp edges, and diffuse depressions described as holes with sloping poorly defined sides. In a two-dimensional image this textural information is conveyed through changes in tone, pit depressions are dark areas surrounded by the lighter tones of the polish between which the transition is abrupt, while the ill-defined edges of diffuse depression have a more gradual tonal transition. Using a digital scanner with an aperture of 50×50 μm on an image at 200 magnifications means that each resolution area, or pixel, represents a square of 0.25×0.25 μm of the flint surface.

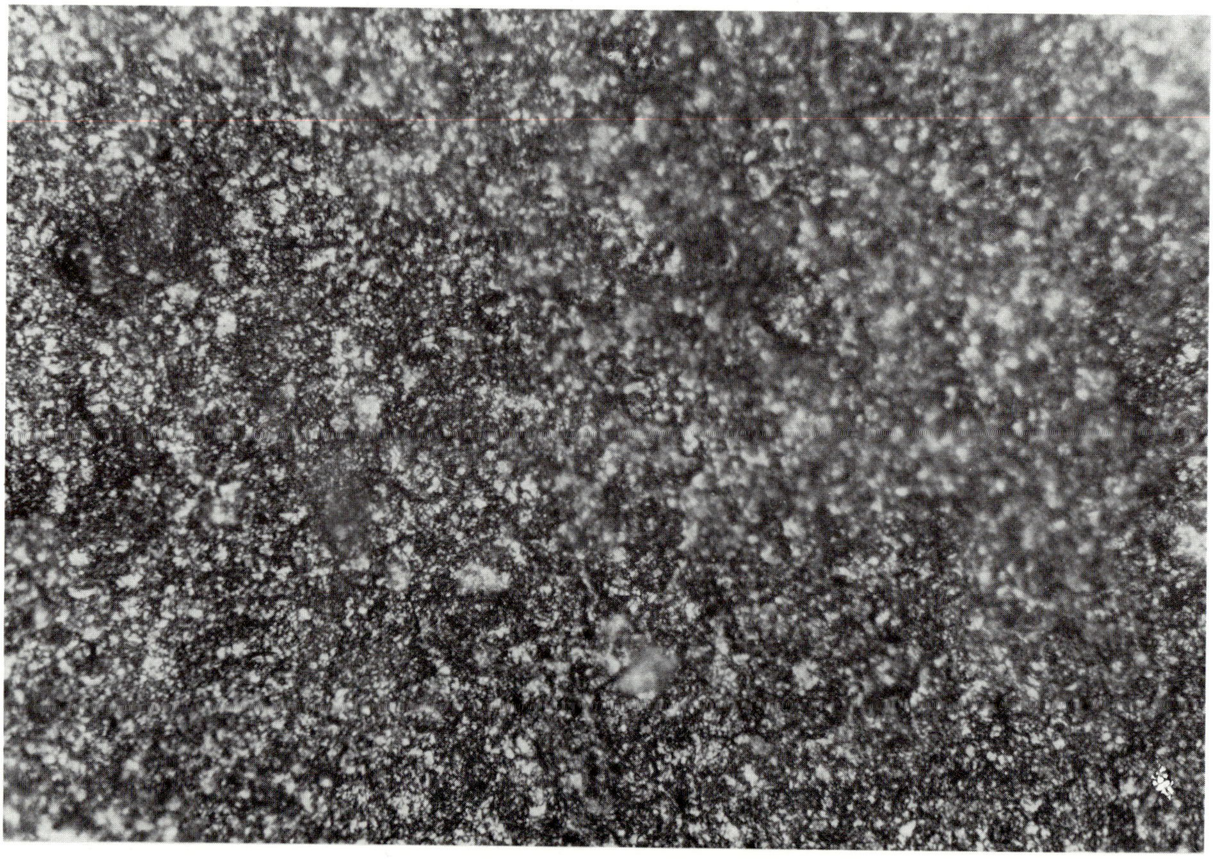

Fig. 8.1. Unused flint.

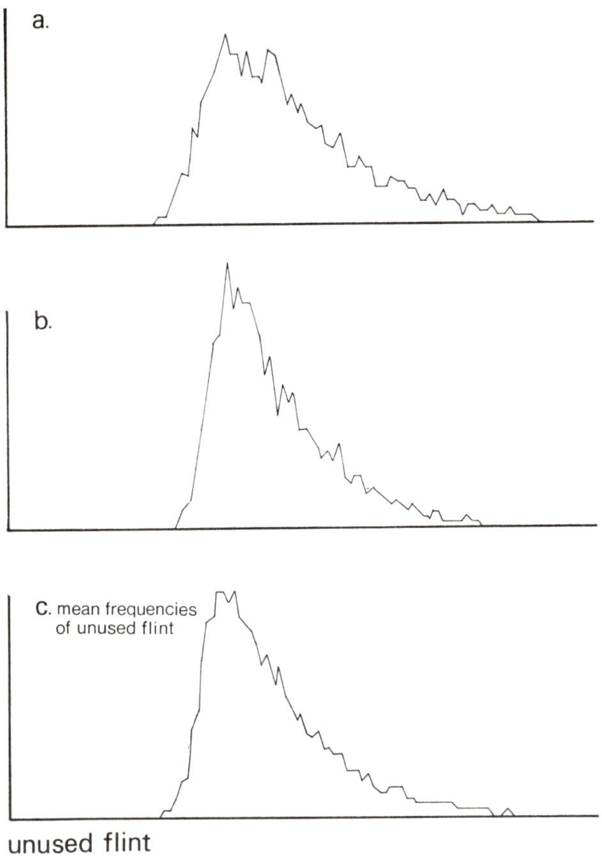

Fig. 8.2. Grey level histograms of unused flint.

With a grey tone scale of 0–255 the difference in tone as reflected by photographic density over a very small area, as between a pit and a diffuse depression, can be detected and quantified because the histogram not only represents the overall spread of tone, and therefore the brightness of an image, but the 'shape' of the histogram indicates the abruptness of tonal transitions.

In this study the histograms were produced from sections of 64×64 pixels of the 512×512 array, giving a total of 4096 pixels. This may be considered a small area from which to characterise a polish, but the histograms of wood polish (Fig. 8.4(*a*)–(*c*)) are from different sections of the same flint surface (Fig. 8.3), and their similarity demonstrates that the relatively small area of 16×16 μm consistently characterises the polish. The advantages of working with a small area are that the amount of computer storage is reduced and the technique can be used when only a small part of the flint surface is polished. Also when a surface contains polish of more than one type they can be selected out by specifying the row and column vector of the array for the origin of the 64×64 pixel section.

Given that grey level histograms can characterise micro-polishes, a statistic that can be extracted is the variance of a polish from unused flint. This is calculated by the square root of the sum of the squared differences between corresponding grey level frequencies divided by the number of grey levels.

$$D^2=\frac{1}{n}\Sigma(f_1-f_2)^2$$

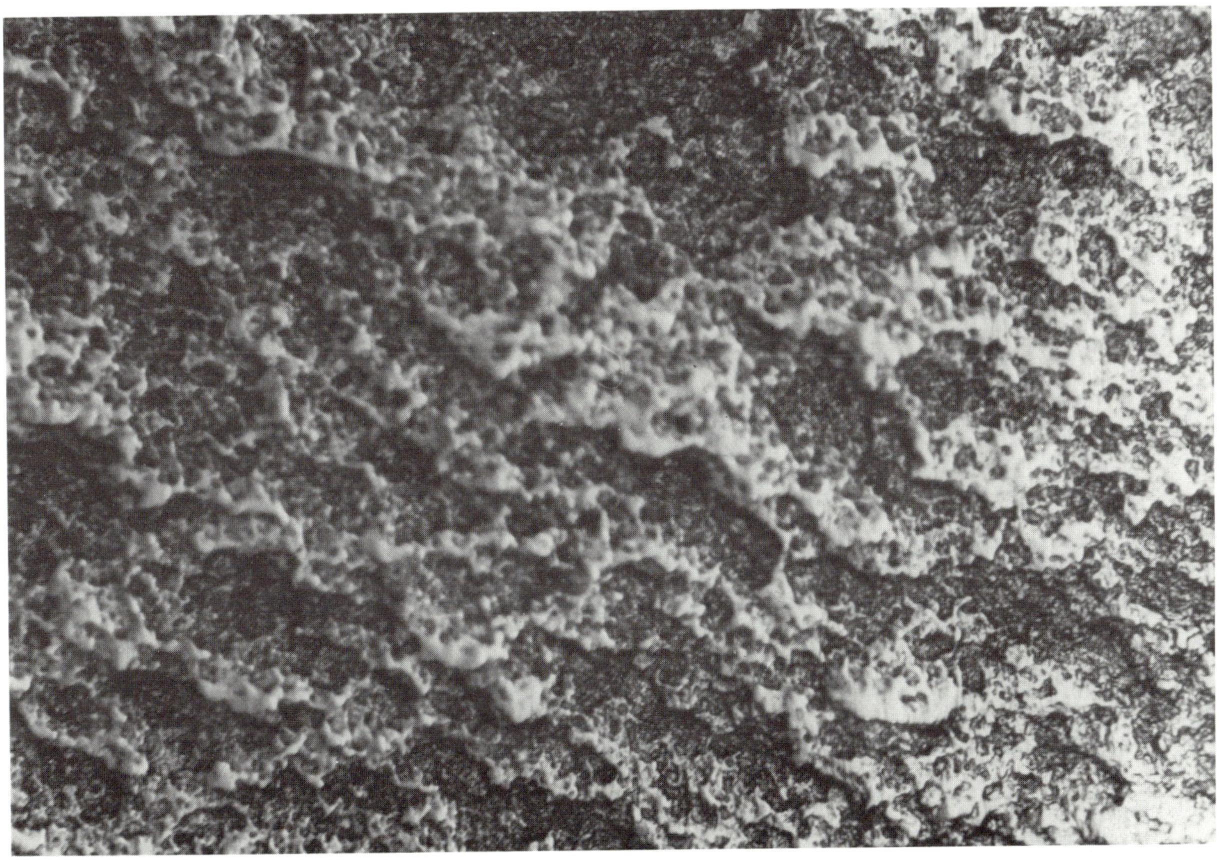

Fig. 8.3. Wood polish.

The way in which this distance statistic can be used in classification is by calculating the mean frequencies at each grey level from a number of histograms of the same texture. These mean frequencies can be subtracted from the frequencies of any sample, resulting in a quantification by the D statistic of the difference between the texture and the sample, the implication being that the lower the value of the D statistic the more probable that the sample polish is of the same type as that for which the means were calculated. Fig. 8.2(*c*) represents the mean frequencies at each grey level of images of unused flint. Using this as a base, the D statistic for a number of polishes can be calculated giving a measure of their difference from unused flint. The values for the different polishes are illustrated as a column in Fig. 8.6(*a*). As expected the values for the D statistic of samples of unused flint are the lowest, differentiating them from the samples having use–wear polish.

A criticism of this method could be that the grey level histograms are unmodified and therefore differences in exposure of the photographs and in processing techniques could be the origin of the differences in the histograms rather than the use–wear polish. The images used in this study have been carefully controlled for consistency in lighting conditions and photographic processing and so the differences in the histograms do reflect polish characteristics. However, when applying the technique to images produced under different conditions, with different films or by different people, some normalising process would be required. To investigate this problem a selection of histograms were standardised by the use of *z* scores, by subtracting the mean and dividing by the standard deviation

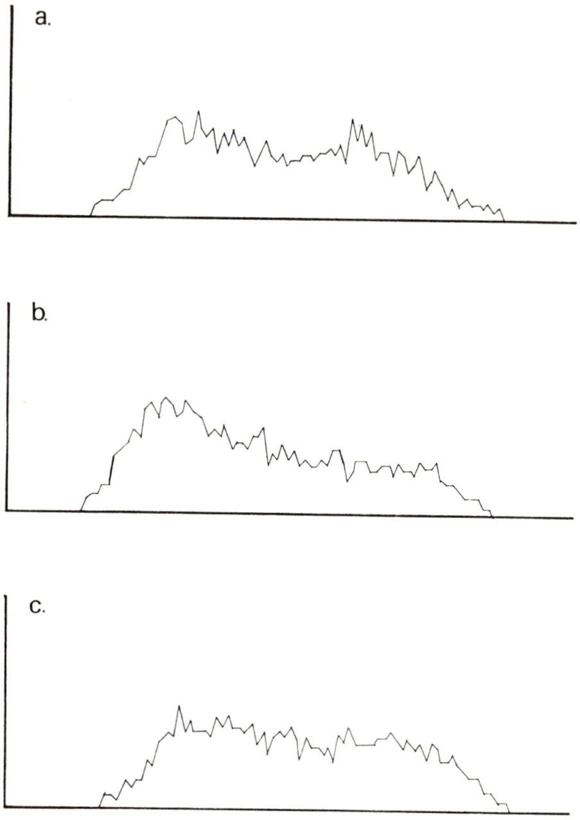

Fig. 8.4. Grey level histograms of wood polish.

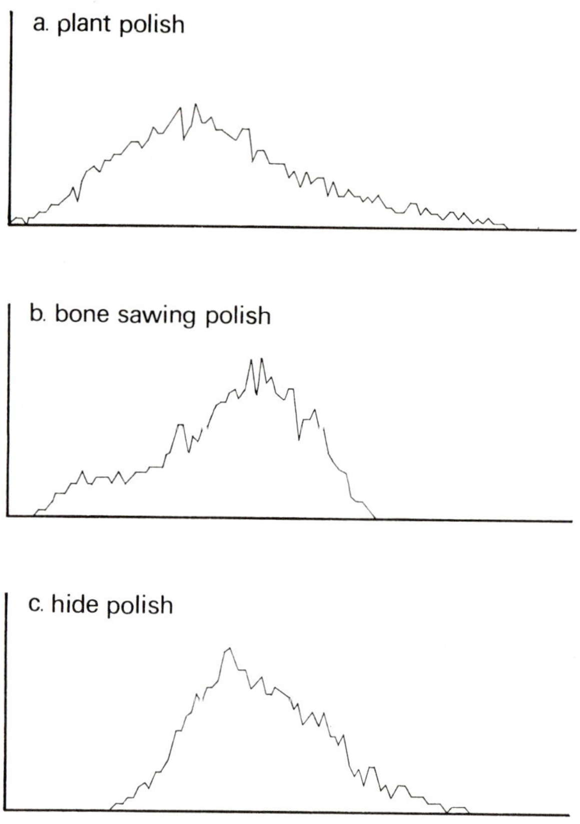

Fig. 8.5. Grey level histograms of use–wear polishes.

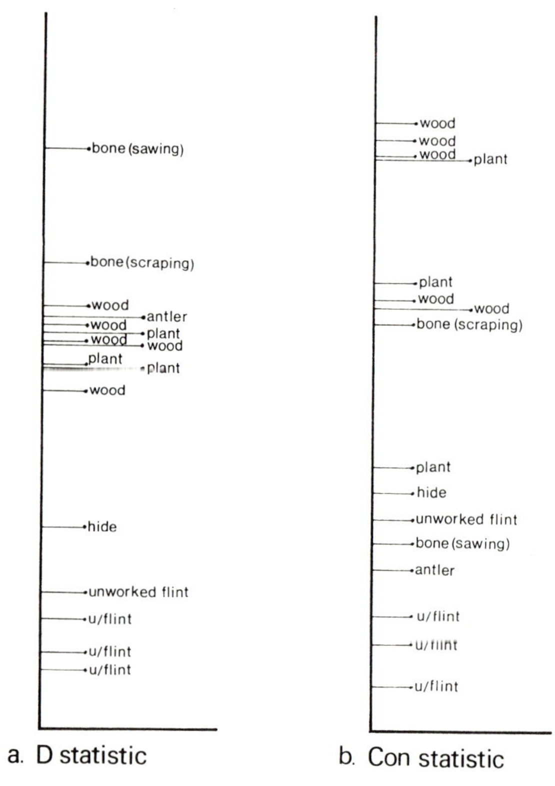

Fig. 8.6. Values of statistics as differences from unused flint.

at each grey level in the histogram. The distributions have the same mean of 0 and unit variance so that any differences in overall brightness caused by different lighting conditions are eliminated. In Fig. 8.7 the frequencies have been grouped to allow direct comparisons to be made and illustrate that the differences inherent in the unmodified histograms remain.

The histograms give textural information but in order to extract information on the spatial arrangement of texture features, co-occurrence matrices can be computed. These are derived from the joint occurrences of grey levels according to parameters of distance and direction in the form of vectors. First the histograms are equalised. The grey levels are combined into groups with as near as possible the same frequency. Equalisation also has the function of normalising the histograms, eliminating the problems discussed above. The resolution areas or cells of the arrays are rescaled from the equalised histogram, the number of grey levels having been reduced in our study to 16. This rescaling has the function of distributing the tonal variation throughout the image and increases the contrast between discrete features (see Pratt, 1975, p. 314). The rescaled array is used as the basis for the production of the co-occurrence matrices. These are calculated from the occurrences of different combinations of grey levels at a set distance and direction from each cell in the array. Fig. 8.8 illustrates a section of an hypothetical array. A matrix is calculated for each specified vector; for cell x in Fig. 8.8, the combination at, say, vector 1,2 (that is, one row down and two columns across, i.e. cell y) is 2,4. Therefore for cell x at vector 1,2 a score would be entered in the 2,4 cell of the matrix. At vector 1,0 (i.e. cell z), a score would be entered in the 2,2 cell of the matrix on the diagonal. This process is repeated for each cell in the array at each specified vector, thus building up the co-occurrence matrix. A statistic that can be derived from these matrices measures the amount of texture present in the image. For example, if all the cells in the array had the same grey level, all the scores in the matrix would be along the diagonal, as all the co-occurrences would be of the same grey level, representing a uniform smooth texture. The Con statistic is a measure of how the high scores concentrate along the diagonal, giving a value of 0 for the above case of a uniform image. An image with more texture will have combinations of different grey levels which will result in the high scores of co-occurrences being distributed throughout the matrix, giving a higher value for the Con statistic.

In this study it was found that the differences between textures of different polishes were most evident at a small scale, and so the Con statistic was calculated for a distance of up to three resolution cells. This represents combinations of grey levels from 0.25 to 1 μm apart on the flint surface. A distance of three cells gives 16 vectors from 0,0 to 3,3 and hence 16 values for the Con statistic from each of the co-occurrence matrices. Following the same procedure as for the D statistic, the mean values of the Con statistic at each vector were calculated for unused flint and these were subtracted from the Con statistic values, and the differences squared, at corresponding vectors for samples of polished flint. Thus a single figure for the difference between unused and used flint surfaces was arrived at, based on the spatial arrangement of different tones in the images. The results are illustrated in Fig. 8.6(b). The Con statistic separates out unused flint but does not clearly differentiate between the

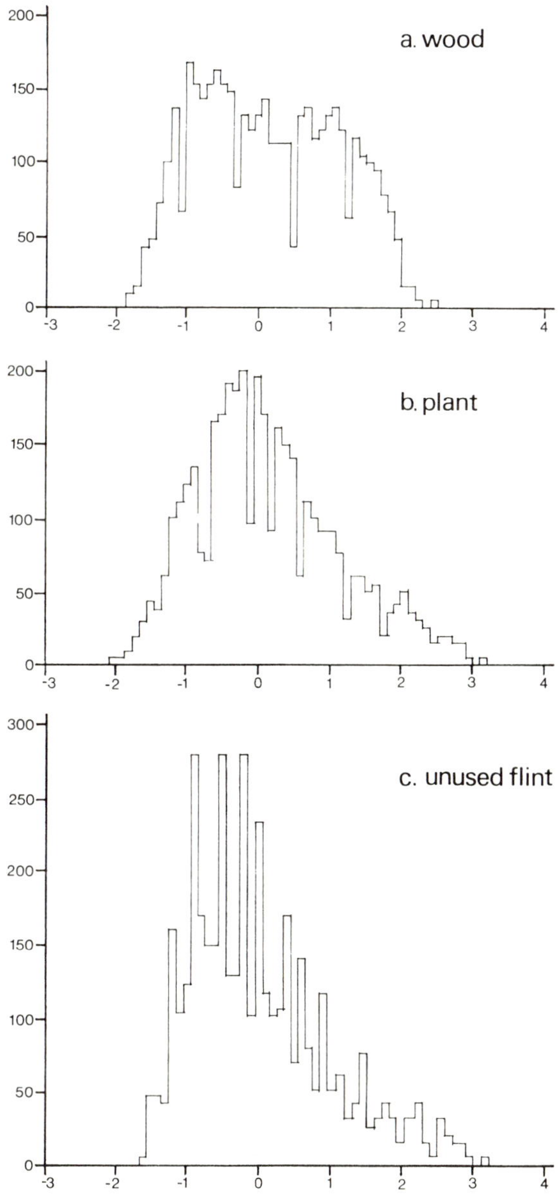

Fig. 8.7. Normalised grey level histograms.

hypothetical array

1	2	2	3
3	2_x	3	4
2	2_z	4	4_y
1	2	3	4

Fig. 8.8. Section of hypothetical grey level array.

polishes; by combining the information from the values of the D statistic and the Con statistic a scatter diagram can be produced in which the axes are the D and Con statistics. As the values of D and Con are the differences of the samples from unused flint, the nearer the origin a point is, the more similar that sample is to unused flint. Fig. 8.9 clearly illustrates that this technique can isolate unused flint. Combining the two statistics in this way also gives better separation between the polished samples. For example, in the D statistic column (Fig. 8.6(*a*)), antler polish is grouped with wood and plant polish, and in the Con statistic column (Fig. 8.6(*b*)), it is grouped with bone polish, while in the scatter diagram it is separated out from all the other samples.

It is interesting to note that in Fig. 8.9 the used surface nearest the origin is that of hide polish. This particular piece was used for only 28 minutes and had relatively little polish, and this is reflected in its position on the scatter diagram, demonstrating the sensitivity of the technique in that the hide polish would be difficult to identify or even separate from unused flint by current visual methods. Compare the hide polish (Fig. 8.10) with the unused flint surface (Fig. 8.1).

While the scatter diagram is an illustrative device to demonstrate the effect of combining two statistics, when a number of statistics that measure different diagnostic features of micro-polish have been produced, they can be combined as a series of attributes that then may be used as classification criteria. As the attributes would be objectively quantified in the form of numbers with scalar values, numerical classification procedures can be adopted. For

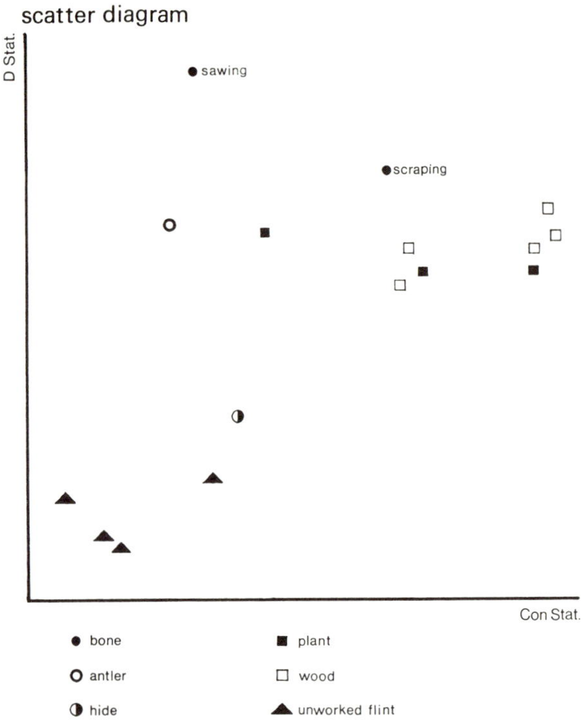

Fig. 8.9. Scatter diagram of differences from unused flint.

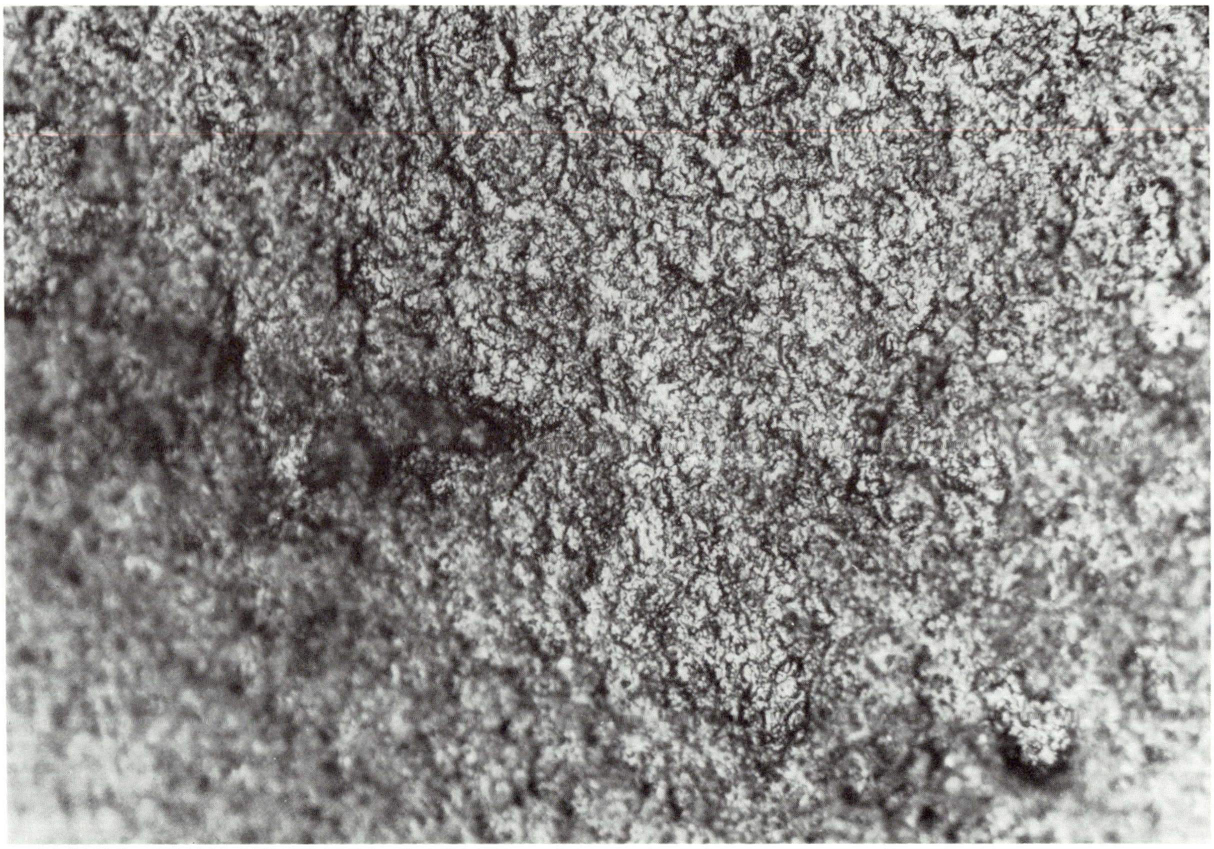

Fig. 8.10. Hide scraping polish.

example, a principal components analysis would determine the relative 'distances' between polish samples according to the different statistical attributes, and discriminant analysis would give a measure of which of the polish statistics are most significant for classification purposes. For multivariate techniques of this kind a greater number of polish samples are required to make the procedures statistically valid. The techniques tested so far are based on a small body of data and the results are presented as examples of the potential of the technique of using image processing. It should be emphasised that at this stage we are testing techniques to discover the nature of polish features rather than presenting a completed classification procedure. However, it is worth noting that using the Con statistic in combination with other texture feature statistics, an accuracy of 89% has been achieved in classifying sandstone textures (Haralick, Shanmugan & Dinstein, 1973). The sandstone textures have less variation among them than exists between micro-polishes produced by use wear, indicating that as good as, if not better, results should be possible with images of micro-polish.

These statistics derived from the arrays produced by digital scanning of photographs of micro-polishes are only examples of the many techniques of image processing that can be used for classification. Research will continue to test the suitability of these techniques, to adapt them to achieve optimum classification of micro-polishes and to develop an auto-classification program that will relieve the microwear analyst of the problems of description and identification. This will speed up the process of use wear identification, enabling larger samples of flint tools to be analysed, reduce the amount of time of apprenticeship, and make different analyses directly comparable by mathematical criteria.

Acknowledgement

The authors gratefully acknowledge the support of the Science and Engineering Research Council.

Bibliography

Brink, J. W. (1978). Experimental study of microwear formation on endscrapers. M.A. Thesis. University of Alberta.

Cahen, D., Keeley, L. H. & van Noten, F. (1979). 'Stone tools, toolkits, and human behavior in prehistory.' *Current Anthropology*, **20**, 661–72.

Darling, Jr, E.M. & Joseph, R. D. (1968). 'Pattern recognition from satellite altitudes.' *IEEE Transactions. Systems Science and Cybernetics*, SCC-4 no. 1.

Hall, E. I., Kruger, R. P., Dwyer, S. J., Hall, D. L., McLaren, L. W. & Hardwick, G. S. (1971). 'A survey of preprocessing and feature extraction techniques for radiographic images.' *IEEE Transaction. Computers*. C-20 no. 9, pp. 1032–44.

Haralick, R. M., Shanmugan, K. & Dinstein, L. (1973). 'Texture features for image classification.' *IEEE Transactions. Systems, Man and Cybernetics*. SMC-3, pp. 610–21.

Keeley, L. H. (1980). *Experimental Determination of Stone Tool Uses. A Microwear Analysis*. The University of Chicago Press, Chicago and London.

Pratt, W. K. (1978). *Digital Image Processing*. Wiley. New York.

Vaughan, P. C. (1981). *Lithic microwear experimentation and the functional analysis of a lower Magdalanian stone tool assemblage*. PhD Thesis. University of Pennsylvania.

9

Microflaking quantification

KAORU AKOSHIMA

Introduction

The problem of appropriate measurement is one of the most embarrassing to microwear analysts. In the processes of recovery, classification, identification, and interpretation, it is often realized that features of micro use–wear are too complex and diversified to be accommodated within our experimental and recording framework where controlled and recorded variables are very limited. Microflaking is no exception. Unlike polishes, which are often uniformly distributed along an edge, microflaking results in assemblages of individual scars which are difficult for us to describe and identify in more objective terms. The 'low power school' has successfully interpreted this category of wear (Tringham *et al.*, 1974; Odell & Odell-Vereecken, 1980), but the criteria of diagnosis have not yet been fully elucidated. The present study is an attempt to describe quantitatively the variability of scar patterns.

Experimental framework

Tohoku University Microwear Research Team has carried out about 400 replicative experiments so far, using shale, chert, obsidian, and sanukite, to make comparative collections of artefacts of known use; to help in the analysis of Japanese Palaeolithic and Jomon artefacts (see, especially, Akoshima, 1981, which is an expanded version of the present paper in Japanese, as well as Kajiwara & Akoshima, 1981; Serizawa, Kajiwara & Akoshima, 1982; Serizawa, 1980*a*, 1980*b*; Kajiwara, 1982*a*, 1982*b*). Seventy-two experimentally made and used tools were put to the present analysis in 1979. The flakes were struck from shale nodules obtained at the river bed of Mogami River, Yamagata. This material is a kind of siliceous hard shale that fractures with conchoidal features, which was widely used in Northeastern District throughout prehistoric times. The working edges remained unretouched in the course of experiments. Independently controlled variables were activity, worked material, and edge angle. The list of experiments is shown in Table 9.1. Such variables as number of strokes, duration of work, place and conditions, shape of tools, prehension and use method, contact angle, distance moved and conditions of worked materials were also recorded, and the resultant flaking scars were photographed. Since microflaking represents a great range of diversity in its characteristics and distribution along the edge, only summary description is possible here. It was found that the edge in fact stabilized (Keller, 1966) after a certain number of strokes. A number of scars are produced at the initial stage of experiment, and the rate of formation gradually decreases after weak points on the edge have already fractured. Thus, the counting of scars was done after the stabilization, so that the 'standard' microflaking characteristics might be identified.

Analysis and results

Microflaking was observed under the stereoscopic microscope at 30× to 60×, mostly 45×. A total of 3840 scars were counted one by one and recorded in terms of shape, size, and termination. Significant correlations exist between these attributes and conditions such as hardness of worked materials and kind of activity.

(a) *Percentages of step flaking*

The cross-section of scars can be precisely detected by making a rubber replica. Figure 9.5 shows typical kinds of microflaking cross section: no. 1 SH92 is a section of deep scalar flaking. Figure 9.5, no. 2 is that of step flaking, the depth of which is about 70 microns (SH92). Figure 9.5, no. 3 is that of vertically overlapping scars (SH93). The percentage of step flaking of all scars except the snapped type is shown in Fig. 9.1. It is evident that (1) harder materials tend to produce a higher percentage of step flaking if the action is constant, and (2) transversal action tends to produce higher percentages than longitudinal action if the worked material is the same. The latter effect is very marked in the case of hard materials when the figure is almost doubled. However, no correlation was recognized between edge angle and the percentage of step flaking.

(b) *Shape of microflaking*

Cumulative graphs of shape of microflaking scars are shown in Fig. 9.2. The range of scars was classified into six categories according to the system used by Tringham *et al.* (1974). One line in the graph represents one edge. It was found that harder materials tend to produce lower percentages of scalar scars, and higher percentages of rectangular and trapezoidal scars than softer materials, if the action is the same; and that transversal action tends to produce lower percentages of scalar scars, and higher percentages of rectangular and trapezoidal scars than longitudinal action,

Table 9.1. *List of experiments*

exp. no.	worked material	activity	strokes	No. of scars	edge width	edge angle	condition *
SH 1	wheat	cut	(15m)	83	67 mm	middle	F. fresh
SH 7	alder	cut	1400	48	42	thin	L. fresh
SH 8	alder	cut	2500	23	30	thin	L. fresh
SH 9	alder	cut	1000	50	45	middle	L. fresh
SH 10	alder	cut	2000	56	62	middle	L. fresh
SH 11	wheat	cut	(25m)	65	68	middle	F. fresh
SH 12	alder	cut	3000	38	40	middle	L. fresh
SH 13	alder	whittle	2000	44	39	middle	L. fresh
SH 14	pine	chop	200	49	45	thin	L. seasoned
SH 15	lamb	cut	500	51	58	thin	L. half-frozen
SH 16	antler	cut	1500	43	40	middle	L.
SH 17	lamb	cut	1000	17	51	middle	L. half-frozen
SH 20	grass (weed)	cut-chop	600	82	52	thin	F. fresh
SH 26	beef	cut	800	36	52	thin	L. fresh
SH 40	reed	cut-chop	3000	75	48	thin	F. fresh
SH 42	reed	cut-chop	2650	67	49	thick	F. fresh
SH 43	rice	cut	3000	27	70	thin	F. fresh
SH 44	pine	saw	2000	32	37	thin	L. fresh
SH 45	rice	cut	3000	59	60	thin	F. fresh
SH 46	pine	whittle	1500	33	37	thick	L. fresh
SH 47	antler	saw	4000	110	50	middle	L. soaked
SH 48	antler	saw	15000	72	45	middle	L. soaked
SH 49	wood	saw	5000	58	58	middle	L. seasoned
SH 54	pork	cut	1300	19	40	thin	L. fresh
SH 55	pork	cut	1600	40	50	middle	L. fresh
SH 56	chicken	butcher	1020	73	41	thin	L. fresh
SH 57	pork	scrape	1100	14	27	thin	L. fresh
SH 66	SUSUKI	cut-chop	800	70	55	thin	F. fresh
SH 68	antler	saw	4300	70	45	middle	L. dry
SH 70	antler	whittle	2000	58	30	middle	L. dry
SH 76	cattle hide	saw	5000	74	47	middle	L. dry
SH 77	gourd	saw	5000	82	40	middle	L. seasoned
SH 79	bamboo	saw	2000	55	37	thin	L. seasoned
SH 80	bamboo	saw	4000	73	35	middle	L. fresh
SH 82	bamboo	scrape	4000	12	20	thick	L. fresh
SH 84	bamboo	scrape	2000	101	41	middle	L. fresh
SH 86	pig bone	saw	3000	69	30	thin	L. fresh
SH 89	pig bone	scrape	1000	39	20	thick	L. fresh
SH 90	pine	scrape	300	73	55	middle	L. seasoned
SH 91	pig bone	saw	500	29	33	middle	L. boiled
SH 92	pig bone	saw	5000	84	38	middle	L. boiled
SH 93	pig bone	scrape	1500	82	36	thick	L. boiled
SH 96	cedar	whittle	1000	77	53	middle	L. fresh
SH 97	duck meat	cut	700	19	20	thin	L. fresh
SH 99	alder	whittle	2000	41	16	middle	L. seasoned
SH 100	pine	whittle	2000	31	15	middle	L. seasoned
SH 101	pig bone	whittle	3000	74	25	middle	L. boiled
SH 108	duck	butcher	–	77	33	middle	L. fresh
SH 111	paulownia	scrape	500	69	40	middle	L. fresh
SH 114	paulownia	whittle	1000	29	25	middle	L. fresh
SH 115	paulownia	saw	1000	42	45	middle	L. fresh
SH 116	pig rawhide	scrape	2000	14	16	thick	L. loam involved
SH 120A	pig rawhide	scrape	2000	79	45	middle	L.
SH 121	pig rawhide	scrape	2000	40	32	thick	L.
SH 122	pig rawhide	scrape	2000	78	24	thin	L.
SH 123	pig rawhide	scrape	2000	73	30	middle	L. loam involved
SH 124	pig rawhide	scrape	800	66	27	thin	L. loam involved
SH 125	pig rawhide	scrape	2000	41	27	middle	L. sand involved
SH 126	pig rawhide	scrape	2000	31	27	middle	L. sand involved
SH 128	pig rawhide	scrape	2000	31	33	thin	L.
SH 129	pig rawhide	scrape	4000	41	32	middle	L.
SH 130	pig hide	scrape	1500	19	14	middle	L. dry
SH 131	pig hide	scrape	1500	34	39	thick	L. dry
SH 140	grass (weed)	cut	2200	53	42	middle	F. fresh
SH 141	SUSUKI	cut-chop	2200	82	45	middle	F. fresh
SH 146	pig hide	scrape	1000	79	51	thick	L. dry
SH 148	pig hide	chop	306	38	35	thin	L. dry
SH 149	pine	scrape	500	55	34	middle	L. seasoned
SH 150	cedar	scrape	500	44	45	middle	L. fresh
SH 151	cedar	scrape	1000	48	30	thick	L. fresh
SH 152	pine	chop	700	44	35	thick	L. seasoned
SH 153	antler	scrape	100	56	42	middle	L. dry

*L. = laboratory, F. = field condition)

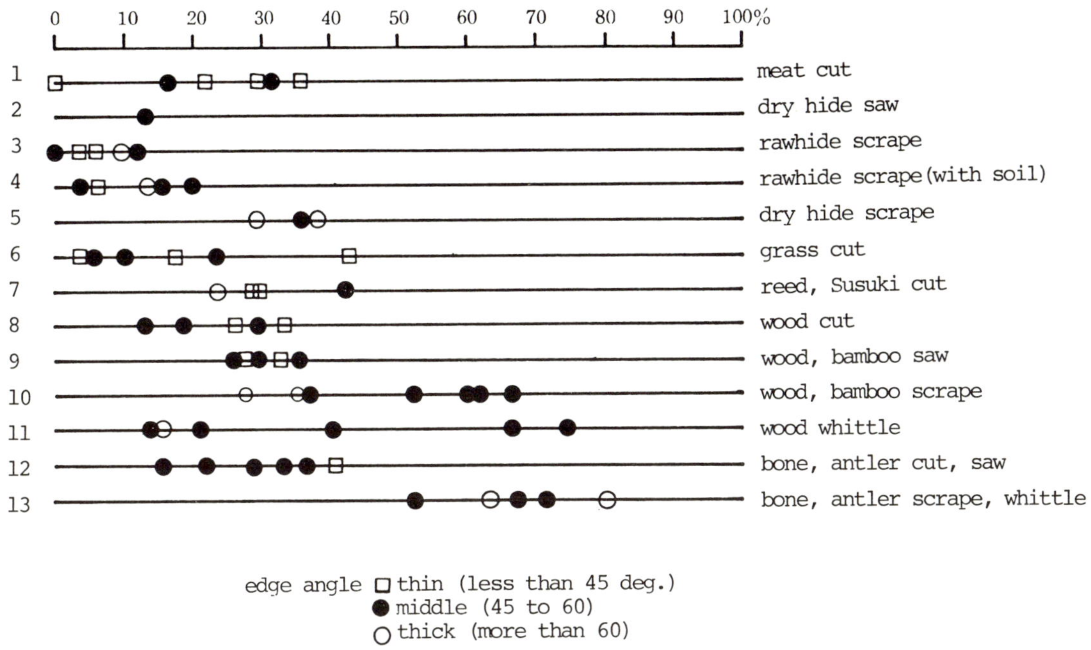

Fig. 9.1. Percentage of step flaking.

if worked material is the same. The results of shape analysis are not inconsistent with the results of termination analysis: there is a tendency to produce scalar scars with feather termination (deep or shallow) by working softer materials in longitudinal actions, while harder materials worked transversely tend to yield rectangular and trapezoidal scars with step terminations (or hinge fractures).

(c) *Size of microflaking*

The width of scars was measured at the edge and divided into Mi (micro, smaller than 0.5 mm), S (small, between 0.5 and 1.0 mm), M (middle, between 1.0 and 2.0 mm), and L (large, larger than 2.0 mm). Percentages of each size category are shown in Fig. 9.3. The tendency for harder worked materials to produce larger scars is indicated here. However, no clear correlation between size and action or edge angle was observed.

(d) *Dorsal–ventral distribution*

The degree to which scars are concentrated on dorsal or ventral aspect of the edge is shown in Fig. 9.4. Only the higher percentage, either dorsal or ventral, is cited. It was reassuring that transversal action tends to yield higher percentages than longitudinal action, but the degree varies according to worked materials. For example, in the case of hard materials (bone and antler), the contrast is very distinctive; (P=percentage<65) to ($58<P<83$). It is interesting that whittling exhibited lower percentages than scraping, if the material is the same.

(e) *Density of microflaking*

The number of flake scars per length (cm) was examined. The density did not show a proportional increase of the number of strokes irrespective of worked material or action, due to the above mentioned stabilization of the edge. The density is dependent on specific conditions of edge and materials, and varies greatly from a few to 25 scars per cm. Transverse action, however, does seem to produce more scars than longitudinal action.

Conclusion

We may conclude that distinctive patterns of flaking scars do indeed exist, and that these patterns can be expressed in quantitative terms. Sometimes the quantitative characteristics allow us to make inferences about tool use. For instance, it is deducible from Fig. 9.1 that if a flake has 10% or less of step flaking it was probably used on soft material in longitudinal action. However, it is really a pity that the correlations or tendencies that have been depicted are *not exclusive* in nature. This leads to a conclusion that a certain scar cannot be a definitive clue to functional determination and the features of flaking scars on the edge *as a whole* should be the unit of analysis and interpretation. This is the inherent limitation of microflaking analysis, because a scar of a specific type (Fig. 9.5) is formed when certain conditions are met at a small portion of the edge at a single moment. It should be emphasized that the usefulness of microflaking analysis lies in considering scars in groups or assemblages, and the distribution pattern must be taken into account. Examples of characteristic types are illustrated in Fig. 9.6.

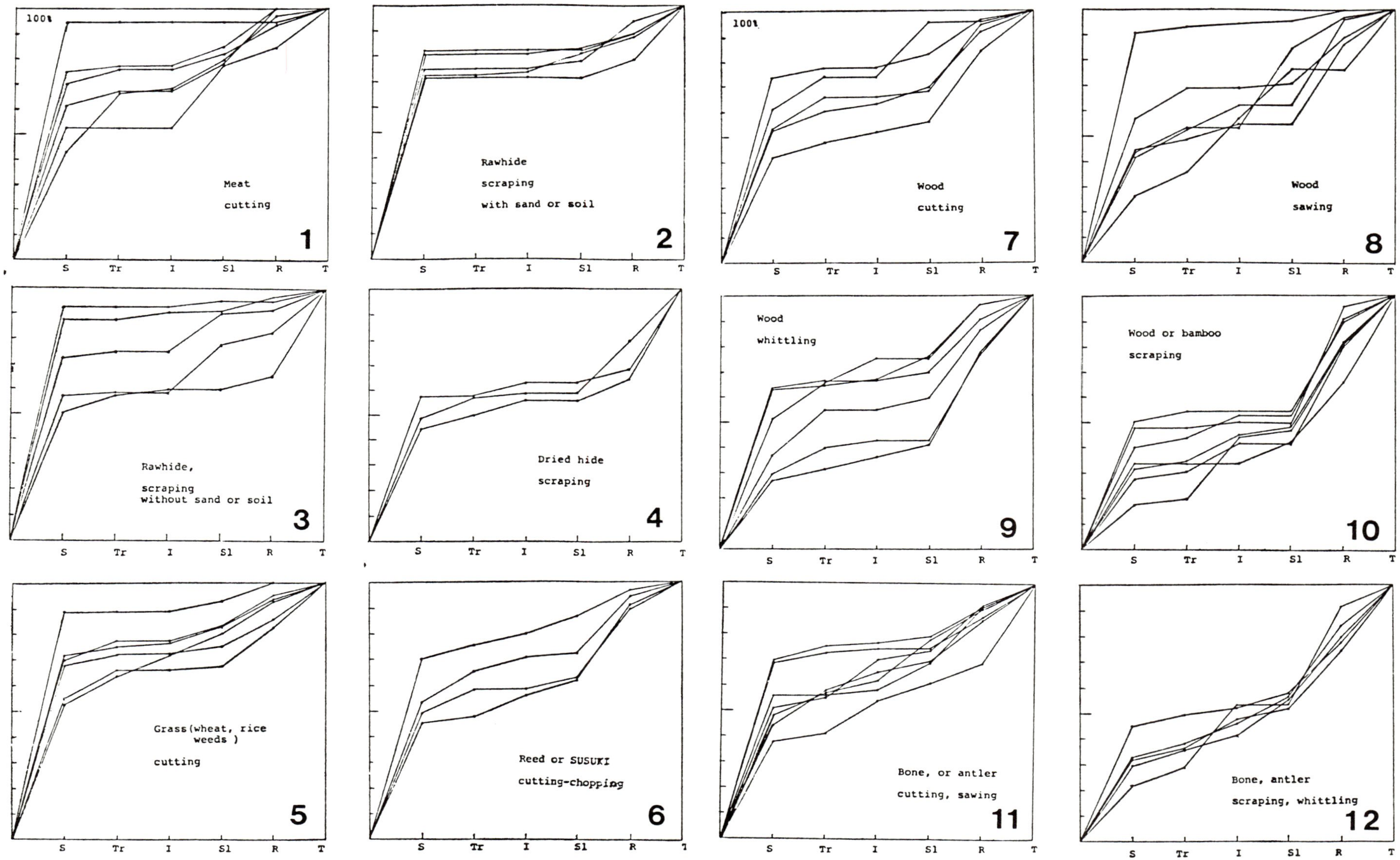

Fig. 9.2. Cumulative graphs of microflaking shapes. Abbreviations (from left S=scalar Tr=triangular I=irregular Sl=sliced R=rectangular T=trapezoidal).

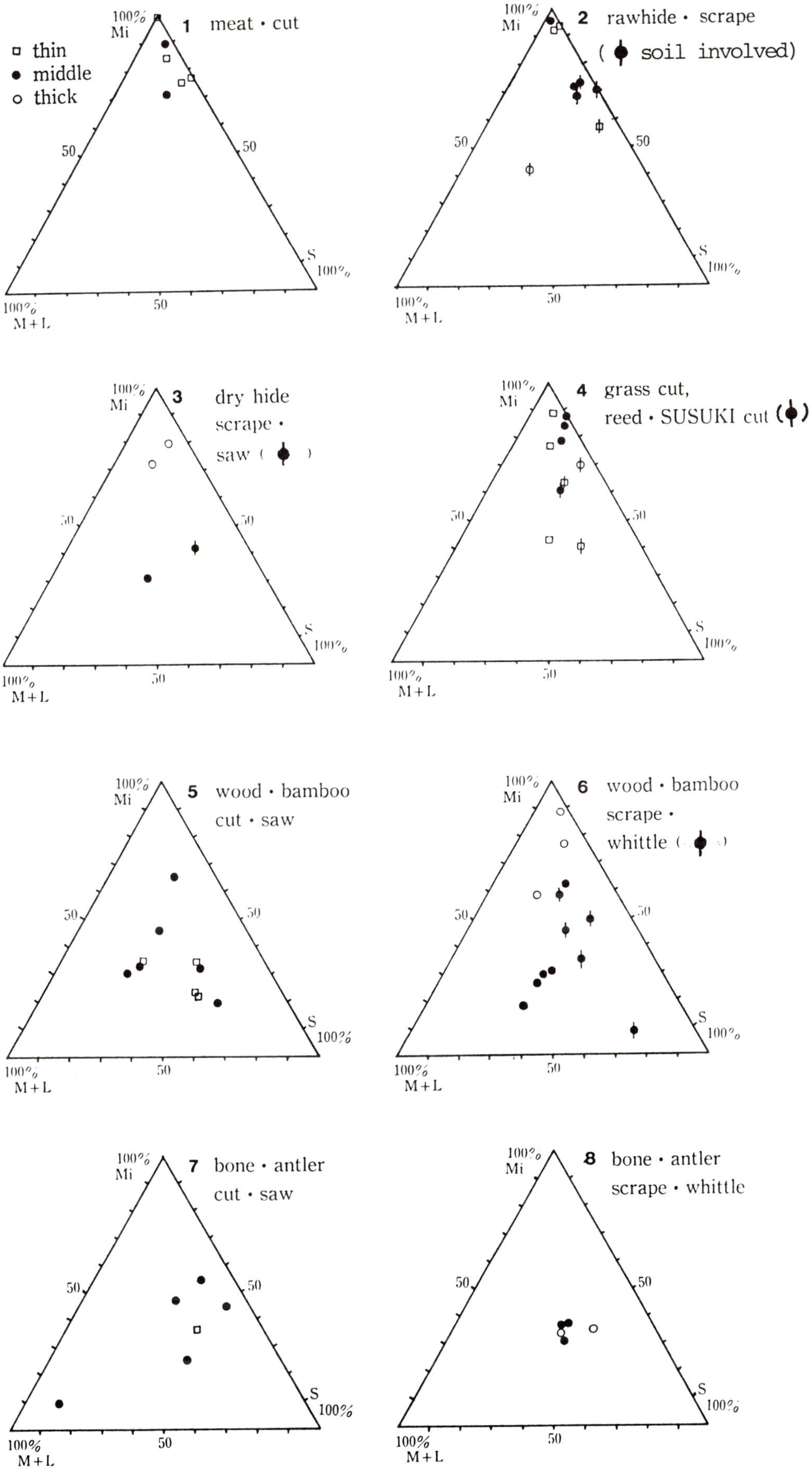

Fig. 9.3. Size of microflaking.

(a) *Soft worked material, longitudinal action*

Diverse scars develop on both aspects of the edge; they are not usually continuous and are small and mainly scalar (Fig. 9.6, nos. 1, 2). When greater force is used as when working hardened hide or half-frozen meat, or when reaping grass (Fig. 9.6, no. 3) scars become larger. Butchering produces more varied types (e.g. large angular) as the edge contacts bone tendons as well.

(b) *Soft worked material, transversal action*

Rawhide scraping produces small scalar scars which are regularly distributed on the rear aspect of the edge (Fig. 9.6, no. 4). When loam or sand is involved, heavy abrasion occurs and the edge is rounded (Fig. 9.6, no. 5), with the enlargement of scars and accentuation of the scalar shape. The presence of loam or sand also increases the number of scars on the frontal aspect. Dry hide tends to produce more step flaking (Fig. 9.6, no. 6).

(c) *Medium worked material, longitudinal action*

Here features were rather experiment-specific. Different kinds of scars are scattered on both aspects of the edge, but are larger than those made by soft materials (Fig. 9.6, nos. 7, 8). Scars are very variable in size and shape and may be continuous or not.

(d) *Medium worked material, transversal action*

Wood scraping produced distinctive microflaking consisting of continuous flat scars (Fig. 9.7, nos. 10, 11). When step fractures form, the scars are mainly rectangular or trapezoidal in plan, and wide but shallow scars are scattered on the frontal aspect. The scar pattern is more varied in the

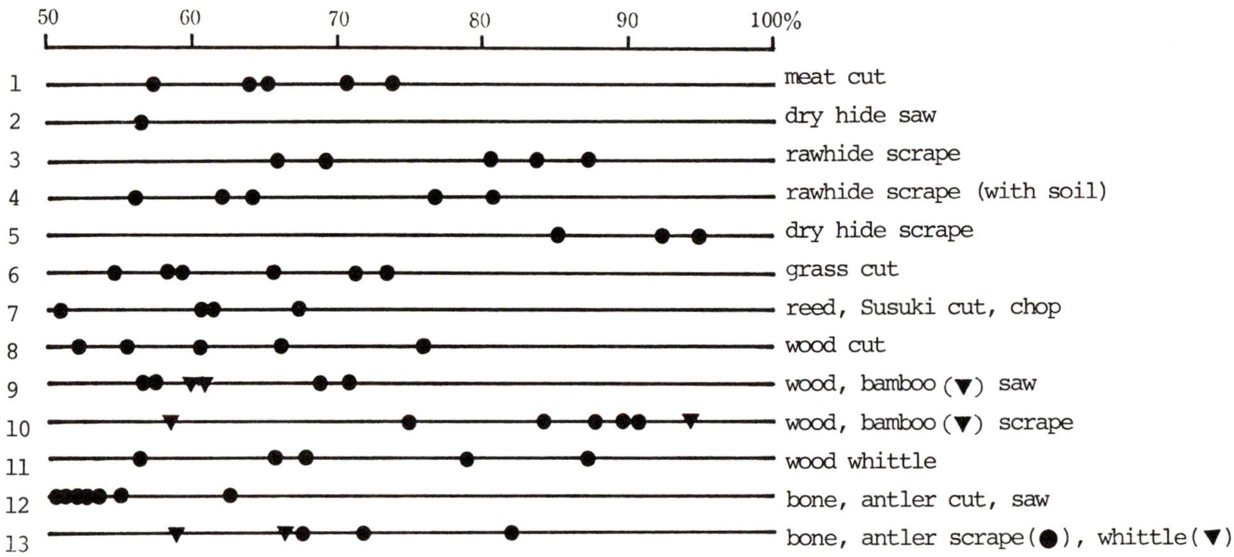

Fig. 9.4. Concentration of microflaking scars on one aspect of the tool.

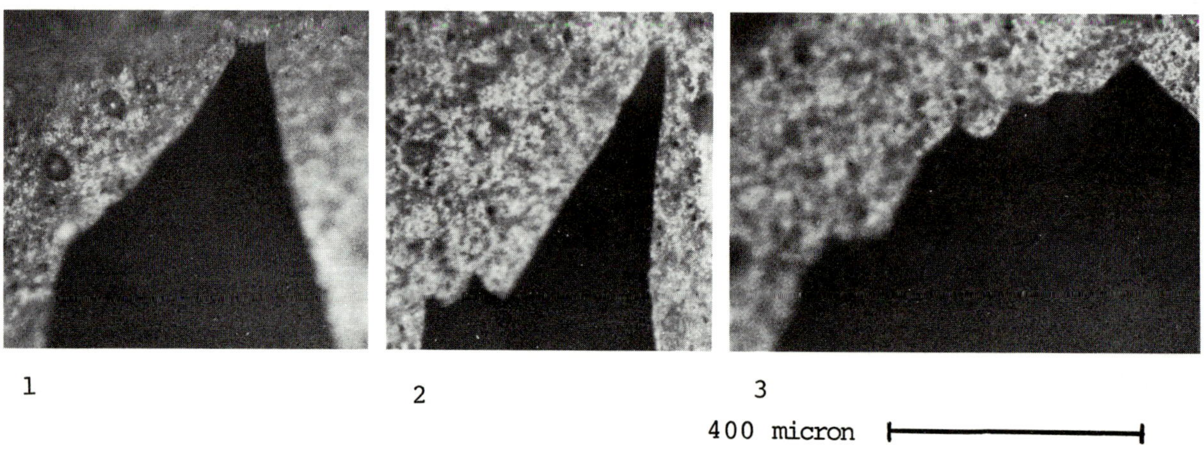

Fig. 9.5. Cross-section of microflaking.

case of bamboo or wood whittling, but sometimes it is similar to that produced by scraping. The axes of scars may be oblique to the edge line (see Fig. 9.7, no. 12 for scars made by chopping). This phenomenon also occurs in wood cutting, meat cutting or bone or antler whittling, when the force is exerted obliquely.

(e) *Hard worked material, longitudinal action*

Sawing hard materials produces very distinctive alternate microflaking patterns (Fig. 9.7, no. 13). When viewed perpendicular to the edge, the pattern is of comparatively angular and large scars. Scars are deep, often overlapping,

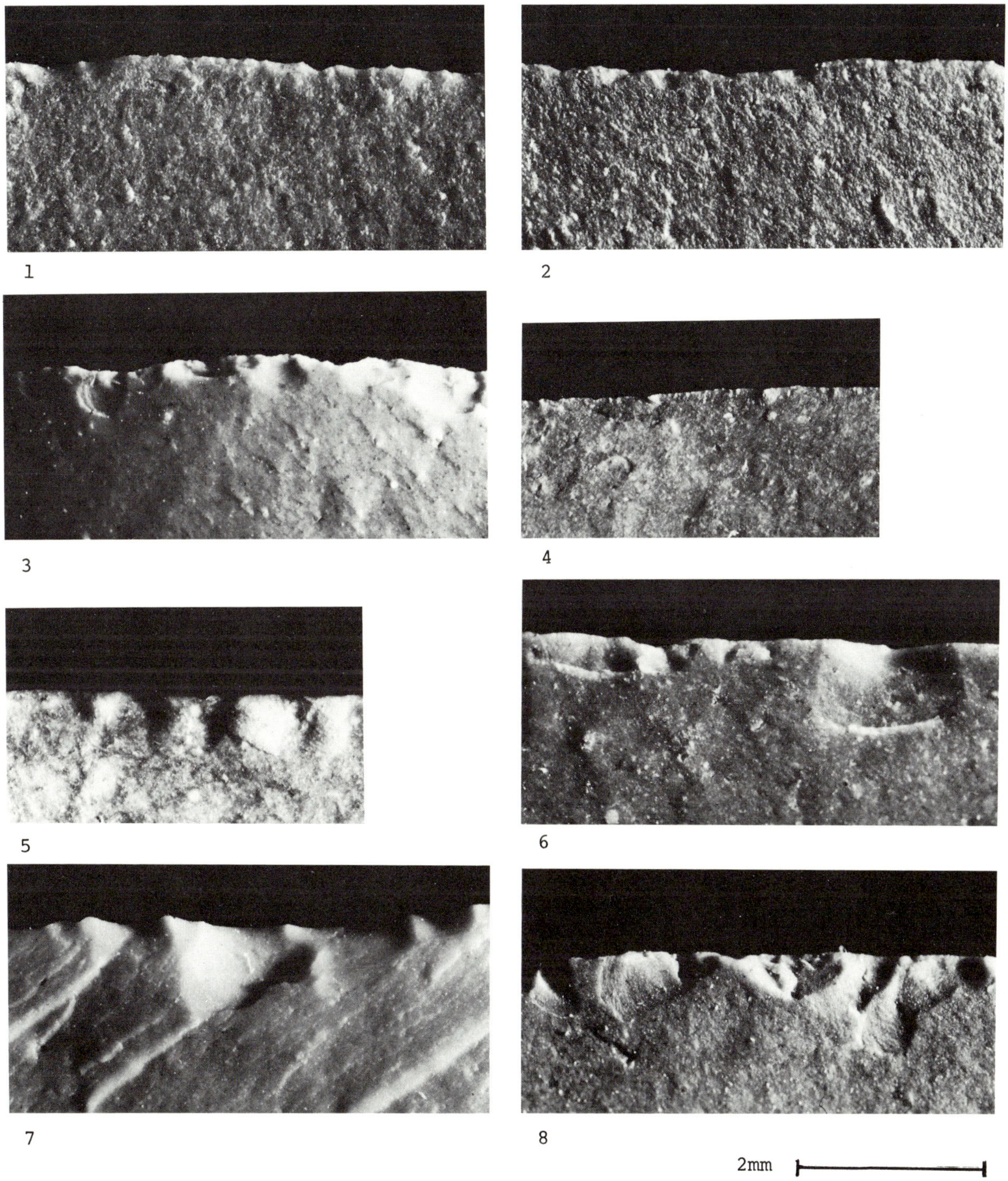

Fig. 9.6. Appearance of microflaking scars. (1) SH55 meat cut. (2) SH1 wheat cut. (3) SH20 grass cut-chop. (4) SH121 rawhide scrape. (5) SH126 rawhide scrape. (6) SH146 dry hide scrape. (7) SH66 Susuki (*Miscanthus sinensis*) cut-chop. (8) SH44 wood saw.

9

3mm

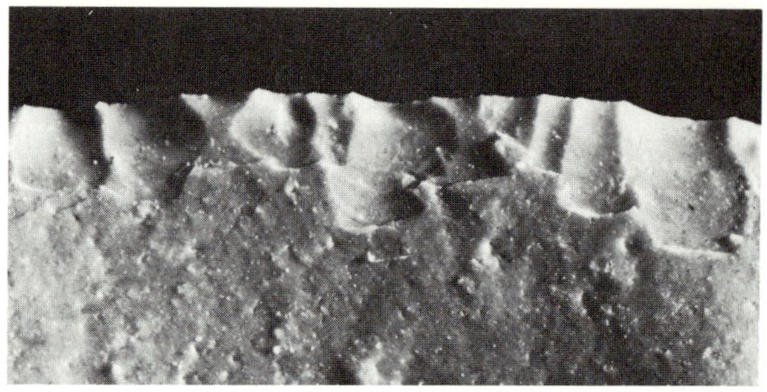

10

3mm

11

12

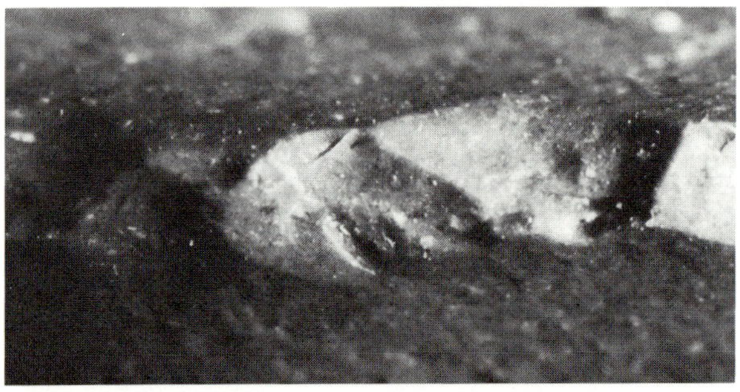

13

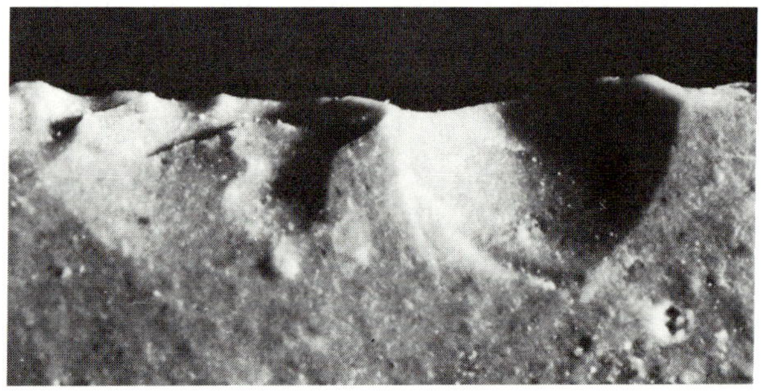

14

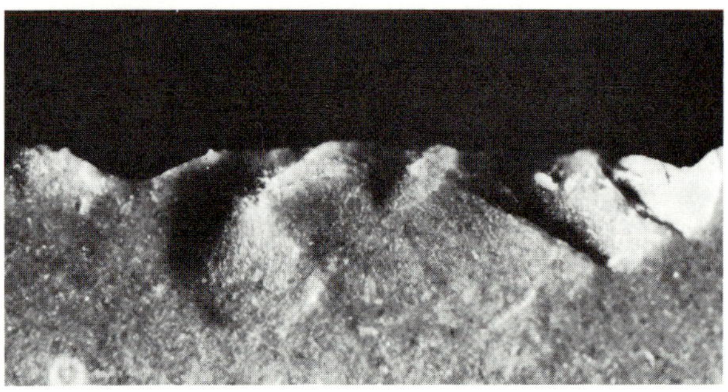

15

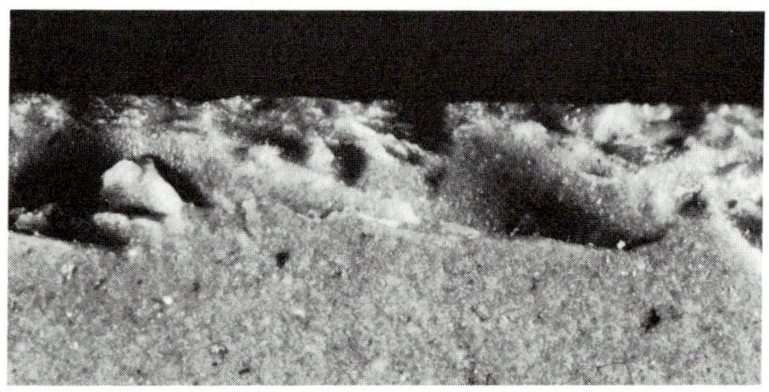

16

17

18

2mm

and are angular when used in dry conditions (Fig. 9.7, no. 14), but when the worked material is soaked, the edge becomes more rounded and scars shallower.

(f) *Hard worked material, transversal action*

Scraping bone or antler produces an overlapping of larger flat scars and shallow step flaked scars which have a crushed appearance (Fig. 9.7, nos. 16, 17). Small scars are scattered on the frontal aspect. In the case of whittling bone or antler, scars are more varied.

Acknowledgements

I would like to acknowledge the continuing guidance and encouragement of Professor Chosuke Serizawa. I am very much indebted to him for drawing my interest to this special field of lithic analysis as well as Palaeolithic studies in general. I owe a special thanks to my colleague Hiroshi Kajiwara, for conducting many experiments with me and for stimulating discussions.

Bibliography

Akoshima, K. (1981). 'An experimental study of microflaking.' *Kokogaku Zasshi*, **66**(4), 1–27 (in Japanese).

Kajiwara, H. & Akoshima, K. (1981). 'An experimental study of microwear polish on shale artefacts.' *Kokogaku Zasshi*, **67**(1), 1–36 (in Japanese).

Kajiwara, H. (1982*a*). 'A microwear analysis of tanged scrapers excavated from Mikamine Site.' *Kokogaku Zasshi*, **68**(2), 43–81 (in Japanese).

Kajiwara, H. (1982*b*). 'Microwear analysis.' In: *Mosanru: Records of Archaeological Material No. 4*, pp. 11–15, 83–95, C. Serizawa (ed.), Laboratory of Archaeology, Faculty of Arts and Letters, Tohoku University (in Japanese).

Keller, C. T. (1966). 'The development of edge damage patterns on stone tools.' *Man*, **1**, 501–11.

Odell, G. H. & Odell-Vereecken, F. (1980). 'Verifying the reliability of lithic use–wear assessment by "blind tests", the low power approach. *Journal of Field Archaeology*, **7**, 87–120.

Serizawa, C., Kajiwara, H. & Akoshima, K. (1982). 'Experimental study of microwear traces and its potentiality.' *Kokogaku to Shizenkagaku* (Archaeology and Natural Sciences). **14**, 67–87 (in Japanese).

Serizawa, C. (1980*a*). *Mukoyama: Records of archaeological Material No. 3.* Laboratory of Archaeology, Faculty of Arts and Letters, Tohoku University (in Japanese).

Serizawa, C. (1980*b*). 'Studies on lithic use–wear.' In: *Kokogaku Bijutsushi no Shizenkagakuteki Kenkyu* (Natural Scientific Studies in Archaeology and Art-History), pp. 461–8 (in Japanese).

Tringham, R., Cooper, G., Odell, G., Voytek, B. & Whitman, A. (1974). 'Experimentation in the formation of edge damage: A new approach to lithic analysis. *Journal of Field Archaeology*, **1**, 171–96.

Fig. 9.7. Appearance of microflaking scars (continued). (9) SH44 wood saw. (10) SH150 wood scrape. (11) SH111 wood scrape. (12) SH14 wood chop. (13) SH68 antler saw. (14) SH68 antler saw. (15) SH86 bone saw. (16) SH89 bone scrape. (17) SH93 bone scrape. (18) SH101 bone whittle.

10

The influence of flint microstructure on the formation of microwear polishes

ROSEMARY BRADLEY AND CHRIS CLAYTON

Introduction

It has always been recognised in the study of use–wear on stone tools that the raw material type is a major controlling factor in the formation and disposition of the traces sustained in use (Wyant & Bayhem, 1976, pp. 232–3; Odell, 1977, pp. 99–100; Seitzer, 1978, p. 6; Semenov, 1964, pp. 33–9; Keeley, 1974). Consequently, in the planning and execution of replicative experiments to form a reference collection of use–wear types for prehistoric pieces, emphasis is given to using the same raw material in the experiments as that represented by the prehistoric artefacts (Brink, 1978, pp. 22–3; Keeley, 1978, p. 73; Broadbent & Knutsson, 1975).

While research has shown that very great differences occur between use–wear formed on different types of stone (e.g. Kamminga, 1978, pp. 82–102; Greiser & Sheets, 1979, p. 289), less attention has been given to the possible variations within a single rock type and, for example, flint is described as 'English fine-grained chalk flint' (Odell *et al.*, 1976). Now that high magnification techniques are being used increasingly to observe the microwear polishes and striations on flint and microfractures and abrasions on damaged edges, it is becoming more important to understand how the flint type influences the development of lithic use–wear. We report here preliminary investigations into the influence of flint microstructure, in which we attempt to combine geological knowledge of the raw material with characteristic features observed in a series of laboratory experiments on microwear formation. In particular we consider whether microwear polishes form in the same pattern and at the same rate on different types of flint.

Microstructure of the upper cretaceous flint

In thin-section, the fine grain-size of flint prevents detailed investigation of the constituents, and this has led to the assumption that flint fabric is homogeneous. However, when etched for one to four minutes in 40% HF at room temperature and viewed in reflected light, typical black flint is seen to consist of at least three main generations of silica (Fig. 10.1): silicified skeletal fragments, spherical quartz 'lepispheres' (which together make up the main framework structure of flint), and an interstitial 'cement' of structurally disordered microfibrous quartz chalcedony (Clayton, 1982).

Silicified skeletal fragments, particularly microfossils such as foraminifera (Fig. 10.1), are generally rare. However, in some areas, these together with fragments of macrofossils such as bryozoans or echinoderms may be major constituents, such as in Turonian flints from Grand Pressigny, France. These represent the first stage of flint formation in chalks and in some cases the fossils preserved may be of some use in source typing studies. Quartz lepispheres are 5–20 μ spherical aggregates of interlocking anhedral quartz grains which typically comprise between 60 and 95% of the structure depending on how densely they are packed together. These were produced initially by precipitation of silica as disordered cristobalite (opal-CT) and then later recrystallised to quartz. Similar objects, but still composed of cristobalite, can be found where the chert has not matured into a quartzose flint (e.g. in the Craie de Villedieu around Tours, Central France: Jarvis *et al.*, 1982). The third type of silica, chalcedony, acts as an interstitial 'cement', infilling voids within the framework of lepispheres and skeletal fragments. It is composed of structurally disordered water-rich silica with sub-microscopic water filled pores and molecular water absorbed on grain and sub-grain boundaries and along lattice defects (Clayton, 1983). The grains, which are generally less than 0.1 μ long, form radiating bundles and rosettes to fill the voids. Since the boundary between the chalcedony and the lepispheres is usually gradual it is probable that the chalcedony grew before the opal-CT/quartz transition, a suggestion supported by the similarity of the oxygen isotope composition of the two (Clayton, 1983).

After the growth and lithification of a flint nodule by precipitation of the above phases, recrystallisation can affect the structurally disordered silica, resulting in local redistribution of the structural and surface bound water into microscopic and sub-microscopic fluid inclusions. These inclusions partially reflect incident light rather than dissipating it, and this produces the characteristic white patches seen in many nodules. Similar recrystallisation within the body of the flint can also be induced by heating the flint, during which structural water migrates from the chalcedony into fluid inclusions (Griffiths *et al.*, this volume). Recrystallisation can also occur on the outer

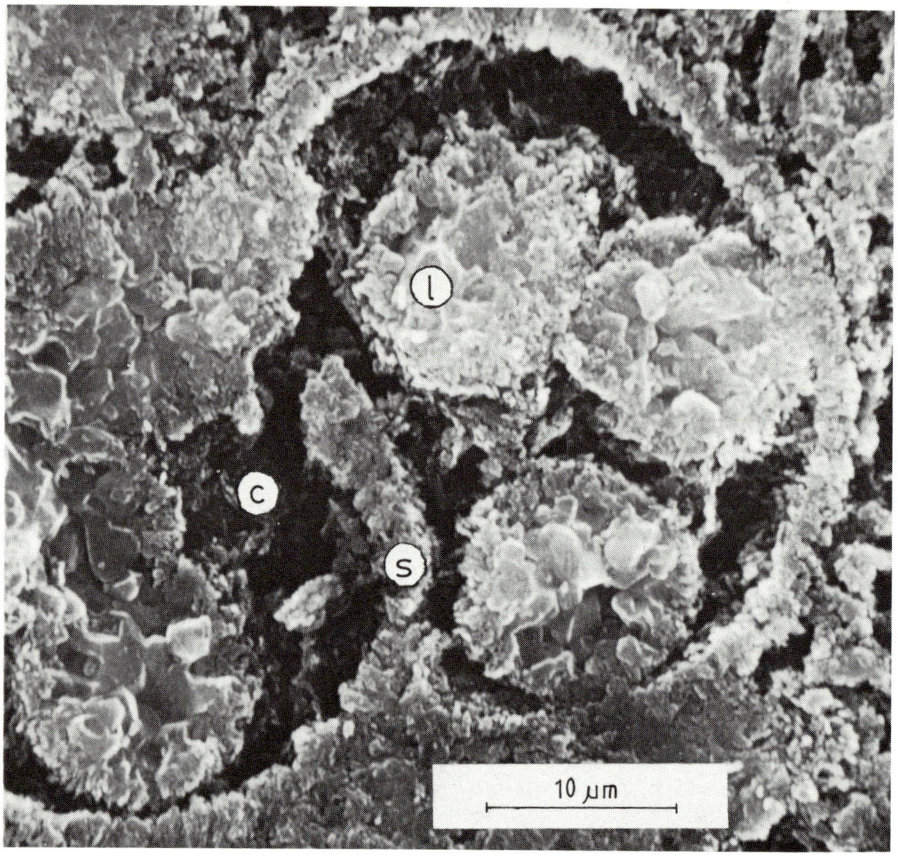

Fig. 10.1. Microstructure of typical blank flint. Acid etched specimen, Brandon, Norfolk (×200), showing (S) framework of silicified skeletal fragments (in this case foraminifera) and (L) lepispheres with interstitial chalcedony (C).

surface of flint in response to the chemical changes in the surrounding pore waters and this leads to development of the familiar white patina (Clayton, 1983). In all cases, the resulting flint is structurally different to the original material and will respond differently to mechanically induced stress (i.e. impact or use–wear).

Fracture and microwear polish formation with respect to flint microstructure

In the flaking of Upper Cretaceous flint which has undergone no recrystallisation the fracture front preferentially breaks the flint through the chalcedony, around the quartz lepispheres (Fig. 10.2(*a*)). The resulting surface is covered with hemispherical domes or craters the size of the lepispheres composing the flint and spaced to reflect the density of packing of the lepispheres in the flint before fracture. This surface roughness must not be confused with the irregularities of the surface which are produced by alterations and deviations in the speed and direction of the propagating fracture front. However, when flint which has undergone recrystallisation is knapped, the susceptability of the chalcedony has been reduced and the fracture surface cuts across the lepispheres (Fig. 10.2(*b*)). This is true whether the recrystallisation has occurred from deliberate heating or not and leads to a much more predictable outcome to the flaking process (see Griffiths *et al.*, this volume). The recrystallisation also results in a much higher proportion of well-crystallised quartz in the structure and since this is more resistant to mechanical and chemical attack it is anticipated that recrystallised flint will show a

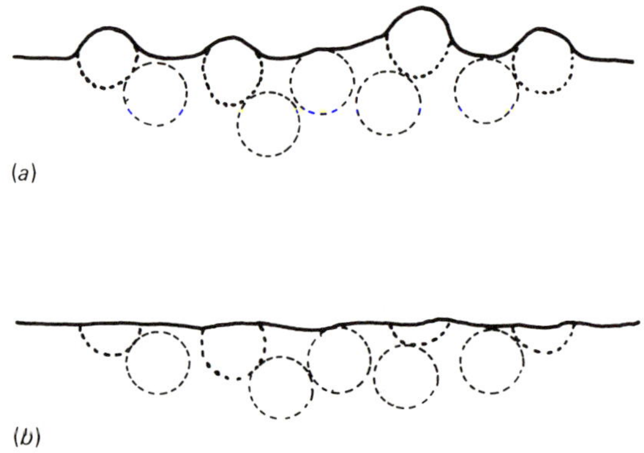

Fig. 10.2. Diagram showing the passage of a fracture front and the effect of recrystallization on fracture morphology. Normal unaltered cretaceous fling (*a*) fractures round lepispheres and through weaker chalcedony. On recrystallization (*b*) the chalcedony becomes stronger and fracturing cuts across lepispheres as well as chalcedony.

greater resistance to microwear polish formation. Conversely, since it is easier to knap, it may exhibit a greater susceptibility to subsidiary microfracturing than unaltered flint and be more susceptible to chipping. In addition, the density of packing of the lepispheres within the flint structure also will influence the form of the traces resulting from use. All of these features, which relate to the microsurface texture, must be considered when examining the variations in microwear formation on different types of flint.

Whether the fracture front has passed around or across the lepispheres, the surface produced shows the plates and grains of which they are composed. Each anhedral grain in a lepisphere is built up from subgrains of quartz which are aligned with reference to the twinning laws of quartz. These in turn compose larger plates which lie close together packed into the sphere. Between the lepispheres is the chalcedony mineral which appears fibrous and matt. Using an optical microscope, even under low magnifications the fresh fracture surface appears grainy and is dull and matt (Fig. 10.3). The SEM also reveals clearly the unused surface of flint with its rough and platey structure (Fig. 10.4). When areas of recrystallisation occur, the microfluid-inclusions which have formed within the structure partially reflect rather than refract the incident light and therefore give a white colour to the flint. Under the optical microscope, and in particular at higher magnifications, such reflections can cause intense glare and lead to problems in observation. Variations in the species of silica present in the flint will influence both the mechanical damage sustained and also that produced by chemical reactions. Those flints with the original structure will not only be more prone to mechanical damage, since they contain a lower proportion of well-crystallised quartz, but also more readily attacked by chemical agents as they are composed of more structurally disordered water-rich silica.

Areas of attack, both mechanical and chemical, will be governed in most cases by the microtopography of the flint. The path of the fracture front, in particular that through unaltered flint, leaves an irregular face with a large surface area and this is especially vulnerable to dissolution and other chemical influences. In addition, the depositional (i.e. precipitation) features described by Anderson (1980) will be situated with reference to surface texture and type. Her research revealed that many microwear polishes had a strong depositional element resulting from the microdissolution of the flint, the formation of silica gels and their subsequent re-deposition on the flint surface. This would be much quicker and more extensive on unaltered flint, since the water-rich poorly ordered silica phase in the matrix around the lepispheres and fossils is more soluble than quartz (Kennedy, 1950, p. 652). By comparison, the far higher proportion of well-ordered quartz found in recrystallised flints renders them more resistant to both mechanical and chemical attack.

Furthermore, some of the features seen on the surface of flint tools after use are interpreted by Anderson as recrystallisation products of the amorphous silica produced in use (1980, p. 183). The interstices between the plates which compose the lepispheres or the deep negative hollows left when a fracture front breaks around a lepisphere represent ideal loci for the accumulation of silica gel and residual deposits such as the phytoliths she describes.

From a close examination of the microstructure of flint, it can be predicted that unaltered lepispheric flint will produce rougher fracture surfaces which are more susceptible to microwear polish formation but slightly less prone to

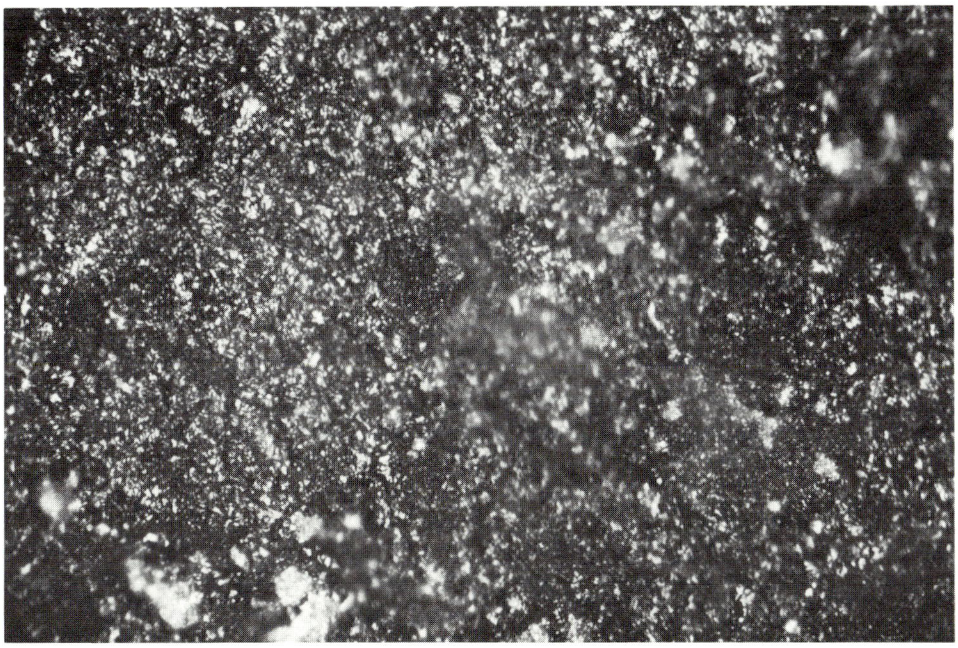

Fig. 10.3. Fresh fracture surface of upper cretaceous Brandon flint as seen under the optical microscope. Showing the grainy and matt texture (×200).

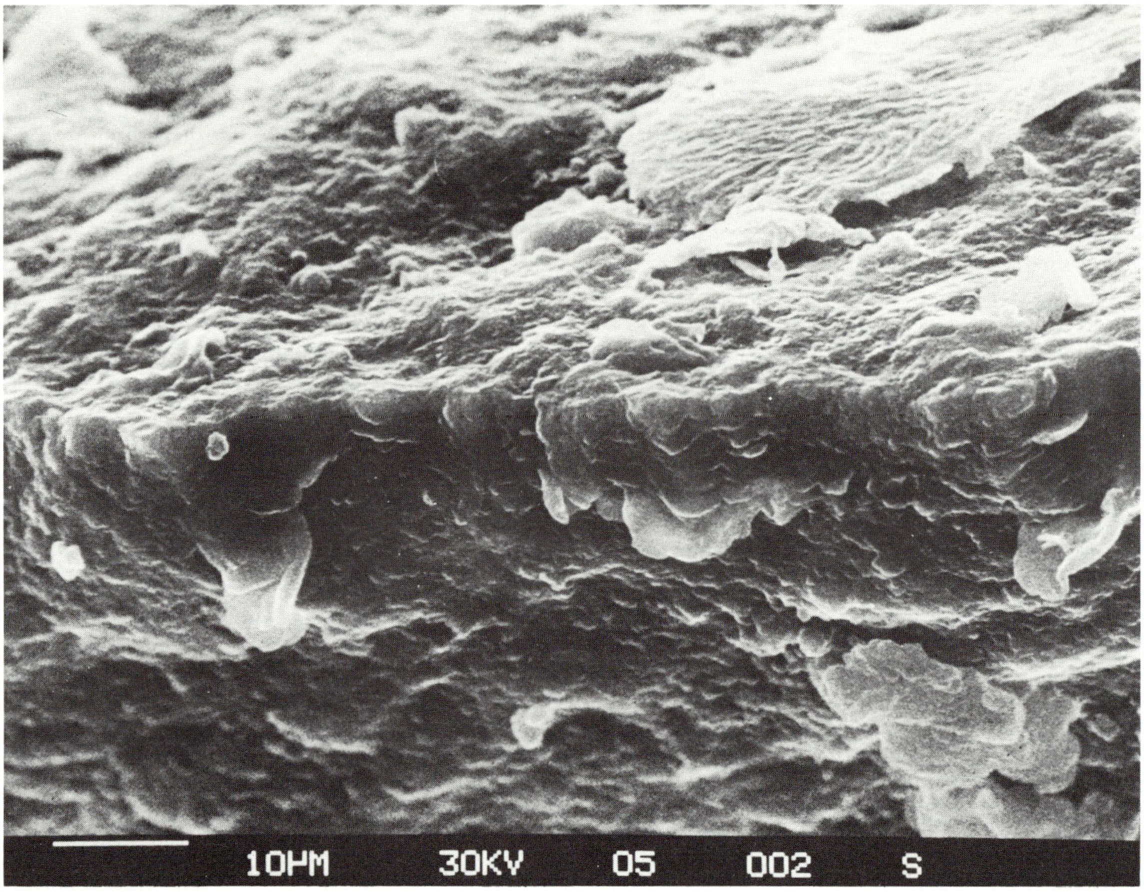

Fig. 10.4. SEM micrograph of fresh flint surface of Upper Cretaceous Brandon flint, showing its rough and platey texture (×1700).

microfracturing than those of recrystallised lepispheric flint. These basic differences affect not only the rate at which microwear damage traces form on a tool during use but also on the disposition of traces when they develop, since the microtopography of, in particular, the initial surfaces, can be very different.

The experimental study

To test empirically the assertion that different types of Upper Cretaceous flint develop use–wear traces in varying forms and speeds in comparable tests, a series of experiments were conducted. Beach pebble flint collected from the north coast of Ireland was knapped and the fracture surface of a flake from each of two nodules was etched in 40% HF at room temperature for one to three minutes to reveal the microstructure of the flint.

Nodule A is a homogeneous pale pink matt flint which etched quite readily to reveal an open lepispheric framework with very few skeletal fragments (Fig. 10.5). The microstructure is very similar to that seen in etched samples of Brandon flint and the lack of recrystallisation indicates the continued presence of chalcedony mineral with its structurally bound water, although the pale colour suggests small amounts of recrystallisation in the chalcedony. It is expected that microwear on this flint will develop in a similar way to that on Brandon flint because both lack extensive recrystallisation and have a similar density of lepispheres in the structure.

Nodule B, which macroscopically is a light grey matt flint, etched more slowly. It is seen to have a coalesced lepispheric network structure with much evidence of later recrystallisation (Fig. 10.6). It contains many foraminiferal fossil fragments and sponge material and also occasional detrital quartz sand grains. The recrystallisation which has occurred has increased the proportion of well-crystallised quartz and fluid inclusions in the structure and it is anticipated that microwear features will form more slowly on this type of flint, as the quartz is less prone to mechanical and chemical attack than the disordered chalcedony.

Keeley & Newcomer (1977, p. 39), in their experiments using flakes of flint on wood, noted that since the polish developed on the elevations of the microtopography first, the original texture of the flint must affect its gross appearance. They also noted that the polished patches which develop occurred on 'domed' high points of the microtopography and our observations have shown that these represent the polished surfaces of upstanding lepispheres. A surface will appear polished when the distances between the irregularities on its surface are less than the

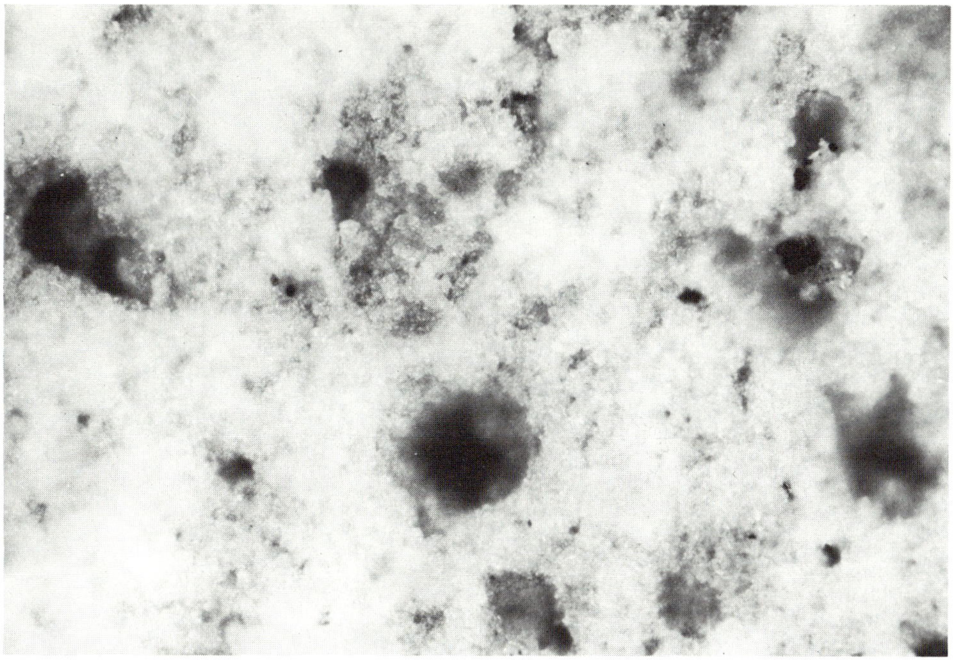

Fig. 10.5. Acid etched surface of nodule A with an open lepispheric framework (×200).

wavelength of the incident light. Therefore as wear occurs during the movement of a flint tool through wood the small quartz plates composing the upstanding lepispheres will be smoothed and deposition may occur to infill the crevices between the asperities. Since wood is not a very yielding material and the lepispheres stand proud of the surface they will sustain the greater part of the initial damage. The production of wood polish contains a strong chemical as well as a mechanical element so its development is relatively rapid and the polished patches on adjoining lepispheres soon coalesce to give the characteristic reticular aspect to the wood polish. Later, sheer polish areas may form as the process continues. The speed at which the polish development occurs depends on how contiguous are the lepispheres and on the state of recrystallisation of the raw material.

A flake each from Nodule A and B were used to cut a piece of fresh wood for a quarter of an hour and after identical cleaning were observed under a Leitz Epivert microscope using Ortholux variable darkfield objectives with magnifications between 50× and 525×. Microphotography on this system proved poor so the photographs presented here were taken using a Nikon microscope.

A fairly extensive, well-developed, characteristic wood polish formed on the flake from nodule A with, in places, isolated polishing of the tops of the lepispheres (Fig. 10.7). Elsewhere more extensive reticular polish developed as the matrix between the lepispheres was polished, with 'craters' in between reflecting the deep negative impressions of lepispheres. The polished surfaces have a fluid aspect with broad, even troughs. This fluidity may indicate areas of silica gel deposition, derived either from local dissolution of the flint or from amorphous silica in the wood itself. In contrast, the surface of the used flake from nodule B was seen to have a very poor polish development (Fig. 10.8). Some of the upstanding lepispheres have been rounded and polished but the wear appears to be much more restricted and slower in forming. Since other factors were kept constant, the variation must be due to the differences in the microstructure of the two flints as revealed by etching.

A similar difference was seen when soaked antler was scraped for 20 minutes with a piece of each flint type. The flake from nodule A developed an extensive polish which has a harder, less fluid aspect than wood polish (Fig. 10.9). It has formed in a reticular fashion and the characteristic pock-marks left between the sheets of polish appear to be negative impressions which lie below the level of polishing and so have remained grainy and matt. The flake from nodule B has an antler polish which, while being characteristic for this material (Fig. 10.10), is considerably less well developed than on the other flake. Wear occurs particularly on the tops of the lepispheres but there is a restriction to its development into sheets. While variations in the angle at which the flake was held to the antler could account for some of the difference, again it appears that the more resistant recrystallised flint matrix was slower to polish.

Finally, the third example illustrates the same phenomenon in the comparison of two flakes, one from each nodule, which cut wet sea-shell for 20 minutes. The surface of the piece from nodule A shows the attributes of shell polish, that is, a very bright, metallic, restricted polish which is heavily sleeked in the direction of use (Fig. 10.11). It appears to form as pedestalled areas surrounded by a surface which has a faint fibrous lustre. The equivalent surface on the flake from nodule B shows smaller, less extensive areas of a similar pedestalled polish (Fig. 10.12) and this variation may be related to the distribution of the lepispheres and fossils in the flint. In addition the flake from nodule B has suffered more edge microscarring, a fact which would be expected from a knowledge of the structure of the flint, and this may have removed some of the polish patches as they developed. Shell is a hard material and it is evident that it flattens upstanding quartz lepispheres and fossils without encroaching on the surrounding matrix or rounding the edges of the polish patches. The differences between the polishes developed on the two types of flint should be less than those seen in the case of wood and antler, where more of the matrix was affected. It is also likely that the redeposition of silica to infill surrounding

Fig. 10.6. Acid etched surface of nodule B with a coalesced lepispheric network (×200).

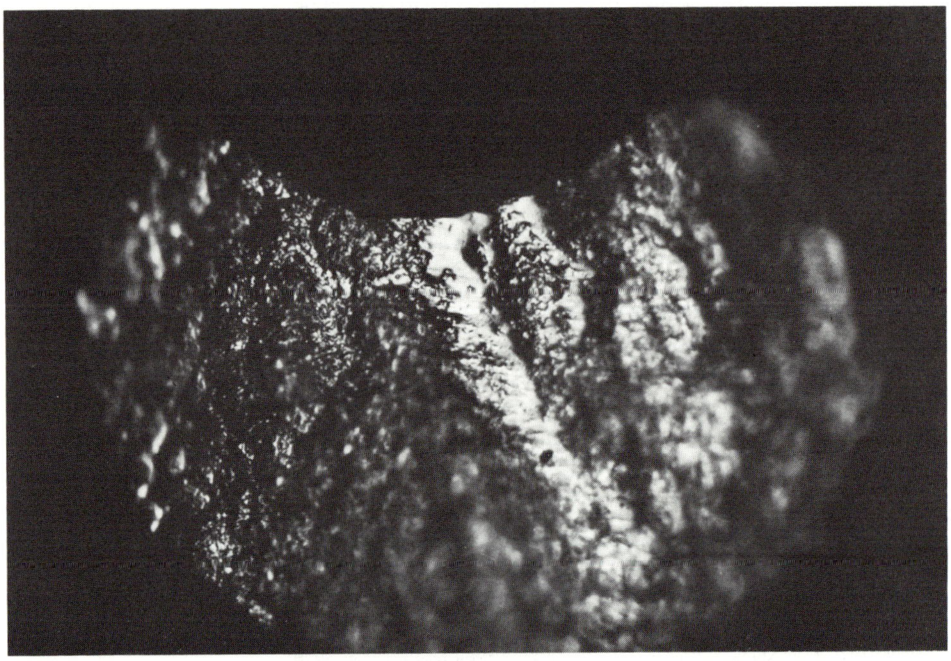

Fig. 10.7. Moderately well-developed sheet of wood polish on tool from nodule A used for 15 minutes to cut fresh wood (×200).

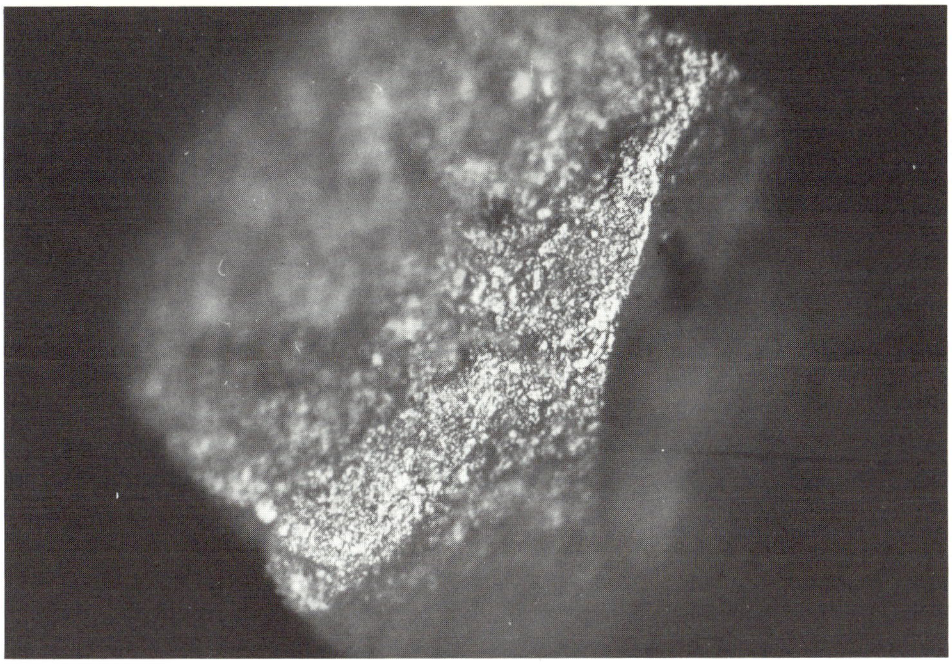

Fig. 10.8. Spots of poorly developed wood polish on tool from nodule B used for 15 minutes to cut fresh wood (×200).

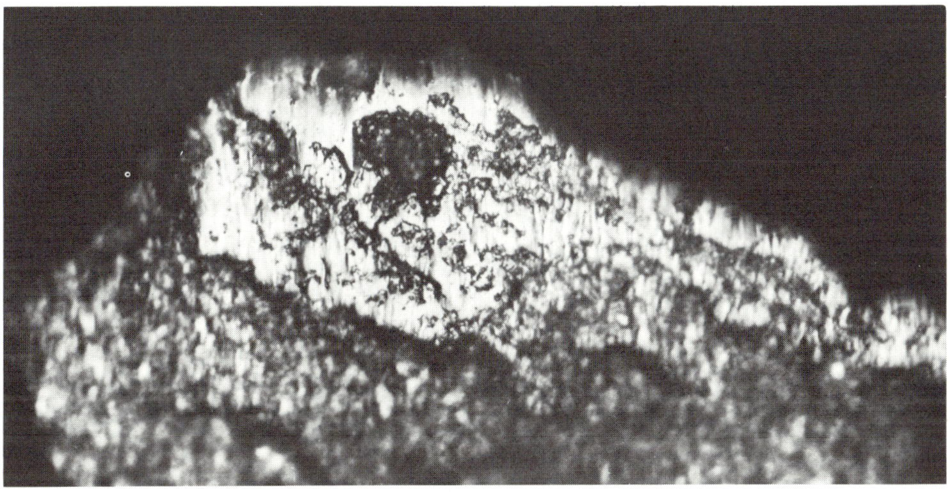

Fig. 10.9. Extensive sheet of polish produced on the tool from nodule A by scraping soaked antler for 20 minutes (×200).

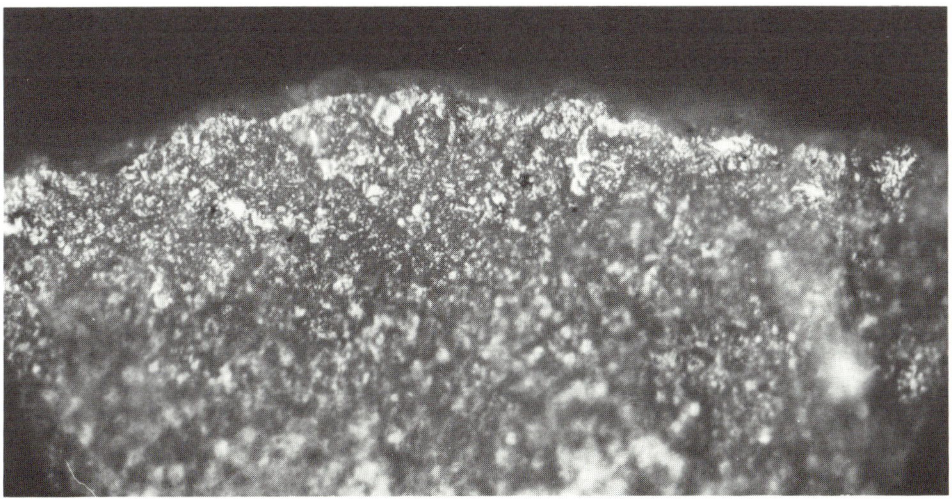

Fig. 10.10. Restricted development of polish on tool from nodule B produced by scraping soaked antler for 20 minutes (×200).

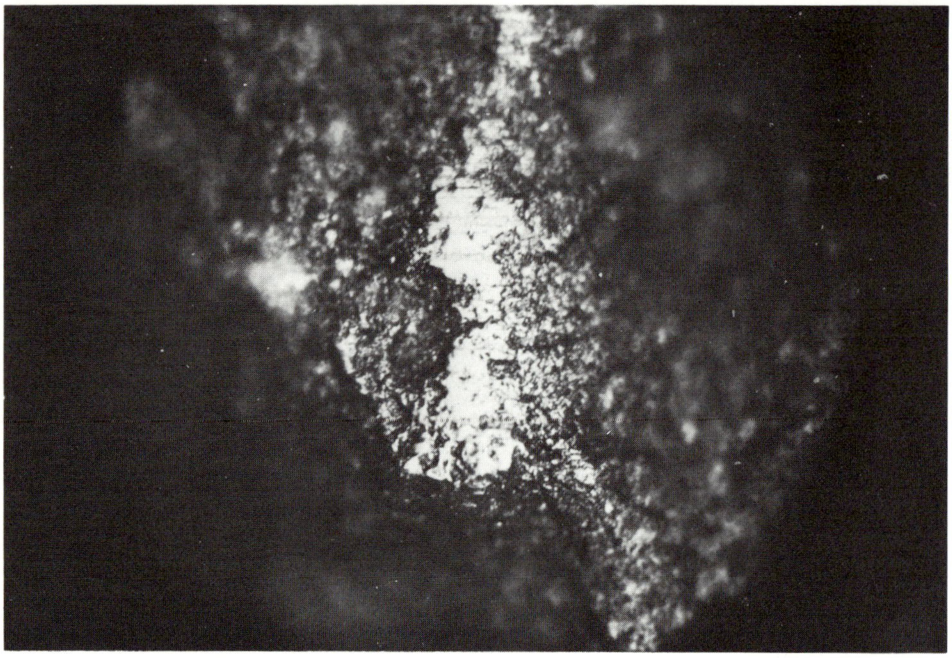

Fig. 10.11. Tool from nodule A with strongly sleeked quite extensive patch of wet shell polish from cutting for 20 minutes (×200).

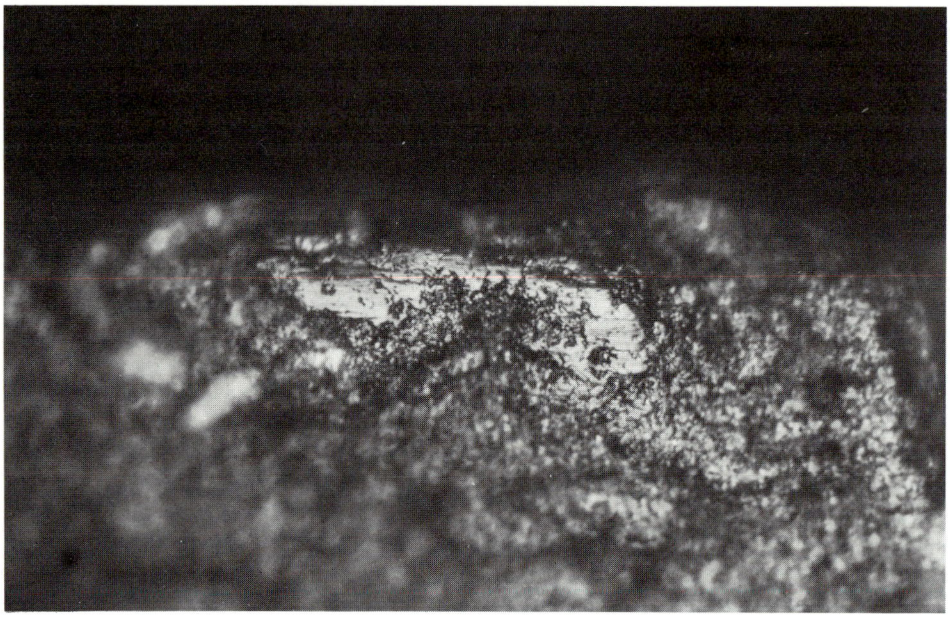

Fig. 10.12. Smaller more restricted polish on tool from nodule B produced by cutting wet shell for 20 minutes (×200).

irregularities is much less in this instance, as less silica would be in solution due to the limited area of contact between the hard shell and the flint.

Conclusion

This short paper has highlighted the importance of variations in raw material as one factor determining the differential formation of use–wear. An examination of the microstructure of Upper Cretaceous flint in etched specimens reveals it to contain a number of different silica types, not usually visible in thin section. An understanding of these has led to the formulation of a series of predictions relating to the behaviour of flint during fracture and microwear polish formation. Experimental evidence has shown that in the cases of wood, antler and shell working these hypotheses were confirmed. It seems that because unaltered Upper Cretaceous flint contains a lower proportion of well-crystallised quartz and a higher amount of chalcedony (structurally disordered fibrous silica) than recrystallised flint, it is more susceptible to both mechanical and chemical damage. The disposition of the resulting microwear traces relates largely to the material worked and the method of use, but they are also influenced by the structure and topography of the flint at a microscopic level. The general appearance of a piece of flint in the hand specimen alone is insufficient to differentiate between structural flint types, which influence microwear in different ways.

The demonstration that microwear polishes form at varying rates under the same conditions of use on flint of different structure has a number of implications for archaeologists. In the study of flint artefacts from assemblages of mixed geological types, comparisons between visually similar groups on the basis of use–wear must be made carefully. It is also possible that certain tools, especially if their utilisation was comparatively brief, and if they were made of recrystallised flint, will carry very slight traces of wear which could be overlooked or considered undiagnostic of a particular use, while their counterparts made of unaltered flint carry classic microwear traces. However, it is hoped that a comprehension of the geological aspects of the microstructure of flint will enable microwear researchers to predict more accurately the varying forms microwear traces may take and better understand anomalies which they observe. The authors remain optimistic.

Acknowledgements

We are indebted to the National Museum of Antiquities of Scotland Artefact Research Unit for the use of their facilities and in particular the Leitz Epivert Microscope and to the SEM Service Facility of the University of Edinburgh at King's Buildings. R.B. would also like to express her thanks to Hugues Plisson of the University of Paris 1 for his help in rephotographing the experimental pieces. Miss A. Grant-Suttie kindly typed the main draft of this manuscript at alarmingly short notice.

Bibliography

Anderson, P. C. (1980). 'A testimony of prehistoric tasks: diagnostic residues on stone tool working edges.' *World Archaeol.*, **12**(2), 181–94.

Brink, J. W. (1978). *An Experimental study of Microwear formation on Endscrapers*. Archaeol. Surv. Canada, Pap. 83, Ottawa.

Broadbent, N. D. & Knutsson, K. (1975). 'An Experimental analysis of quartz scrapers. Results and applications.' *Fornvannen*, **70**, 113–28.

Clayton, C. J. (1982). 'Growth history and microstructure of flint.' *IAS 3rd EUR. MTG. Copenhagen*, Abstr.

Clayton, C. J. (1983). Geochemistry of chert formation in Upper Cretaceous chalks. Unpubl. PhD Thesis University of London.

Greiser, S. T. & Sheets, P. D. (1979). 'Raw Materials as a functional variable in use–wear studies.' In: Hayden, B. (ed.), *Lithic Use–Wear Analysis*, London, pp. 289–96.

Jarvis, Ian, Gale, Andrew & Clayton, Chris (1982). 'Litho and biostratigraphical observations on the type sections of the Craie de Villedieu Formation Upper Cretaceous (Western France). *Newsl. Stratigr.*, **11**(2), 64–82.

Kamminga, J. (1978). Journey into the microcosms: A functional analysis of certain classes of prehistoric stone tools. PhD Thesis. Univ. Sydney.

Keeley, L. H. (1974). 'Technique and methodology in microwear studies: a critical review.' *World Archaeol.*, **5**, 323–36.

Keeley, L. H. (1978). 'Preliminary microwear analysis of the Meer assemblage.' In: van Noten, F. (ed.), *Les Chasseurs de Meer*, Brussels, pp. 73–86. (=Dissertationes Archaeologicae Gandenses 18).

Keeley, L. H. (1980). *Experimental Determination of Stone tool uses: A Microwear Analysis*, London.

Keeley, L. H. & Newcomer, M. H. (1977). 'Microwear analysis of experimental flint tools: a test case.' *J. Archaeol. Sci.*, **4**, 29–62.

Kennedy, G. L. (1950). 'A Portion of the system silica–water.' *Econ. Geol.*, **45**, 629–53.

Odell, G. H. (1977). The application of microwear analysis to the lithic component of an entire prehistoric settlement: methods, problems and functional reconstructions. PhD Thesis, Univ. Harvard.

Odell, G. H., Tringham, R., Roberts, M., Voytek, B. & Whitman, A. (1976). 'Microwear analysis.' *J. Fld. Archaeol.*, **3**, 239–40.

Seitzer, D. J. (1978). 'Forms vs function: microwear analysis and its application to Upper Palaeolithic burins.' *J. Medd. Lunds. Univ. Hist. Mus.*, n ser. 2, pp. 5–20.

Semenov, S. A. (1964) *Prehistoric Technology. An experimental study of the oldest tools and artefacts from traces of manufacture and wear.* London.

Wyant, J. R. & Bayhem, F. E. (1976). 'An experimental study of lithic use–wear.' *Arizona State Mus. Archaeol. Ser.*, **10**, 231–43.

11

Problems encountered in a high-power microwear study of some Egyptian predynastic lithic artefacts

D. L. HOLMES

Introduction

Judging from some of the literature on microwear analysis of stone tools, it might be supposed that it is a fairly straightforward process to investigate use–wear on artefacts made on a variety of siliceous lithic materials, particularly anything which can be called chert or flint. It was with this kind of expectation that I started looking for use–wear traces (in the form of polishes and striations readily observable at high-magnification) on some Egyptian Predynastic lithic artefacts from excavations in the Nagada area near Luxor in Upper Egypt, organized by Fekri Hassan of Washington State University. The raw material in question here is flint, or chert, from the local Eocene limestone cliffs (Thebes Formation). To familiarize myself with wear traces on this Egyptian flint, I did a number of experiments using flakes and tools made of this material. Also, knowing that it might not always be possible to obtain Egyptian flint for experimental use, I made and used some pieces made on British Upper Cretaceous Chalk flint to see whether the results would be comparable. (Later, I also looked at some experimental tools on French Upper Cretaceous flint made and used by Mark Newcomer.)

Two main problems, however, became apparent. One was that, although the Predynastic pieces selected for examination appeared fresh, unpatinated, and in suitable condition for microwear analysis, many showed what I have termed post-depositional surface modification or '*pdsm*', which is generally very obscuring of any use–wear traces that might be present. What causes *pdsm* is unclear, but one, or a combination, of a natural chemical adherent, light wind gloss, or trample are likely possibilities. The other problem was that polish did not seem to form as readily on the experimental pieces of Egyptian flint as on those of European Upper Cretaceous flint. Moreover, although both lithic materials looked very fine-grained to the naked eye, and fairly similar at low magnification (50×), at high-magnification (400×) the surface microtopographies of the two kinds of flint were generally quite different from each other.

Some experimental observations

This section describes a preliminary study investigating the apparent difference in the development of wear traces on Egyptian flint compared with European Upper Cretaceous flint. In addition, some experimentally trampled pieces are briefly discussed.

I had started to build a reference collection of experimental pieces made on Egyptian flint, which I had used on a variety of different materials for various lengths of time. I had also used some pieces made on British Upper Cretaceous chalk flint. Finding that polish seemed to develop more readily on specimens made on the British flint, I attempted some paired experiments to verify this apparent difference. One piece of each pair made on one flint type would be used to work the same material for the same length of time and in the same way as an experimental specimen made on the other flint type. I also re-examined my experimental collection and tried to assess the amount of use–polish on each piece. On that basis, I tried to predict whether it might be possible to deduce that each specimen had been used, and if so, on what material, if each piece was imagined to be (a) experimental (but with unknown details), and (b) ancient (allowing for some minor *pdsm*). This sort of prediction may be criticized, and it may be considered desirable to have a blind test approach, where a specimen's use (or if used at all) is unknown to the microwear analyst. It is my view, however, that knowing a specimen was used will have led me to be a little over-optimistic about my predictions. This bias, if anything, should tend to decrease the apparent difference between the two varieties of flint.

Since my sample of experimental pieces made on British flint was small, Mark Newcomer loaned me some tools he had used which were made on French Upper Cretaceous flint. The French flint was found to be very similar to the British flint in terms of its polish development capacity, although its microtopography is generally slightly different from the British material. Overall, it is more similar to British flint than to Egyptian flint, and so the two are considered together as European flint.

Table 11.1 shows that if the experimental pieces made on Egyptian flint are imagined to be ancient artefacts, 57.8%

would be considered unused, compared with 15.4 per cent for the pieces made on the European flint. The difference between the totals for the experimental and the ancient predictions arises from the fact that it is easier to be certain that traces on experimental pieces are the result of use, than on ancient specimens, where trample or other *pdsm* have to be considered, and which can confuse any use–wear traces that might be present.

Table 11.2 shows the number of observations in each 'amount of polish' category. The categories themselves are crude and judged by eye, but the results do support the apparent difference between the two kinds of flint. Figures 11.1 and 11.2 indicate what is meant by the terms diffuse and lightly developed, etc. It may be seen that the pieces made on Egyptian flint more often show no, or very rare, traces of use (23.4%) compared with the pieces on European flint (3.8%). Conversely, for developed traces there is a greater percentage for the European flint than for the Egyptian. The results may be limited quantitatively, but they do confirm the impression of a difference between the raw materials.

Considering the paired experiments (see Tables 11.3 and 11.4) the results again show a difference between the flint types. Figures 11.3 and 11.4 indicate the kind of difference that can result when artefacts made on different flints are used on the same material for the same amount of time and in the same manner.

The reason for the difference between the raw materials appears to be indicated by the surface microtopographies. At high-magnification (400×), it seems that Egyptian flint (Figure 11.5) is slightly coarser and different in microstructure from European flint (Figure 11.6).

When in the field in Egypt during the summer of 1981, I conducted a trampling experiment. I put down 30 flakes

Table 11.1. *Predicted conclusions if some used experimental specimens are imagined to be (a) experimental (but with unknown details), and (b) ancient*

	Flint							
	Egyptian				European			
Prediction	(a) Frequency	%	(b) Frequency	%	(a) Frequency	%	(b) Frequency	%
used	30	46.9	15	23.4	19	73.1	12	46.2
uncertain	14	21.9	12	18.8	5	19.2	10	38.5
unused	20	31.3	37	57.8	2	7.7	4	15.4
	64	100.1	64	100.0	26	100.0	26	100.1

Table 11.2 *Frequency of observations in each 'amount of polish' category for experimentally used pieces*

	Flint type			
	Egyptian		European	
Amount of polish	Frequency	%	Frequency	%
none, rare	15	23.4	1	3.8
diffuse	44	68.8	16	61.5
light–medium development	4	6.3	7	26.9
abundant development	1	1.6	2	7.7
	64	100.1	26	99.9

Table 11.3. *Predicted conclusions if the paired experimental specimens are imagined to be (a) experimental (but with unknown details), and (b) ancient*

	Flint							
	Egyptian				European			
Prediction	(a) Frequency	%	(b) Frequency	%	(a) Frequency	%	(b) Frequency	%
used	4	33.3	4	33.3	8	66.7	5	41.7
uncertain	2	16.7	1	8.3	3	25.0	6	50.0
unused	6	50.0	7	58.3	1	8.3	1	8.3
	12	100.0	12	99.9	12	100.0	12	100.0

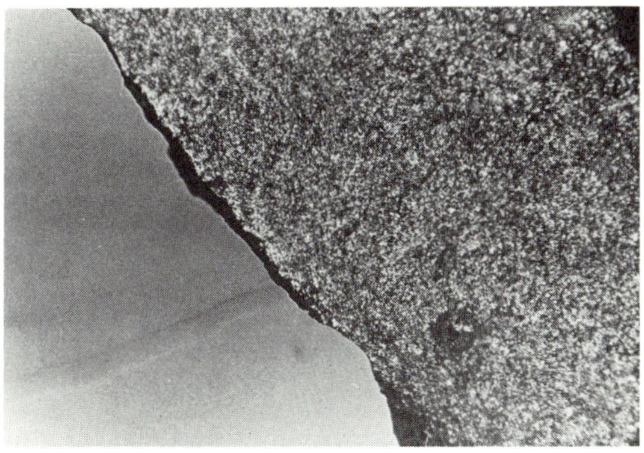

Fig. 11.1. A minor diffuse polish at the edge of a scraper used to scrape a hard wood for 45 minutes (50×), Egyptian flint.

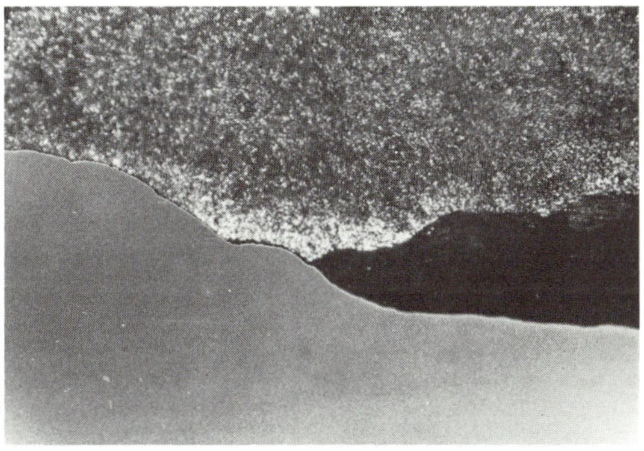

Fig. 11.2. A light-to-medium developed polish at the edge of an unretouched flake used to cut bone for nine minutes (50×), British flint.

Fig. 11.3. Maximal traces on a scraper made on Egyptian flint used to scrape a hard wood for 105 minutes (200×). Note: polish limited to edge.

Fig. 11.4. Maximal traces on a scraper made on British chalk flint used to scrape a hard wood for 105 minutes (same experimental conditions as for scraper in Fig. 11.3) (200×). Note: polish extends back from the edge.

Fig. 11.5. Microtopography of fresh, unused flint surface, Egyptian flint (400×).

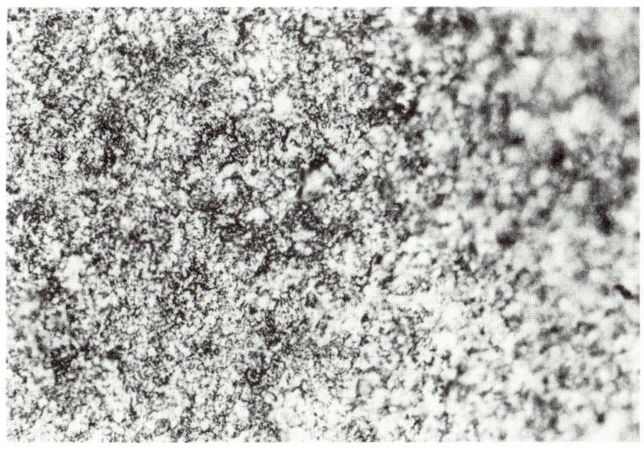

Fig. 11.6. Microtopography of fresh, unused flint surface, British chalk flint (400×).

Table 11.4 *Frequency of observations in each 'amount of polish' category for the paired experimental specimens*

	Flint type			
	Egyptian		European	
Amount of polish	Frequency	%	Frequency	%
none, rare	5	41.7	1	8.3
diffuse	6	50.0	8	66.7
light–medium development	1	8.3	3	25.0
abundant development	0	0.0	0	0.0
	12	100.0	12	100.0

made on the local flint on the desert scree surface (consisting mostly of sand with some stones) in a metre square area in front of the camp's mess tent entrance for two weeks. In that period, about 15 people would go in and out of the tent at least five times a day (for meals and tea-breaks). The flakes were placed so that they were not in contact with each other. Once they had been mapped the string used to mark the metre square area was removed so that there was nothing to remind anyone that the flakes were on the ground, and since the flint was a beige colour the flakes blended in with the ground surface.

Microwear analysis revealed traces which often appear similar to presumed non-use traces, or *pdsm*, on archaeological specimens, particularly diffuse polish as in Figure 11.7, and some striations, e.g. Figure 11.8. Other trampling traces comprise 'asterisks' as in Figure 11.9, and 'Stapert polish' (Moss, pers. comm; see Figure 11.10).

Observations on some predynastic lithic artefacts

My study of microwear traces on Predynastic artefacts began by looking for used flakes among samples of unretouched flakes from a settlement site in the Nagada area, but without much success. Some 50×50 cm areas of 10–20 cm depth were left for me to excavate, using a soft brush for fear that screening might cause confusing damage

Fig. 11.7. Diffuse polish on an experimentally trampled flake (50×).

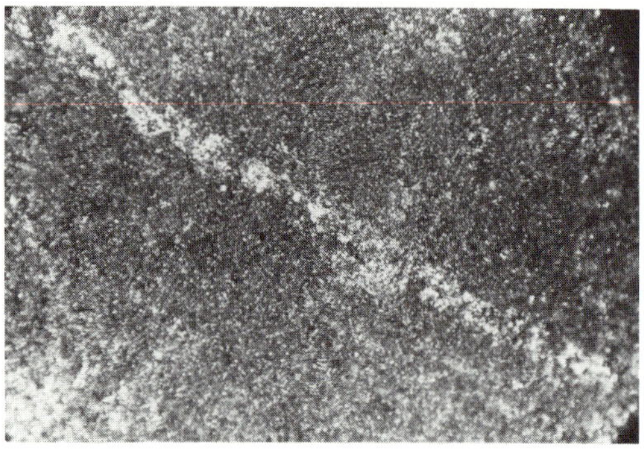

Fig. 11.8. Striation on an experimentally trampled flake (50×).

Fig. 11.9. 'Asterisk' on an experimentally trampled flake (200×).

Fig. 11.10. 'Stapert polish' on an experimentally trampled flake (200×).

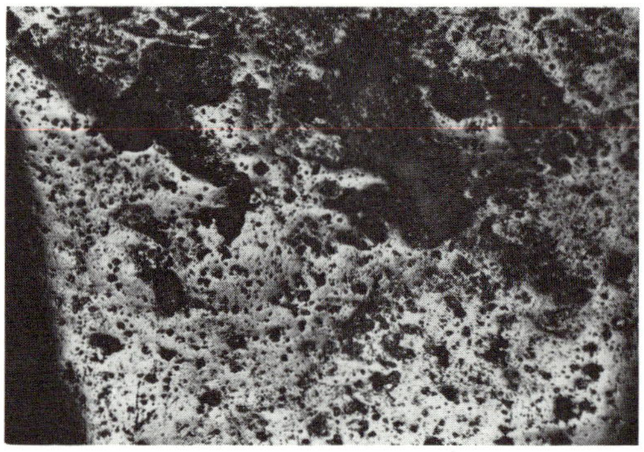

Fig. 11.11. Sickle gloss on the tooth of a Predynastic sickle blade from South Town, Nagada (100×).

Fig. 11.12. Hide (?) polish on a scraper from North Town, Nagada (50×).

on artefacts intended for microwear analysis. (Normally, all artefacts were recovered by screening the sediment.) As I uncovered each potential microwear specimen, I bagged it separately. Such a careful excavation technique, however, seems to have been unnecessary. I put some experimental flakes into a metal bucket containing some desert scree sediment, tipped the contents of the bucket into a sieve with a metal mesh and sieved with a vigorous side-to-side motion. I then bagged each flake separately. Yet such treatment seems to have caused no traces in most cases and only very minor traces in others. Traces caused by artefacts being bagged together, however, seem to be more important and cause for more caution.

More recently, I have examined some more formal tools from various settlement sites in the Nagada area. On the sickle blades that I have looked at, there was no trouble in seeing use–polish (Figure 11.11). Problems arose, however, when I looked at a scraper. Though appearing to the naked eye in suitable condition for microwear analysis, I found it to be covered with *pdsm*, although it had probably been used. The use–polish was impossible to 'disentangle' from all the other traces. Another scraper had only minor *pdsm* and had use–wear traces which could be interpreted as probable hide polish (Figure 11.12). Two bifaces examined had, in one case, no *pdsm* and, in the other, minor *pdsm*, and both had probable wood polish. One had traces at two ends, and the other, at one end. A motion perpendicular to the utilized edges was indicated in each case.

Discussion and conclusion

This preliminary study indicates that microwear analysis has problems which are of a nature generally not discussed in microwear literature.

It may be noted, however, that other microwear analysts have found differences in the rate of polish development due to slight differences in the nature of the raw material. Unger-Hamilton, cutting reeds with blades hafted in a sickle mount, found it took 200 strokes to get a visible-to-the-naked-eye polish with blades made on fine-grained British chalk flint from Brandon, Suffolk, compared with 2500 strokes and 5000 strokes with blades made on relatively medium- and coarse-grained Syrian flint, respectively (Unger-Hamilton, pers. comm.). Gysels & Cahen (1982, p. 223) also note that polish develops most rapidly on fine-grained flint rather than coarser varieties. Vaughan (1981, p. 90) and Sala (1982, p. 21) too have found that grain size affects polish development, and Beyries (1982) has found that polish varies with flint type as well as rock type, having used pieces made on various French flints, quartzite and an igneous rock type. These and the results described in this paper, indicate that microwear analysts need to exercise caution with different flint types. An experimental programme to see how time of use correlates with amount of polish and flint type would be of great interest. A mineralogical investigation to find how the structures of different flint types vary and how they affect polish development would also prove valuable. My initial impressions, however, have been that, despite differences in the rate of polish development with different kinds of flint, the polishes within each polish category (bone, wood, etc.) are, broadly speaking, similar when well developed. Vaughan (1981, p. 90) also seems to have found this.

My comment on the investigation of use–wear traces on archaeological specimens made on Egyptian flint is that it should be undertaken with much caution and patience. Only pieces that under the microscope (at 50×) show, at most, only minor *pdsm* should be considered for analysis. Also, it is probably advisable to look at formal tools only. One should not expect to find evidence of casual, short period use on artefacts such as unmodified flakes. The proportion of used flakes in a sample of unretouched flakes is quite likely to be small, and if they have only been used for a short time, say less than 30 minutes, then many, or even all, may not have sufficient traces for the microwear analyst to diagnose those pieces which were actually used, as used. Also, experiments suggest that flakes usually have shorter efficient lifetimes than many formal tools. A flake may be blunt after a few minutes cutting wood, but I have used a scraper on wood for a period of 105 minutes, and found it to be still as efficient at the end of the experiment as it was at the beginning. Formal tools, i.e., pieces that have been deliberately modified by retouch, however, may be expected to have been used and probably for long periods of time, since it may be assumed that people do not make tools if they have no function; and a tool may be used until, at least, a job is completed, or the tool is no longer functional. Exceptions to this generalization may be tools such as backed bladelets, where the functional part is an unretouched edge, or arrowheads used once, penetrating their quarry in approximately one second. The predominant tool types on the Nagada sites, however, are scrapers, notches, denticulates, burins and bifaces (Hassan, Hays, Holmes & Demuynck, in preparation; Hassan & Holmes, in preparation) which can remain functional for long periods.

Acknowledgements

I wish to thank Dr Mark Newcomer of the Institute of Archaeology, London and Dr Fekri Hassan of Washington State University, USA for their help and advice in my research. The Nagada Project is supported by grants to Fekri Hassan from the National Science Foundation and the Smithsonian Institution, USA.

Bibliography

Beyries, S. (1982). 'Comparaison de traces d'utilisation sur différentes roches siliceuses.' *Studia Praehistorica Belgica*, **2**, 235–40.

Gysels, J. & Cahen, D. (1982). 'Le lustre des faucilles et les autres traces d'usage des outils en silex.' *Bulletin de la Société Préhistorique Française*, **79**, 221–4.

Hassan, F. A., Hays, T. R., Holmes, D. L. & Demuynck, M. A. 'Lithic artefacts of the Nagada Predynastic sites: I, typological, attribute, and morphometric analyses of non-bifacial tools and debitage.' In: F. A. Hassan, ed., *Prelude to Civilization – the Predynastic of the Nagada Region*, in preparation.

Hassan, F. A. & Holmes, D. L. 'The flint "axes" of Nagada.' In: F. A. Hassan, ed., *Prelude to Civilization – the Predynastic of the Nagada Region*, in preparation.

Sala, I. (1982). Carinated tools or cores: a microwear study of material from Ksar Akil, unpublished BA thesis, Institute of Archaeology, University of London.

Vaughan, P. (1981). 'Microwear analysis of experimental flint and obsidian tools.' In: F. H. G. Engelen, ed., Third International Symposium on Flint, *Staringia*, **6**, 90–1.

12

Mount Sandel microwear: a preliminary report

JOHN V. DUMONT

Introduction

This report presents the conclusions concerning the function of particular artefact types derived via the use of the medium magnification microwear method (see Keeley, 1980; and also Cook & Dumont, this volume). No attempt has been made here either to formulate statements about the activities represented at the site by the use-episodes found on the tools, or to assess the typological composition of activity tool-kits (e.g. hideworking or woodworking tools). These topics, in addition to the supporting individual microwear interpretations, will be addressed in the author's dissertation (Dumont, 1985). The limits of space have prevented discussions of the methodology employed in this study and the theoretical foundation of the functional interpretations. With respect to the methodology, I have stressed the importance of the recognition, identification and distribution of microwear polishes and linear features, with the evidence of the edge-damage being utilised in only a very restricted manner (Dumont, 1985, and Cook & Dumont, this volume). The theoretical basis of the functional interpretations can be found in Dumont, 1982; 1985; and Cook & Dumont this volume.

A total of 273 artefacts were examined during this study. Eighty-four of these artefacts exhibited functional and/or potential hafting traces while 65 exhibited technological traces. The total number of artefacts with microwear traces (including functional, hafting and technological traces) is 127. The functional and potential hafting traces are summarily presented, in units of artefacts within each type, in Tables 12.1 and 12.2. Table 12.3 provides a summary of the functional interpretations relating to each type, though now the data is presented in units of use-episodes. Figures 12.1 through 12.8 are examples of the types of microwear polishes found on the Mount Sandel artefacts. With the exception of the type 'edge-damaged/retouched flakes and blades', I have followed Woodman's 1978 typology (Woodman, 1978, pp. 33–40).

The artefacts

Core axes, core axe resharpening flakes and flake axes

Twenty-eight resharpening flakes were examined and 14 exhibit functional and/or technological traces. Of the 14 flakes with traces, 12 exhibit functional traces while the remaining two appear to exhibit traces (stone-on-stone polish, linear arrangements of 'stone' polish, striations and scattered areas of wood polish) related to the initial manufacture of the axes. The functional traces consist of wood polish, hide polish, linear arrangements of wood polish, edge-damage, and edge–/*arête*-rounding. Wood polish was the most common trace, occurring on 11 of the 12 utilised flakes, followed by edge-damage (present on nine), linear arrangements of wood polish (on seven), striations (on four), and edge–/*arête*-rounding that was restricted to the only flake exhibiting hide polish. Of the 11 flakes with functional wood polish, seven (*ca.* 64%) also had linear arrangements of this trace and nine (*ca.* 82%) exhibited edge-damage. The presence of hide polish on the working edge of one resharpening flake (MU2124) is surprising. Of the three core axes examined, only one exhibited traces: wood polish, striations, linear arrangements of wood polish and edge-damage. Both ends of this core axe appear to have been utilised for woodworking.

All of the examined flake axes (six in number) exhibit traces that are considered to be related to their use. These traces include wood polish, linear arrangements of wood polish, striations and edge-damage. Although stone-on-stone polish was present on the surfaces of two axes, its distribution suggests a technological rather than functional origin. Two flake axes (MU2264 and MU2287) exhibit areas of wood polish across both surfaces (dorsal and ventral) in a distribution that is more easily explained by inferring the former presence of a wooden haft than by 'woodworking' (see Fig. 12.9). Of the functionally significant traces, wood polish was the most common and occurred at the working edge of each of the six axes. Edge-damage, of an apparent functional origin and primarily consisting of snap-like fractures (Keeley's half-moon breakages (1980, p. 25)), was also present at the working edges of at least four of the axes. Linear features were surprisingly uncommon and occurred on only two of the six axes. Only one axe exhibited linear polish arrangements, while two exhibited striations. This frequency of occurrence is less than half (*ca.* 44%) of the frequency of linear features found on the demonstrably utilised core axe resharpening flakes: eight of the 11 resharpening flakes used on wood (*ca.* 74%) exhibited striations and/or linear polish arrangements. This difference in frequency is only apparent at the typological level of analysis (i.e. resharpening flakes *vs* flake axes); it disappears when one examines the frequency of linear features with respect to the manner of tool use: seven of the working edges used for chopping wood exhibit linear features, while five of those used for planing/adzing wood also exhibit linear features. The difference in frequency with respect to each tool type is likely to be merely an artefact of the small size of the examined sample.

A total of 21 distinct use-episodes have been identified on 20 working edges of the 19 demonstrably utilised core-axes, resharpening flakes and flake axes. With the exception of a single use-episode on a resharpening flake, all of the functions are related to woodworking; the exception, MU2124, was used to work hide. Two principal manners of tool use have been identified on the basis of the distribution and orientation of the available traces: chopping and planing/adzing. The distinction between the two manners of use is one of tool orientation and the nature of the tool–wood contact. Chopping involves a low to oblique orientation of

Table 12.1. *Summary of the recognised functional and technological microwear traces per artefact type: in units of artefacts*
Key: s-o-s, stone-on-stone polish; ?, unidentified polishes or deposits; LAP, linear arrangements of polish or deposits; IF, impact fractures; –, not applicable.

Artefact type	Polishes or deposits						Linear traces					Edge alterations		
Class of trace	wood	bone	antler	hide	meat	functional haematite deposits	s-o-s	?	LAP	striae	damage*	rounding	crushing	IF
Core axe resharpening flakes														
functional	11	0	0	1	0	0	0	0	7	4	9	1	0	0
technological	0	0	0	0	0	–	1	0	1	1	0	0	0	0
Core axes														
functional	1	0	0	0	0	0	0	0	1	1	1	0	0	0
Flake axes														
functional	6	0	0	0	0	0	0	0	1	2	4	0	0	0
technological	0	0	0	0	0	–	2	0	0	0	0	0	0	0
Borers														
functional	1	3	0	2	0	1	0	2	1	6	4	5	4	0
technological	0	0	0	0	0	–	3	0	3	0	0	0	0	0
Edge-damaged/retouched flakes and blades														
functional	2	4	0	0	2	12	0	0	3	10	17	10	0	0
Variously backed blades														
functional	0	1	0	1	0	2	0	1	0	4	4	3	0	0
technological	0	0	0	0	0	–	1	0	0	0	0	0	0	0
Truncated blades														
functional	1	0	1	1	0	0	0	1	1	3	3	0	0	0
technological	0	0	0	0	0	–	1	0	1	0	0	0	0	0
Scrapers														
functional	0	1	0	0	0	0	0	0	0	0	1	0	0	0
Picks														
technological	0	0	0	0	0	–	1	0	0	0	0	0	0	0
Burins														
technological	0	0	0	0	0	–	0	0	0	1	0	0	0	0
Needles														
functional	0	2	0	1	1	0	0	0	2	0	0	0	0	0
technological	0	0	0	0	0	–	1	0	0	0	0	0	0	0
Triangles														
functional	0	1	0	2	0	0	0	0	0	0	2	1	0	0
technological	0	0	0	0	0	–	3	0	0	0	0	0	0	0
Obliquely blunted points														
technological	0	0	0	0	0	0	3	0	0	0	0	0	0	0
Misc. microliths														
technological	0	0	0	0	0	0	2	0	0	0	0	0	0	0
Rods														
functional	0	2–4	0–1	2–3	0–1	0	1	0	3	2	11	2	0	3–5
technological	0	0	0	0	0	–	47	0	0	0	0	0	0	0

* only the edge-damage that can be confidently related to other microwear traces, on the affected edge, is listed here.

Table 12.2. *Summary of the potential hafting traces per artefact type: in units of artefacts*
Key: LAP, linear arrangements of polish

Artefact type	Polish or deposit: wood	bone	antler	unidentified deposits[1]	biased distribution of pedogenic iron[2]	Linear traces: LAP	striae
Flake axes	2	0	0	0	0	0	0
Borers	1	0	0	1	0	0	1
Picks	1	0	0	0	0	0	1
Needles	1	0	0	0	0	0	0
Rods	2	1–2	1	1	2	1	0

[1] Perhaps a hafting adhesive.
[2] Perhaps reflecting the distribution of a former adhesive.

first one, and then the opposing, edge-aspect with respect to the wood. The roughly alternate contact of the two edge-aspects at similar orientations should produce a roughly similar distribution of traces corresponding to the similar areas and direction of tool–wood contact on both aspects. Chopping also involves the use of impact; i.e. the edge strikes the wood after a swing of varying length and strength. Although the use of a haft would no doubt improve the efficacy of 'chopping', there is no *a priori* reason to assume the necessary presence of such an aid. The microwear analysis has, unfortunately, shed no light on the question of the hafting of core axes.

Planing and adzing have been conflated into a single category because of the difficulty in defining criteria that could confidently discriminate one action from the other on the basis of the distribution and orientation of the microwear traces. In both instances the tool is oriented at a low to oblique angle, against the wood, with respect to predominately a single edge-aspect. This orientation causes a greater area of surface contact of one aspect against the wood when compared to the opposing aspect. The distribution of the traces across both aspects reflects the differential tool–wood contact. The difference between adzing and planing rests on the initial contact or entry of the edge into the wood. Like chopping, adzing requires that the edge impacts the wood after a swing of varying length and strength. Planing has no such requirement; the edge can be pushed through the wood from an initial state of rest. In a sense, then, chopping and adzing are related in that the contact between the wood and the edge causes a rapid deceleration of the tool, while planing (also whittling and scraping) involves an acceleration of the edge or tool through the wood. In both cases a portion of the kinetic energy is dissipated by producing work against the wood, damage to the edge (including microwear polishes) and heat.

Of the 15 use-episodes on the core axe resharpening flakes and core axe, nine are 'chopping' (60%) while six are 'planing/adzing' (40%). All of the use-episodes represented on the flake axes, six in number, are of planing/adzing. With regard to the manner of tool use, all nine of the chopping use-episodes are associated with core axe resharpening flakes, while the 12 planing/adzing use-episodes are evenly divided between the group consisting of the core axe resharpening flakes (including the core axe) and the flake axes. In more simple terms, the core axes (represented by the resharpening flakes) were utilised for both chopping and planing/adzing, with chopping being exclusively associated with this type of axe; and although the flake axes were exclusively used for planing/adzing, this same function was also performed by the core axes (represented by the resharpening flakes). There is, then, a direct correlation between tool type and function: within the limits of this sample, all flake axes were used for planing/adzing while chopping was restricted to the core axes. It is also clear that the ancient tool users considered core axes suitable for planing/adzing. There are no significant differences in either the spine or edge angles of those edges used for chopping as opposed to those used for planing/adzing (see Dumont, 1985).

If the haft reconstructed for flake axe MU2287 is accurate (see Fig. 12.9) and common to all or most of the flake axes, then the proximity of the hafting sleeve to the working edge may well have made the hafted tool totally unsuitable for chopping. One could argue that the haft would impede the entry of the working edge into the wood and, in so doing, cause damage to the haft itself. If the core axes were hafted then it is likely that the haft would have been designed to avoid any interference with the worked material. A core axe that is suitable hafted for 'chopping' could, of course, be used for planing/adzing as well.

Borers

Thirty-two borers (microawls) were examined as part of this study. Of these 32 artefacts, 12 exhibited traces but only nine had functionally significant microwear. An additional two artefacts exhibited traces that may have been caused by hafting, while another merely exhibits technological traces. The possible hafting traces consist of wood polish and striations on MU109, and an unidentified deposit (perhaps a hafting adhesive) on MU(Y61). The technological traces consist of linear arrangements of stone-on-stone polish that occur at the edges of three artefacts where their orientation and position, on the edge-aspects opposite those that show retouch scars, strongly suggest that they were caused by hard hammer retouch (Keeley's hammerstone smears (1980, p. 28; see also Fig. 12.8). The functional traces consist of polishes, deposits, linear arrangements of deposits, striations, edge–/*arête*-rounding and edge-damage/crushing. Of the polishes, bone polish was the most common and occurred on the working tips of three borers. Wood polish was present on two borers though only one is thought to be derived from the use of the tool; the polish on the other (MU109) is more likely the result of hafting (see the discussion of MU109 in Dumont, 1985). Hide polish was also observed on the working tips of two borers. Two more artefacts exhibit haematite deposits that by their distribution and form suggest an origin other than through natural causes. Striations were found on six artefacts in apparent association with wood polish (two instances), hide polish (one instance), bone polish (one instance), linear

Table 12.3. *Summary of the functional interpretations per artefact type: in units of recognised use-episodes*

Artefact type	Manner of use									
Worked material	chopping	planing/ adzing	scraping/ planing	boring or piercing	sawing/ cutting	whittling	projectile components	unidentified	totals[2]	totals[3]
Core axe resharpening flakes										
wood	9	3	0	0	0	0	0	0	12	
hide	0	1 or	1	0	0	0	0	0	1	13
Core axes										
wood	0	2	0	0	0	0	0	0	2	2
Flake axes										
wood	0	6	0	0	0	0	0	0	6	6
Borers										
bone	0	0	0	3	0	0	0	0	3	
hide	0	0	0	2	0	0	0	0	2	
wood	0	0	0	1	0	0	0	0	1	
haematite[1]	0	0	0	1	0	0	0	1	2	
unidentified	0	0	0	1	0	0	0	0	1	9
Edge-damaged/retouched flakes and blades										
wood	0	0	0	0	1	1	0	0	2	
bone	0	0	2	0	0	0	0	0	2	
meat	0	0	0	0	2	0	0	0	2	
haematite[1]	0	0	19–20	1	6–7	0	0	6	33	
unidentified	0	0	0–1	0	0–1	0	0	1	2	41
Variously backed blades										
bone	0	0	1	0	0	0	0	0	1	
hide	0	0	1	0	0	0	0	0	1	
hide and haematite	0	0	0	0	1	0	0	0	1	
haematite[1]	0	0	1	0	0	0	0	0	1	
unidentified	0	0	0	0	1	0	0	0	1	5
Truncated blades										
wood	0	0	0	0	0	1	0	0	1	
antler	0	0	1	0	0	0	0	0	1	
hide	0	0	1	0	0	0	0	0	1	
unidentified	0	0	1	0	0	0	0	0	1	4
Scrapers										
bone	0	0	1	0	0	0	0	0	1	1
Needles										
–	0	0	0	0	0	0	2	0	2	2
Triangles										
–	0	0	0	0	0	0	3	0	3	3
Rods										
stone	0	0	0	1	0	0	0	0	1	
hide	0	0	0	0	1	0	0	0	1	
wood	0	0	0	0	1	0	0	0	1	
–	0	0	0	0	0	0	10	0	10	13

[1] or a material bearing this mineral.
[2] per worked material per artefact type.
[3] all use-episodes per artefact type.
[2] nr=not recorded.

arrangements of haematite deposits (one instance) and with edge-damage/crushing and haematite deposits (one instance). Edge- and/or *arête*-rounding occurred with both hide (two instances) and wood (one instance) polishes. Rounding also affected the edges and/or *arêtes* of two other artefacts that exhibited no polish. Edge-damage or crushing was observed on the working tips of at least four and up to eight borers; the uncertainty being caused by the difficulty in resolving damage from retouch scars. Of the certain instances of edge-damage, only two were 'associated' with polish (bone). Three of the possible cases of edge-damage occurred with hide, bone and wood polishes. The remaining uncertain instance and the two certain instances occurred in 'association' with no polish.

Nine use-episodes have been identified on the nine utilised borers. By the evidence of the polishes, bone, wood, hide, haematite and an unidentified material were worked by the borers. Although the working of bone, wood and hide with this tool type is not surprising or counter-intuitive, the rationale of working haematite with a borer is not clear. It is of course possible that lumps of haematite were perforated for beads but such artefacts were not found at the site; hence the suggestion remains unsupported. It is perhaps more likely that at least one of the borers exhibiting

(*a*)

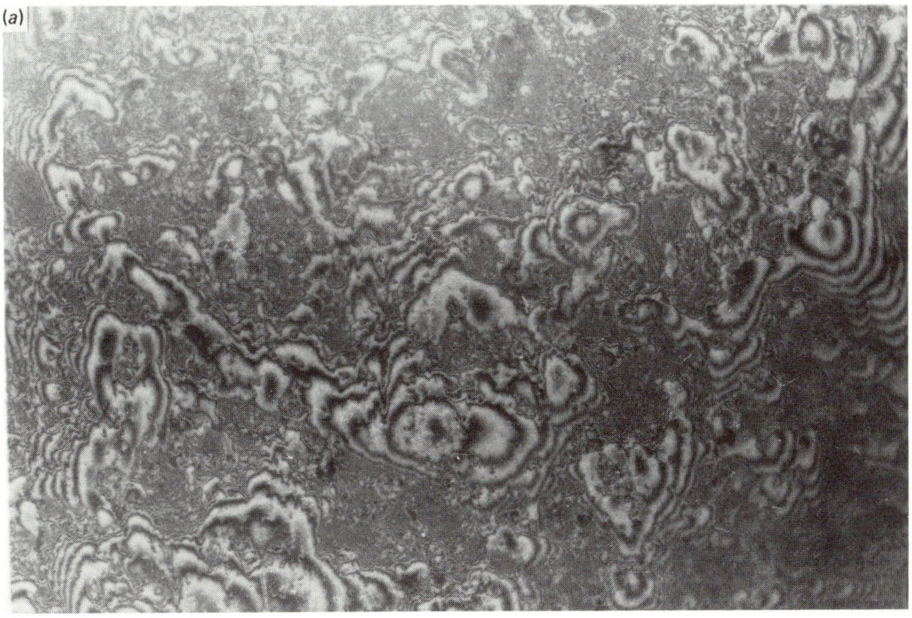

(*b*)

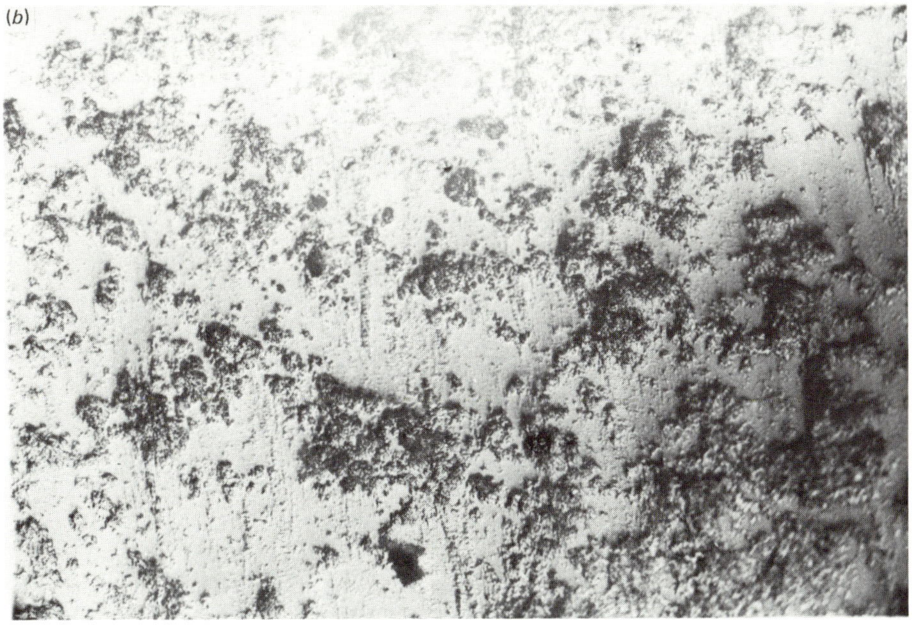

Fig. 12.1. (*a*) is the interference pattern produced by the wood polish shown in (*b*). A full description of these two plates can be found in Dumont (1982). Both photomicrographs were taken at a magnification of 200 diameters. Length of the photomicrographs: 0.566 mm.

haematite may have acquired this trace by working a material, such as hide, that had been treated with haematite for the purposes of colouration or preservation. In the absence of hides recovered from the site, this uncommon suggestion must also remain speculative. The working of the other materials (bone, hide and wood) is less difficult to assert in the absence of the final product.

With respect to the manners-of-use of the borers, two methods have been inferred from the orientation and distribution of the observed traces: boring and piercing (conflated into one category in Table 12.3). The manner-of-use of one artefact, though, remains entirely unknown. Boring, a rotary motion around the long axis of the tool, was the most common manner-of-use and is represented by six or seven of the nine use-episodes. Piercing, a motion parallel to the long axis of the tool, is represented by at least one of the use-episodes. The distribution of the traces has allowed a resolution of the rotary boring action into clock-

Fig. 12.2. An example of bone polish. The photomicrograph was taken at a magnification of 320 diameters. Length of the photomicrograph: 0.354 mm.

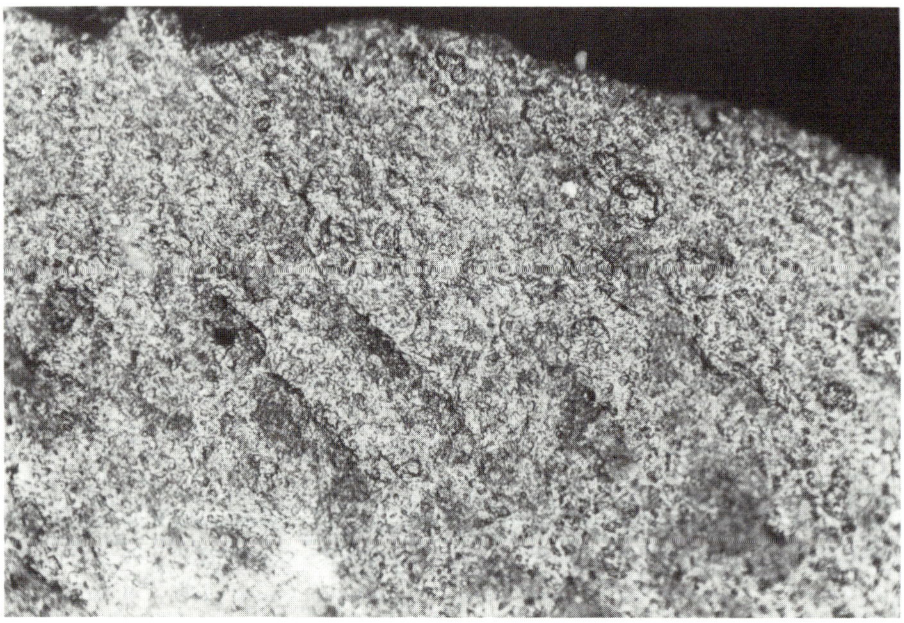

Fig. 12.3. An example of meat polish. The photomicrograph was taken at a magnification of 200 diameters. Length of the photomicrograph: 0.566 mm.

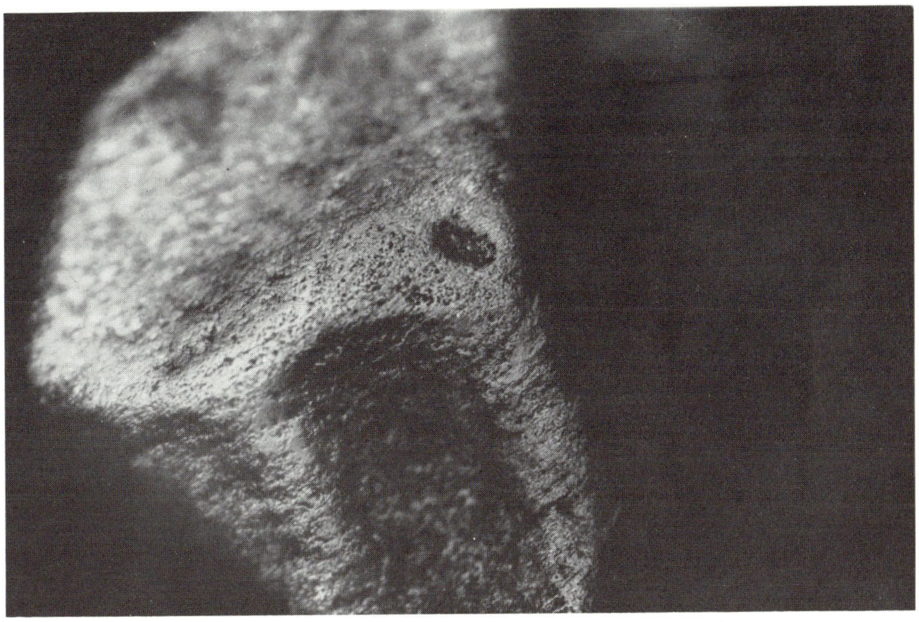

Fig. 12.4. An example of hide polish (possibly dry hide polish). Note the presence of extensive edge rounding and the micro-potlid fracture. The photomicrograph was taken at a magnification of 200 diameters. Length of the photomicrograph: 0.566 mm.

wise and anti-clockwise motions in 4 instances (clockwise: three examples; anti-clockwise: one example); the remaining three boring use-episodes are either undifferentiated or represent a reciprocal action.

The longitudinal extent of the traces back from the working tip of each borer and the greatest cross-sectional dimension of the tip at the limit of their extent provide an approximation of the size of the holes produced in the worked materials. The inferred diameter of the holes represents the minimum value of the true greatest diameter of each hole produced by the tool at the stated depth of penetration. The least greatest diameter of the holes, then, ranged from *ca.* 1.7 mm to 5.7 mm with a mean diameter of 3.84 mm (standard deviation: 1.26 mm).

The microwear analysis has not shed much light on the hafting of the Mount Sandel borers. Only two tools exhibit traces that may relate to the former presence of a haft but the evidence is far from compelling. On the one hand there is certainly no *a priori* reason to assume that the tools were necessarily hafted for they are quite easily manipulated by hand. On the other hand the absence of conclusive, and the paucity of suggestive, hafting traces does not, of course, indicate that hafts were not employed.

Edge-damaged/retouched flakes and blades. Thirty-one edge-damaged or casually retouched flakes and blades were examined as part of this study. A total of 39 utilised edges or edge-segments were identified on 20 of the artefacts. The microwear traces found on these artefacts consist of polishes, haematite deposits, linear arrangements of polish/deposits, striations, edge-/*arête*-rounding and edge-damage. Pedogenic iron compounds were also found on six of the artefacts exhibiting functional traces. For the purposes of the present discussion, the haematite deposits will be considered a 'polish' because although it is a deposit, its presence conveys the same sort of information as the polishes *per se*; i.e. the partial identity of the worked material. The polishes (including the haematite deposits) were the most common of the microwear traces and occurred on 38 of the 39 utilised edges/edge-segments (hereafter abbreviated to edge-segments). The vast majority of this group consists of edge-segments exhibiting the functional haematite deposits: 31 of the 38 edge-segments with polish showed this trace (*ca.* 82%).

Bone polish occurred on four of the edge-segments with one in apparent 'association' with meat polish. Meat polish also occurred on another edge-segment giving a total, then, of two occurrences. Wood polish was found on the remaining two edge-segments exhibiting polish. Edge-damage occurred on 36–38 of the utilised edge-segments. Of the 22 edge-segments exhibiting edge-/*arête*-rounding, all but one also exhibited functional haematite deposits; the exception being 'associated' with only edge-damage. Although virtually all of the rounding instances were 'associated' with the functional haematite deposits, only *ca.* 68% of the edge-segments exhibiting haematite (21 of the 31 segments) were affected by rounding. With respect to the edge-damaged/retouches flakes and blades, then, edge-/*arête*-rounding is virtually exclusively 'associated' with the working of haematite (or a material bearing this mineral) but it is not a necessary effect of such use.

Striations occurred on 15 edge-segments. Of these 15 occurrences, 12 were 'associated' with functional haematite deposits, one with wood polish, one with bone polish and the remaining instance on an edge with both bone and meat polishes. All of the edge-segments exhibiting both functional haematite deposits and striations were also affected by edge–/*arête*-rounding. Linear polish arrangements were only observed on three edge-segments: two of bone polish and one of haematite.

Forty-one use-episodes have been identified on the 39 edge-segments of the 20 utilised artefacts. The majority of these use-episodes (33 of the 41, *ca.* 80%) were concerned with the working of either haematite directly or the working of a material bearing this mineral. Of these 33 'haematite-working' use-episodes, 19 to 20 consisted of scraping/planing (*ca.* 58–61%), six to seven of sawing/cutting (*ca.* 18–21%), one of boring (*ca.* 3%) and six of an unidentified manner(s)-of-use (*ca.* 18%). The remaining eight use-episodes are evenly divided amongst wood-, bone- and meat-working, and an unidentified worked material. With respect to woodworking, sawing/cutting and planing/whittling are each represented by a single use-episode, while boneworking is represented by two scraping/planing epi-

Fig. 12.5. An example of antler polish. The photomicrograph was taken at a magnification of 200 diameters. Length of the photomicrograph: 0.566 mm.

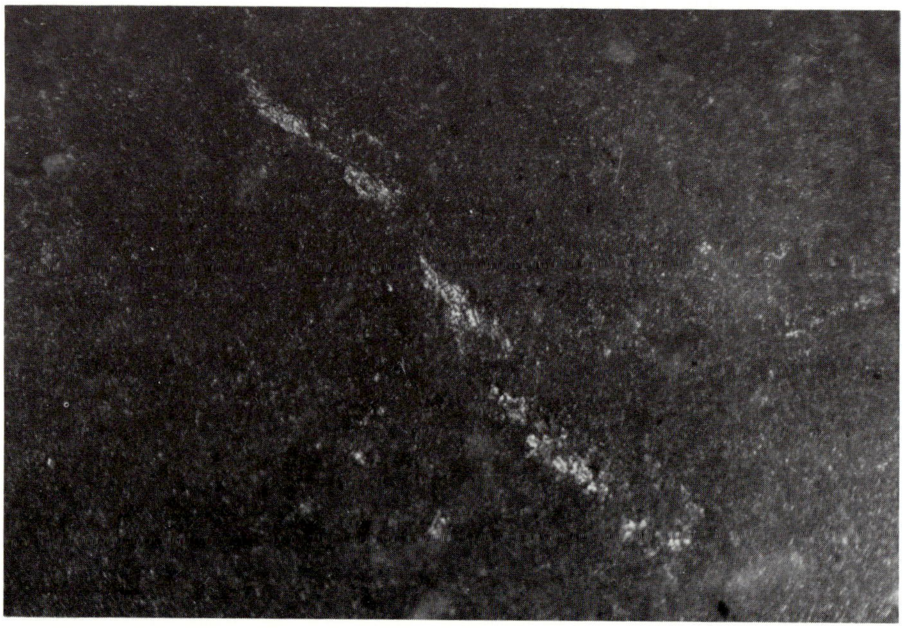

Fig. 12.6. A linear arrangement of wood polish. The photomicrograph was taken at a magnification of 100 diameters. Length of the photomicrograph: 1.15 mm.

sodes. Although meat-working or butchering is only represented by two episodes of sawing/cutting, I would not be surprised if this figure is under-representative of the true frequency of meat-cutting with these artefacts. With respect to the unidentified worked materials, one episode of either sawing/cutting or scraping/planing is represented while another completely unidentified manner-of-use is also represented. With respect to all the use-episodes, then, the frequency of the various functions can be ranked as follows: scraping/planing haematite (or a material bearing this mineral) (*ca.* 46–49%), sawing/cutting haematite (or a material bearing this mineral) (*ca.* 15–17%), unidentified uses against haematite (or a material bearing this mineral) (*ca.* 15%), scraping/planing bone (*ca.* 5%), sawing/cutting meat (*ca.* 5%), sawing/cutting wood (*ca.* 2.4%), planing/whittling wood (*ca.* 2.4%), boring haematite (or a material bearing this mineral (*ca.* 2.4%), either sawing/cutting or scraping/planing an unidentified material (*ca.* 2.4%) and, finally, an unidentified use against an 'unknown' material (*ca.* 2.4%). It is interesting to note that the majority of the use-episodes (21–23 of the 41, *ca.* 51–56%) were concerned with scraping/planing: a frequency that is perhaps unexpected of artefacts of this type.

In order to investigate the potential morphological–functional differences within this group of tools, I have divided the artefacts (or edge-segments) into several groups (see below) and compared the newly derived mean values of their metrical attributes (see Dumont, 1985). Potential differences were tested for significance through the use of the Fisher Exact Probability Test (FEPT) and the Mann–Whitney U-Test (MWUT). The groupings consist of (1) all of the utilised artefacts, (2) all of the artefacts interpreted to have been used against haematite or a material bearing this mineral, (3) all of the artefacts utilised on materials other than haematite, (4) all of the examined artefacts except those used on haematite, (5) all of the artefacts devoid of traces and, finally, (6) all of the examined artefacts. The attributes that were compared consist of: length, maximum width, maximum thickness, length/width ratio, thickness/width ratio, weight, and the spine and edge angles of the lateral edges. After applying the above statistical tests to the metrical attributes of the different groups, it became evident that significant differences only exist between the width, thickness and weight of 'all the utilised artefacts' *vs* 'the apparently non-utilised' groups of artefacts.

With respect to the width of these two groups, the demonstrably utilised artefacts tend to have higher values than those with no functional traces (mean width of 17.5 mm compared to 13.0 mm, respectively). The FEPT confirmed this apparent tendency by giving the result that the probability of this distinction being present within a single uniform population would be 2.51%: i.e. the probability of the presence of two distinct but overlapping groups is 97.49% ($p=0.0251$, $n=30+1$ median value). The MWUT provides a similar probability of greater than 97.83% ($n_1=20$, $n_2=11$, standard normal deviate (SND)=2.02). Similarly, the thickness of the demonstrably utilised artefacts is also greater than the apparently non-utilised artefacts (mean thickness of 5.3mm compared to 4.2mm, respectively). Both the FEPT and MWUT confirm that the difference is significant. The FEPT gives a probability of 2.75% that a single population is represented: i.e. the probability that two distinct though overlapping groups are actually represented is 97.25% ($p=0.0275$, $n=31$). The MWUT gives a similar probability of greater than 95.64%

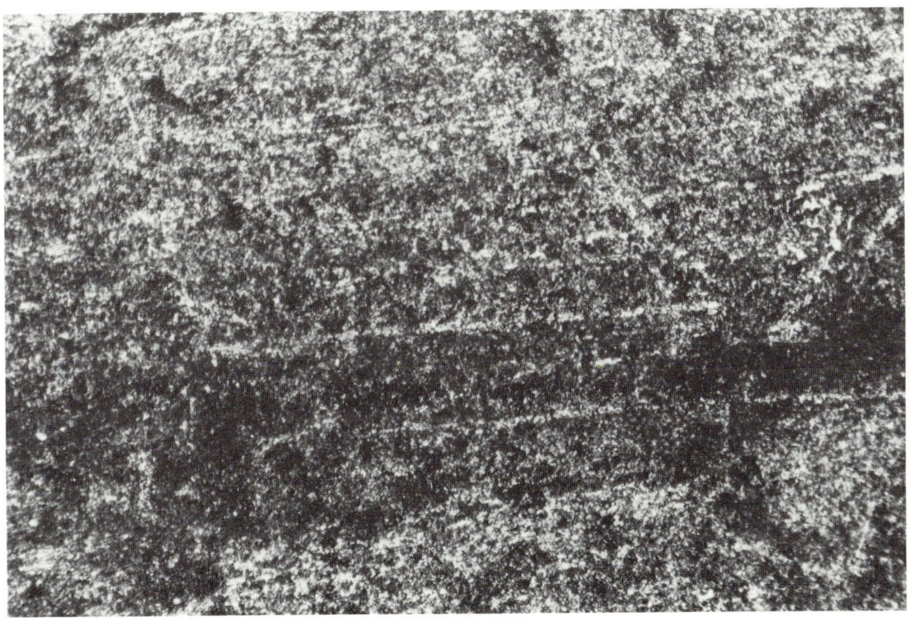

Fig. 12.7. Linear arrangements of haematite deposits near and parallel to the working edge of an edge-damaged/retouched blade. The photomicrograph was taken at a magnification of 100 diameters. Length of the photomicrograph: 1.15 mm.

(n_1=20, n_2=11, SND=1.71) for the presence of two distinct groups. With respect to the weight of the two groups, the demonstrably utilised artefacts tend to be heavier than those without functional traces (a mean weight of 5.3gr compared to 3.2gr, respectively). The FEPT and MWUT both confirm this apparent difference with the probabilities of 97.05% (FEPT: p=0.0295, n=31) and greater than 96.56% (MWUT: n_1=20, n_2=11, SND=1.82).

These results indicate that the ancient tool-users at Mount Sandel were selecting the wider, thicker and heavier blades and flakes, per given length, for use against the materials represented by the polishes and deposit. The absence of observable microwear traces, aside from edge-damage, on thc apparently non-utilised flakes and blades does not, of course, necessarily indicate that these artefacts were not utilised. All or some of the apparently non-utilised tools may have been used against materials, such as meat, that produce a slowly forming and unobtrusive polish or, perhaps, they may have been used against other materials for a time period too short to produce an observable polish. These speculative comments should not be read as an argument for the presence of unrepresented activities, but rather as a caution against the *a priori* dismissal of their possible former presence.

Variously backed blades

Eight variously backed blades (three naturally backed, two partially backed and three fully backed) were examined during this study. Five utilised edges were identified on four of the eight artefacts. The traces found on these tools consist of polishes/deposits, striations, edge–/*arête*-rounding, edge-damage and stone-on-stone polish. Polish/deposits were observed on all five identified working edges: two instances of hide polish (on a single artefact), two of functional haematite deposits, one of bone polish and one instance of an unidentified deposit. Striations were observed on four of the five working edges; the exception being the edge exhibiting hide polish. Edge–/*arête*-rounding also affected all but one of the working edges; no rounding was observed in 'association' with the bone polish. Edge-damage occurred on all five working edges while stone-on-stone polish was restricted to the edge exhibiting the unidentified deposit.

The five use-episodes represented on the five working edges were concerned with the working of hide, bone, haematite (or a material bearing this mineral) and an unidentified material. One artefact (MU2203) may well have been used to work haematite impregnated hide, thus giving rise to the observed juxtaposition of hide polish and haematite deposits along one edge. With respect to the two hide-working use-episodes, one consisted of scraping/planing while the other was of a sawing/cutting action. The remaining two scraping/planing episodes were against bone and haematite (or a material bearing this mineral). The remaining sawing/cutting episode was directed against the unidentified material. It is interesting to note that the majority (three of the five) 'manners-of-use' are of scraping/planing; a frequency that is, at least to this writer, somewhat unexpected in light of their typology.

Although I am reluctant to make any generalisations based on statistics, because of the small size of this sample, concerning the utilised and apparently non-utilised backed blades, the following informal comments can be made based on the metrical attributes of each individual tool. Each of the utilised artefacts are longer, wider, heavier and, with one exception, thicker than the apparently non-utilised artefacts: i.e. they are simply larger. Furthermore

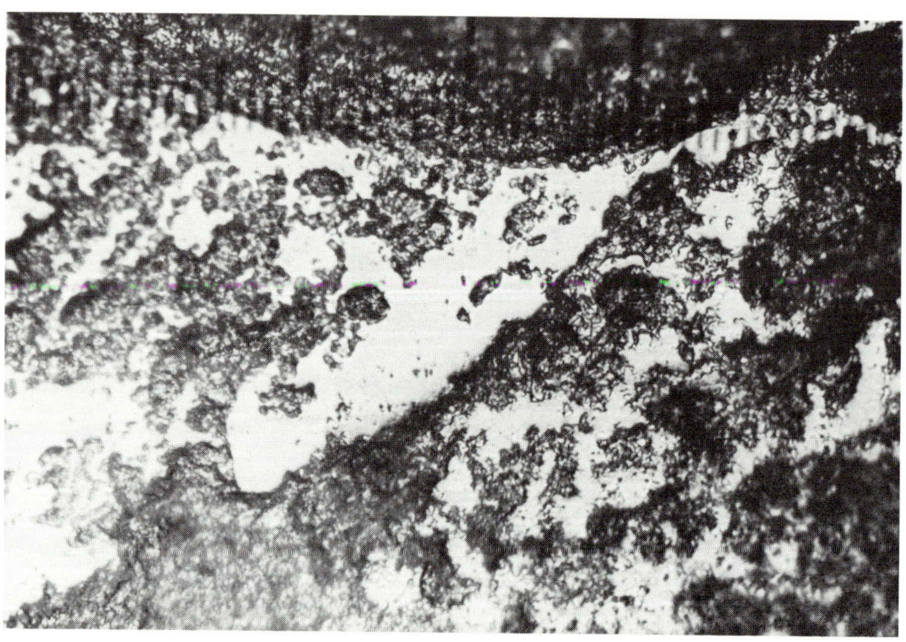

Fig. 12.8. An example of stone-on-stone polish. The photomicrograph was taken at a magnification of 200 diameters. Length of the photomicrograph: 0.307 mm.

each of the utilised artefacts have a lesser length/width ratio value than the 'non-utilised' artefacts: i.e. the demonstrably utilised artefacts are less 'blade-like', in final form, than the others. The absence of observed microwear traces on the apparently non-utilised though intentionally manufactured artefacts does not necessarily indicate that they experienced no use. It is possible, but not demonstrable, that they were used either against softer materials, such as meat, that are not conducive to rapid and obtrusive polish development, or perhaps they were used for too short a period of time for a polish (or other traces) to be recognisable on their surfaces.

Truncated blades

Four truncated blades were examined during the course of this study. Four demonstrably utilised edges were identified on three of the four artefacts. The traces found on these tools consist of polishes, linear arrangements of polish, striations, edge-damage and stone-on-stone polish. Neither edge- nor *arête*-rounding was observed on these tools. Polishes were found on each of the four identified working edges: one instance each of antler, wood, hide and an unidentified polish. Linear arrangements of wood and stone-on-stone polish were only observed on a single working edge. Striations and edge-damage occurred on three of the edges; the remaining edge (with hide polish) was free of these traces.

Four use-episodes, corresponding to each of the four working edges, were identified on the three artefacts and were concerned with the working of antler, wood, hide and an unidentified material. Three of these use-episodes consisted of scraping/planing while the woodworking episode was of whittling. With one exception, the working edges of these artefacts were the lateral sides.

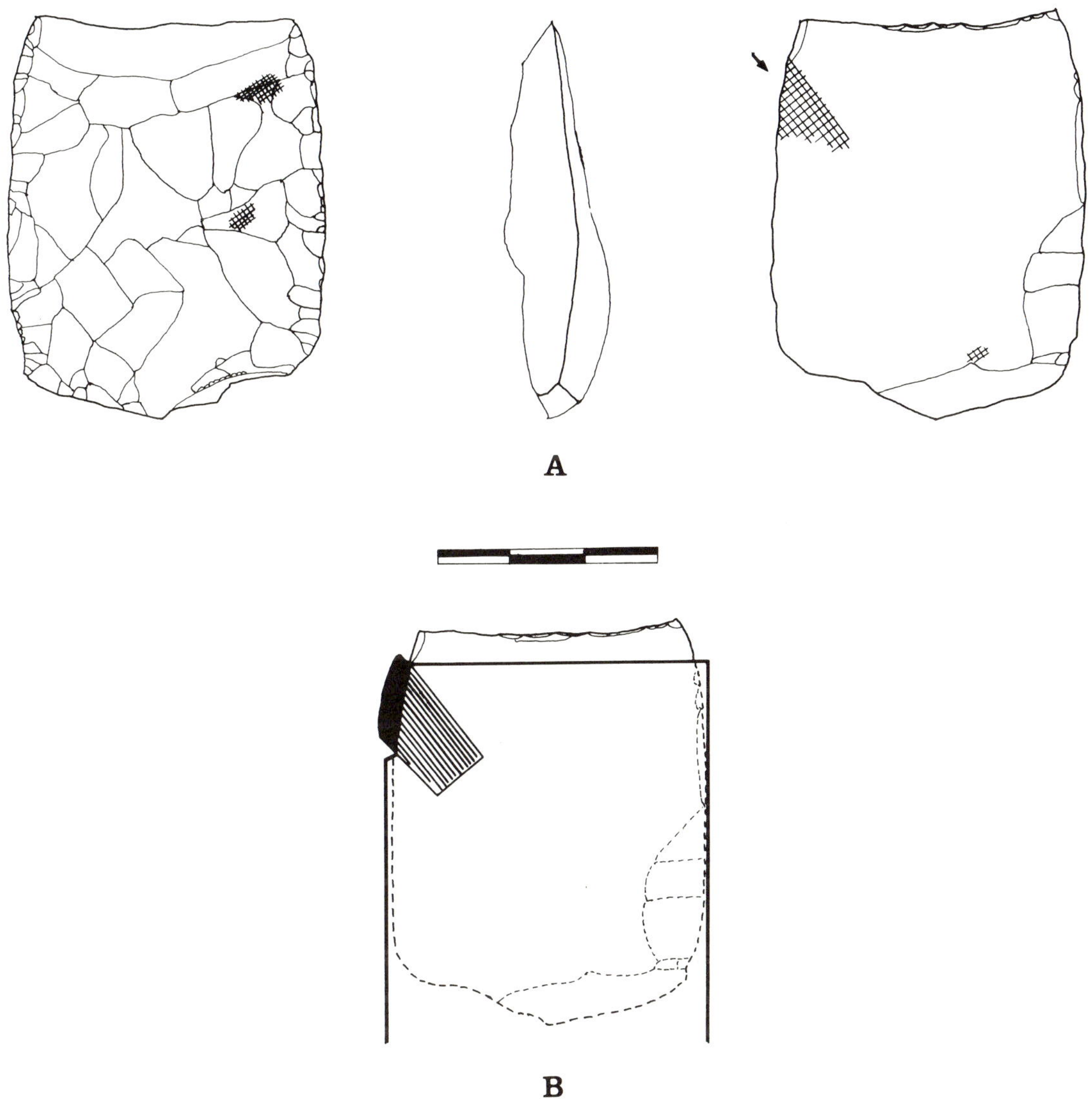

Fig. 12.9. (A) is a schematic drawing of three views of flake axe MU2287 that shows extensive and well-developed areas of wood polish (areas marked by cross-hatching) which the author suggests were caused by the use of a wooden sleeve-like haft (B) that was periodically tightened through use of a wooden wedge (hatched and solid areas of (B)). The scale is in centimetres.

Scrapers, picks and burins

Two scrapers were examined and one exhibited microwear traces: bone polish and edge-damage at the steeply retouched working edge. The function of this tool has been interpreted as 'scraping bone'. Only one pick was examined. The surface of this tool exhibited potential hafting (wood polish) and technological traces; the latter presumably caused during its manufacture. The single examined burin only exhibited traces that are thought to relate to the preparation and removal of one of the burin spalls (see Dumont, 1985, for a full description of these artefacts).

Needles

Of the 34 needles examined during this study, only two exhibited functional traces while a third exhibited traces that may have been caused by a haft. The traces observed on these three artefacts consist entirely of polishes: one instance of wood polish that may be a hafting trace; a single instance each of hide and meat polishes that may well be functionally related; two instances of linear arrangements of bone polish; and probably technological stone-on-stone polish. Both of the linear arrangements of bone polish appeared on the two needles with likely functional traces. The identity and distribution of the meat and hide polishes, and, the location and orientation of the linear arrangements of bone polish, are consistent with the suggestion that these two tools were utilised as projectile tips.

The low percentage of needles with functional or hafting traces (*ca.* 8.8%) appears remarkably small when compared to the rates achieved in the other artefact types from Mount Sandel and Star Carr (with the exception of the Star Carr microliths; see Dumont, 1983; and Dumont, 1985). It would be unwise to argue that the three *identifiably* utilised needles represent the only tools of this type that actually experienced use. If it is the case that all or most of the needles were used as weapon tips, the types of polishes likely to be produced on their surfaces are exactly those that are the most difficult to recognise in their early stages of formation. The recognition of hide and meat polishes on two of the needles indicates that these particular tools were in energetic contact with meat and hide for relatively long periods. The extended duration of contact could have been achieved by the building-up of a palimpsest of repeated uses of the weapon containing the armatures, or the armatures themselves in different shafts, in successive hunting episodes. One might expect that the weapons were reused until such time that a poorly directed impact caused contact with the prey's skeleton resulting in damage to some component of the weapon. Such instances could be represented by the formation of the linear arrangements of bone polish visible on two of the needles. The absence of recognisable functional traces on the remaining 32 needles, then, might only indicate that these armatures experienced insufficient duration of contact (hunting episodes?) rather than the absence of use.

Triangles

Microwear traces were observed on four of the nine triangles examined from Mount Sandel. Three of these artefacts exhibit stone-on-stone polish that is thought to be related to the tools' manufacture rather than to use. Functionally significant traces were observed on three artefacts: hide polish (two instances), bone polish (one instance), edge-rounding (one instance) and edge-damage (three instances). Although edge-damage occurred on each of the three utilised triangles, only in two instances can it be confidently described as resulting from use.

The microwear analysis has not provided unequivocal functional interpretations for these tools. In each case the functional interpretation has had to rely on the morphology of the tool and the functional preconceptions associated with this particular tool type. In a manner of speaking, then, the analysis has only served to confirm, or at least not refute, the traditional assumption that these microliths were weapon armatures without actually enabling this interpretation to be derived directly from the type and pattern of traces observed on each tool.

Obliquely blunted points

Seven obliquely blunted points (obliquely trimmed points) were examined but none exhibited functional or potential hafting traces. Three of these microliths, though, did show traces (stone-on-stone polish) that are thought to relate to their manufacture.

Miscellaneous microliths

Six variously backed and/or trimmed microliths, that do not fall easily within the recognised types, were also examined but none exhibited functional traces. Two of these, though, exhibited stone-on-stone polish at locations consistent with a technological origin.

Rods

Of the 101 rods examined as part of this study, only 19 had traces that could be argued to have resulted from either the use of the tool or the former presence of its haft. Eleven of the 19 rods exhibited traces (stone-on-stone polish and striations) that could be ascribed to the hard hammer percussion utilised in their manufacture. In addition to these 11 rods, a further 36 rods exhibited only technological stone-on-stone polish at or near their abruptly retouched edge. The presence, then, of technological stone-on-stone polish on 47 of the 101 rods (*ca.* 46%) indicates that hard hammer percussion was a common method of abruptly retouching these tools.

The traces that are thought to relate to the actual use of the rods rather than to their hafting or manufacture are composed of polishes, striations, 'impact fractures', edge–/*arête*-rounding and edge-damage. Polishes (including linear arrangements) were the most common functional trace and were observed on 11 artefacts: areas of bone polish were present on two to four rods; linear arrangements of bone polish on three; hide polish on two or three; functional stone-on-stone polish on one; meat polish on perhaps one; and perhaps antler polish on another. Striations were observed on two rods: one with an impact fracture and one with functional stone-on-stone polish. Although the majority of the rods exhibit a snap at one or both ends (77, representing *ca.* 76% of the examined rods), only those breaks that are 'associated' with a pseudo-burin facet (or other 'associated' flake removal originating from the break, see Barton & Bergman, 1982, for a discussion of experimentally produced impact fractures) and also other corroborative functional and/or potential hafting traces are here considered as impact fractures. Impact fractures were observed on three rods: two along with potential hafting traces and one with striations. Impact fractures may also be present on two other rods but no other corroborative traces were observed on their surfaces. Edge–/*arête*-rounding affected the leading edge of two rods and, in each instance, hide polish was present as well.

Traces that are, by virtue of their distribution, more easily interpreted as having resulted from hafting, either from contact against the haft itself or from the former presence of a hafting adhesive, were observed on nine rods. These traces were found either along the backed edge or differentially across the surface(s) at one end of the rods. The following polishes and deposits have been interpreted as evidence of hafting: areas of bone polish on one or two rods; linear arrangements of bone polish on one; wood polish on two; antler polish on one; suggestively biased distributions of pedogenic iron compounds (perhaps reflecting the distribution of a former hafting adhesive) on two; and an unidentified deposit on one rod.

Of the 19 rods with functional or hafting traces, 13 have been assigned functional interpretations of varying but mostly diminished confidence. The remaining six rods exhibited either no functional traces or traces that were too ambiguous for even speculation. In the interpretation of the microwear traces, two different methods were required: (1) the comparison of the observed traces to a model of the *presumed* function of the tools and (2) the more cautious and stronger argument from the nature, distribution and orientation of the observed traces themselves. The latter method was only possible with three rods which in my opinion exhibited sufficient microwear data: (1) boring holes in stone; (2) cutting hide; and (3) sawing/cutting wood. Although this approach was not allowed by the traces on each of the other ten tools, the nature and location of the available traces (including the impact fractures and hafting traces) were, in each case and as a group, consistent with the assumption of their use as projectile tips (three instances), barbs (six instances), or, a tip or barb (a single instance).

With respect to the hafting of the presumed projectile components, only three rods (two tips and one barb) exhibited relevant traces. Both of the inferred projectile tips showed a biased distribution of pedogenic iron that has been interpreted as reflecting the extent of a former hafting adhesive; while the barb exhibited antler polish at its backed edge. This suggests that the barbs were set into slotted antler shafts while the tips were mounted at the fore of the shaft and nearly entirely encased within some type of adhesive. Of the three rods that have been interpreted as *not* being projectile components, two exhibited traces that are thought to relate to wooden hafts; no hafting traces were found on the third. Of the four other rods whose functions remain unclear, three appear to have been mounted within slotted bone hafts and one perhaps exhibits traces of a hafting adhesive. If these four rods were in fact projectile components, then bone as well as antler shafts were utilised. As usual, the demonstrably utilised rods need not and are not likely to be the only utilised rods within the examined collection of artefacts. The absence of functional and/or hafting traces on the other 82 rods merely reflect the fortune or skill in either or both the hafting or use of the presumed projectile components or other forms of tools.

Conclusions

The remarks presented above indicate that many of the Mount Sandel artefacts were subjected to a variety of uses and do not simply represent the waste and/or final products of a lithic artefact production area. Although the presence of particular activities at this site, as elucidated by the microwear analysis, has not been addressed in this paper, it would be certainly perverse to believe that all of the use-episodes found on the tools represent off-site activities. It has yet to be determined by this study, though, if spatially distinct areas within the site contain artefacts of related function that could, in turn, define specific activity loci. These points and others, such as the relative frequency of particular functions and the intriguing use of haematite, will be fully discussed in the final report of this study (Dumont, 1985).

Bibliography

Barton, R. N. E. & Bergman, C. A. (1982). 'Hunters at Hengistbury: some evidence from experimental archaeology.' *World Archaeology*, **14** (2), 237–48.

Dumont, J. V. (1982). 'The quantification of microwear traces: a new use for interferometry.' *World Archaeology*, **14** (2), 206–17.

Dumont, J. V. (1983). 'An interim report of the Star Carr microwear study.' *Oxford Journal of Archaeology*, **2**(2), 127–45.

Dumont, J. V. (1985). 'A microwear analysis of selected British Mesolithic sites.' D. Phil. Thesis, University of Oxford.

Keeley, L. H. (1980). *Experimental Determination of Stone Tool Uses: A Microwear Analysis.* Chicago University Press.

Woodman, P. C. (1978). *The Mesolithic in Ireland: Hunters and Gatherers in an Insular Environment.* Oxford: British Archaeological Reports, British Series no. 58.

13

Wear analysis of a Lower Magdalenian flint assemblage from southwestern France

PATRICK VAUGHAN

Chipped-stone tools from prehistoric sites are usually so abundant that the use–wear analyst is able to study only a small portion of the excavated lithic material. To counter this sampling problem the author chose for his dissertation research a small cave site in southwestern France which in seven excavation seasons had yielded a lithic assemblage of sufficiently restricted size that it could be studied in its entirety.

The site of Cassegros (near Trentels, Lot-et-Garonne) is in a long karstic cave located near the top of a limestone butte, at an altitude of 140 m above the alluvial plain of the Lot river (Le Tensorer, 1981). Investigations at Cassegros, conducted since 1973 by Prof. Dr J.-M. Le Tensorer,* have revealed cultural materials principally within the cave's first chamber measuring *ca.* 50 m^2 and also on the terrace area in front of the cave entrance (Fig. 13.1). Although Mousterian artifacts have been recovered from deeper sondages, the majority of the lithic material exposed to date by meticulous three-dimensional recording of objects within metre squares and sedimentary levels dates from towards the end of the Würm glacial period. The greatest number of finds come from two levels with stone tool assemblages which are classified on typological grounds as Lower Magdalenian, specifically '0' (zero) and *I*.

The flints from the Magdalenian '0' level were chosen for use–wear analysis because of their relatively restricted number (*ca.* 850 pieces from 16 m squares excavated 1973–78) and because of a distinct stratigraphical demarcation of the level due to heavy staining of its sediments from red ochre, which is not the case of the adjacent levels. Sedimentological studies situate the Magdalenian '0' level, which varies in thickness from 5 to 25 cm, at the beginning of the Würm IV glacial period (17th millennium BC) and during a time of very cold climatic conditions in the region. Study of the horizontal distribution of the remains has indicated a denser concentration of finds inside the first chamber as compared to the terrace, a hearth and scattered hearth debris just within the cave entrance, concentrations of pure ochre, and an abundant spread of splintered bones from cold climate fauna, in particular reindeer. Bone artifacts from the level are few but include a manufactured 'polisher', four needles or fragments thereof, and various polished, grooved, or pierced bones.

From all the excavated flints, both retouched and unretouched, of the 1973–79 seasons, those larger than 1 cm were examined from microscopic traces of use at 280×

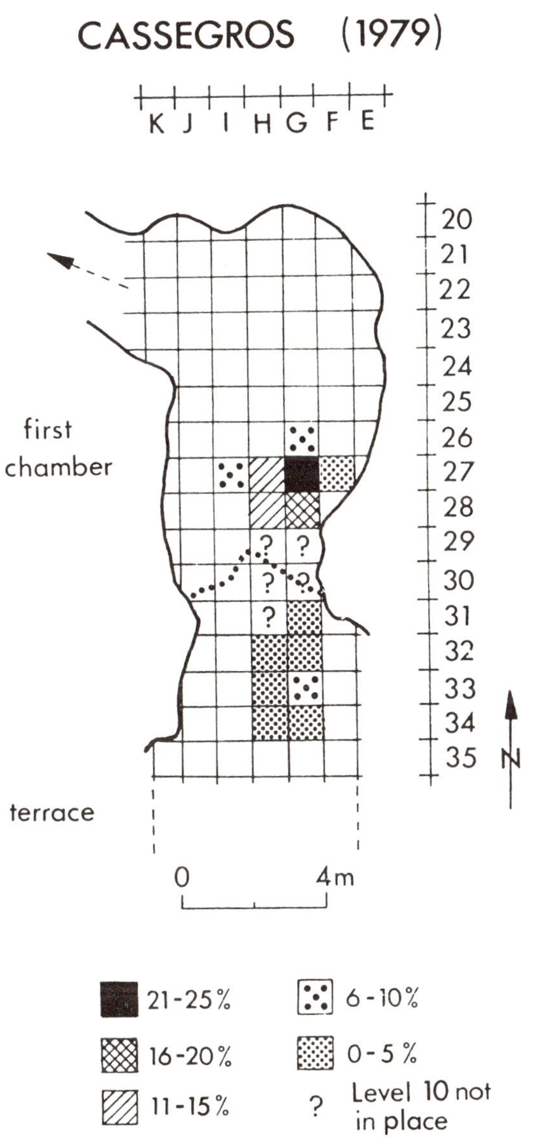

Fig. 13.1. Plan of Cassegros (Lot-et-Garonne, France), 1973–79 excavations. Code indicates the relative percentage of flints with use traces from the Magdalenian '0' (level 10).

* Seminar für Ur- und Frühgeschichte der Universität Basel. The author is most grateful to Prof. Le Tensorer for facilitating study of the Cassegros material.

magnification. By employing the 'high-power' or micropolish method of wear analysis (Keeley, 1980; Vaughan, 1985), the author observed systematically the attributes of use polishes, striations, and edge rounding on the flints so as to reconstruct their functions. These interpretations were also correlated with the typological classifications of the tools and were likewise related to the positions of the implements within the excavated portions of the site. The total analyzed sample amounted to 532 pieces, of which only one-third (*N*=189, or 36%) were retouched tools.

Approximately one-third of the analyzed flints from the Magdalenian '0' at Cassegros (*N*=158, or 30%) were found to bear microscopic traces which could be related to intentional usage by the cave's occupants. The 158 flints with use traces document 283 cases of specific intentional utilization of the edges or ridges of the flints, giving an average of about two separate uses per tool. In actual fact, though, only about half the utilized flints (*N*=73, or 46%) displayed more than one usage. As to the materials which the tools worked, approximately 60% of the individual used edges and ridges bore traces from dry hide or leather, 13% from wood, 11% from antler and bone, 9% from unspecified harder materials, and 7% from miscellaneous or unidentifiable substances. The majority of the used edges (*ca.* 60%) involved a transverse action such as scraping, planing, or whittling, whereas only 35% of the edges bearing use traces attest to longitudinal actions such as cutting, slicing, sawing. Thus preparing dry hides seems to be the major activity that is reflected in the use traces on the flints.

Interesting relationships were established between the types of tools and their functions. Although *un*retouched flints comprised nearly two-thirds of the studied pieces (*N*=343, or 64%), only 24% of the unretouched pieces (*N*=82) bore use traces, and of those only 11% (*N*=9) showed three or more separate usages. The retouched flints, on the other hand, were fewer (*N*=189, or 36%) but exhibited a much higher proportion of utilized pieces in general (40%, *N*=76) and of intensively used tools showing three or more uses (27%, *N*=21). Thus the retouched tools received more attention not only in their preparation but also in usage. Nonetheless, it is also apparent that ordinary unretouched flints were put to good use and that a comprehensive functional study of a stone tool assemblage should not be limited to the retouched pieces alone.

Each group of specialized tool types – such as end-scrapers, burins, side-retouched blades, splintered pieces, side-scrapers, etc. – bore use traces from *several* different kinds of worked substances, with dry hide usually the predominant contact material. In other words, no tool type group was found to be associated exclusively with one function, but there were varying degrees of preference amongst the tool groups for one or more functional possibilities. The strongest correspondence between form and function was seen between the end-scraper group (Fig. 13.2) and use traces from dry hide (77% of usages). Likewise, there was a very strong correlation between the function of hide-working and the tool groups such as end-

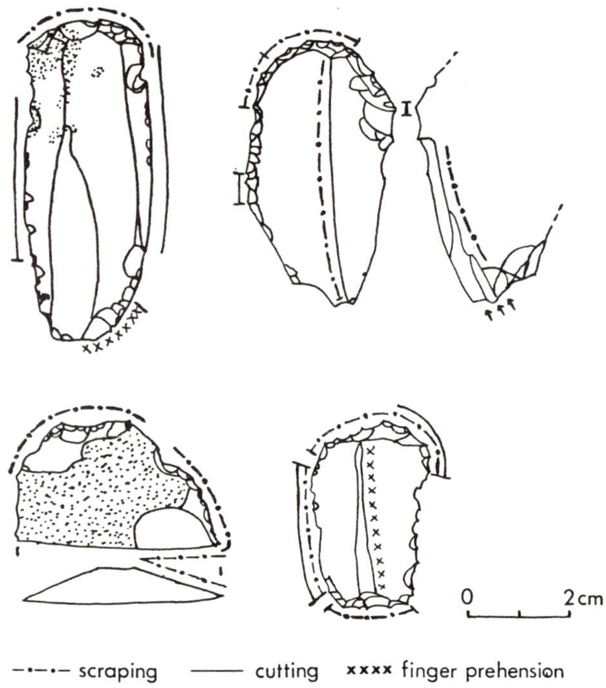

Fig. 13.2. Examples of intensively utilized end-scrapers from Cassegros. With the exception of bone-working traces on the right edge of the burin/end-scraper (upper right), the utilized zones were involved in scraping and cutting dry hide.

scrapers, splintered pieces, side-scrapers which displayed the greatest proportion of retouched tools which were used, once again retouched or otherwise remodified, and then reused again. By contrast, the 13 used burins displayed a lack of preference for one contact material, in that the used edges bore a wide variety of use polishes from bone, antler, wood, an unidentifiable harder material, and dry hide. In addition to the surprising use of burin tools to work hide, the Cassegros burins bore wear traces only rarely (one case) on the burin tip or bevel. Rather, on eight of the used burin tools the *sides* of the burin facets were used; and on the four remaining used burins, edges other than those formed by burin removals bore the use traces.

When the tools' functions are related to their proveniences (Fig. 13.1), it is clear that the greatest majority (81%) of the flints with use traces come from the excavated area within the cave's first chamber, just as do virtually all (93%) of the more intensively used flints, that is, those pieces displaying three or more uses. If the find-spot of a tool is at least a general indication of the area (but perhaps not the precise spot) where it was utilized, then it is clear that most of the activities involving stone tools took place inside the cave's first chamber and not on the terrace in front of the cave.

According to the microwear traces, then, the major emphasis in the use of stone tools at Cassegros during the Magdalenian '0' seems to have been directed toward the working of dry hides. Yet ethnographic evidence from recent stone tool-using cultures in North America and Africa shows that various implements and accessories made from bone, wood, and antler also played a major role in the processing of animal skins, particularly in the fresh state before the hides were tanned or dried (Gallagher, 1979; Steinbring, 1966; Yellen, 1977). Cassegros has so far yielded remains of a few bone tools which correspond to ethnographically known bone implements that serve in various stages of hide-preparation, namely a bone 'polisher' which could have served to dehair the fresh hides and bone needles for sewing together pelts. Since stone tools would have been used to fabricate and maintain any wooden, bone, or antler implements used in hide-preparation, the presence on various flints of use traces from wood, bone, and antler is not incongruous with the wide variety of tasks which are actually involved in preparing finished hides and furs from fresh animal skins. Furthermore, the use of bone and antler tools to work the hides while fresh would also explain the absence of use traces from fresh hides on the flints. Thus, the entire range of hide-processing activities on fresh, dried, and tanned pelts may well have been carried out at Cassegros by means of a tool kit which included flint, bone, antler, and wooden implements.

The very cold climate which sedimentological analyses have documented for the Magdalenian '0' at Cassegros would understandably have created the need for an abundance of hides and furs for clothing, bedding, and shelter. The distribution of tool functions at the site points overwhelmingly to a localization of the hide-preparation activities within the first chamber of the cave, where the hearth and its swept-out debris were also found. One may conclude, then, that in a severe periglacial climate an enclosed working area which could be heated would have been a prime location for the long and arduous activities involved in hide-preparation.

Bibliography

Gallagher, J. (1979). 'Contemporary stone tools in Ethiopia: implications for archaeology.' *Journal of Field Archaeology*, **4**, 407–14.

Keeley, L. H. (1980). *Experimental Determination of Stone Tool Uses: A Microwear Analysis.* Chicago: University of Chicago Press.

Le Tensorer, J.-M. (1981). *Le Paléolithique de l'Agenais.* Cahiers du Quaternaire, no. 3. Paris: Editions du CNRS.

Steinbring, J. (1966). 'The manufacture and use of bone defleshing tools.' *American Antiquity*, **31**, 575–81.

Vaughan, P. (1985). *Use-Wear Analysis of Flaked Stone Tools.* Tucson, Arizona: University of Arizona Press.

Yellen, J. (1977). *Archaeological Approaches to the Present.* New York: Academic.

14

The transport and abrasion of flint handaxes in a gravel-bed river

P. HARDING, P. L. GIBBARD, J. LEWIN, M. G. MACKLIN AND E. H. MOSS

Summary

An experiment to investigate the movement and resulting modification of flint handaxes is described. Modern facsimile handaxes were placed in the Afon Ystwyth, a gravel-bed river in mid-Wales. The weighed and numbered handaxes were recovered after having been moved by flood water, each individual being reweighed to determine weight loss. Selected recovered handaxes were examined to determine the effects of abrasion arising from different distances of transport. A sequence of progressive alteration was found: polish appearing at 8.9 m, notches and extensive edge damage at 38 m and considerable edge damage by 57 m. Edge damage stabilised and edge polish became developed at 150 m. These results are discussed in the light of previous interpretations of flint implement abrasion.

Introduction

Throughout much of Europe the greatest frequencies of Lower Palaeolithic flint artefacts are found in deposits of fluvial gravels and sands. In many cases these deposits were laid down by rivers flowing under conditions rather different to those prevailing at the present day. The gravels and sands originated from high concentrations of debris, derived from destruction of local rocks, and deposited in river channels. High flow velocities are necessary to move such materials, although they need have only occurred on certain occasions during the year to mobilise the coarser components. Most artefacts have probably been incorporated into river sediment by channel migration, overbank deposition or by burial possibly following recycling by mass movement on slopes. In active rivers artefacts might be expected to occur in channel and bar gravels alike as a result of transport. The amount of transport undergone by any individual axe will have depended on where and how it originally entered the fluvial system, how quickly it became buried and on the preservation of the deposits from reworking.

This paper describes an experiment to record the dispersal and subsequent movement of handaxes in a river. The characteristics and development of the abraded surfaces which are often present on these tools is also discussed. The study and classification of handaxes using typology (Bordes, 1961; Roe, 1964, 1968; Wymer, 1968), technology (Newcomer, 1971; Bradley & Sampson, 1978) and function (Jones, 1980; Keeley, 1980) is well known, and has contributed greatly to the understanding of the Lower Palaeolithic period. Abrasion has been used, together with associated finds, as an indication of the distance which handaxes have moved from their abandonment site. Handaxes with sharp edges, associated with flakes, and vertebrate bones were thought not to have moved a great distance, whereas those with heavily abraded edges probably have (e.g. Clark, 1974, p. 103; Clark & Kiroshina, 1976, pp. 163, 167; Singer *et al.*, 1973, pp. 27–8; Wymer, 1968, pp. 31–3, 1976). Various attempts have been made to classify these conditions (Wymer, 1968, 1976; see Clark, 1974; p. 103). Wymer (1968, pl. XI) adopted a range of visual attributes to describe wear which were based on the reduction in height of flake scar ridges. The terms he applied were: mint, sharp, slightly rolled, rolled, and very rolled. Shackley (1974) subsequently undertook a series of experiments from which she recognised seven stages of abrasion based on microscopic examination of ridge structures. She noted four factors which affect the rate of abrasion: the hardness of the flint, the velocity of the water current, the shape of the artefact and the grain size and lithological composition of the river bed. She also suggested that two types of abrasion occurred in rivers. Firstly, the chipping of the implement resulting from clast collisions and, secondly, a 'slow grinding down' of upstanding ridges by the abrasive effect of finer debris in transport (i.e. a blasting effect by sand and silt grains). She concluded that the examination of handaxe condition was important in the investigation of artefact assemblages. Despite the recognition and importance of abrasion very little work has been presented to establish both how it occurs and the rate of its occurrence. Similarly, the dispersal of artefactual material and its behaviour within a river has received little attention. Isaac (1967, 1977, pp. 81–4) has reported on the behaviour of handaxe replicas in abraided streams under tropical conditions and described a possible stream mechanism for concentrating handaxes from a wide catchment area. Schick (1980, pp. 129–30), also working in East Africa, experimented with the destruction of a chipping floor by water action and showed that the recovery of flakes was biased in favour of the larger pieces.

In addition, certain experiments have been carried out with the behaviour of mammalian bone refuse both in the flume chamber and in stream environments (Gifford, 1978; Hanson, 1980). None of these experiments considered wear or abrasion of the artefacts and, interestingly enough, the results of Isaac's experiments seem somewhat at variance to those reported on below.

In the light of these views on the evidence of fluvial sedimentology and flint handaxe abrasion, a number of questions arise concerning the occurrence of handaxes in coarse sediments:

1 To what extent are concentrations of handaxes an artefact of the rivers' sedimentary processes and in what part of the river are they most likely to occur?
2 How far can handaxes travel and do they travel different distances from different subenvironments of the river?
3 How much transport is necessary to abrade flint artefacts, how is weight lost, how long does it take and how is abrasion expressed?

In an attempt to answer some of these questions an experiment was initiated. This involved the placing of replicas of flint handaxes in a river which as closely as practically possible approximated to those thought to have deposited implement-bearing Pleistocene gravels and sands. For this purpose the Afon Ystwyth in Dyfed, Wales was selected, where two of the authors (J.L. and M.G.M.) were conducting experiments on clast transport. During the study the preferred conditions of handaxe movement, the distances moved by individual pieces, their weight loss and the influence of their shape and size were considered. In addition, associated damage and microwear traces arising from the transport were investigated to determine both their expression and development.

Handaxes

The handaxes were made over a period of approximately seven days in 1982 (by P.H.). Originally 100 were prepared, from which 60 were selected for the experiment. The flint was obtained from the Grimes Graves area in Norfolk. A variety of flint types was used, ranging from good quality flat nodules, through large poor quality pieces to rounded cobbles derived from local gravel sources. Some handaxes were made from single nodules while others were made from large flakes. No attempt was made to standardise handaxe quality, type or size, so that all forms from lanceolates to refined cordiforms and ovates (Bordes, 1961) were produced. The purpose was to manufacture a complete range of useable handaxes. All were made using the well-known technique described by numerous writers (Warren, 1921; Newcomer, 1971; Bradley & Sampson, 1978; Newcomer & Sieveking, 1980) in the production of large bifacial tools. This involves (i) rough bifacial flaking around the edge with a hard stone hammer to achieve basic shaping, (ii) thinning and shaping, and (iii) finishing using a red deer antler. All stages were carried out by direct percussion and striking platforms were prepared, especially in stages (ii) and (iii), by 'turning the edge' to strengthen them before percussion. Some handaxes were produced from the standing position by holding the piece in the left hand and flaking with the right, while others were made from a seat, resting the flint on a knee pad. In all cases such a tool could be made in 10–20 minutes. Some broken handaxes were retained when the piece was still functional (e.g. handaxe no. 50).

Study site

The Afon Ystwyth is a gravel-bed river in mid-Wales, at Llanilar (SN 628754), where the catchment area is 152 km^2 and valley gradient is approximately 4 m km^{-1}. The channel has been artificially straightened though without bank protection works. The channel is rapidly shifting with developing gravel bars and active transport of bed materials. Material up to 0.40 m intermediate diameter has been observed to move during floods, but median grain size of the sandy gravels is generally around 16–32 mm. Finer sedimentary units are also found as bar tail deposits, as discontinuous sand bodies on bar tops, and on the flood plain surface. The gravels are entirely derived from local Palaeozoic shales and gritstones, with a high proportion of disc- and blade-shaped clasts. River discharges, gauged 4 km below the study site, average 5.5 m^3s^{-1} but may rise to over 300 m^3s^{-1} in extreme floods. Mean annual flood is 90 m^3s^{-1}, with gravel sediments being in motion on several occasions in the year at well below this discharge. Because of this rate of sediment-transporting activity taking place on discrete occasions when river levels rise and fall again within a few days, an opportunity for experimental work on handaxe transport in rivers is usefully provided.

To some extent these conditions resemble those that prevailed in England under cold climate conditions when the coarse sediments underlying the finer sedimentary units of present floodplains were actively in transport. There are differences: in this study the gradient is steep, the size and discharge of the small river are not those of larger English counterparts, bed material lithology is different, and the storm–rainfall dominated river regime is not the same. But for a first experimental study, this actively developing coarse-sediment reach does provide an accessible natural gravel-river environment where the movement and modification of handaxes can be observed, taking place alongside the transport of appropriately sized natural materials in an eroding and depositing river.

The nature of the experimental site is shown in Fig. 14.1. A footbridge conveniently spans the 30 m wide channel. There is a well-developed left bank bar extending some 110 m downstream with a pool alongside. The low flood channel crosses to the left bank across a shallow riffle, and bars then occur alternately on the left and right banks. Inner channels are also to be found between bar and bank; these become reactivated at intermediate flows, and at high (handaxe transporting) flow bars, pools and riffles are all submerged. Further downstream the channel becomes more braided in appearance with more than one active channel, even at low flows during the study period.

Fig. 14.1. View of the reach of the Afon Ystwyth used for the handaxe experiments.

Methods

Before the handaxes were placed in the monitored channel reach, their size and shape parameters were measured and recorded. The 60 handaxes were weighed (to ±0.1 g) and long (A), intermediate (B) and short (C) prime axes were measured in mm; from these maximum-projection sphericity (ψ_p, Sneed & Folk, 1958) was calculated for each handaxe. In addition, identifying numbers (ranging from 1 to 60) were painted on, and all handaxes were photographed.

On 31 January 1983, during a flood with an estimated peak discharge of 68 m^3s^{-1}, two handaxes were put into the Afon Ystwyth from each of 30 points 1 m apart on the footbridge spanning the river. Surveying of water surface gradient at the flood peak and measurement of water depth across the channel allowed values of average bed shear stress to be estimated. These ranged from 34 to 61 N m^{-2} (Fig. 14.2A). The computed maximum values of bed shear stress were high enough to move all the handaxes (Costa, 1983, Fig. 4, p. 995), the largest of which had an intermediate (B axis) diameter of 123 mm.

In later weeks, under lower stream flow conditions, the experimental channel reach was repeatedly searched and the locations (distance travelled in metres from the footbridge) of handaxes were recorded. At the beginning of April 1983 all the handaxes that could be found were collected and their distances from the footbridge were again noted. In July 1983 two more handaxes came to light and were recovered. All relocated handaxes were reweighed and any weight loss was recorded. A subsample of seven of the recovered handaxes, whose distance transported and weight loss was known, were sent for microwear analysis.

Results

Of the original 60 handaxes placed in the channel, 59 (98%) moved, 27 (45%) had their new locations recorded and 19 (32%) were recovered and reweighed. The results are tabulated by size classes (see Table 14.1). Although values for maximum distance moved and size class support the general relationship that distance travelled is inversely related to particle size, there is no straightforward correlation simply between the size of individual handaxe intermediate axes and distance moved (r=0.042). If, however, a stepwise multiple regression is carried out with distance as the dependent variable and handaxe A, B and C axes, weight and shape (ψ) as the independent variables, a multiple correlation coefficient (r) of 0.656 is obtained (significant at p=≤0.01). Most of the improvement in the correlation coefficient is attributable to the inclusion of handaxe long axis (A) and weight as independent variables. Both are, as expected, negatively correlated with distance moved. A plot of distance travelled with reference to initial handaxe insertion point (Fig. 14.2B) suggests that susceptibility to movement and distance transported is a function of local bed shear stress. Axes in the low-flow and inner channels at the edge of the river (Fig. 14.1) did not

Table 14.1. *Distance moved and weight loss data for handaxes*[1]

Size class (mm)	Initial number of handaxes	Number of handaxes moved	Number of handaxes found	Number of handaxes lost	Maximum distance moved (m)	Average distance moved (m)	Minimum distance moved (m)	Maximum weight loss (g)	Average weight loss (g)	Minimum weight loss (g)
52.0–56.1	2	2	0	2	nr[2]	nr	nr	nr	nr	nr
56.2–71.3	7	6	4	3	44.7	20.5	0.0	0.3[3]	0.3[3]	0.3[3]
71.4–86.5	24	24	13	11	150.0	37.1	1.0	5.8	1.4	0.0
86.6–101.7	19	19	7	12	120.0	47.2	2.6	2.4	1.25	0.0
101.8–116.9	6	6	1	5	46.0[3]	46.0[3]	46.0[3]	2.7[3]	2.7[3]	2.7[3]
117.0–123.0	2	2	2	0	25.0	15.0	5.0	2.0	1.2	0.4

[1] Size class units are ±1, 2, 3 standard deviations from the mean value of handaxe intermediate (B) axis – 86.55 mm.
[2] NR=not recorded.
[3] Values for only one handaxe recorded.

move far, whereas those on the gravel bar moved furthest or were lost, probably by burial. The greatest recorded distances of movement were achieved by handaxes inserted on the margins of the gravel bar; these were transported from the bar through the pool and then deposited towards the head of the next bar downstream.

It is important to appreciate that the distance an axe is transported does depend not only on its size but also on where it was dropped. The axes behave in the same way as the local mobile sediment: if placed on an active gravel bar they may move and become buried, but if placed in slacker water they will not. In an environment of dynamic bed-forms, transport is an episodic process involving a phase of movement, and then a prolonged phase of burial which may be followed by renewed transport if those sediments are re-eroded again. Handaxe wear, even in an active transporting environment, is achieved intermittently. In these conditions, it should be noted that several handaxes were transported more than 100 metres in a few months, probably as a result of a relatively small number of episodes of movement, each lasting no more than a few days. This can be compared with 75 m of handaxe movement recorded by Isaacs (1967) over two years in an ephemeral, braided stream in Kenya. Nevertheless, despite contrasts in hydrological and depositional regime between Isaac's sand-bed stream and the perennial, gravel-bed Afon Ystwyth, handaxe transport and abrasion in both fluvial environments is interrupted by prolonged periods of within-channel sediment storage.

Under flood conditions high suspended sediment concentrations obscure the stream bed and prevent *direct* observation of handaxe movement. None the less, it is likely that handaxes were transported as bedload, moving largely by sliding and rolling. Movement by sliding would probably abrade the facets of handaxes, while rolling could result in breakage and wear along edges (Fig. 14.7).

For all the handaxes that could be recovered and reweighed, weight loss incurred during transport was calculated. Although no systematic relationship between weight loss and handaxe size emerged (Table 14.1) there is a strong, positive correlation between weight loss and distance transported (r=0.826, significant at $p=\leq 0.001$). Figure 14.3 shows weight loss in grammes against distance moved in metres, and has a simple linear regression line fitted to the data. The slope coefficient (B_1) 0.0274, indicates the rate of handaxe weight loss per unit distance travelled; for this flood 0.0274 g m^{-1} seems appropriate. The maximum recorded weight loss of 5.8 g represented 2.52% of the original weight of handaxe number 16. It is interesting to note that this amount of weight loss occurred to an artefact that would still be classified as fresh by archaeologists using the Wymer System (see below p. 120 and Fig. 14.6).

To assess whether handaxe size and shape parameters influenced weight loss during transport a stepwise multiple regression was carried out. Inclusion of handaxe weight, A, B and C axes, and handaxe shape (ψ) as independent variables, in addition to distance moved, did slightly increase the correlation coefficient, giving a multiple r value of 0.886 (significant at $p=\leq 0.001$). Most of the improvement in explanation of the regression model was, however, due to the inclusion of handaxe weight, with heavier/larger handaxes generally losing more weight per unit distance transported.

The wear analysis of the handaxes

Little is known about the microscopic, depositional alterations to flint implements, and this experiment provided an opportunity to examine one of the many variables affecting the preservation of Palaeolithic artefacts. Since the wear analyst (E.H.M.) joined the project late, there was no opportunity to view the handaxes before the experiment began. However, two handaxes made later (by P.H.) were examined with attention to the traces of manufacture.

Some literature on use–wear states that microwear polishes are variable in appearance due to a variety of causal factors, including temperature, pressure, moisture, the chemical and physical properties of the stone upon which they form and of the abrasive, or polishing agent (Anderson, 1980; Mansur-Franchomme, 1983; Moss, 1983, pp. 15–18). Moss (1983) has shown that the polish created on stone by stone is highly variable and that when abundant water is present, the formation of both polish and striations

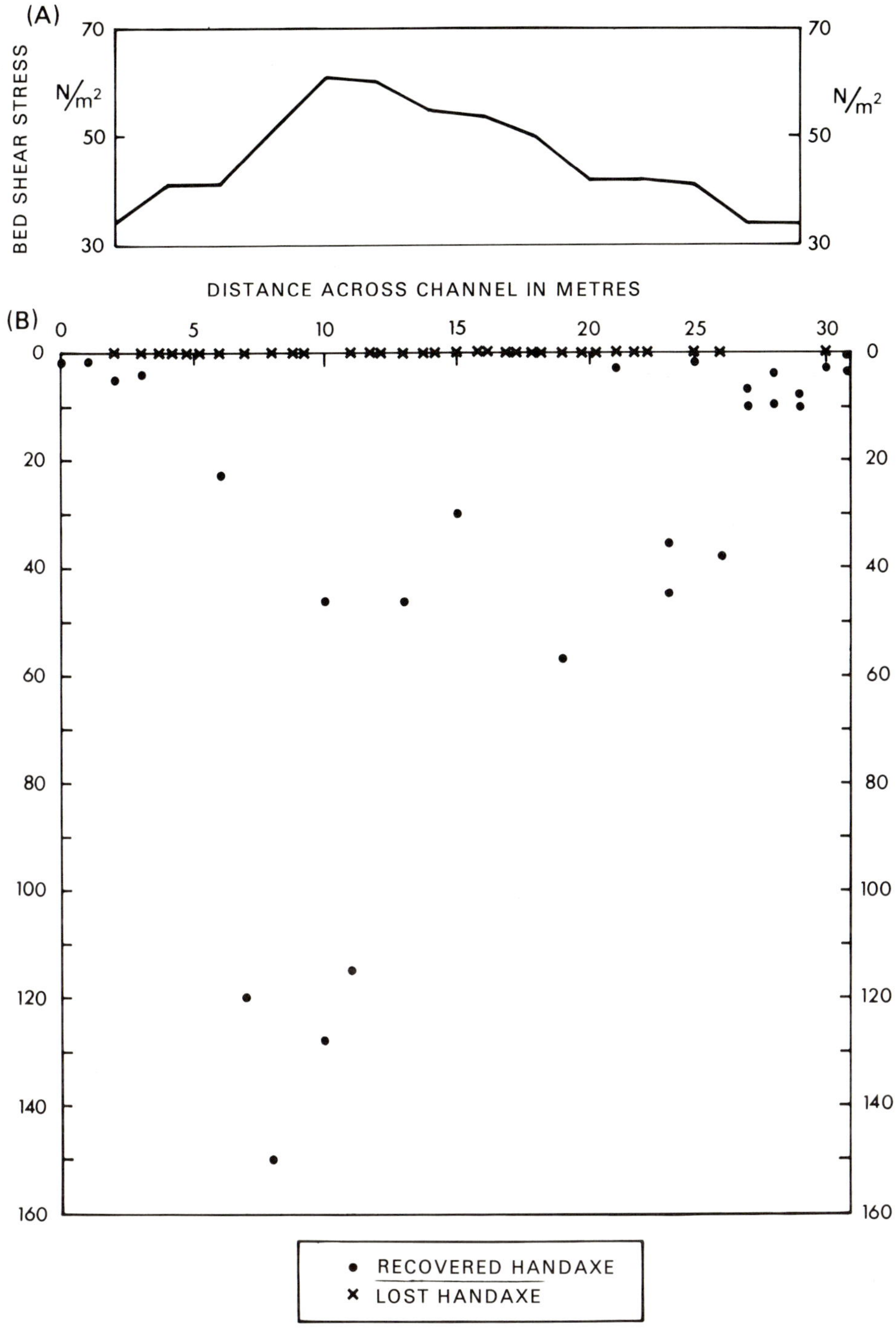

Fig. 14.2. (*A*) Average bed shear stresses across the channel at the insertion point for the 31 January 1983 flood. (*B*) Plot of distance travelled by recovered handaxes from initial insertion point.

seems to be inhibited. In a pristine state, the handaxe should exhibit some flint-on-flint polish present on the faces from the manufacturing process (Keeley, 1980; Moss, 1983). The two handaxes made later for the microwear study had more than the usual amount of this polish due to manufacture because these had been laid on the knapper's knee while they were made, thus increasing the pressure of flint against flint as the flakes were struck off the handaxes (Fig. 14.4(*a*) and (*b*)). When a handaxe is in a newly manufactured state, polish is largely absent from the edges. Preparatory abrasion to the striking platform is removed by the flaking process. In the present study, handaxes, when placed in the river, are assumed to have retained a normal amount of polish due to manufacture.

The seven experimental handaxes selected for study were examined in detail at magnifications of 56×, 140× and 280×, using a microscope equipped with incident light and bright field. By no means all of any handaxe surface area was covered in the examination, but an attempt was made to sample the whole axe, i.e. flat surfaces, concavities, protrusions and edges, on both sides of each handaxe.

Examination showed that the wear caused by the river transport seems to follow a progressive series of alterations:

Handaxe 59 (travel distance 2.6 m). This axe looks essentially fresh. The surfaces still have no more polish than could be attributed to that of manufacture (Fig. 14.4(*c*)).

Handaxe 56 (travel distance 8.9 m). Much of the axe is still untouched by polish, but parts, especially ridges and flat facets are beginning to display an increase in quantity of polish (Fig. 14.4(*d*)).

Handaxe 50 (travel distance 38.2 m). The edges are still largely without polish. Notches and other edge damage which were not part of the manufacturing process are beginning to be prominent; any edge polish was undoubtedly carried away during the formation of these flake scars (Fig. 14.4(*e*)). Some facets show the formation of a heavy polish, which resembles that produced by antler, but whose distribution is entirely different from use–wear known to the analyst (Fig. 14.4(*f*)). The distribution is random on the convexities, ridges and flat surfaces.

Handaxe 38 (travel distance 57 m). After travelling some distance down river, some edges are now dramatically affected (Fig. 14.5(*a*)). In addition, more of the surface has been altered by the creation of the same kind of heavy polish shown in Fig. 14.4(*f*) (Fig. 14.5(*b*)), though some slight concavities remain untouched.

Handaxe 14 (travel distance 120 m). This axe has even more surface area covered by polish (Fig. 14.5(*c*)) than handaxe 38, but the polish itself remains similar to that on handaxe 50 (Fig. 14.4(*f*)). Figure 14.5(*d*) shows the presence of some striations, which are never very common in this series.

Handaxe 16 (travel distance 150 m). On this handaxe, the edge damage seems to have stabilised and as a result, polish which has been forming at the edges is not removed (Fig. 14.5(*e*)). The polish resembles that in Figs. 14.4(*f*) and 14.5(*d*), but it is even more extensive than on handaxe 14. This axe is also the thinnest and one of the two lightest of the seven. The quantity of polish may have been affected by the morphology of the piece and its susceptibility to the hydrodynamics of the river.

Handaxe 33 (travel distance 184 m). This handaxe conforms to the pattern of increasing edge stabilisation and increasing development of polish at the edges. However, all aspects of wear resemble handaxe 38 rather than the two immediately preceding which have travelled further (Fig. 14.5(*f*)).

Though preliminary record photographs were taken of the handaxes, there was no opportunity to record macro- and microscopic edge flaking in detail before the handaxes were put in the river and little can be said concerning the incidence of edge damage arising from river transport. Beginning with handaxe 50 in the series, the edges begin to be duller to the touch, and therefore edge damage is known to have taken place; the polish does not begin to be retained at the edges until edge damage has ceased. Points of percussion are visible on some surfaces. Polish is distributed equally on both surfaces. No breaks occurred, and the larger edge damage which occurred during river transport (Figs. 14.6 and 14.7) could in other circumstances be attributed to poor manufacture, but not in this experiment.

Handaxe 16 has ridges which could be termed 'sharp' using Wymer's system (Wymer, 1976, p. 329). Microscopically, these ridges are beginning to lose some of their polish through the continued microflaking (battering) of the ridge. By extension of these findings to, hypothetically, more heavily 'rolled' handaxes, there should be rounded edges and ridges with little polish, and heavily polished facets. It is almost impossible to imagine the analysis of use wear traces on such implements. Keeley's analysis of implements from Hoxne, Clacton and Swanscombe showed that many pieces were too affected by post-depositional natural factors to be analysed for use–wear; 201 out of 267 handaxes at Swanscombe were affected (Keeley, 1980, pp. 120–1). As

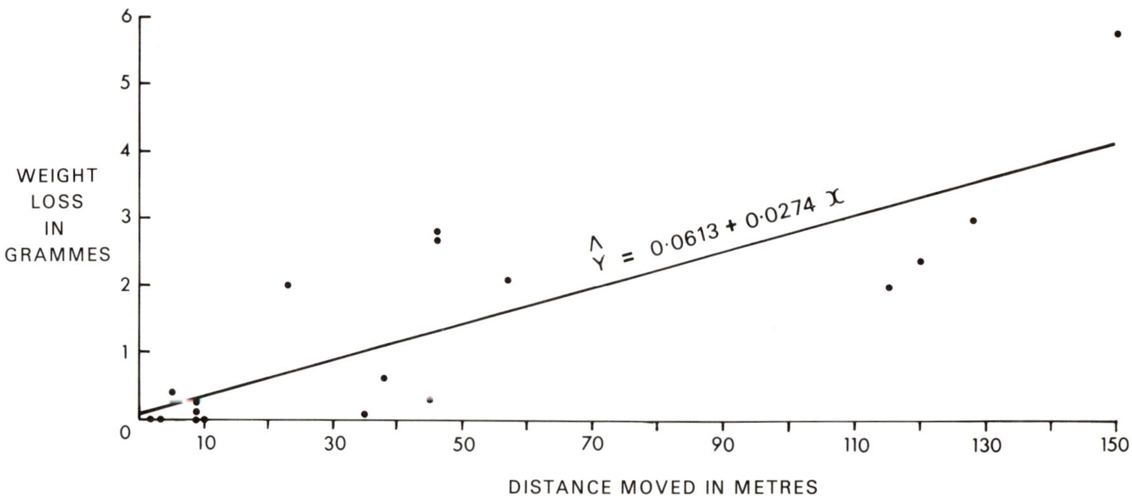

Fig. 14.3. Plot of handaxe weight loss against distance transported.

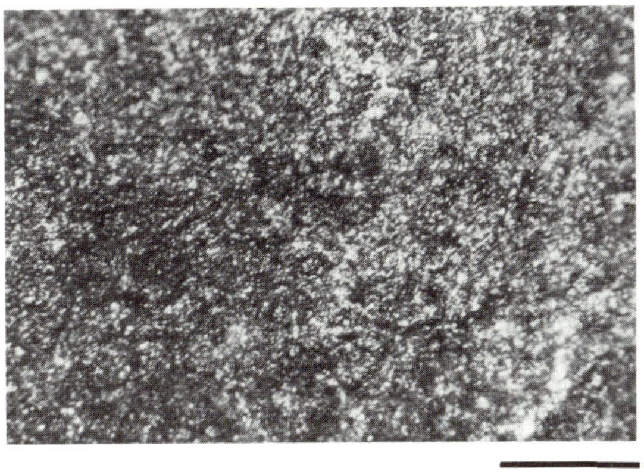

(a)

(b)

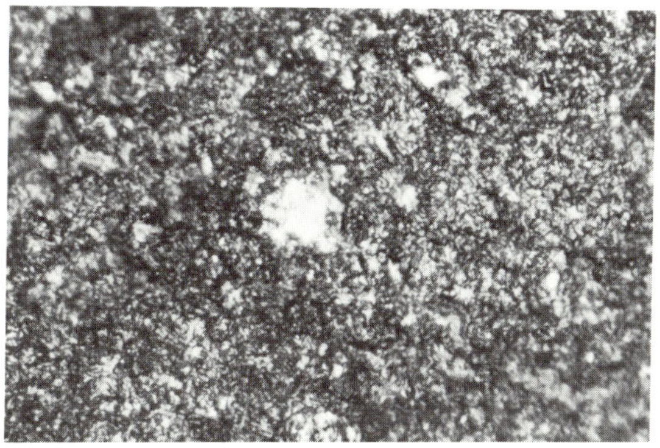

(c)

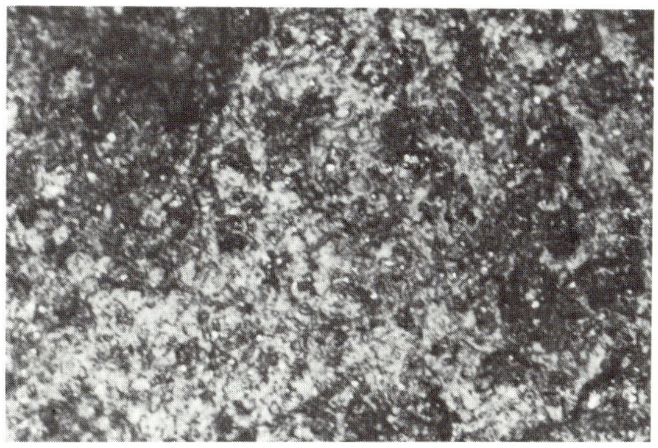

(d)

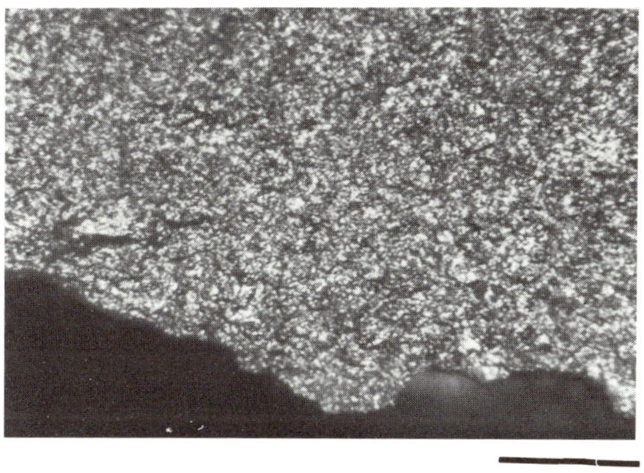

(e)

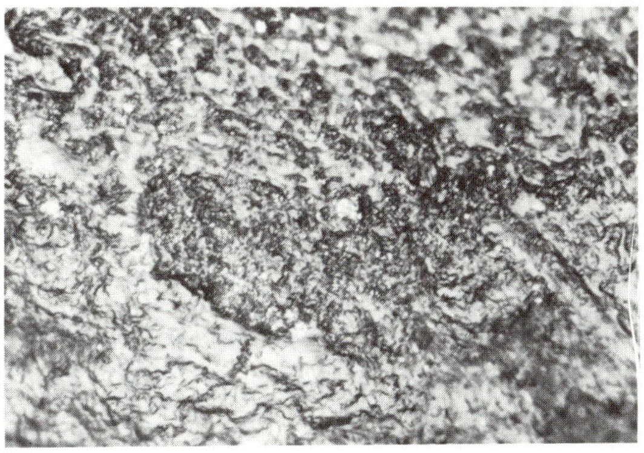

(f)

Fig. 14.4. (*a*) Surface of a freshly made handaxe, 140×. (*b*) Surface of a freshly made handaxe, 280×. (*c*) Surface of handaxe 59 (travel distance 2.6 m), 280×. (*d*) Surface of handaxe 56 (travel distance 8.9 m), 280×. (*e*) Edge of handaxe 50 (travel distance 38.2 m), 140×. (*f*) Surface of handaxe 50, 280×. The scale=0.10 mm at 140× magnification. Other photographs were taken at 280×, for which the same scale=0.05 mm.

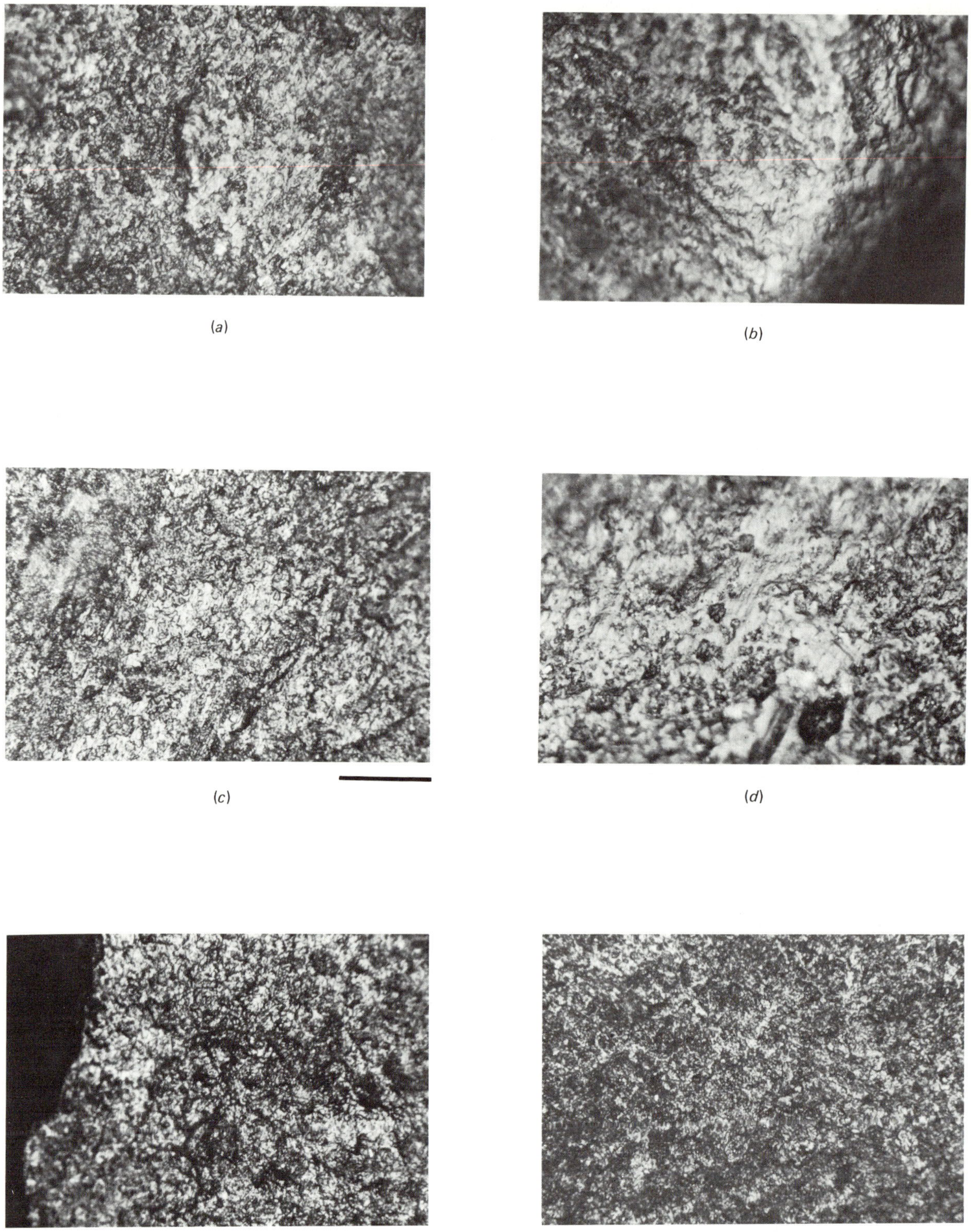

Fig. 14.5. (*a*) Edge of handaxe 38 (travel distance 57 m), 280×. (*b*) Ridge of handaxe 38, 280×. (*c*) Surface of handaxe 14 (travel distance 120 m), 140×. (*d*) Surface of handaxe 14, 280×. (*e*) Edge of handaxe 16 (travel distance 150 m), 140×. (*f*) Surface of handaxe 33 (travel distance 184 m), 140×.

this author states: 'The vast majority of British handaxes have been recovered from secondary or "geological" contexts, usually river gravels (*op. cit.*, p. 160).' This implies that their depositional environments were similar to those in this experiment.

It was not possible to establish any firm *rates* of wear experimentally. While distance travelled seems to be an important factor in the experimental handaxe wear, this supposition may not be directly applicable to the wear on redeposited Lower or Middle Palaeolithic handaxes from cold stage coarse sediments, since other possible wear causes have to be taken into account.

Stapert has extensively discussed possible macroscopic and microscopic post-depositional phenomena which may affect Lower and Middle Palaeolithic artefacts. Keeley's pl. 15 showing a ' "water-rolled" surface on an implement from Clacton' (1980, p. 205) and Stapert's fig. 5 (1976*a*) look similar, but Stapert states that the polish could have been formed either mechanically or chemically or both (Stapert, 1976*a*, p. 14). Singer & Wymer attribute one cause of lustre to 'the action of fine suspended sediment in flowing water' (1976, p. 17; Keeley, 1980, p. 128). Stapert has also put forward arguments for the presence in certain cases of cryoturbation retouch (1979), wind gloss (1979), solifluction traces (1979) and gelifluction scratches (1976*b*). Importantly, he attributes some edge rounding not to fluviatile action but to 'solution processes in the soil' (Stapert, 1979, p. 133). While in the case of the present series of experiments most of these non-fluviatile agents of abrasion and gloss formation do not apply they must in many cases have affected the condition of contemporary artefacts found in Pleistocene sediments.

In summary, microscopic examination of the experimental specimens shows that stone polish accumulates first on the flat facets and ridges of the handaxes, and that it tends to develop in intensity and extent of distribution with the distance travelled in the river. The polish does not begin to be visible at the edges until most of the available macro- and microscopic striking platforms have been struck at angles likely to produce flakes and most edge damage has ceased

Fig. 14.6. Degree of edge removal due to river transport. The photograph is of handaxe 16 after the experiment. The line drawing shows the original outline and was drawn from a photograph of handaxe 16 taken before the experiment.

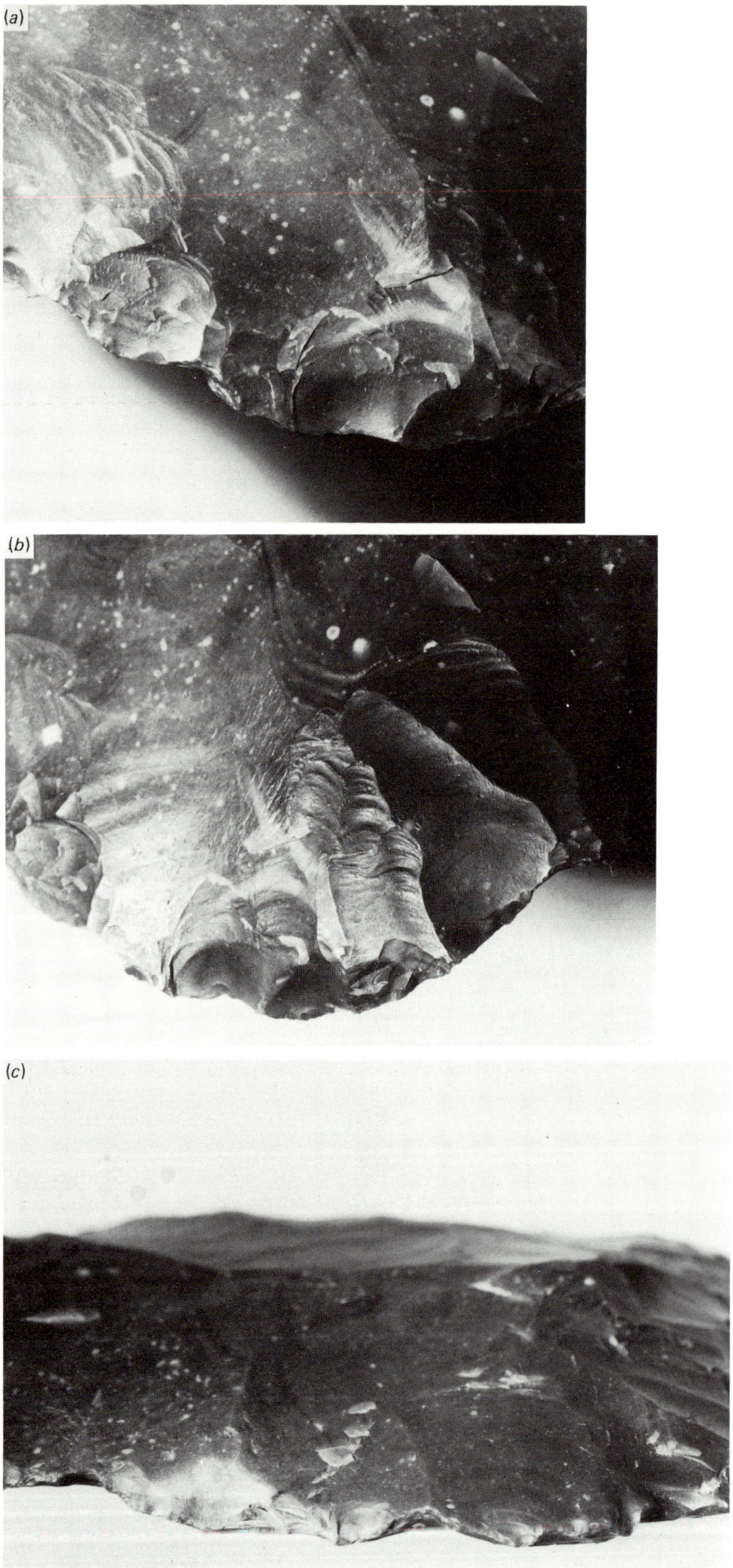

Fig. 14.7. Handaxe 16. Detail photomicrographs showing irregular damage to prepared edge of implement after river transport, ×3 magnification (photo. Botany School, Cambridge). (*a*) and (*b*) Top views of adjacent part of handaxe edge showing abrupt damage scars superimposed on flaking. (*c*) Side view (from low angle) showing small scars caused by edge battering from both sides.

to form. The time it takes for an edge to stabilise depends not only upon edge morphology, but upon the weight and shape of the handaxe, the strength of the river flow in relation to river bed morphology, and the grain size of the river sediments. Stabilisation in our experiments may be correlated with distance travelled. However, the smaller amount of polish on handaxe 33, which travelled the furthest, is not surprising in view of the number of variables at work.

The polish itself most resembles that produced by antler because the formation of striations, which would ordinarily indicate stone wear, has probably been inhibited by the presence of water. (See Keeley, 1980, pl. 53; Moss, 1983, pl. 6.3.) However, the distribution of the polish is unlike that of use–wear.

Shackley (1974) has used similar magnifications to those in the figures used in the present work, as well as incident light, but only her fig. 2 seems comparable with polish present on a ridge. It is disappointing that she has not described her experiments, since both the 'stress cracks' of her fig. 2 and the "braided" ridge' of her fig. 3 (*op. cit.*, p. 501) seem to be phenomena present on freshly struck flint. Our sample was too small to clarify Shackley's four factors which affect rate of alteration, although common sense attests to their relevance. The length of time the implement remains in the current may be of additional significance.

It is clear from what is already known concerning microscopic polish formation and sedimentation that few Lower Palaeolithic artefacts which have been subjected to river transport will remain unmodified by other factors. Post-depositional changes to the microscopic condition of these pieces are to be expected, particularly those arising from thousands of years of burial under varying climatic conditions. Microscopic examination of the experimental handaxes has demonstrated certain progressive changes arising from transport by the river. These changes are marked and varied enough to provide several points of reference for future work.

Conclusions

This study of the transport and abrasion of flint handaxes in the fluvial environment is at an early stage. Nevertheless, some interesting patterns are already emerging. The limited amount of data available at present does include some variation. However, work is continuing to support and clarify the results already obtained.

In general it appears that handaxe weight loss increases with distance travelled and that the heavier the handaxe the greater the loss in weight. How the shape of individual pieces influences transport distance is not yet clear. However, there is some evidence that ovate forms may be moved further than more elongate types. The distance travelled by the handaxes is strongly determined by their original starting position within the river. Current velocity and bed topography are particularly important in this respect.

Study of retrieved handaxes has shown that surface polish develops rapidly during this type of fluvial transport. After only 38 m of travel both edge wear and polish are well developed, although the rate of development of these features may be dependent upon river clast lithology. The polish is found initially on flat facets and ridges of the implements. In general, it increases in both intensity and distribution with distance travelled.

This modern analogue study represents a first attempt to understand some of the processes affecting the movement and modification of stone implements after abandonment by man. This type of study is closely akin to taphonomic investigations of the formation and preservation of fossil assemblages from living communities (e.g. Hanson, 1980; Shipman, 1981). Together they will inevitably provide much new and valuable information which will permit greater precision in our understanding of artefact and fossil assemblages in various environments.

Acknowledgements

We would like to thank G. de G. Sieveking and Dr M. H. Newcomer for encouraging us to publish this work. M.G.M. thanks I. Heywood and S. Openshaw for help in computer analysis of the handaxe data. P.L.G. gratefully acknowledges NERC and the Leverhulme Trust for financial support.

Bibliography

Anderson, P. C. (1980). 'A testimony of prehistoric tasks: diagnostic residues on stone working edges.' *World Archaeol.*, **12**, 181–93.

Bordes, F. (1961). *Typologie du Paleolithique Ancien et Moyen.* Delmas: Bordeaux.

Bradley, B. & Sampson, C. G. (1978). 'Artifacts from the Cottages site.' In: (ed., C. G. Sampson), *Paleoecology and archaeology of an Acheulian site at Caddington, England.* Southern Methodist University: Dallas, pp. 83–137.

Clark, J. Desmond (1974). *Kalambo Falls Prehistoric site*, volume II, Cambridge University Press, England.

Clark, J. D. & Kiroshina, H. (1976). 'New Plio–Pleistocene archaeological occurrences from the plain of Gadeb, Ethiopia. In: (eds., J. D. Clark & G. Isaac), *The Earlier Industries of Africa.* Colloque V, IX Congress of the International Union of Protohistoric and Prehistoric Sciences. UNESCO Paris, pp. 159–216.

Costa, J. E. (1983). 'Paleohydraulic reconstruction of flash flood peaks in the Colorado Front Range.' *Bull. Geol. Soc. Am.*, **94**, 986–1004.

Gifford, D. P. (1978). 'Ethnoarchaeological observations of natural processes affecting cultural materials.' In: (ed., R. A. Gould), *Explorations in Ethnoarchaeology.* University of New Mexico Press: Albuquerque, pp. 77–103.

Hanson, C. B. (1980). 'Fluvial taphonomic processes: models and experiments.' In: (eds., A. K. Behrensmeyer & A. P. Hill), *Fossils in the making.* University of Chicago Press: Chicago, pp. 156–81.

Isaac, G. (1967). 'Towards the interpretation of occupation debris. Some experiments and observations.' In: *Kroeber Anthropological Society Papers*, Vol. 37. Berkeley, California, pp. 31–55.

Isaac, G. L. L. (1977). *Olorgesailie: Archaeological Studies of a Middle Pleistocene lake basin in Kenya.* University of Chicago Press: Chicago.

Jones, P. (1980). 'Experimental butchering with modern tools and its relevance for Palaeolithic Archaeology.' *World Archaeol.*, **12**, 153–65.

Keeley, L. H. (1980). *Experimental determination of stone tool uses: a microwear analysis.* University of Chicago Press: Chicago.

Mansur-Franchomme, M. E. (1983). 'Scanning electron microscopy of dry hide working tools: the role of abrasives and humidity in microwear polish formation.' *Jl. archaeol. Sci.*, **10**, 223–30.

Moss, E. H. (1983). *'The functional analysis of flint implements – Pincevent and Pont d'Ambon: two case studies from the French final Palaeolithic.'* Brit. Archaeol. Rep.釉ternl. Ser. No. 177, Oxford.

Newcomer, M. H. (1971). 'Some quantitative experiments in handaxe manufacture. *World Archaeol.*, **3**, 85–94.

Newcomer, M. H. & Sieveking, G. de G. (1980). 'Experimental flake scatter patterns: a new interpretative technique.' *Jl. Field Archaeol.*, **7**, 345–52.

Roe, D. A. (1964). 'The British Lower and Middle Palaeolithic: some problems, methods of study and preliminary results.' *Proc. Prehist. Soc.*, **30**, 245–67.

Roe, D. A. (1968). 'British Lower and Middle Palaeolithic handaxe groups.' *Proc. Prehist. Soc.*, **34**, 1–83.

Schick, K. (in Bunn, H. *et al.*) (1980). F×Jj50: an early Pleistocene site in northern Kenya.' *World Archaeol.*, **12**, 109–36.

Shackley, M. L. (1974). 'Stream abrasion of flint implements.' *Nature*, **248**, 501–2.

Shipman, P. (1981). *Life History of a Fossil: An Introduction to Taphonomy and Paleoecology.* Harvard University Press: Cambridge, USA.

Singer, R., Wymer, J., Gladfelter, B. G. & Wolff, R. G. (1973). 'Excavation of the Clactonian industry at the golf course, Clacton-on-Sea Essex.' *Proc. Prehist. Soc.*, **39**, 6–74.

Singer, R. & Wymer, J. (1976). 'The sequence of Acheulian industries at Hoxne, Suffolk.' In: (ed., J. Combier), *L'evolution de l'Acheuléen en Europe.* Colloque x, ix Congress of the International Union of Protohistoric and Prehistoric Sciences. UNESCO: Paris, pp. 14–30.

Sneed, E. D. & Folk, R. L. (1958). 'Pebbles in the lower Colorado River, Texas, a study in pebble morphogenesis.' *Jl. Geol.*, **66**, 114–50.

Stapert, D. (1976*a*). 'Some natural surface modifications on flint in The Netherlands. *Palaeohistoria*, **18**, 8–41.

Stapert, D. (1976*b*). 'Middle Palaeolithic finds from the northern Netherlands.' *Palaeohistoria*, **18**, 44–72.

Stapert, D. (1979). 'The handaxe from Drouwen (Province of Drenthe, The Netherlands) and the Upper Acheulian.' *Palaeohistoria*, **21**, 128–42.

Warren, S. H. (1921). 'Excavations at the stone-axe factory of Graig-Lwyd, Pennaenmawr.' *Jl. R. Anthrop. Inst.*, **51**, 165–99.

Wymer, J. J. (1968). *Lower Palaeolithic Archaeology in Britain.* John Baker: London.

Wymer, J. J. (1976). 'The interpretation of Palaeolithic cultural and faunal material found in Pleistocene sediments.' In: (eds., D. A. Davidson & M. L. Shackley), *Geoarchaeology.* Duckworth: London, pp. 327–34.

15

Prehistoric flint mining: the history of research – a review

GÜNTER SMOLLA

Flint mines were among the first remains of prehistoric man's activities to have been observed, described and sometimes excavated.

Already, in the 1860s and 1870s several ancient mines and workshop sites were described, including places such as Grimes Graves and Cissbury, Rijkholt, Grand Pressigny, Spiennes and Obourg and several others in Britain, France and Belgium. Even as far east as Upper Silesia the pits of Makau (Makow) were first described in the late 1870s. In the nineteenth century, these sites were not the subject of heated discussion, as were the hand-axes of Boucher de Perthes, the first cave excavations, the megalithic monuments, the Swiss pile dwellings, the Danish Kjöken Möddinger or the spectacular finds of Hallstatt and La Tène. Perhaps this was because these mines were discovered and described by engineers or by empirically oriented scientists and others who avoided sophisticated interpretations, or perhaps because of their obviously industrial character.

As far as I know, the first comprehensive description of the most important sites in western Europe was that published in Washington DC in 1899, in the report of the US National Museum for the year 1897. The author, Thomas Wilson, was curator of the division of prehistoric archaeology in this museum after 1887; but in his earlier years he had been US consul in the city of Ghent in Belgium and his descriptions, especially those of the site of Spiennes and of the Brandon Flint knappers, appear to be first hand not compilations. That this chapter on 'flint mines and quarries in Western Europe and the United States' has been generally overlooked is surprising, but who would expect it to be included in a volume entitled *Arrowpoints, Spearheads and Knives of Prehistoric Times?*

The first monograph on prehistoric mining was Julius Andree's book *Bergbau der Vorzeit*, published in 1922. Andree was a trained geologist, who later became a specialist in Palaeolithic archaeology. He was a very industrious and thorough collector of reports and his book remains a useful work of reference.

It is interesting to see that Andree needed 28 pages for the description of flint mining, as opposed to 21 pages for copper mining, eight for tin and six for salt mining. Forty-eight years later, the mining engineer Robert Shepherd (1980) included 122 pages in his summary of publications on flint and chert mining, and only 47 pages for copper mining and 22 pages for other mining activities. This illustrates the comparative development of these subjects in the intervening period.

A considerable number of flint mines were excavated and studied between the two world wars. Research continued in the best-known flint mining centres in western and northern Europe and spread to other countries – especially to Poland. In western Europe, Grahame Clark & Stuart Piggott's *The Age of the British Flint Mines* (1933) was perhaps the most valuable report of this period. It summarised what was known of European mining and its cultural significance before the development of radiocarbon dating, and was a stimulus for further research.

After the last war, Martin Jahn published the next compilation after Andree's, but with a new paradigm. Born in 1888, Jahn was a trained archaeologist whose main interest was in Iron Age problems. In his later years he concentrated on economic questions in prehistory. His paper 'Gab es in der vorgeschichtlichen Zeit bereits einen Handel?', published in 1956, emphasises the importance of understanding the organisation of mining activities and the distribution of special raw materials from known sources in order to reconstruct neolithic economic structure.

Of course Jahn was not the first to recognise these problems and he never claimed to have been (he was the most modest archaeologist I ever met). Subsequently he went further and published his paper *Der älteste Bergbau in Europa* (1960) (at this time, he was 72), the first archaeological typology of neolithic mines.

Since Jahn, a number of other archaeologists including the Maastricht group (Felder, 1981) Elisabeth Schmidt (1973), Leonhard Fober & Gerd Weisgerber (1980), Robert Shepherd (1980), Gale Sieveking (1979), Jazec Lech (1980) have discussed mining techniques and developed typologies, though these unfortunately differ between each other. Perhaps it would be a good idea if a commission of experts from various countries – including experts of mining activities other than flint – could produce a paper on technical nomenclature in several languages!

However, despite some unsolved problems, the present state of knowledge concerning the technical aspects of mining research seems to me so far advanced – thanks mostly to the Maastricht group and our Polish colleagues – that we can enlarge the definition of mining research to include other closely related questions. The first such topic, the question of distribution and exchange of mined materials has been studied almost since the time of the first exploration of the mines themselves. A good deal of work has been done on distribution studies (Sherratt, 1976) in recent years, often in close cooperation with specialists in analysing the different raw materials and their sources (Sieveking *et al.*, 1972; Cowell, 1981; Phillips *et al.*, 1977).

In the past few months several publications have appeared, mostly in German, on this topic and I would like to mention these briefly.

Christoph Willms (1982) study on the neolithic flint and obsidian trade in Europe gives some basic models for further discussion on this subject. F. R. Van Itersen-Scholten's (1981) paper on the question of transport in the Netherlands is also stimulating, because it emphasises the importance of water transport which, though not entirely neglected previously, has generally been underestimated.

On the other hand, Manfred Korfmann (1982), in his monograph on the Tilki Tepe on the shore of Lake Van in Eastern Turkey, has included a chapter on the obsidian trade in the ancient Middle East, which gives us some information on another ancient and widely used method of transport, that of simply carrying the material on one's back. If one carrier can bear 40–45 kg about 36 km a day – and Korfmann gives good arguments for such a belief – then this underlines the significance of Willms' statement (1982), that the heaviest hoard of flint slabs he knows had a weight of 32 kg.

In relation to these suggestions, we must also accept that both forms of transport (by water and by carrier) probably were used in combination. The ethnographic model for this could be the system of 'portages' between different river systems as practised by North Americans and the similar *Wologs* of the later Middle Ages in eastern Europe, which made possible the traffic between the Baltic and the Black and Caspian Seas. It seems to me that if we accept this combination of water and land transport, then the traffic patterns in prehistoric Europe from neolithic times onwards becomes much more comprehensible.

Andreas Zimmermann has shown, in a paper published in 1982, that neolithic settlements differ statistically in the amounts of flint implements they produce or use and that we have to recognise some sort of 'site-hierarchy' even as early as the Linearbandkeramik LBK. This parallels the evidence for a similar organisation of the distribution of obsidian in the Near East put forward by Renfrew and others (1969). It seems that this hierarchical model should be of considerable use in analysing our distribution evidence. Recent studies, including those published in this volume, have documented the distribution of material from flint mines in many European countries. When Bernd Engelhardt's excavation of mining sites of *Platten Silex* (tabular flint) in northern Bavaria are published, only a few gaps will remain in the system of mining sites in most parts of Europe and the dispersal of flint material from them.

The most important result of the most recent studies, in my opinion, is that the use and the distribution of special sorts of flint material – with few exceptions – are seen to be not only limited to closely defined chronological periods but are also related to cultural or archaeological units, primarily defined by ceramic styles such as Bandkeramik, Michelsberg pottery and Funnel beakers. This was not always as evident as it is today. It was only 26 years ago that Ulrich Fischer (1956) first put forward satisfactory evidence that the distribution of pottery styles was identical with particular burial customs! Today most of us will agree, I think, that archaeological units restricted in time and space in some way reflect social units and also, as we heard from Professor Zeitlin, mirror political units in a broader sense. So it is possible that, in the future, we will become more and more familiar with the dynamics of neolithic social and political development, with the help of such distribution research.

But for this we also need better chronological control. We need this for several reasons. Firstly, our symposium programme has shown that the theatre of research is expanding from Europe to the Near East and Egypt.

As Andreas Redlich emphasises in his book '*Studien zum Neolithikum Mittelasiens*' (1982), in his chapter 'silex-Gruben und-Werkstatten', no less than 23 papers on flint mining and manufacturing in central Asia were published by our Soviet colleagues between 1956 and 1972. Palaeolithic mining pits have also been recorded since 1945, in river gravels in Angola, published by H. Breuil and J. Janmert (1950) and confirmed by Desmond Clark (1963), and there is also recent evidence from eastern Asia, North America and Australia, that has to be taken into account.

But in addition we must also look at the chronological and technical relationship between flint mining and the other activities. Gerd Weisgerber has described (page 131) the relations between flint and copper mining techniques on the one hand, and between mining techniques and those for digging out graves and grave-systems in rock formations, which can be traced at least as far back as the Ghassul–Bersheba culture, in the second half of the 4th millenium BC (Fig. 15.1).

Technically, there is little difference between digging out artificial caves and digging tunnels in more or less solid rock formations for graves or for mining-systems, so that the working tools are likely to be similar. It seems likely that the lateral galleries at Monte Tabuto in Italy and possibly also the mines at Lisboa-Rocio in Portugal are technically closely related to the Mediterranean artificial rock cut tombs. Does the pattern of distribution between southern France and Kleinkems of closely similar grooved tools, presumably connected with special mining methods, give us a way to test this hypothesis?

The other relationship that must be tested is the technical similarity between the digging of mining shafts and water wells. Neolithic wells are known as early as the elder LBK in Moravia (J. Luning, pers. comm.) and are equally early in Bulgaria (H. Todorova, 1982, p. 92). They are also known in Neolithic China. These narrow cylindrical shafts are a considerably technological achievement and the methods of construction could be closely related to those of similar shafts in the flint mines.

Many of these activities are associated with the industrial

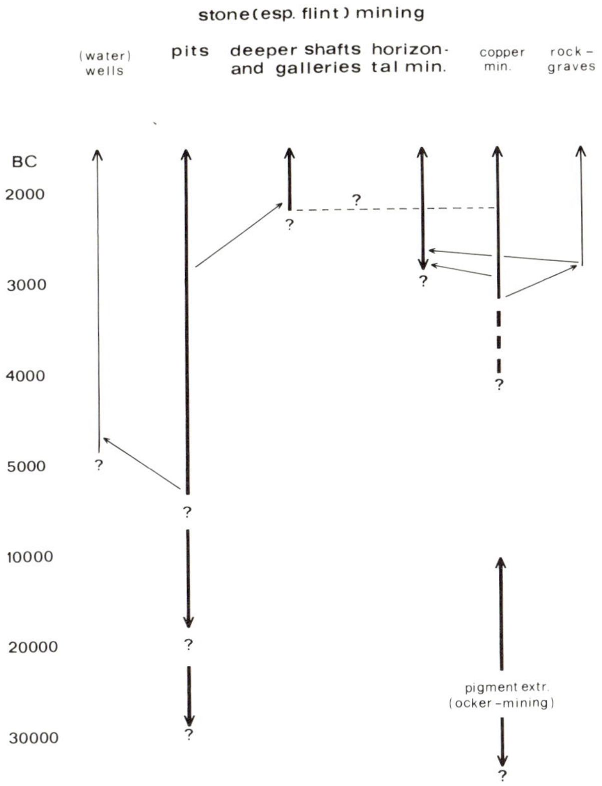

Fig. 15.1. The chronology of flint mining.

and agricultural changes and the population expansion originally described by Childe as the *Neolithic Revolution*. Though Childe's concept is no longer generally acceptable in its original form, there is no doubt that the changes in scale of activity that accompany the expansion of early agriculture in Europe and the Near East do constitute a real technical revolution.

Flint mining was an especially important part of this process and the remains of this activity are much better preserved and more suitable for excavation than most other Neolithic activities.

The role of technology and technicians is one of the most exciting themes in the social and political history of mining and miners in Neolithic times, and one can look forward to new and exciting developments in this field.

Bibliography

Andree, I. (1922). *Bergbau in der Vorzeit*, I. Bergbau auf Feuerstein, Kupfer Zinn und Salz in Europa. Leipzig.

Breuil, H. & Janmart, J. (1950). 'Les limons et graviers de l'Angola du North-Est et leur contenu archéologique.' *Diamang Publ. culturais*, No. 5, Lisboa.

Clark, J. Desmond (1963). 'Prehistoric cultures of north east Angola and their significance in Tropical Africa.' *Diamang Publ. culturais*, No. 62, Lisboa.

Clark, J. G. D. & St Piggott, (1933). 'The age of the British flint mines.' *Antiquity*, **7**, 166–83.

Felder, P. J. (1981). 'Prehistoric flint mining at Ryjckholt–St Geertruid (Netherlands) and Grimes Graves (England).' *Proceedings of the Third International Symposium on flint*, (*Staringia*, No. 6), Netherlands, pp. 57–62.

Fischer, U. (1956). 'Die Gräber der Steinzeit im Saalegebiet.' *Vorgeschichtliche Forschungen*, Bd. 25, Berlin.

Fober, L. & Weisgerber, G. (1981). 'Feuersteinbergbau – Typen und Techniken.' In: *5000 Jahre Feuersteinbergbau*. 2. Aufl. Bochum, pp. 32–4.

von Itersen-Scholten, F. R. (1981). 'Late neolithic flint transport in the Netherlands.' *Staringia*, **6**, 115–17.

Jahn, M. (1956). 'Gab es in der vorgeschichtlichen Zeit bereits einen Handel? Abhandlungen der Sächsischen Akademie der Wissenschaften zu Leipzig.' *Phil.-hist.*, Kl. Bd. 48, Berlin.

Jahn, M. (1960). 'Der älteste Bergbau in Europa. Abhandlungen der Sächsischen Akademie der Wissenschaften zu Leipzig.' *Phil.-hist.*, Kl. 52, H. 2, Berlin.

Korfmann, M. (1982). 'Tilkitepe, Die ersten Ansätze prähistorischer.' *Forschung in der östlichen Türkei. Istanbuler Mitteilungen*, Beiheft 26. Tübingen.

Lech, J. (1980). 'Flint mining among the early farming communities of central Europe.' *Przeglad Archaeologizny*, **28**, 5–55.

Phillips, A. P., Aspinall, A. & Feather, S. W. (1977). 'Stages of "Neolithisation" in southern France: supply and exchange of raw materials.' *Proceedings of the Prehistoric Society*, **43**, 303–16.

Redlich, A. (1982). 'Studien zum Neolithikum Mittelasiens.' *Antiquitas*, Reihe 3, Bd. 25. Bonn.

Renfrew, C. (1975). 'Trade as action at a distance: Questions of integration and communication.' In: Sabloff J. & Lamburg-Karlofsky C. (eds.), *Ancient Civilization and Trade*. Albuquerque. University of New Mexico Press, pp. 3–59.

Schmid, E. (1973). 'Reviere urgeschichtlichen Silex Bergbaus in Europa.' *Der Anschnitt*, 25, H. 4, 12–15.

Shepherd, R. (1980). Prehistoric Mining and Allied industries. London, Academic Press.

Sherratt, A. G. (1976). 'Resources technology and trade: an essay in early European copper metallurgy.' In: Sieveking, G.de G., Longworth, I. H. & Wilson, K. M. (eds.), *Problems in Economic and Social Archaeology*, London. Duckworth, pp. 557–81.

Sieveking, G. de G. *et al.* (1972). 'Prehistoric flint mines and their identification as sources of raw material.' *Archaeometry*, **14**(2), 151–76.

Sieveking, G. de G. (1979). 'Grimes Graves and prehistoric European flint mining.' In: Crawford, H. (ed.), *Subterranean Britain*. London. Baker, pp. 1–44.

Todorova, H. (1982). 'Kupferzeitliche Siedlungen in Nordost-Bulgarien.' *Materialien zur Allgemeinen und vergleichenden Archäologie*, Vol. 13, München.

Willms, Chr. (1982). 'Zwei Fundplätze der Michelsberger Kultur aus dem westlichen Münsterland, gleichzeitig ein Beitrag zum neolithischen Silexhandel in Mitteleuropa.' *Münstersche Beiträge zur Ur- und Frühgeschichte* 12. Hildesheim.

Wilson, Th. (1899). 'Arrow points, spearheads and knives of prehistoric times.' *Report of the US National Museum for 1897*, pp. 811–988, Washington.

Zimmermann, A. (1982). 'Zur Organisation der Herstellung von Feuersteinartefakten in Bandkeramischen Siedlungen. Internationales Koloquium Nové Vozokany 17–20. Nov. 1981.' *Siedlungen der Kultur mit Linearkeramik in Europa*. Nitra, pp. 319–23.

16

The technological relationship between flint mining and early copper mining

G. WEISGERBER

Introduction

Archaeologists are often of the opinion that copper mining preceded sophisticated flint mining. The techniques necessary for complex flint mining are presumed to be dependent on technology previously developed in mining copper. The conclusion is that the existence of a flourishing flint mining industry of central and western Europe is due to stimulation provided by methods taken over from copper mining.

The purpose of the present paper is to illustrate that this is not the case, from a study of the attendant technology (Jovanovic, 1981).

Mining and processing of copper can be postulated for the 5th millennium BC and has been proved to exist in southeastern Europe in the 4th millennium BC (Jovanovic, 1982; Cernych, 1972, 1978). Flint mining is associated with the Funnelbeaker culture of north and eastern Europe and the Lengyel culture of Austria and Hungary, which are in close contact with the Copper Age cultures (such as the Tiszapolgar) of southeastern Europe. In Poland the processes of flint mining are known to have been carried out in the Mierzanowice culture when copper in small quantities is already in use. Therefore one could reach the conclusion that the late Neolithic cultures were stimulated to mine flint because copper mining was already being carried out in southeastern Europe. This implies that the idea of mining flint by the most developed mining technology was not possible, except under the influence of technology previously developed for copper mining.

The relationship between geology and mining

(a) Flint deposits

From the point of view of mining technology geological deposits of flint may be classified as either primary or secondary. Primary deposits such as those found at the Grimes Graves (GB 13,[1] Spiennes (B 12) or Krzemionki (PL 6) flint mines, are those where the flint layers are more or less horizontal in the bed-rock. Such deposits are usually exploited by systematic shaft-mining, although other techniques are also used. For example drift tunnels for exploiting primary flint beds are known at the mines of Kleinkems (D 1, Pleigne (CH 1), Monto Tabuto (I 1) and probably those at Lisbon/Rocio (P 1). Among shaft mines the types range from simple workings ('Duckelbau') to large scale, systematic mines such as Ryckholt (NL 1) or Krzemionki (PL 6). The existence of desirable materials in layers and seams and the necessary stability of the adjoining bed-rock provided the basis on which extensive mine works could be carried out at an economically feasible level. Apparently there was a considerable demand for materials of fine quality.

However, quality of material was not the only factor in determining whether open-cast or underground mining was carried out, as is shown by the examples of mines at Vienna (Mauer) (A 1) and both Sümeg (H 2) and Tata (H 3) in Hungary. In these areas, material of poor quality was mined from which large implements could not be produced. Vienna is an underground mine with entrance shafts, while the two Hungarian sites are open quarries. The material was mined since nothing of better quality was available within a reasonable distance.

But there is no doubt that the earliest flint mining was that undertaken in secondary deposits. Three types of flint in secondary deposits are recognised: 'Maas Egg' and 'Baltic Flint' are alluvial deposits derived from marine sediments or from glacial gravel deposits. Flint nodules, weathered out of less resistant rock and deposited in residual beds through erosion, are typical of such sites as Grand-Pressigny (F 12) or Aubel (B 9). The exploitation of marine strand terraces in Denmark (DK 5.6) seems to be a local variation of gravel exploitation. Deposits of alluvial/glacial/marine and weathered materials were mined in open-workings, such as open diggings and pits. No underground mining has been found in Europe which is related to these types of deposit.

(b) Copper deposits

Copper is often found in deposits similar to those of flint.[2] Flint and copper deposits are both found in 'seams', and

[1] Site numbers such as GB 13 are taken from Weisgerber *et al.* (1980).

[2] For the purpose of this paper, copper deposits which were mined from the Middle Bronze Age are not taken into consideration. These deposits were first worked after easily available deposits had been fully exploited in the early Bronze Age, and no more ores could be mined. The history of metallurgy illustrates how people became adept at exploiting increasingly poorer ore deposits.

also as concentrations formed as a result of weathering and erosion processes. In Europe, however, both types of deposits only played a minor role in the Copper Age, as far as the frequency and amount of recovered raw material are concerned. Most of the deposits mined in the Copper Age were found in steeply inclined veins in the rock, which are easily distinguishable by the greenish colour of the carbonate ores. Such veins were generally worked out down to ground water level.[3]

Early copper mining

Timna, Israel

The research on copper mining at Timna has been studied by Conrad & Rothenberg (1980). At this site copper ores occur as impregnated zones in horizontal layers as a result of the 'Red Bed' type of mineralisation. The copper has been precipitated onto organic materials and forms small lumps and nodules similar to flint nodules. These may occur irregularly throughout a layer or may be concentrated in horizons within the white Nubian sandstone. The nodules may also be weathered out of the sandstone and exposed by wind abrasion. Such deposits occur as concentrations of copper within weathered sediment.

Exploitation of the site at Timna began in the 4th millennium BC when the ore nodules were simply collected from the surface; later, simple pits were dug to obtain the nodules in lower layers. Some simple open-quarry pits were later enlarged into underground works in the 3rd millennium (Rothenberg, 1979). This sequence is characteristic of early copper mining. As the ore 'seams' at Timna are more irregularly deposited than flint deposits, systematic mining techniques are not applied at the very beginning. The forms and size of mine adits are irregular as they follow the concentration of ore deposits. In locations where there is a high concentration of ore, larger mine systems develop. When the lode peters out, an unsystematic search is carried out for further concentrations. Timna is perhaps an unusual example of early copper-mining exploitation and should not be regarded as typical of Copper Age mining procedures.

Rudna Glava, Yugoslavia

A single example of Vinca culture copper-mining technology dated to 3500 BC has been found at Rudna Glava in Yugoslavia (Jovanovic, 1982). It may be presumed that many other mines of the same type will be found when a systematic search is made for them. At Rudna Glava, copper carbonate is deposited in fissures in limestone bedrock. Mining technology developed here – as in the case of Timna – in response to conditions prevailing at the site and no systematic exploitation was undertaken. Where the ore outcropped on the surface, it was simply dug out in slanting pits irregularly sited over the area. The 'shafts' simply followed the steeply inclined veins of ore down to ground water level. In mining terminology, such works would not be termed 'shafts', as the inclined tunnels were not constructed to provide access to underground galleries.

There is a major difference in technology between Rudna Glava and systematic flint mining such as that carried out at Spiennes (B 13) and the former can be compared only with undeveloped flint mines such as Monte Tabuto in Italy (I 1) or Hov, in Denmark (DK 1).

Veshnoveh, Iran

Another early Copper Age mine at Veshnoveh (near Qom) in Iran is similar in its technology to Rudna Glava; the ore seams have been worked by more or less horizontal tunnels, or, where the seams are inclined by almost vertical tunnels, leading to cavities similar to those of Rudna Glava (Holzer & Momenzadeh, 1971).

At Veshnoveh the same type of tools were used for mining the so-called 'grooved stone mauls' as in Rudna Glava. This type of implement is also common in mines exploiting flint in hard bed-rock (D 1, F 1.4).

Timna and Rudna Glava are not the only types of mining technology known in the early Copper Age cultures. On the basis of geological conditions it seems likely that the most common type of mining works must have been similar to those of Aibunar, Bulgaria (Cernych, 1972, 1978) (see Fig. 16.1), and also near Huelva in southern Spain (Rotherberg & Blanco, 1980, 1982) where steeply inclined seams of quartz and copper ore were simply quarried from the surface. In Bulgaria the remains of these workings were up to ten metres deep and many kilometres long, interrupted by bridges of sterile rock. There are mines of similar size in Oman, whereas the examples in Spain and Ireland (Jackson, 1980) are very much smaller.

At these sites the seams vary from less than a metre to several metres in breadth, so that mining technology is again governed by geological conditions. Planned exploitation did not take place. Since the mining pits were usually no deeper than five metres, there were no special technical problems to solve, as is the case when depths are greater.

We have shown that from the viewpoint of mining technology it is not possible to derive the flint mining system in central and western Europe from the contemporary or earlier copper-mining technology of southeastern Europe. Early copper mines do not have the highly developed systems of galleries and shafts present in flint mines and these could not have developed from any known examples in copper mines. The technology developed for deep-shaft mining works of flint was far superior to the technology of copper mining. A comparable technology for copper mining was achieved, for instance, in Timna, at a far later period beginning about 1200 BC in the Late Bronze Age (Conrad & Rothenberg, 1980).

[3] The terms 'primary' and 'secondary' deposits used when referring to flint, cannot be applied in the same way to copper deposits, due to the more complicated chemical and physical weathering processes of the copper ores.

Fig. 16.1. Ai Bunar, Bulgaria. This copper mine is a good example of trench mining in more or less vertical ore veins.

Chronology

A comparison of dates for flint mines and copper mines shows that chronologically the one type of technology (flint) cannot be derived from the second type (copper).

The available dates (Appendix 1) show that mining technology for flint has been carried out in Europe since Palaeolithic times, although the forms are very simple. Flint mining in eastern Europe starts in Upper Palaeolithic times PL 1, 12, 17, 19), but systematically constructed mines (Krzemionki PL 6) start at the middle of the 4th millennium BC and at the same period in southeastern Europe (Vienna, A 1) and at the Belgian mine of Jandrin-Jandrenouille (B 12) in western Europe. The early 3rd millennium BC reveals an increase in shaft mining (at the sites GB 3, 4; B 1, 4). If we use the German terminology (see Appendix) we can say that 'Duckelbaue' date from both 4th and 3rd millennia (at sites B 2, 11; DK 1, 2; F 23 and GB 1). 'Pingenbaue' (GB 2, H 2, 3, PL 2, 5, 13, 15, 19) have also been recognised from the same periods and from much earlier times. These correspond in type to the less systematically developed copper mines of the same age. The copper mines are plainly not older than the fuller developed flint mines, they are of the same age.

Thus the aspects of mining technology, chronology and areal distribution all show that flint-mining technology was not developed subsequently to copper-mining technology.

However, one cannot deny that, during the Copper Age, the general economic and social organisation was different from that of the early Neolithic (Lichardus, 1980). An increased demand for high quality copper and flint was satisfied as each supplier developed the most efficient mining technology for the material and types of site available. During this period flint and copper were materials of equal economic importance. For many purposes flint was still the most used raw material – for example, for sickles – and competition between the two materials really occurred much later. Competition produced the well-known excellent accomplishments by the 'flint-smith', which themselves would not have been possible without the high quality raw materials obtained by sophisticated mining techniques.

Acknowledgements

The author is grateful to Mrs A. M. Martin, Bochum, for translating this paper.

Appendix 1: dated flint mines in Europe

The following table taken from the catalogue of the 1980 flint mining exhibition at the German Mining Museum (Weisgerber *et al.*, 1980) lists the dates given for flint mines in Europe. Also included are dates referred to in Burleigh (1981). The mining terms are given in German, as these are more precise than those used in English. Approximate translations for the German terms are as follows:

German	*English*
Schachtsysteme	exploitation by shafts and a system of gallery networks
Schächte	pits
Duckelbaue	rounded pits with irregular cavings at the bottom, similar to bell pits
Hangpingen	diggings along a slope and into a slope
Tagebaue	open-air mining
Stollen	drift, adit
Gräberei	shallow digging
Kuhle	rounded pit deeper than a 'Gräberei'
Pinge, Pingenbau	irregular open digging larger than a 'Kuhle'

	Place	Type of mining	Culture	Radiocarbon dates (non-calibrated)
A1	Vienna	Schächte	Lengyel	3500
B1	Spiennes	Schachtsysteme	Michelsberg	from 3000
2	Obourg	Duckelbaue	from end of Bandkeramik	before 2000
3	Flénu	Schächte		
6	Avennes	Schächte	Michelsberg	
11	Mesvin	Duckelbaue		Lv 65 3270±150 215 3390±150 BM417 3881±123
12	Jandrain–Jandrenouille	Schachtsysteme		IRPA367 3495±260
CH1	Pleigne	Hangpingen	(Cortallod?)	B2057 2990±240 2050 3070±100 2601 3260±100
D1	Kleinkems	Tagebau und Stollen		2000
3	Aachen	Tagebau		KN2662 2630±140
DK1	Hov	Duckelbau	Trichterbecher C	3000
2	Bjerre	Duckelbau	Trichterbecher C	3000
3	Aalborg	Duckelbau	Late Neolithic	1500
F23	Hardivillers–Troussencourt	Duckelbau		C-14 2500–2000
43	Etables	(Schacht)	Cerny	before 3000
46	St. Mihiel		Seine–Oise–Marne	C-14 2200–2000
GB1	Blackpatch	Duckelbau		BM290 3140±130
2	Church Hill	Kuhlen und Pingen		BM181 3390±150
3	Cissbury	Schachtsysteme		BM185 2780±150 BM183 2770±150 BM184 2700±150
GB4	Harrow Hill	Schachtsysteme		BM182 2980±150
9	Easton Down	Kuhlen		BM190 2530±150
13	'Grimes Graves'	Schachtsysteme		BM 2340–1750
H2	Sümeg	Pingen		A-246 2570±160 2220±160
3	Tata	Pingen		HV-1770 1860± 65
NL1	Ryckholt–St Geertruid	Schachtsysteme		GrN-4544 3120± 60 GrN5549 3050± 60
PL1	Orońsko	Pingen	Late Palaeolithic	10 000–9000
2	Tomaszów	Pingen	Linearbandkeramik–Mierzanowice	GrN-7050 3945± 40 7590 3750± 70 7591 4195± 70 7592 3720± 65
4	Polany–Kolonie II	Duckelbau	Mierzanowice	GrN-6834 2040± 40 6833 2055± 35
5	Polany II	Pingen	Mierzanowice	1541± 81
6	Krzemionki	Schachtsysteme	Trichterbecher Kugelamphoren Mierzanowice	3500–2500 2500–1800 1800–1600
7	Ruda Kóscielna	Schachtsysteme (?)	Mierzanowice	(1800–1600)
8	Borownia	Schachtsysteme (?)	Mierzanowice	(1800–1600)
9	'Korycizna'	Schachtsysteme(?)	Mierzanowice	(1800–1600)
11	Ożarów		Mierzanowice	(1800–1600)
12	Świeciechów–Lasek	Gräberei–Kuhlen	Palaeolithic to Mierzanowice	
13	Maków	Pingen	Linearbandkeramik to Trichterbecher	
14	Jerzmanowice–Dąbrówka	Kuhlen	Lengyel	4000

	Place	Type of mining	Culture	Radiocarbon dates (non-calibrated)
15	Sąspów	Pingen	Linearbandkeramik to Trichterbecher	GrN-7692 3750±135 7693 3625± 75 Bln-1462 3375± 60 GrN-7052 C 3375± 90 BM-1128 3096±102
16	Bębło	Kuhlen	Lengyel–Polgar	4000
17	Wołowice	Kuhlen	Magdalenien	15 000–9000
19	Brzoskwinia	Pingen	Late Palaeolithic	GrN8519 9555±400
S1	Kvarnby	Pingen und Duckelbau	Trichterbecher	BM410 2900±115

Bibliography

Burleigh, R. (1975). Radiocarbon dates for flint mines. *Staringia,* **3**, 89–91.

Černych, E. N. & Radunceva, A. (1972). 'Les anciennes mines de cuivre des environs de Stara Zagora.' *Archeologia*, **14**, 61–73.

Černych, E. N. (1978). 'Aibunar, a Balkan copper mine of the fourth millennium BC (investigations of the year 1971, 1972 and 1974).' *Proceedings of the Prehistoric Society*, **44**, 203–18.

Conrad, Hans G. & Rothenberg, B. (1980). 'Antikes Kupfer im Timna-Tal.' *Der Anschnitt, Beiheft* 1 (Veröffentlichungen aus dem Deutschen Bergbau-Museum, Bochum, No. 20).

Fober, L. & Weisgerber, G. (1980). 'Feuersteinbergbau – Typen und Techniken.' In: Weisgerber *et al.*, pp. 32–47.

Holzer, H. F. & Momenzaden, M. (1971). 'Ancient copper mines in the Veshnoveh area, Kuhestan-E-Qom, west-central Iran.' *Archaeologia Austriaca*, **49**, 1–22.

Jackson, J. S. (1980). 'Bronze Age copper mining in Counties Cork and Kerry, Ireland.' In: 'Scientific studies in early mining and extractive metallurgy.' In: P. T. Craddock (ed.), *British Museum Occasional Paper*, **20**, 9–30. London.

Jovanovic, B. (1981). 'Flint mining and early copper mining in the south-east Europe.' *Staringia*, **6**, 63–4.

Jovanovic, B. (1982). *Rudna Glava. Der älteste Kupferbergbau* im Zentralbalkan. Bor/Beograd (Museum für Bergbau und Hüttenwesen/Archäologisches Institut, Besondere Ausgaben, No. 17.

Lichardus, J. (1980). 'Zur Bedeutung der Feuersteingewinnung in der jüngeren Steinzeit Mitteleuropas.' In: Weisgerber *et al.*, pp. 265–70.

Rothenberg, B. & Weyer, H. (1979). *Sinai*. Bern.

Rothenberg, B. & Blanco-Freijeiro, A. (1980). 'Ancient copper mining and smelting at Chinflon (Huelva, S.W. Spain).' In: Craddock, pp. 41–62.

Rothenberg, B. & Blanco-Freijeiro, A. (1981). *Ancient Mining and Metallurgy in South-West Spain* (*Metal in History 1*). London.

Weisgerber, G., Slotta, R. & Weiner, J. (eds.) (1980). *5000 Jahre Feuersteinbergbau – Die Suche nach dem Stahl der Steinzeit. Deutsches Bergbau-Museum, Bochum*, (Veröffentlichungen aus dem Deutschen Bergbau-Museum, No. 22).

17

The exploitation of chocolate flint in central Poland

ROMUALD SCHILD

The prehistoric exploitation of the central Polish chocolate flint began to be studied by archaeologists shortly after World War I, when Stefan Krukowski pointed out that most of the terminal Palaeolithic and Mesolithic assemblages were made of this excellent and well-characterised raw material.

Because of the colour of the vast majority of its varieties the flint was, from the first, described as chocolate coloured and has since become generally known as the chocolate flint.[1] Krukowski was able to predict the position of the natural outcrops of this flint from the locations of the terminal Palaeolithic and Mesolithic sites (Krukowski, 1920, p. 194). Later, in 1922, he and Samsonowicz identified in the field the first chocolate flint outcrops, in the northeastern foothills of the Holy Cross Mountains (Krukowski, 1923; Samsonowicz, 1923). Surveys in subsequent years resulted in the recognition of several flint extraction sites and test excavations and surface collecting was undertaken at some of these (Krukowski, 1939).

During the mid 1960s, the Institute of the History of Material Culture returned to the chocolate flint belt, to carry out studies of flint mining, distribution and use, as part of a long term project. This included a systematic regional survey to locate all recognisable flint extraction sites and the excavation of selected sites. Plans for associated research included chemical analysis, flint distribution studies and studies of the raw material economy of the prehistoric societies utilising this particular flint type.

In 1968, a detailed survey of the northeastern slopes of the Holy Cross Mountains was undertaken. An area measuring more than 1000 km^2 was prospected and 17 potential flint extraction sites were recorded (Fig. 17.1), including some already known from the previous investigators (Schild, 1971).

Excavations at selected flint extraction sites began in 1971. Three sites have so far been studied, Polany II and Polany Kolonie II in 1971 and 1972, and Tomaszów I, between 1973 and 1975.

Several flint processing workshops associated with nearby mines were also excavated at Polany Kolonie I and at Tomaszów II. More recently, salvage excavations directed by Młynarczyk and Lech have been undertaken at the mine of Wierzbica-Zele.

[1] The term chocolate flint is traditional. Correctly, the stone should be classified as chert since it occurs in non-Cretaceous formations.

Research approaches and aims

In the prehistoric communities, flint mines and flint distribution can be thought of as part of a larger system of raw material procurement and use. As such, mining and the distribution of the raw material are necessarily influenced by the types of mining technique in use and by the character of the stone technology: that is, the total system of core and tool preparation and use, repair and conservation. A dynamic approach is required to both mining and processing, to take these factors into account. Micro-stratigraphical analysis is used to reconstruct the cutting and exploitation of the flint mines and the stages and dynamics of their refilling. Estimates are made of the size of quarries and their raw material output, as well as of the size and type of products turned out by the flint processing workshops; and the artefact assemblages at these workshops are compared with assemblages found on settlements using chocolate flint at various distances from the raw material sources.

The chocolate flint

In the foothills of the Holy Cross Mountains (Góry Świętokrzyskie), the limestones and karstic clays containing chocolate flint occur in a northeast – southwest belt some 50 km long and 3–6 km wide, lying between the Radomka and Kamienna Rivers.

The outcrops form three main groups: eastern, central and western (Fig. 17.1). Each includes several flint extraction sites. Some 20 km to the southeast from the eastern group are a few small outcrops of the chocolate flint near Gliniany. The flint nodules and slabs of the eastern and central groups occur in limestone, while those of the western group are chiefly associated with pre-Quaternary karstic clays.

The limestone was initially classified as belonging to the Upper Astartian (Samsonowicz, 1923, 1932, 1934; Pożaryski, 1948; Dembowska, 1953; Karczewski, 1960). Later changes in the formal sub-division of the Upper Malm merged the Upper Astartian with the Late Oxfordian (Kutek, 1967; Malinowska, 1967). However, the lithostratigraphic position of the oolithic and pellithic flint-bearing limestones allow them to be assigned to the *Sutneria palatynata* zone, within the Lower Kimmeridgian of the Paris Basin (Kutek, 1961).

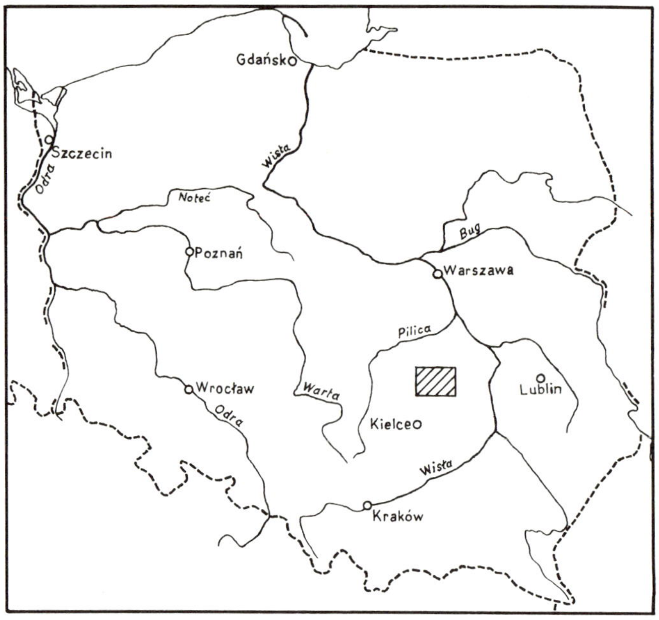

Sketch map of Poland

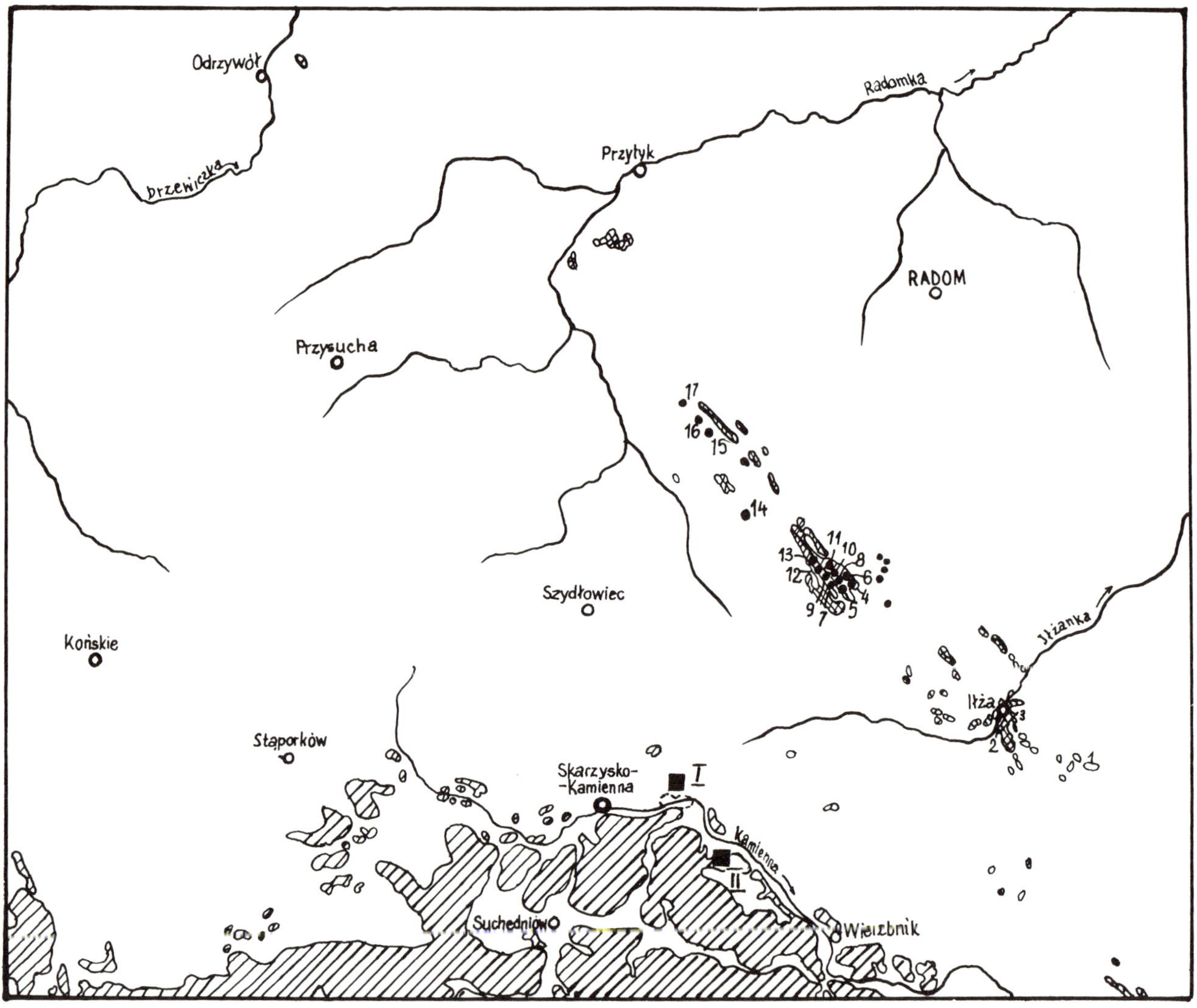

Chocolate flint locality map

A bed of Quaternary sediments of variable depth overlies both limestone and karstic clays. In the areas where prehistoric extraction sites occur, the limestone and clay is relatively close to the surface, and seldom covered by more than 4 m of sediment.

The chocolate flint is a rock of unusual appearance. It is generally very easy to recognise on the basis of simple inspection, in spite of its considerable local variability. The main petrographic and physical characteristics of this flint are tabulated in the Appendix.

An attribute classification based on a large number of samples collected at every identified extraction site gave 11 main groups of flint, each containing a few sub-varieties. Four groups are the most common and apparently were the most important in the past.

The mines

The oldest quarry pits to be identified appear to be those found by Krukowski before World War II at Orońsko II2, in the Voivodedom of Radom (Krukowski, 1939). The pits of this site yielded a rich lithic assemblage containing, among other pieces, blade and flake cores with opposed and single platforms and a crude dihedral burin, as well as a single, typical, pointed arch-backed blade. This inventory suggests that at least one of these small, narrow but deep pits is associated with the Final Paleolithic Arch Backed Piece complex (ABP) dated elsewhere in Poland to the Late Alleröd and to the early Younger Dryas. It is interesting to note that two rich assemblages of the ABP complex at the site of Całowanie (Layer IV), some 100 km to the northeast, are made of Orońsko flint.

In addition to tools typical of the ABP complex the mine of Orońsko II also contain traces of exploitation from later prehistoric periods.

Quarried chocolate flint was also extensively used by the Masovian communities dating from the Younger Dryas and the early Preboreal. No mining pits have yet been associated with this taxonomic unit. However, a Polany Kolonie I (cutting I) part of a large Masovian pre-core preparation workshop occurred in a secondary position in a karstic sinkhole, clearly suggesting processing of flint extracted nearby during this period (Schild, 1980).

Though no quarry pits of Mesolithic age have so far been identified, chocolate flint was apparently mined at the site of Tomaszów (Voivodedom of Radom), since a (charcoal) radiocarbon date, GrN-7051, 6555±45 years, BP, was obtained from one of the numerous Mesolithic workshops near the mine which specialised in pre-core preparation (Figs. 17.2 and 17.3).

Four flint mines exploiting the chocolate flint during the Neolithic and Bronze Age have so far been investigated. These can be described briefly.

The large mine of Tomaszów I was excavated between 1973 and 1975. Four cuttings, totalling 385 m^2, were stripped exposing 23 pits of various ages.

The flint bed at Tomaszów I occurs in pre-Quaternary karstic clays, which are heavily involuted by glacitechtonics, and covered by fine-grain washes and a glacial till. Because of the glacitectonics the flint bed lies at various depths, ranging from 1 m to more than 10 m below the surface. The deepest quarry pits (Pit 6) lie in the vicinity of 4 m. The varying depth of the flint bed limited exploitation and resulted in a number of abandoned, blank, pits, and also caused spatial crowding of the quarries in the more promising areas of the mining field.

The quarry pits investigated are conical to tubular in cross-section and cut through the Quaternary overburden (Fig. 17.4). Their diameters vary from 1.1 m to 3.6 m, while an oval pit of the early Bronze Age measures nearly 9 m by 4 m. Several pits have a bank at mid-depth to facilitate communication and transport.

It is believed that the total size of the mining field at Tomaszów I is in the vicinity of 9000 m^2. The number of pits is estimated at between 579 and 604.

A number of flint processing workshops are located around the pits. During the early and Middle Neolithic, the main production of the workshops appears to be associated with the preparation of pre-cores for single platform cores for blades (Fig. 17.5), while in the early Bronze Age rough-outs of bifacial celts were the most popular product.

A number of radiocarbon dates have been obtained from the pits at Tomaszów I. The dates are as follows: Pit 6 – 6145±70 years BP (GrN-7591); Pit 10 – 5990±110 years BP (GrN-7592E, alkali extract) and 5715±65 years BP (GrN-7592R, alkali residue); Pit 1 – 5895±40 years BP (GrN-7050); Pit 4 – 5700±70 years BP (GrN-7590).[2]

These radiocarbon dates place the pits within the Linear and Lengyel Pottery taxonomic units, which is corroborated by their artefactual content. No radiocarbon dated quarry pits at Tomaszów I are associated with the Middle or Late Neolithic, although some of the pits excavated contain flint artefacts suggesting such an age.

Another large Neolithic mine being extensively excavated is that at Wierzbica-Zele, in the Voivodedom of Radom (Młynarczyk, 1981, 1982). This is a large site, nearly 12 hectares in extent. Here the chocolate flint was exploited by means of large shafts, varying in depth from 4.5 m to 6.14 m, cut through Quaternary sediments into the flint-bearing karstic clays or limestone.

Very extensive flint mining operations in the chocolate belt date from the Early Bronze Age and from the transitional period between the Neolithic and the Bronze Age.

Fig. 17.1. Chocklate flint exploitation sites and upper Jurassic (*b*) and Triassic (*a*) outcrops; position of the location map is indicated on schematic map of Poland. Key: 1. Prędocin; 2. Iłża I; 3. Iłża II; 4. Polany II; 5. Polany I; 6. Polany III; 7. Polany kolonie III; 8. Polany kolonie IIa; 9. Polany kolonie IV; 10. Polany kolonie II; 11. Polany kolonie I; 12. Wierzbica–Krzemienica; 13. Wierzbica–Zele; 14. Tomaszów I; 15. Orońsko II; 16. Guzów; 17. Chronów kolonia; I. Haematite quarry of Rydno (Nowy Młyn); II. Haematite quarry of Wielka Wieś. (After Schild (1971).)

[2] All of the dates from Groningen were measured under the supervision of Professor Willem G. Mook to whom the author is greatly indebted.

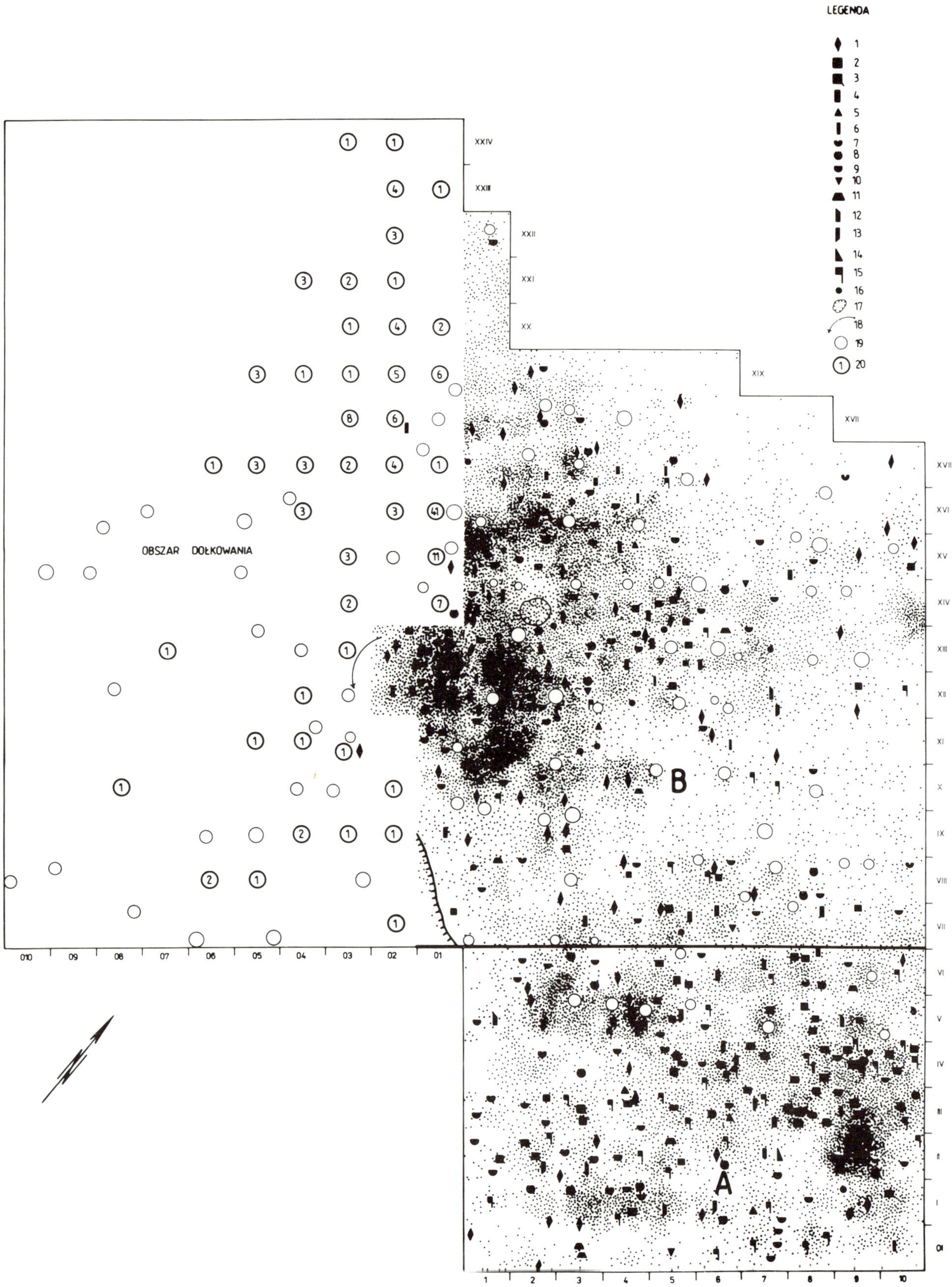

Fig. 17.2. Tomaszów II, Cut II; Scatter pattern of Vistulian, Late Mesolithic flint processing workshops showing density of artefacts. Key: 1. Cores and pre-cores; 2. Scrapers; 3. Raclettes; 4. End-scrapers; 5. Burins; 6. Borers; 7. Notches; 8. Dencticulates; 9. Retouched pieces; 10. Triangles; 11. Trapezoids; 12. Truncations; 13. Komornica points; 14. Stawinoga backed bladelets; 15. Microburins; 16. Varia; 17. Post-depositional pits; 18. Articulations; 19. Trees; 20. Borings. (Prepared by M. Marczak.)

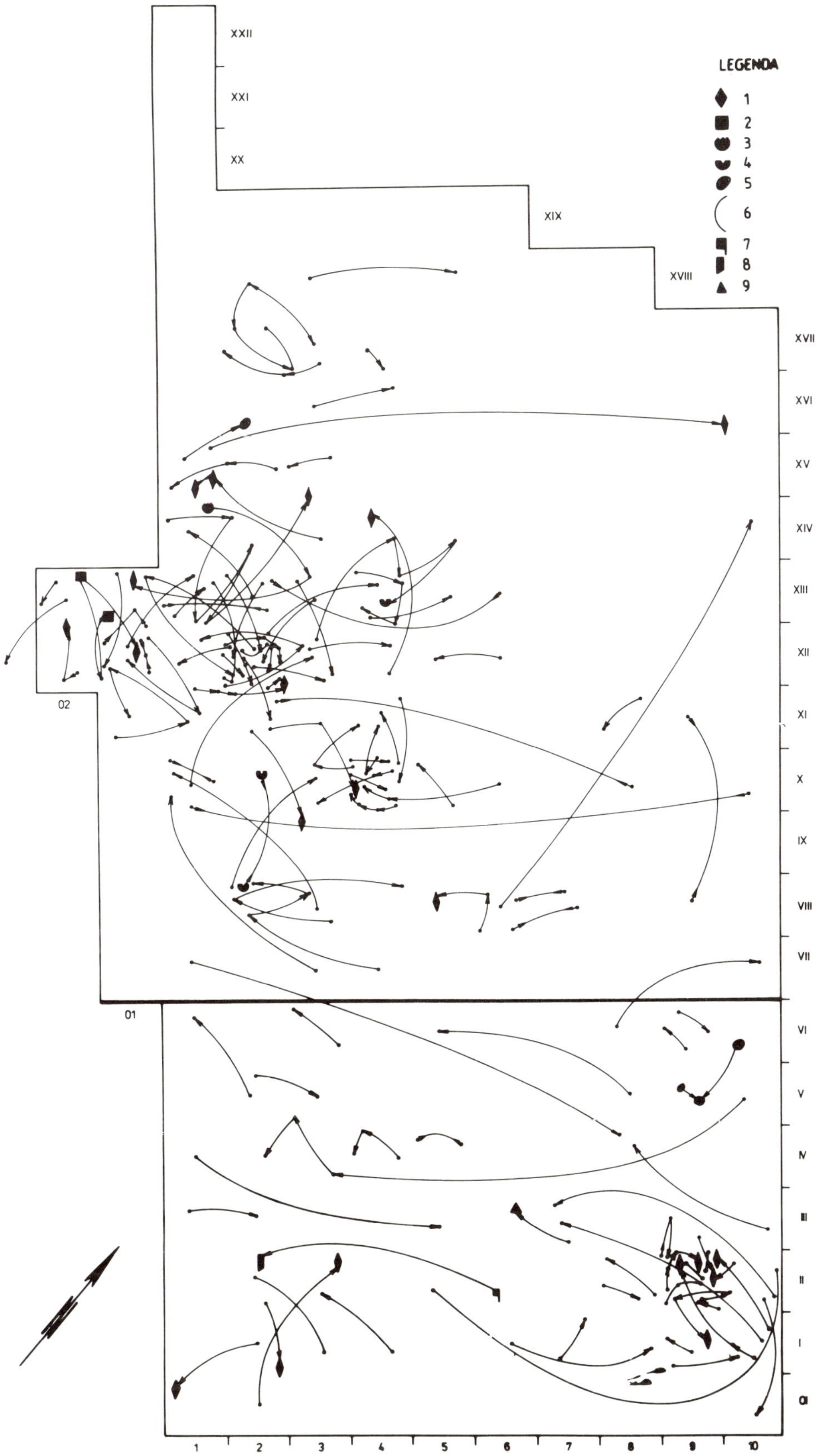

Fig. 17.3. Tomaszów II, Cut II; Network of refitting artefacts. For key see Fig. 17.2. (Prepared by M. Marczak.)

At least three mines were in use (Tomaszów I, Polany II and Polany Kolonie II).

The flint mine of Polany Kolonie II was a small enterprise, measuring about 1500 m² and containing 39±19 pits (p=0.05) of which eight were entirely or partially excavated (Schild, Królik & Mościbrodzka, 1977). The pits are cut into the limestone and have short, dead-ended, narrow galleries, with niches and cavities extending radially from the base of each shaft, at a depth of about 2.5 m (Fig. 17.6). The galleries and niches follow a thin bed of flattish nodules which was found particularly suitable for the production of bifacial celts, sickles and points. Nodules of this flint were extracted from the limestone by the use of levers, hammers and picks made of red deer antler (Fig. 17.7). Flint processing workshops surrounding the pits produced rough-outs of celts, points and sickles.

Two radiocarbon dates have been obtained from Pit 1 (Mook, 1977). The first date of 4005±45 years BP (GrN-6833) is from the *Basal Hearth* and must be contemporaneous with the operation of the shaft. The second, that of 3990±40 years BP (GrN-6834) is from the *Upper Hearth*, post-dates the exploitation of Pit 1, but must certainly be contemporaneous with the use of other pits.

An estimate of the volume of the output has been calculated, based on the recorded technology of manufacture, the amount of waste recovered, and on estimates of productivity of the bed and of the volume of limestone excavated. It appears that the number of rough-outs obtained from each pit fluctuated between 270 and 510 pieces (see Schild *et al.*, 1977). The small number of pits, as well as the limited volume of the yield, indicates that the mine at Polany Kolonie II could have been operated by a single community, a village or a kinship group.

The latest mine recorded in the chocolate flint belt is that of Polany II (Chmielewska, 1973, 1980). The limited excavations carried out at this apparently very large site record the presence of several pits cut through shallow Quaternary sediments down into the limestone. A radiocarbon date of 3491±81 (BM-1235) years BP (Chmielewska, 1980) indicate that this flint extraction site was associated with the Mierzanowice culture, a fact corroborated by the lithic material recovered.

The distribution

The first studies on the distribution of chocolate flint to settlement sites preceded the first discoveries of its outcrops and research on its mining and processing (see Krukowski, 1920, 1922). It appears that chocolate flint was most popular during the Final Paleolithic and early Neolithic. Of

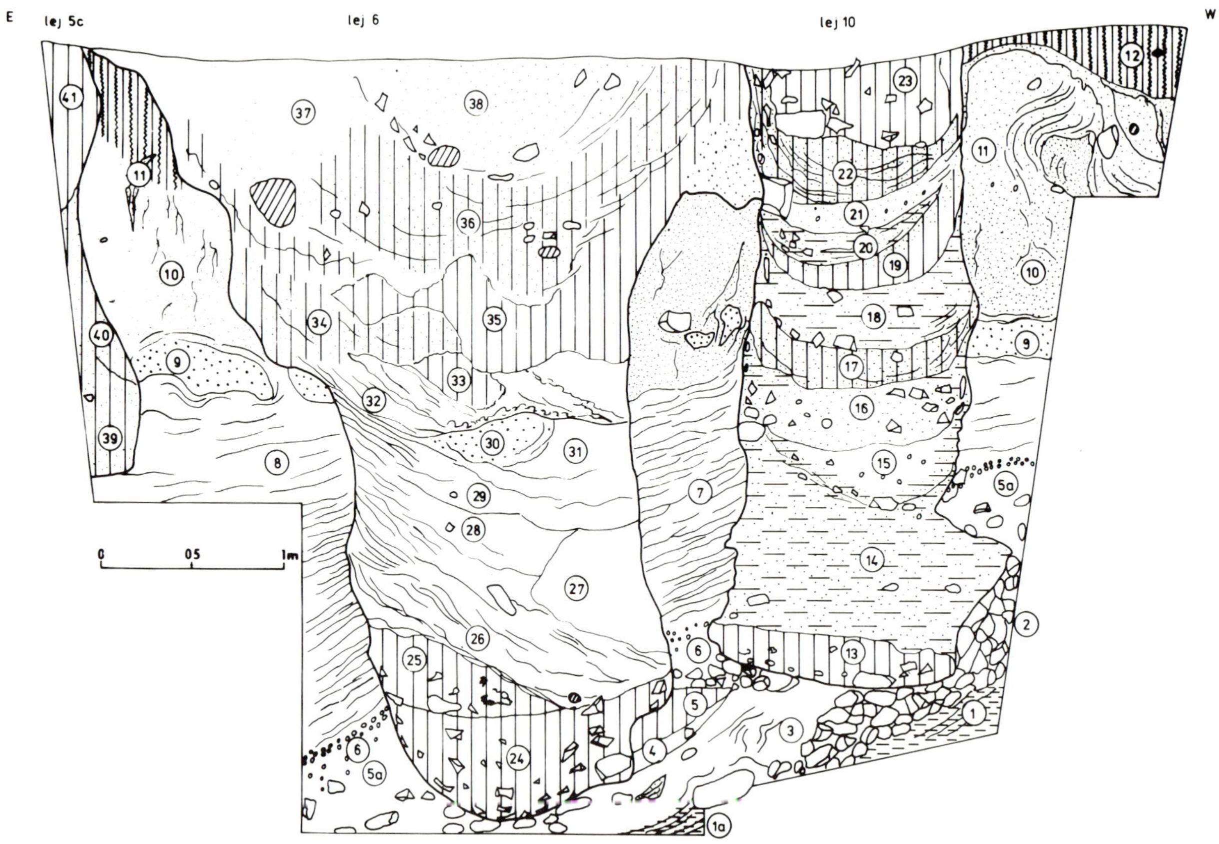

Fig. 17.4. Tomaszów I, Cut III/74. Cross-section of Pits 5c, 6 and 10. Key (simplified): 1. Karstic clay; 1a. Fine sand with marl gravel; 2. Flint nodules in clayey matrix; 3. Karstic clay; 4–6. Karstic clays, some with flint (5), and marl gravel; 7 and 8. Slope sands; 9–11. Cryoturbated glacial, slope sands; 12. Boulder clay; 13–23. Elements of the fill of Pit 10; 24–38. Elements of the fill of Pit 6; 39–41. Elements of the fill of Pit 5c.

these two periods the former is much better understood, while the intricacies of the distribution during the Neolithic are at an early stage of research (see Kaczanowska & Lech, 1977; Lech, 1979, and Lech this volume).

No doubt, chocolate flint was already known to the societies of the Middle Paleolithic, although there is some reason to believe that only the erratic nodules were used at that time (Kazłowski, 1972). During the Upper Palaeolithic, chocolate flint was exploited at several settlements. Most of these occur near Cracow, relatively far from the outcrops of the chocolate flint. It has not been established whether the nodules come from quarried or erratic sources of this flint.

During the Final Paleolithic, chocolate flint was extensively used by the communities belonging to the ABP complex, as well as those of the Masovian (or Swiderian). Among a total of 70 assemblages associated with the ABP complex, at least 35 contain quarried chocolate flint. A preliminary quantitative study of their distribution has demonstrated that the percentage of sites containing over 50% of the chocolate flint decreases rapidly with distance from the source area. The major decline, the so-called fall-off effect, occurs at a radius of some 100 km from the source (Schild, 1976).

About half of the 300 or so Masovian and related assemblages recorded in Poland contain varying quantities of chocolate flint; in most cases this can be shown to belong to mined varieties. The distribution of these assemblages shows that in the northern zone (excluding the Vistula Valley) the fall-off in the number of sites containing over

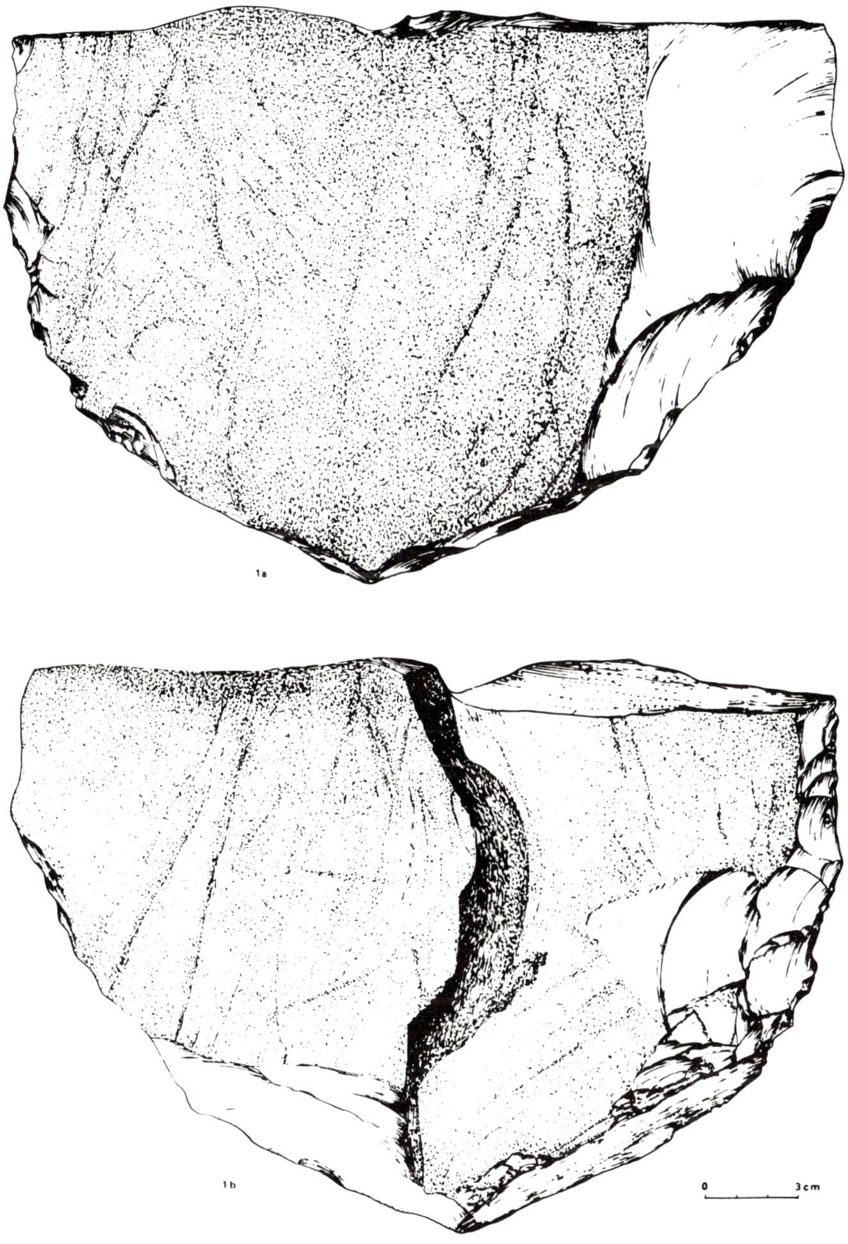

Fig. 17.5. Pre-core for blades. Pre-striking platform on top, prepared, wedge-like pre-flaking surface, bottom right.

50% of chocolate flint occurs at a distance of about 200 km from the source area. In the southern zone, including the Vistula Valley, the fall-off effect is seen at about 100 km from source. But in the Vistula Valley, north of the source, fall-off does not begin until well beyond the 200 km radius (Schild, 1976, p. 165). Sporadic occurrences of quarried chocolate flint are noted as far as 400 km from the source, e.g. at Pomorsko, on the Oder, and at Kašetos and Duba, near Lida in Lithuania.

It is believed that such regularities in the flint distribution largely reflect mean ranges of nomadic movement of prehistoric communities during the Alleröd oscillation and the

Fig. 17.6. Polany kolonie II, Pit 1 after excavation, note niches at the bottom of pit.

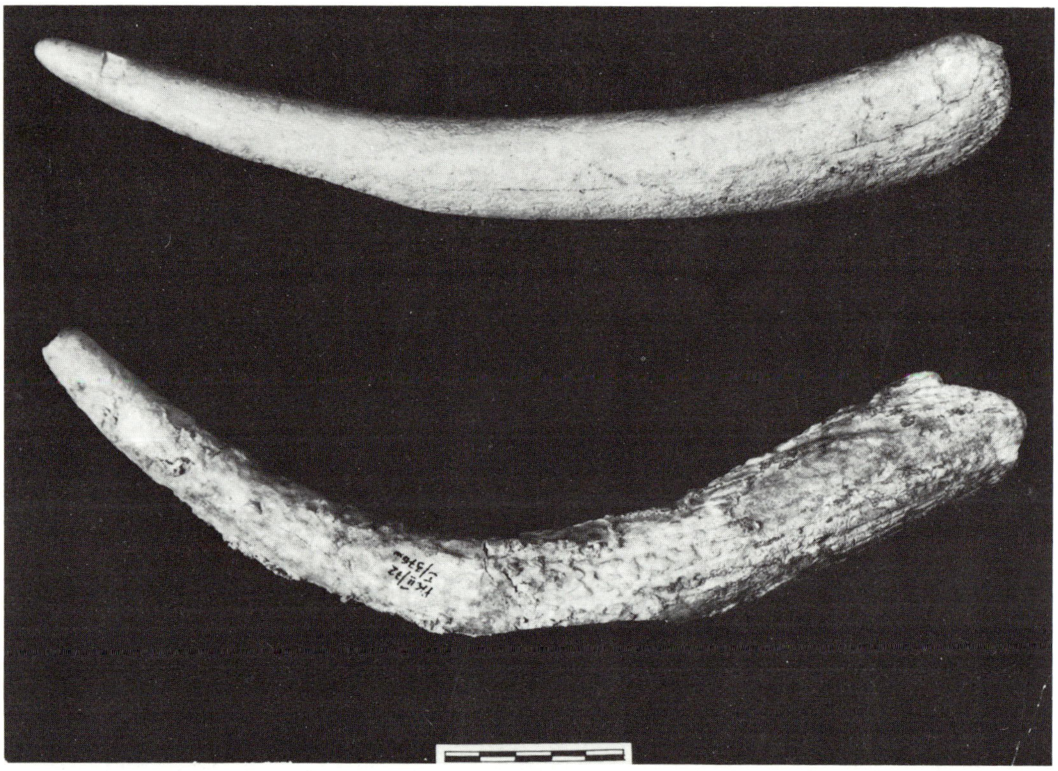

Fig. 17.7. Polany kolonie II, antler levers from Pit 1. Scale in cm.

Younger Dryas and the competition of the southern Cracow–Wieluń Jurassic flint sources.

Changes in the structure of Final Palaeolithic assemblages made of chocolate flint can also be related to the distance from the source area of the raw material (Fig. 17.8). The assemblages from workshops located at the quarries show heavy stress on pre-core preparation and minimal indices of tools. Flint processing workshops placed at some distance from source have high indices of core exploitation and core preparation. Living sites occurring at considerable distance from the source area are characterised by a much higher tool index, a falling index of core preparation as well as a relatively high index of core exploitation, while assemblages located at a very great distance from the place of extraction are characterised by high indices of advanced core exploitation and very high indices of tools. Elements of core preparation are almost non-existent.

Recent studies of raw material use during the Mesolithic (Cyrek, 1983) indicate that chocolate flint was only used sporadically in the Early Mesolithic, except on sites located close to the source. Its utilisation increased during the Late Mesolithic and is mainly associated with Jerzmanowice or Vistulian groups. Quantitative differentiation of various types of artefacts in relation to the distance from source is similar to that of the Masovian, although some differences are evident (see Fig. 17.9).

Chocolate flint, definitely acquired by quarrying, is well represented at sites of the Linear Pottery and the early or middle Lengyel–Polgar Pottery complex in Little Poland and the Sandomierz region (Kamieńska, 1964; Kozłowski, 1968; Kaczanowska & Kozłowski, 1971; Kaczanowska & Lech, 1977; Więkowska, 1971). Considerable quantities of this stone also occur in Kujavia (Gabałówna, 1962; Kaczanowska & Lech, 1977; Balcer, 1983) and among Danubian sites in Slovakia (Kaczanowska & Kozłowski, 1971), while sporadic occurrences have been noted in Upper Silesia (e.g. Balcer, 1977*a*, 1983). The most distant sites containing chocolate flint come from as far away as *ca.* 500 km from source (Fig. 17.10).

In the early Funnel Beaker settlements of Kujavia raw material indices of almost 90% chocolate flint have been recorded (Chmielewski, 1952, p. 24; Balcer, 1983, p. 124). On the other hand, the communities of the late phase of the same complex almost entirely abandoned the use of this raw material (Balcer, 1983, p. 132) and turned in preference to Turonian flint of Swieciechów and of similar outcrops (Balcer, 1975, 1976). This change is most probably to be related to technology: i.e. to the type of tools manufactured.

Slightly higher values of chocolate flint are found at sites of the same period belonging to the Globular Amphorae complex; however, here chocolate flint is always second in the importance to the famous banded flint of Krzemionki Opatowskie (Balcer, 1983, p. 208). In the Corded Ware complex, too, chocolate flint is dominant in the region located close to the source (Balcer, 1983, p. 227) and rare beyond the 50–100 km radius.

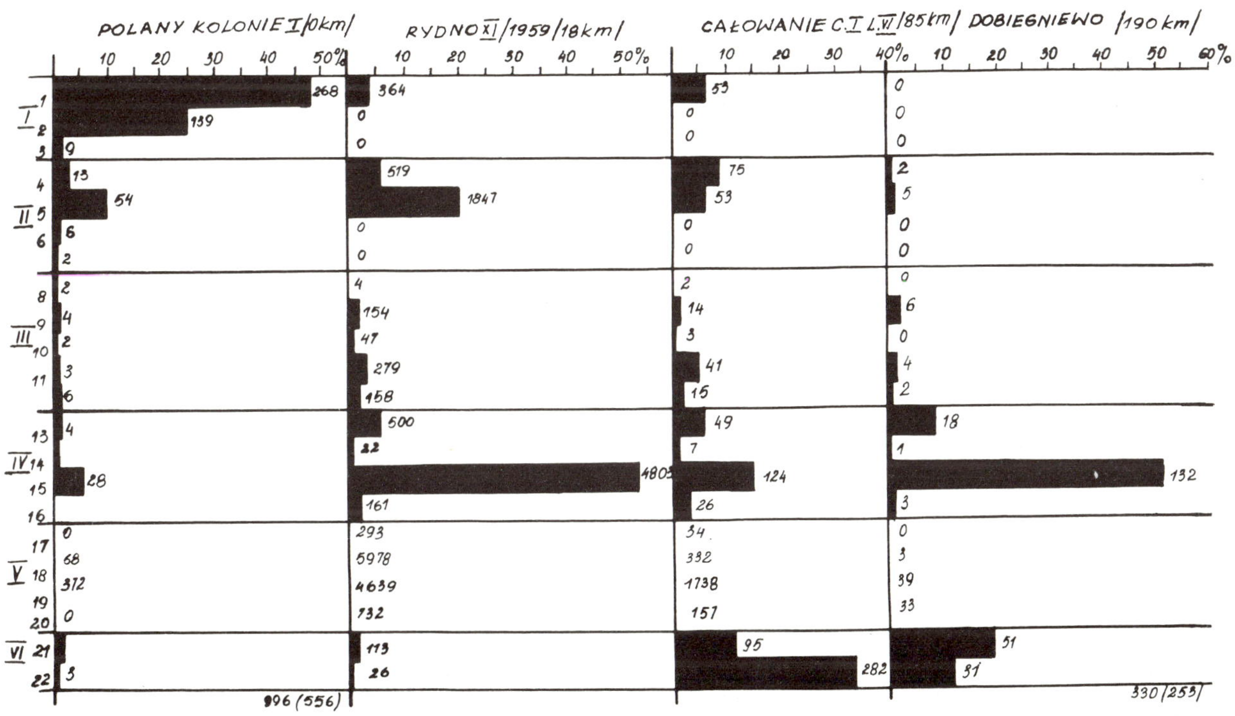

Fig. 17.8. Restricted technological structure of Masovian assemblages made of chocolate flint in relation to the distance from the source. Distance from source indicated in column heading. Key (simplified): I. Group of early core preparation; II. Group of advanced core preparation; III. Group of early core exploitation; IV. Group of advanced core exploitation; V. Group of non-diagnostic waste; VI. Group of tools and tool production waste. (After Schild (1980).)

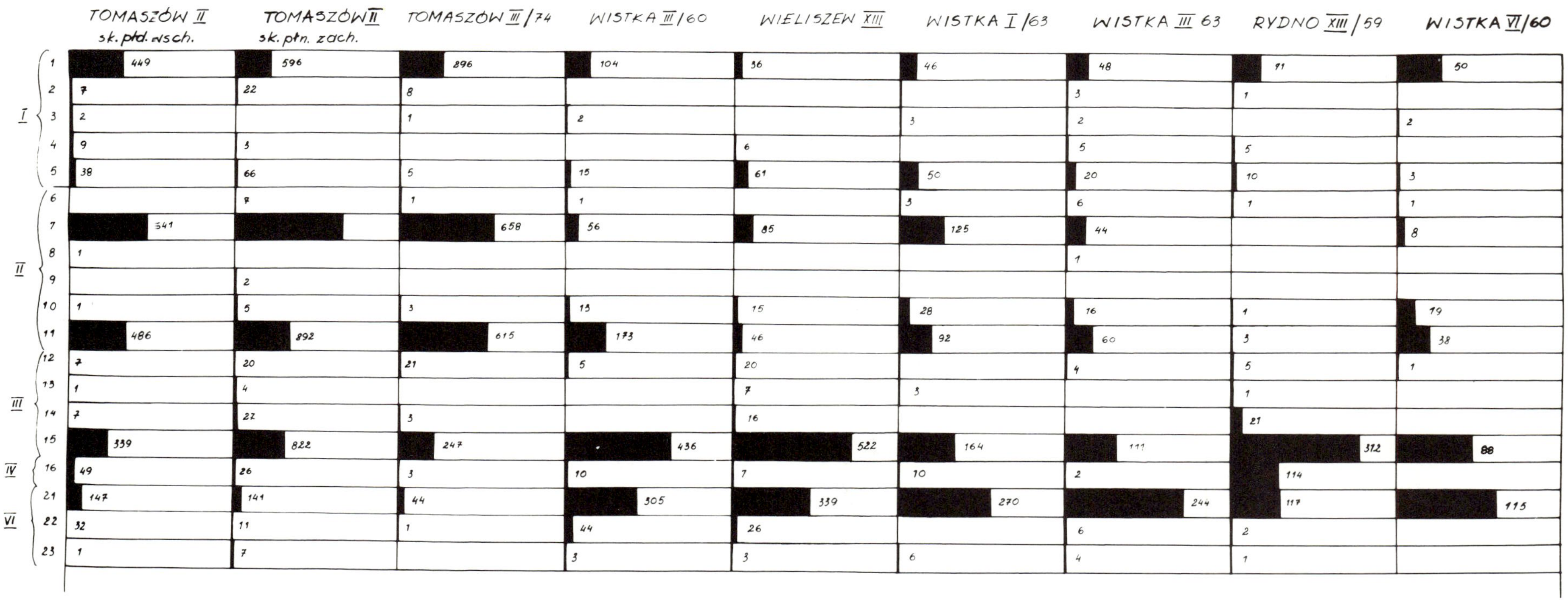

Fig. 17.9. Restrictec technological structure of Vistulian, Late Mesolithic assemblages made of chocolate flint. Position of sites in columns is taxoncmic. Distance from source is as follows: Tomaszów II – 0.5 km; Rydno XIII/59 – 25 km; Wieliszew XIII – 125 km; Wistka I/63, III/63 and VI/60 – 195 km. Key (simplified): I. Group of core preparation and initial exploitation; II. Group of flake production; III. Group of blade production; IV. Group of core rejuvenation; VI. Group of tools and tool production waste; Group V, non-diagnostic, omitted.

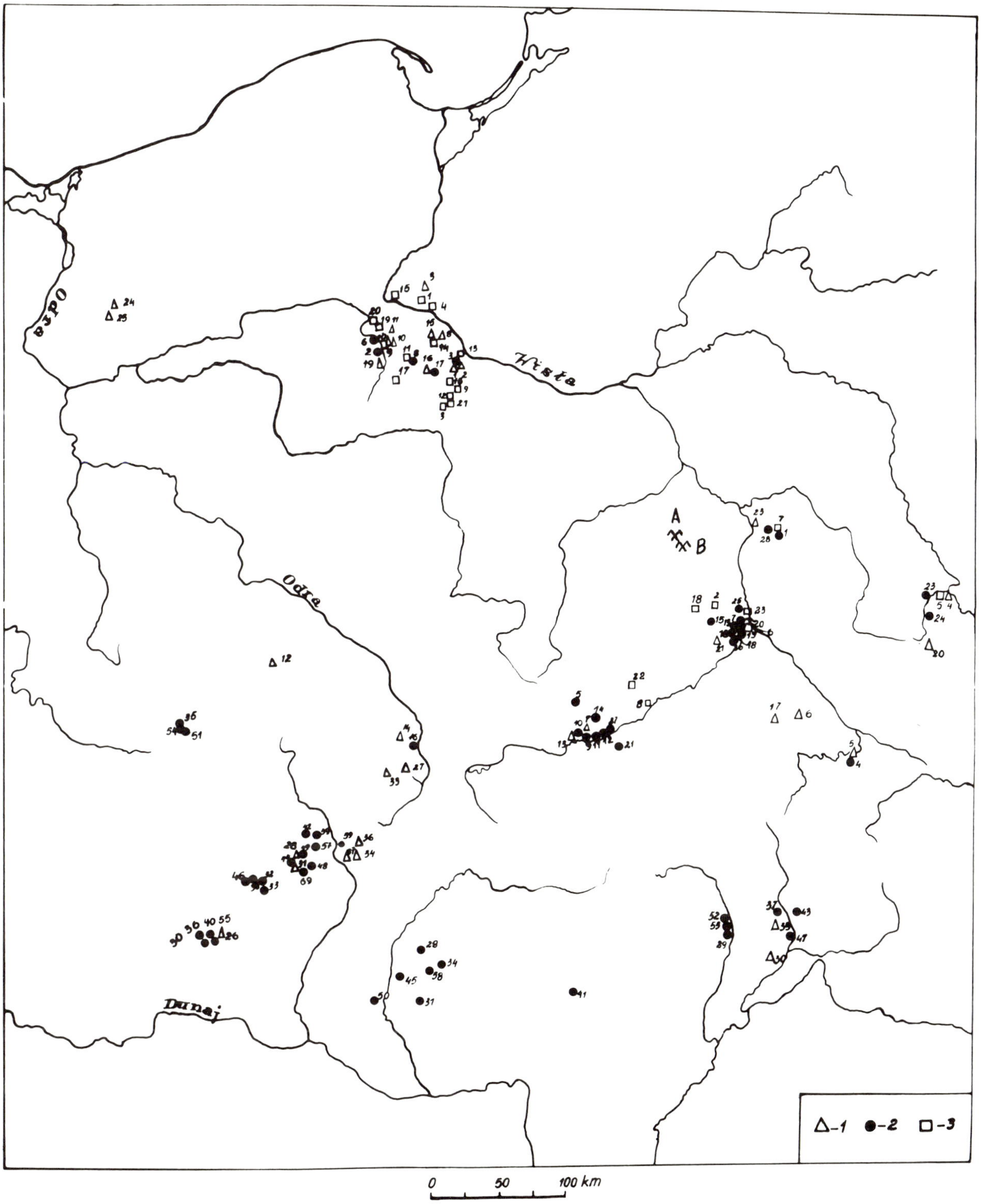

Fig. 17.10. Distribution of Early and Middle Neolithic sites containing chocolate flint. Key: 1. Linear Pottery; 2. Lengyel-Polgar Pottery; 3 Funnel Beaker Pottery; A. Mine of Tomaszów I; B. Mine of Wierzbica-Zele. (According to quoted literature and personal communications of B. Balcer, L. Domańska, J. Lech and A. Zakościelna.)

The distribution of chocolate flint at the turn of Neolithic and Bronze Age seems to be very limited and is largely confined to settlement sites located close to its source (see Machnik, 1967; Balcer, 1977*b*). This appears to contradict data on the intensity of mining at the same period at Tomaszów I, Polany Kolanie II and Polany II. The explanation of this apparent contradiction lies, most possibly, in the fact that expanded consumption of chocolate flint was limited to small geographical zones near the mines. On the other hand, this could have been caused by strict community ownership of the mines and a significantly reduced traffic in stone raw materials.

Appendix: chocolate flint petrology and physical characteristics

Colour: black (10YR 2/1); very dark brown (10YR 2/2); very dark gray to dark brown (10YR 3/1) – (10YR 4/4); rarely, one group only, dark reddish brown (5R 3/2). Some groups may also contain additional black (10YR 2/1) or grayish (10YR 7/1) – (10YR 4/1) patches, lenses and under/cortical bands.

Lustre: distinct waxy/rare/to dull and dull waxy.

Composition: chalcedony with additional admixtures of opal and accessory components as pyrite, quartz, ferrum-oxides, ferro-hydroxides, phosphates, etc.

Textures: homogenous to crypto homogenous, more or less crypto-crystalline and mixed crypto-cristalline-spherulithic, rarely spottic. Microcrystalline unit of chalcedony are usually under 0.01 mm, rarely larger, up to 0.05 mm to 0.1 mm. Microfossils range from 0.03 mm to 0.01 mm and are elongated or globular, made up of silica, collophane, bituminous substance and ferro-hydroxides. The colour of the flint depends mainly on the microfossils.[3]

Cortex: light gray or slightly brownish gray, calcareous, usually thin (less than 0.5 mm).

Transparency: in most cases good to medium, rarely poor to very poor.

Shape: round, regular, flattish nodules and thin to medium-thick slabs. Usually, the nodules and slabs reach no more than 20 cm in length and 10 cm in thickness.

Bibliography

Balcer, B. (1975). *Krzemień świeciechowski w kulturze pucharów lejkowatych. Eksploatacja obróbka i rozprzestrzenienie.* Wrocław: Ossolineum.

Balcer, B. (1976). 'Position and stratigraphy of flint deposits. Development of exploitation and importance of Świeciechów flint in prehistory.' *Acta Archaeologica Carpathica*, **16**, 179–99.

Balcer, B. (1977*a*). 'Z badań nad krzemieniarstwem neolitycznym w dorzeczu górnej Odry.' *Przeglad Archeologiczny*, **25**, 5–51.

Balcer, B. (1977*b*). 'Osada kultury mierzanowickiej na st. 1 w Mierzanowicach.' *Wiadomości Archeologiczne*, **42**, 175–212.

Baker, B. (1983). *Wytwórczość narzędzi krzemiennych w neolicie ziem Polski.* Wrocław: Ossolineum.

Chmielewska, M. (1973). 'Badania stanowiska Polany II w pow. Szydłowiec, w 1971 r.' *Sprawozdania Archeologiczne*, **25**, 29–37.

Chmielewska, M. (1980). 'Polany, Fundstelle II, Wojw. Radom.' In: 5000 Jahre Feuersteinbergbau.' *Veröffentlichungen aus dem Deutschen Bergbau-Museum Bochum*, **22**, 583–6.

Chmielewski, W. (1952). *Zagadnienie grobowców kujawskich.* Lódź: Muzeum Archeologiczne.

Cyrek, K. (1983). 'Surowce krzemienne w mezolicie dorzeczy Wisły i górnej Warty.' In: *Człowiek i środowisko w pradziejach*, eds., J. K. Kozłowski & S. K. Kozłowski, pp. 106–13. Warszawa: PWN.

Dembowska, J. (1953). 'Górna jura między Radomiem i Jastrzębiem.' *Biuletyn PIG*, **218**, 31–46.

Gabałówna, L. (1963). 'Slady osadnictwa kultur z cyklu wstęgowych w Radziejowie kujawskim.' *Prace i Materiały Muzeum Archeologicznego i Etnograficznego w Łodzi. Seria Archeologiczna*, **9**, 25–122.

Kaczanowska, M. & Lech, J. (1977). 'The flint industry of Danubian communities north of the Carpathians.' *Acta Archaeologica Carpatica*, **17**, 5–28.

Kaczanowska, M. & Kozłowski, J. K. (1971). 'Materiały krzemienne z osady neolitycznej i eneolitycznej w Nowej Hucie Mogile (stanowisko 48).' *Materiały Archeologiczne Nowej Huty*, **4**, 67–100.

Kamieńska, J. (1964). 'Osady kultur wstęgowych w Samborcu, pow. Sandomierz. In: 'Studia i materiały do badań nad neolitem Małopolski.' *Prace Komisji Archeologicznej*, **4**, 77–189.

Karczewski, L. (1960). *Ślimaki astartu i kimerydu północno-wschoniego obrzeżenia Gór Świętokrzyskich. Prace Instytutu Geologicznego*, Vol. 32, pp. 1–68. Warszawa: Institute of Geology.

Kozłowski, J. K. (1968). 'Materiały krzemienne pochodzące z zespołu jam otaczających chatę lendzielską w Nowej-Hucie Mogile (stanowisko 62).' *Materiały Archeologiczne*, **9**, 149–54.

Kozłowski, J. K. (1972). 'Archaeological materials.' In: 'Studies on Raj Cave near Kielce (Poland) and its deposits.' *Folia Quaternaria*, **41**, 61–117.

Krukowski, S. (1920). 'Pierwociny krzemieniarskie górnictwa, transportu i handlu w holocenie Polski, część 1.' *Wiadomości Archeologiezne*, **5**, 55–74.

Krukowski, S. (1922). 'Pierwociny krzemieniarskie górnictwa, transpotu i handlu w holocenie Polski, część 2.' *Wiadomości Archeologiczne*, **7**, 34–57.

Krukowski, S. (1923). 'Sprawozdanie z działalności państwowego konserwatora zabytków prehistorycznych na okręg kielecki w r. 1922.' *Wiadomości Archeologiczne*, **8**, 64–84.

Krukowski, S. (1939). 'Paleolit.' In: *Prehistoria ziem poslkich*, ed., J. Kostrzewski. Kraków: Akademia Umiejętności.

Kutek, J. (1961). 'Kimeryd i bonon Stobnicy.' *Acta Geologica Polonica*, **11**, 103–56.

Kutek, J. (1967). 'Z zagadnień stratygrafii górnej jury w Polsce.' *Biuletyn Instytutu Geologicznego*, **203**, 97–105.

Lech, J. (1979). 'Krzemieniarstwo w kulturze spełeczności ceramiki wstęgowej rytej w Polsce. Próba zarysu.' In: *Poczatki neolityzacji Polski potudniowo-zachodniej*, ed., W. Wojciechowski. Wrocław: Uniwersytet Wrocławski.

Machnik, J. (1967). *Stosunki kulturowe na przełomie neolitu i epoki brazu w Małopolsce.* Materiały do prahistorii ziem polskich, Part 3. Warszawa: Instytut Historii Kultury Materialnej.

Malinowska, L. (1967). 'Uwagi o stratygrafii górnej jury pozakarpackiej w Polsce.' *Biuletyn Instytutu Geologicznego*, **203**, 115–20.

Młynarczyk, H. (1981). 'Wierzbica, woj. radomskie, stanowisko "Zele".' *Informator Archeologiczny 1981*, pp. 22–3.

Młynarczyk, H. (1982). 'Wierzbica, woj. radomskie, stanowisko "Zele".' *Informator Archeologiczny 1982*, pp. 66–7.

Mook, W. G. (1977). 'Datowanie C14 próbek węgla z wnętrza szybu 1 w Polanach koloniach II. In: *Kopalnia krzemienia czekoladowego z przełomu neolitu i epoki brazu w Polanach*

[3] The composition and texture of samples were determined by Dr Olech Juskowiak.

Koloniach, eds., R. Schild, H. Królik & J. Mościbrodzka, pp. 117. Wrocław: Ossolineum.

Pożaryski, W. (1948). *Jura i kreda miedzy Radomiem, Zawichostem i Kraśnikiem.* Biuletyn Instytutu Geologicznego 46. Warszawa: Instytut Geologiczny.

Samsonowicz, J. (1923). 'O złożach krzemieni w utworach jurajskich północno-wschodniego obrzeżenia Gór Świętokrzyskich. *Wiadomości Archeologiczne*, **8**, 17–24.

Samsonowicz, J. (1932). 'Przebieg i charakter granicy między jurą a kredą na północno-wschodnim zboczu Łysogór.' *Sprawozdania Państwowego Instytutu Geologicznego*, **7**, 169–226.

Samsonowicz, J. (1934). *Objaśnienia arkusza Opatów ogólnej mapy geologicznej Polski.* Warszawa: Państwowy Instytut Geologiczny.

Schild, R. (1971). 'Lokalizacja prahistorycznych punktów eksploatacji krzemienia czekoladowego na północno-wschodnim obrzeżenu Gór Świętokrzyskich.' *Folia Quaternaria*, **39**, 1–61.

Schild, R. (1976). 'Flint mining and trade in Polish prehistory as seen from the perspective of the chocolate flint mining of central Poland. A second approach.' *Acta Archaeologica Carpathica*, **16**, 147–77.

Schild, R. (1980). 'Introduction to dynamic technological analysis of chipped stone assemblages.' In: *Unconventional Archaeology*, ed., R. Schild, pp. 57–85. Wrocław: Ossolineum.

Schild, R., Królik, H. & Mościbrodzka, J. (1977). *Kopalnia krzemienia czekoladowego z przełomu neolitu i epoki brazu w Polanach Koloniach.* Wrocław: Ossolineum.

Więckowska, H. (1971). 'Materiały krzemienne z psad kultury ceramiki wstęgowej i trzcinieckiej w Opatowie.' In: *Z polskich badań nad epoko kamienia*, ed., W. Chmielewski, pp. 103–83. Wrocław: Ossolineum.

18

Flintmining in Sweden during the Neolithic period: new evidence from the Kvarnby–S. Sallerup area

ELISABETH RUDEBECK

Introduction

Flint mines have been documented in the Kvarnby–S. Sallerup area, 10 km east of Malmö, in the south of Sweden on several separate occasions. The mines have always been discovered in connection with the commercial exploitation of chalk in the area (Fig. 18.1).

It was the Swedish State Geologist N.-O. Holst, who in 1904 first discovered mines in the vicinity of Ängdala farm in S. Sallerup (Fig. 18.1). Holst visited the chalk quarries of that time in order to find an explanation to the many worked antlers found in the chalk by the quarry workers. He discovered vertical shafts dug through 0.5–2 m of till and through 1–3 m of underlying flintbearing chalk. The shafts were about 2 m wide at the mouth and horizontally extended at the bottom (Holst, 1906).

In 1906–7 four similar shafts were investigated in chalk quarries nearby by B. Schnittger (1910).

In 1951 shafts were investigated by C.-A. Althin. He interpreted the shafts as chalk extraction pits from the Roman Iron Age or later (Althin, 1951).

In 1968, 1975 and 1977–83 Malmö museum carried out several investigations of flint mines, mainly near the Ängdala farm in S. Sallerup (Fig. 18.1). These investigations have all been caused by commercial chalk quarrying and paid for by the chalk quarrying company (Malmökrita AB). The Swedish law states that the exploiter/landowner has to pay for the archaeological excavations before exploitation. As the excavations have been rescue excavations they have been carried out under severe time pressure and with insufficient funds. Many features (mines and settlement remains) have only been partially excavated or documented before their destruction. The mine shafts have been partially excavated by hand (mainly the top and bottom layers) and partially with the aid of machinery.

Until just recently a total of about 60 mine shafts had been documented in the Kvarnby–S. Sallerup area. In the autumn of 1982, however, about 400 mine shafts (and settlement remains) were discovered as the top soil was removed northeast of the Ängdala farm (Figs. 18.1+18.4). These mines are not going to be excavated now, but only documented in plan, re-covered with top soil, and preserved. The settlement remains adjacent to the mines have been partially excavated in 1983 with funds from the Swedish Central Board of Ancient Monuments.

Geology and raw material

The quarried Maastrichtian chalk in the south of Sweden is not primary bedrock, but consists of large, ice-transported slabs embedded in clayey till. The slabs vary in size between less than 100 m in length to almost 1 km. They are found in an area 4.5 km long by 7–800 m wide in southwest Scania (Fig. 18.1). The slabs are divided by thick layers of clayey till and are covered with glacial gravel, glacial clay and till of varying thickness. In some places the chalk reaches the surface. The maximum thickness of the slabs is about 30 m. Recent investigation of the origin of the chalk slabs has been carried out by State Geologist B. Ringberg of the Swedish Geological Survey. Ringberg's conclusion is that the slabs were transported to their present location by an ice movement from the south at the beginning of the final phase of the latest glaciation. They originate from the bottom of the Baltic Sea. Through repeated ice melting and ice advance, sediments and till were deposited on top of and between the slabs (Ringberg & Rudebeck, 1982).

The flintbearing chalk has a soft and crumbly structure due to the ice transportation. The flint occurs in the chalk as nodules concentrated in more or less coherent and sloping layers.

The dark shiny flint is of excellent quality, although patches of lighter gray, coarser material can occur in the nodules. The cortex is even and smooth and there is a sharp boundary between it and the flint (Rudebeck *et al.*, 1980). The nodules are of varying size, between 5 and 40 cm long, the larger ones being 20 cm wide and 20 cm thick. It has been observed that the quantity of nodules in the chalk varies locally. In some places the chalk surrounding a mine shaft only contains a few scattered nodules and in other places several nodules are concentrated at a certain level. One test pit has been dug in the chalk close to a mined area in order to try to estimate the size and quantity of flint nodules in the upper parts of the chalk (the mined level). The test pit was 2.5×3 m wide and 1.8 m deep and 127 nodules were obtained from about 50% of the mass volume of the pit (weight: 145.5 kg). Of these, 34 nodules (74.5 kg) were as long as 15–25 cm (but 23 of the 34 nodules were of a very irregular shape, with outgrowths and cavities, which would have made them unsuitable for the production of large tools such as axe blades or daggers. Only 11 nodules (23 kg) were 15–25 cm long and of a shape suitable for the production of large tool types. The majority of the nodules, 95 pieces (79.5 kg), were 10–20 cm long and of irregular shape.

It is necessary to do similar investigations concerning the size and quantity of the nodules before any general estimates can be made about the flint output. However, it is interesting to note that the shafts investigated so far have (perhaps with the exception of the open-cast area (Fig. 18.1)) been scattered on what seems to be the periphery of the recently discovered central area with approximately 400 shafts. It is possible that the chalk in this central area

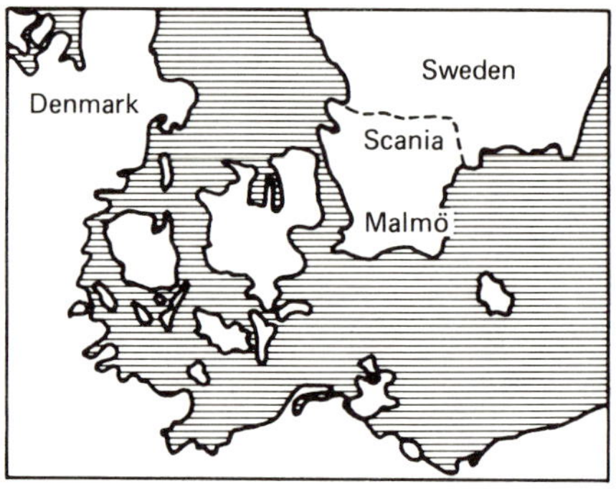

(*a*)

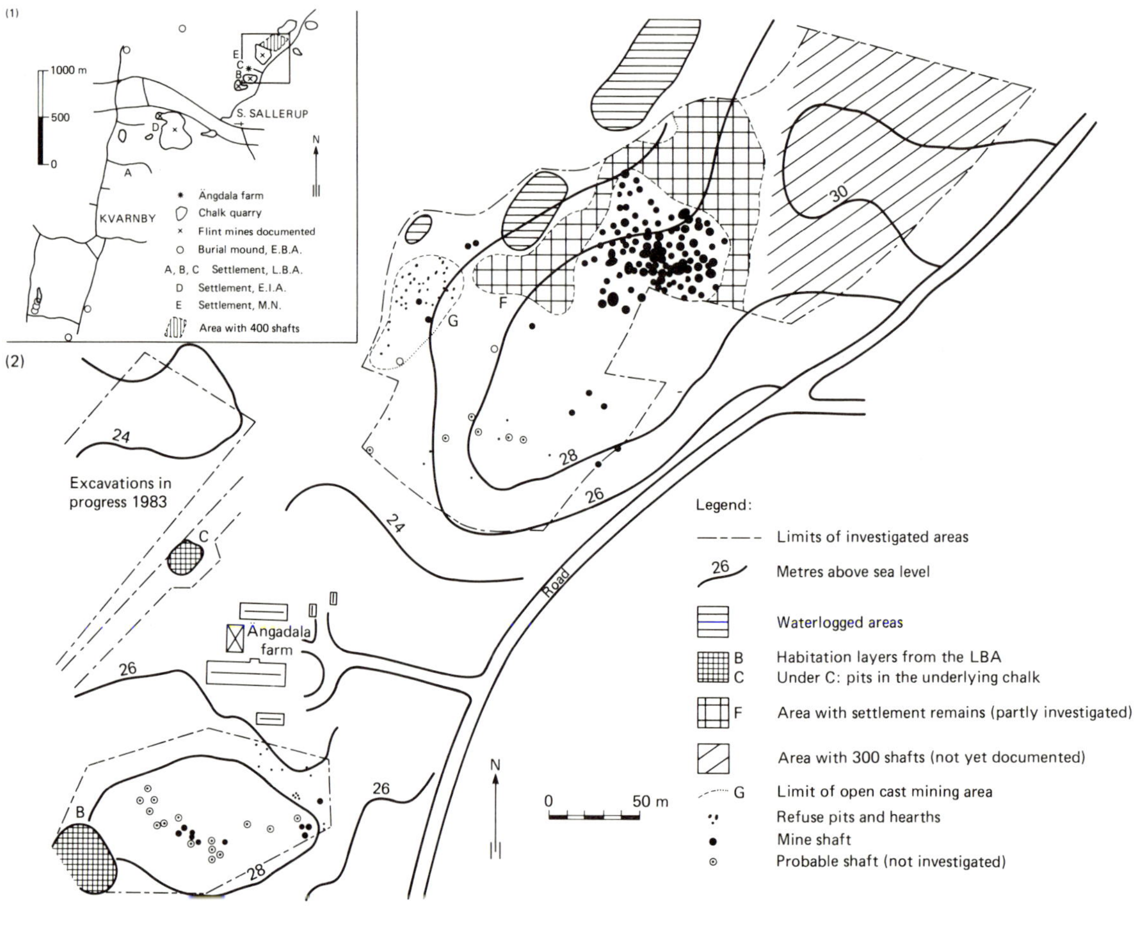

(*b*)

Fig. 18.1. (*a*) Malmö, Sweden. (*b*) Maps showing: (1) The Kvarnby–S. Sallerup area in the south of Sweden where flint mines have been documented. (2) The area around Ängdala farm where flint mines have been investigated between 1977 and 1983. Abbreviations used in Fig. 18.1. are: EBA (Early Bronze Age); LBA (Late Bronze Age); EIA (Early Iron Age); MN (Middle Neolithic).

contains larger quantities of flint nodules and/or larger nodules than the surrounding chalk.

Mines

The flint nodules have been extracted from:

a Surface pits varying in depth between 0.5 and 2 m. These pits have been dug in areas where the chalk lies between 0 and 1 m below the till (Fig. 18.2(*a*)). They occur in interconnected pit systems over large open cast areas and as separate pits.
b Open shafts varying in depth between 2 and 7 m. The thickness of the overlying till varies between 1 and 5 m (Fig. 18.2(*b*)–(*e*)). The diameter at the mouth of the shafts varies between 2 and 9 m.

There is no evidence of true underground mining in galleries, but only of side workings at the bottom of the shafts. The side workings are at the most 2 m long and 1 m high and they occur both in the shallow shafts and in the deeper ones (Fig. 18.2(*b*), (*d*) and (*e*)). The soft structure of the chalk makes it very liable to collapse and dangerous to undermine. There is in fact evidence of the collapse of the thin chalk roof in the side working of a shallow shaft (Fig. 18.2(*b*)).

The shafts have probably been largely refilled very soon after the flint extraction finished. There is evidence of a knapping floor at the bottom of only one shaft. The filling in the lower part of the shafts consists of densely packed chalk and the limit between the chalk filling and the floor of the shaft can be very difficult to ascertain. The reason for this may be that much of the chalk dump was never brought up to the surface, but was piled aside at the bottom of the shafts as the extraction continued. The bulk of the refill consists of clayey till, more or less mixed with sand, gravel and chalk. The upper parts of the filling often contain chalk, charcoal-mixed layers and refuse pits with pottery bone, and antler.

Large quantities of chipping waste are often found in the top layers of the refill, but sometimes also throughout the entire filling.

The surface pits have often been used as refuse pits or hearths and their contents are similar to those of the top layers of the deeper shafts (Fig. 18.2(*a*)).

It is important to note that the top layers of the refill may well be secondary to the mining period and therefore of a much later date than the shafts themselves.

Mining tools and technology

There is evidence of platforms or some other kind of wooden construction in some shafts. Post holes, both vertical and horizontal, have been observed in the chalk walls (Fig. 18.2(*c*)). In one instance it was observed that a horizontal post hole went through the wall of one shaft into the adjoining shaft.

The impression of a diagonally standing post has been observed in a few shaft fillings. Branches were cut off and the posts were probably used as crude ladders and left behind as the shafts were refilled (Fig. 18.2(*d*)). In many shafts there were minor wood impressions in the filling. These impressions vary in diameter and length and some could be the remains of branches without any particular function, whereas others could well be the remains of posts used in some construction or the remains of wooden wedges or levers (Fig. 18.2(*b*)–(*e*)).

One wood impression which was found at an acute angle to a fragmentary piece of antler was interpreted as the remains of a wooden handle of a composite tool (Fig. 18.3(*a*)).

Two red deer antler picks have been found in the mines (Fig. 18.3(*b*)). Marks in the chalk walls of a few shafts have obviously been made by the use of picks of this kind. Beside these picks, about 15 pieces of antler have been found. Some of these antlers have probably been used as levers. A few fragments of scapulas from some large mammal have been found in the top layers of some shafts and it is possible that these scapulas were used as spades or shovels.

Oblong, roughly chipped and pointed flints, 10–15 cm long, have been found in the mining areas and could have been used as picks for the extraction of the flint nodules.

Tool production and dating

Many flint knapping loci are found in the mining areas. Some of them mainly contain protuberances and irregularities removed from the nodules, together with large primary and secondary flakes (with the cortex more or less preserved on the dorsal surface of the flakes). Other knapping loci contain larger amounts of smaller tertiary flakes (without cortex) and chips, indicating a production stage beyond the rough shaping of a blank.

D. Seitzer-Olausson of the Institute of Archaeology at the University of Lund has studied some preforms and waste material from the mining areas. Her conclusion is that at least three major types of tools were produced at the mines: thick-butted square axes, flake axes and daggers. There is also evidence of the production of small tools such as scrapers, drills and arrow heads, but these tools are regarded as spin-off products from the production of larger tools (Rudebeck *et al.*, 1980).

The flake axes, which are of varying size and form, occur both as blanks, rough-outs and finished tools and it is probable that they were hafted and used at the site. Most of the thick-butted axes probably left the mining area as blanks, although there are fragments of polished and repolished axes in the finds. The daggers were probably not fully finished at the site, as there is no evidence of parallel pressure flaking in the collection (Rudebeck *et al.*, 1980).

The above-mentioned tool types belong to the Scandinavian Middle Neolithic (thick-butted axes) and Late Neolithic/Early Bronze Age (daggers). The flake axes are generally considered to belong to the Late Mesolithic/ Early Neolithic, but in the Kvarnby area they have been found at sites from the Early Neolithic to the Early Iron Age. It is possible that the flake axes of the later deposits

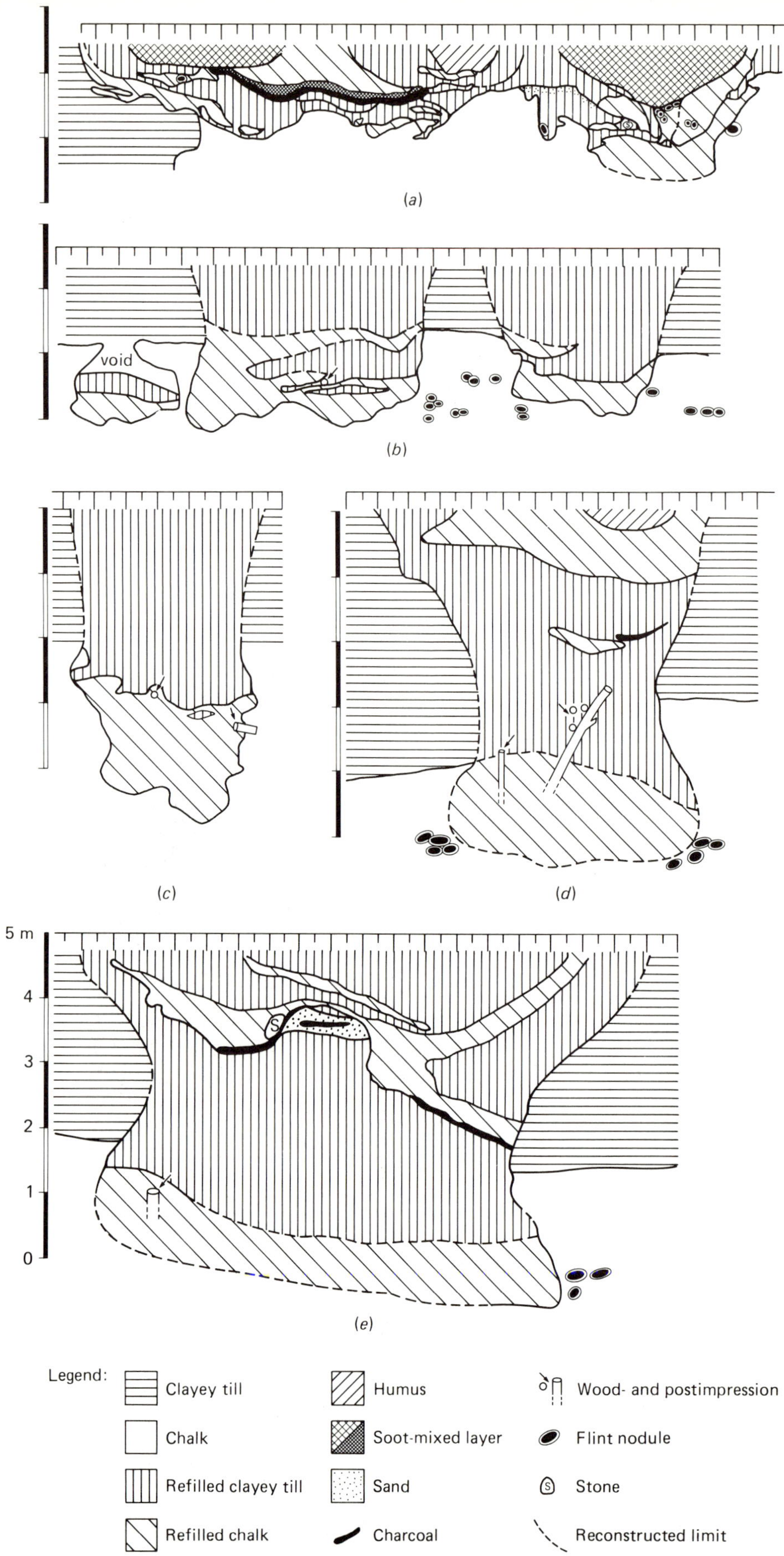

Fig. 18.2. Sections of (*a*) Surface pits in the superficial chalk. (*b*) Open shallow shafts. The shaft to the left with a sideworking. Note the collapsed chalk roof of the side working. (*c*) Open shaft without side workings. Note the horizontal post hole in the chalk wall to the right. (*d*) Open shaft with minor side workings. Note the postimpression in the refill. (*e*) Open shaft with wide diameter and side workings.

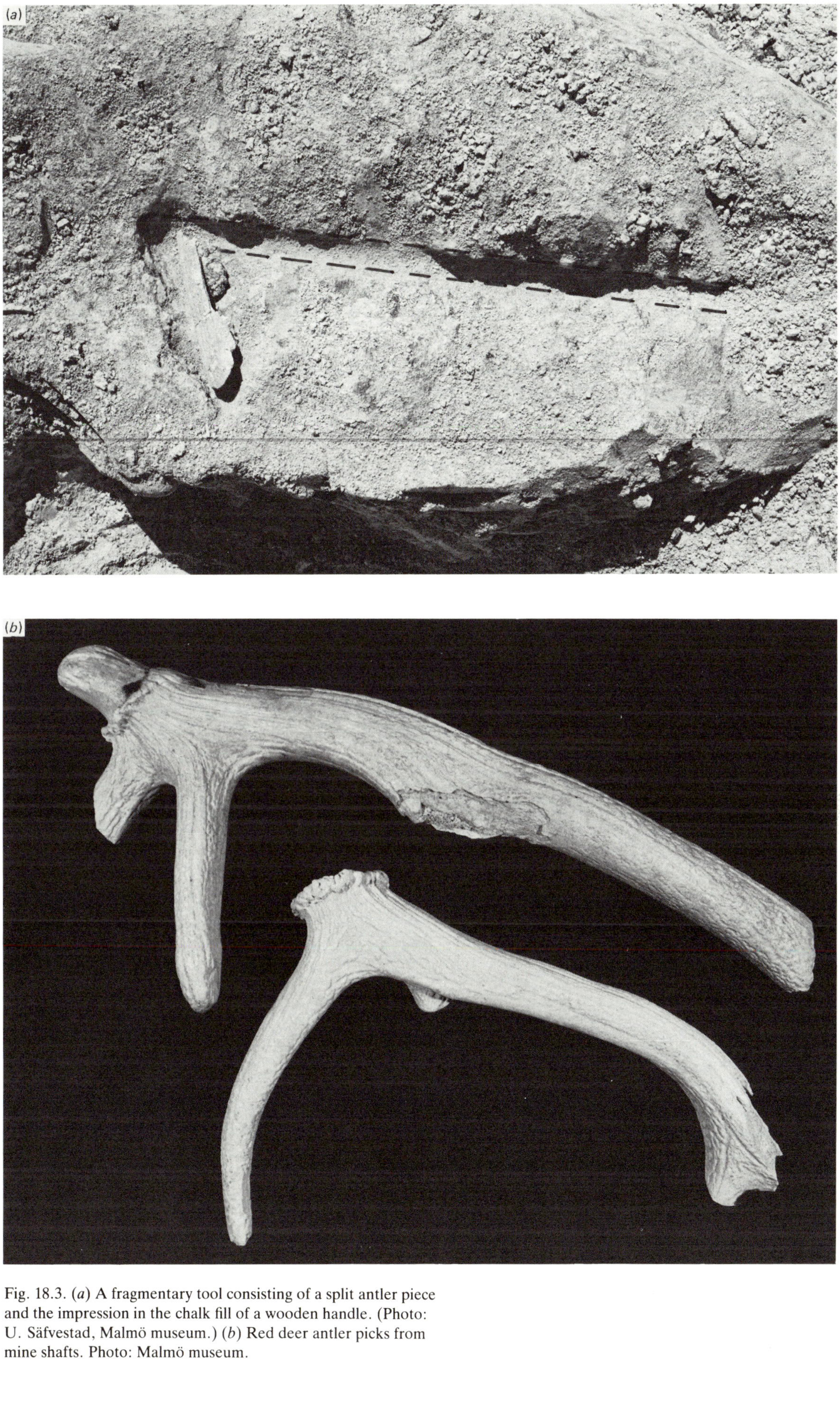

Fig. 18.3. (*a*) A fragmentary tool consisting of a split antler piece and the impression in the chalk fill of a wooden handle. (Photo: U. Säfvestad, Malmö museum.) (*b*) Red deer antler picks from mine shafts. Photo: Malmö museum.

are secondary, but there is also the possibility of the use of this multi-functional and easily made tool continuing throughout this entire time span (especially in the flint-rich area of Kvarnby–S. Sallerup) (Rudebeck *et al.*, 1980).

Four fragmentary thin-butted flint axes have been found in the mining area. These axes belong to the Early Neolithic. None of the axes seem to be made from the mined flint, but from a coarser or lighter flint type. It is interesting to note that rough blanks, roughly chipped picks and flakes of a coarser flint have been found in the mining area. This flint could well have been extracted from the till or the gravel covering the chalk.

Pottery has been found in the top layers of the filling of a few shafts and in the open cast area. Most of the pottery can be dated to the Early Neolithic TRB Culture, but there is also pottery from the Late Bronze Age (ridge decorated pottery). A pair of bronze tweezers from the Early Bronze Age have been found in the top layers of one shaft.

Radiocarbon dates from four shafts and two open cast pits fall between 5080±65 BP and 4560±70 BP and place the mining activity within the Early Neolithic. However, it is obvious from the previous account of tool production that this dating is not conclusive and that flint extraction probably continued throughout the Neolithic period.

Burial mounds from the Early Bronze Age and settlement remains from the late Bronze Age are common in the Kvarnby–S. Sallerup area. It is clear from the excavations of some of these that Kvarnby flint was used for many small tools, such as scrapers, drills and knives (Widholm, 1974), but it is not clear whether the flint was actually mined during these periods.

The extension of the mining area

The chalk has been quarried in the Kvarnby–S. Sallerup area since the late 19th century and the quarries cover an area of about 200 000 m² (not including the quarries opened after 1977). Most of the quarries were opened without archaeological investigations prior to the quarrying and an unknown number of mines and other features have obviously been destroyed; it is therefore difficult to estimate the original extension of the mining area. Through field observation of flint debris and chalk mixture in the top soil it should be possible to roughly delimit the mining area. However, as the intensity of mining obviously varies even in areas where the chalk lies close to the surface (see Fig. 18.1), it is as yet impossible to know whether the area with 400 mine shafts is the focal area or only one of several areas with the same mining intensity.

Fig. 18.4. Aerial photo of the 400 mines north east of Ängdala farm (see Fig. 18.1). (Photo: L. Wilhelmsson, Malmö museum.)

Bibliography

Althin, C.-A. (1951). 'The Scanian flint mines.' *Meddelanden från Lunds Universitets Historiska Museum*, pp. 139–58.

Holst, N.-O. (1906). 'Flintgrufvor och flintgräfvare i Tullstorpstrakten.' *Ymer*, 2.

Ringberg, B. & Rudebeck, E. (1982). 'Geologi och Industri, Skandinaviens äldsta gruvdrift.' *Varv*, No. 3.

Rudebeck, E., Seitzer-Olausson, D. & Säfvestad, U. (1980). 'Die südschwedischen Feuersteingruben – Ergebnisse und Probleme.' *5000 Jahre Feuersteinbergbau, Die Suche Nach Dem Stahl Der Steinzeit.* Bochum.

Schnittger, B. (1910). 'Förhistoriska flintgrufvor och kulturlager vid Kvarnby och S. Sallerup i Skåne.' *Antikvarisk Tidskrift För Sverige*, **19**, 1–101.

Widholm, D. (1974). 'En bronsåldersboplats vid Kvarnby.' *Malmö Museums Årsbok*, 1973–4, Malmö.

19

A flint quarry in the Hambledon Hill Neolithic enclosure complex

R. J. MERCER

Introduction

Nine seasons of excavation at Hambledon Hill have now completed the investigation of this hilltop enclosure within the framework of our present understanding of the Middle Neolithic in southern Britain (Mercer, 1980). The hill itself, a chalk boss, is set in the confluence of the River Stour which runs south eastwards to the sea at Christchurch and its minor tributary the River Iwerne running north–south along the western edge of Cranborne Chase to meet it. The hill, therefore, commands the 'Stour Gap' in the upland massif that stretches from the Devon/Dorset border to Salisbury in central Wiltshire. It also lies at the junction of two economic and geographical units – the upland massif and the lowland region of the Vale of Blackmoor lying between Shaftesbury, Wincanton, Sherborne and Blandford along the Somerset/Dorsetshire border. At around 3000 bc the chalk uplands of southern England may well have been covered with soils altogether more stable and fertile than the truncated profiles that we observe there today. Little is known of the likely landscape or of its degree of utilisation in the lowland areas but Barker & Webley (1978) have suggested that in such a setting the river margins may have been extensively exploited for cattle grazing.

That enclosure construction has been in progress in Britain since the middle centuries of the fourth millennium bc, is demonstrated by recent radiocarbon determinations from the excavations at Briar Hill in Northampton (Bamford, 1979). In the south west of England also the fortified and defended promontory enclosure at Hembury in Devon (Liddell, 1931–35) yielded three radiocarbon dates, apparently related to timbers of its constructional phases, which range from 3300–3000 bc. At Hambledon Hill itself construction appears to have commenced in the early centuries of the third millennium bc, some 700 radiocarbon years after the initial appearance of farming communities in southern Britain. The sequence of activity on the site is still only sketchily perceived, but it would appear that two discrete enclosures were built on the hill: one of 8.5 ha (the Main Enclosure) and one of about 1 ha (the Stepleton Enclosure) – although not necessarily at the same time. Later than the construction of these, a massive series of outworks were put up to enclose the whole hilltop, an area of some 75 ha. This great outwork enclosure, one of the very largest neolithic enclosures in Europe, was defended by a timber reveted rampart of 'box' construction, which appears to have been reconstructed on at least one occasion. On its southern flank the *enceinte* appears to have been reinforced towards the end of its occupation by one, or sometimes two, outer lines of ramparts, converting the enclosure to a multi-vallate structure. Shortly after this conversion, very large parts of the great enclosure appear to have been destroyed by fire of such intensity as to burn timbers deeply into their earthfast sockets. This destruction would appear also to have ended the use of the smaller Stepleton enclosure and a number of intact human skeletons were retrieved in this area; one with a leaf-shaped arrowhead contained within its thoracic cavity. It is suggested that this evidence points to the destruction of the site occurring as a result of intercommunal conflict, a termination noted at several other enclosure sites in Britain in this period – Crickley Hill in Gloucestershire, Carn Brea in Cornwall (Mercer, 1983), and at Hembury in Devon. Clear indications of pit excavation in the neolithic period for the extraction of flint were found immediately behind the great outwork enclosure rampart on the steeply sloping south west flank of the site (see Fig. 19.1). This 'flint mine' is diminutive in character and quite distinct from the altogether more massive enterprises with which archaeologists are familiar, and which relate to the same general period, such as Harrow Hill, Cissbury and Blackpatch in Sussex, and others elsewhere in Britain.

Flint mine complex 1

To understand the evidence of flint mining at Hambledon it is important to appreciate the role that erosion and solution has played in producing the weathered chalk sub-soil surface of the present day. It is likely that five major phases of cultivation have taken place at Hambledon Hill since the neolithic period. This probably included some 1500 annual cycles of cultivation, and the prevailing conditions of solution and sub-aerial erosion on the hilltop has lowered the level of the original soil cover and the natural chalk surface by as much as a metre, as is shown by the truncation of ditch and other feature fillings. The present truncated 'sub-soil' is a discoloured and broken chalk surface of variegated texture and composition (quite as common a phenomenon as hard white chalk on southern English chalk-land archaeological sites). It is deeply riven with fissures filled with clay with flints and other secondary natural deposits.

On this type of surface, on a steep slope some 15° from horizontal behind the outwork rampart at the south west extremity of the site, occurred a series of patches of what appeared to be discoloured and broken chalk. The chalk was very hard and no artefactual material was found on these patches. In circumstances of large area excavation, with extensive cross-referencing of sub-soil textures, and allowing the scrupulous inspection of all sub-soil anomalies, it was possible to recognise that these were archaeological structures, and to show with excavation that they were flint quarries. Under different circumstances they could not have been recognised. Nothing whatever was visible on the surface before excavation and only with hindsight is it possible to associate these features with very faint and somewhat indefinite anomalies on aerial photographs of the site. The initial lesson to be drawn from the exercise is clear. Such flint quarries may be an altogether more frequent phenomenon than has been established archaeologically to date.

Though no finds were visible on the surface, the commencement of routine test excavation of these 'sub-soil anomalies' produced relatively small quantities of flint-flaking débris and a few fragments of bone and antler. The archaeological status of the anomalies thus became clear and excavation proceeded in the normal way.

The fill of the main quarry (FQ1, 2 & 3) comprised shallow, truncated tip lines of near sterile material, introduced from the west side, and dipping downwards to the east. Very few artefacts were contained within this deposit, which had the appearance of a deliberate backfill. A few unabraded fragments of neolithic bowl pottery were present, however, and provided a *post quem* date for excavation and backfilling of the structures. The floor of the quarry and the lower part of the walls were of hard tabular

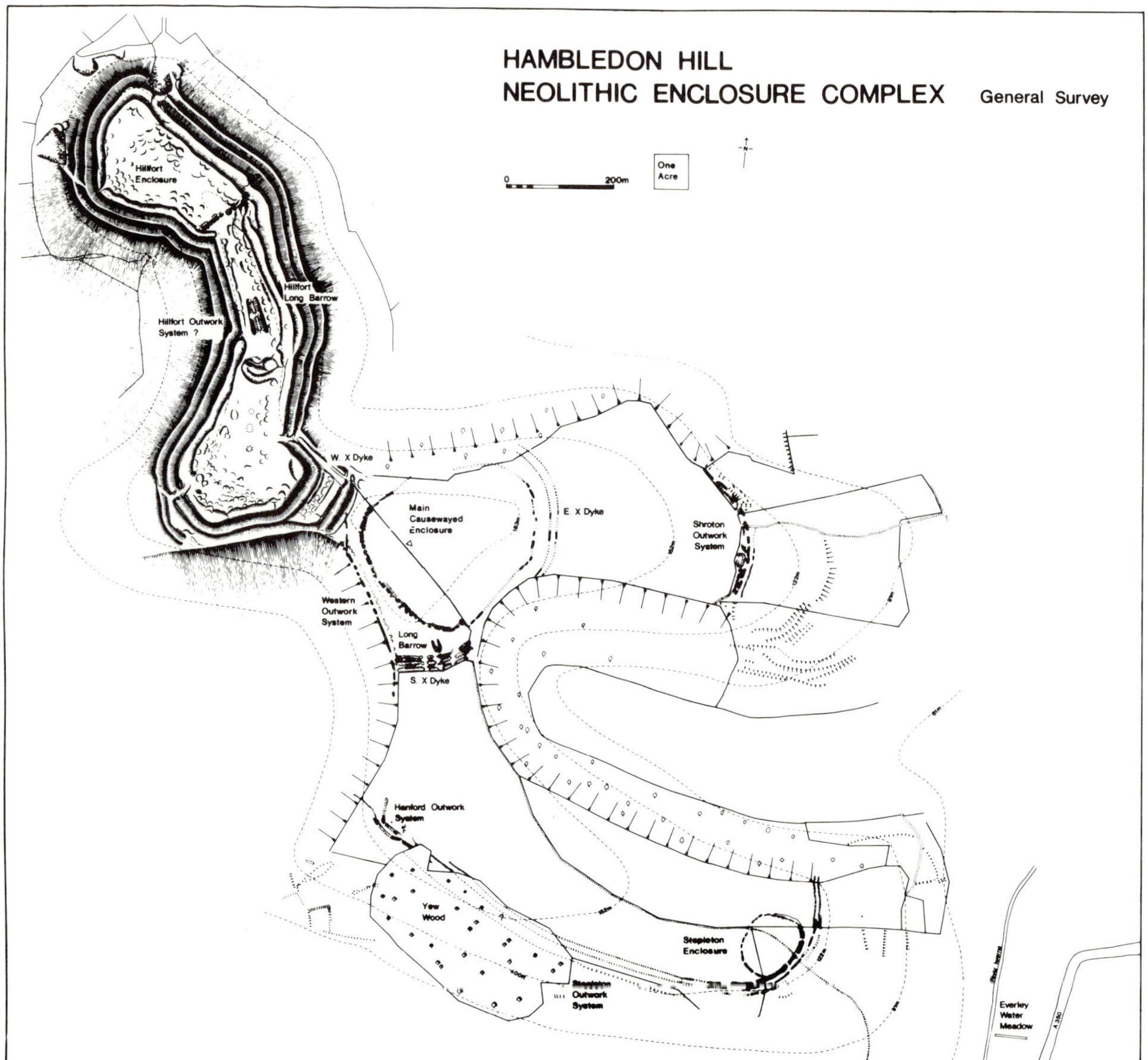

Fig. 19.1. Plan of enclosures at Hambledon Hill. The flint mine complexes are south of the Hanford outwork system.

chalk, but the product of quarrying such solid bedrock material was not to be found in the backfilling. It would appear, therefore, that the quarried bedrock blocks had either totally weathered down before backfilling began or was quite deliberately not reintroduced into the quarry. Little in the way of weathered silting was encountered in the base of the quarry, which would suggest that it was not left open for any very extended period of time. It did, however, remain unfilled for long enough for very considerable deposits of small rodent bones to form in the base of the shaft. The writer has experienced in another context the depressingly regular propensity of the Short-tailed vole (*Microtus agrestis*) to rush head-long into flint mine shafts and feels that the deposits in question may also indicate no more than the passage of a season prior to backfilling. The short period over which the quarry seems to have been left open, therefore, cannot be held to account for the weathering down of the chalk rubble matrix extracted from it and must be assumed that this material was disposed of on another part of the site. It would seem reasonable to suggest that it might well be that the chalk blocks were incorporated in the timber-framed rampart situated 10 m to the west of flint quarries. It follows that it may be possible to tie the period of quarry-working, and the use of the flint from the quarry, very closely to the period of outwork construction at Hambledon. The need for flint supplies, close at hand, during the construction process renders this suggestion attractive. The material that was backfilled into the quarry itself would appear to have been scraped up from around the quarry. Patches of material of quite high humic content appeared within the filling, which in turn accounts for the incorporation of a relatively thin scatter of struck flakes and pottery.

The reason for this apparently quite deliberate backfilling was presumably to render the ground fit for subsequent use, although on such a slope it would appear inherently unlikely that human occupation would figure in this later development. However, if the great outwork enclosure at Hambledon was used for corralling cattle, such backfilling may be regarded as understandable and necessary.

The structure of the flint mines

If we examine the plan of the flint quarries, we can see that galleries or undercuts extended from the base of the first flint mine complex (FQ1) in all directions following the flint seams and linking FQ1 to FQ2 and 3. A gallery to the south of FQ1, only broad enough and high enough at its opening to allow extraction of the desired raw material by a man lying on his side, breaks through into what appears to be a further quarry pit – FQ4. In the opening of this gallery from FQ1 an antler 'rake' was located. Little other *in situ* material was found on the floor of the quarry.

FQ4 is a quite separate substantial pit, dug to the south of the main FQ 1/2/3. It is roughly circular in form. Once again it is clear that the hard block chalk material that must have been extracted from the pit was not included in the backfilling and the same conclusion is suggested for this anomaly as in FQ1, 2 and 3. This common factor, however, is the only resemblance between the filling of FQ4 and that of the larger quarry complex to the north, for in this instance the initial silting of the pit to a depth of about 10 cm was followed by a much richer deposit of organically stained deliberate backfilling containing considerable quantities of pottery, antler and bone. Except in its conjunction with the quarry complex to the north, this pit in every way resembles the domestic rubbish-filled pits common throughout southern Britain at this period (Field, Matthews & Smith, 1964), perhaps suggesting further complexity in the more general interpretation of neolithic pits, at present covered by the functional *label* 'grain-storage pits'. The floor of FQ4 certainly coincides with a seam of tabular flint some of which is, however, left *in situ*; furthermore a recess had been dug into the south wall of the pit, a 'mini-gallery', presumably to allow the extraction of a flint nodule, so that it may have had a multi-purpose history including flint extraction.

The two adjacent pits, FQ5 and 6, are respectively to the south west and south east of FQ4. They are somewhat shallower pits and may represent initial soundings for flint quarrying activity. They appear to be deliberately backfilled and contained no artefactual material. In both cases the pits appear to have been filled up and then recut from the surface and the recut backfilled with ashy material – a procedure that is difficult to explain in rational terms. These pits are only included with the flint quarry complex by virtue of their proximity and because of the isolation of the whole complex from further pits on the site. Of course, it may be that these pits, though associated with the quarrying activity, did not actually function as quarries themselves.

The last pit, FQ7, lies to the south of FQ5 and is a somewhat deeper pit than FQ6. It clearly did encounter a seam of flint and was certainly backfilled straightforwardly with chalk rubble.

Flint mine complex 2

Twenty-five metres to the north west of flint mine complex 1 is a secondary quarry site, partly examined during the 1982 excavations. The limited excavation of this complex revealed a tip-line orientation denoting deliberate backfilling similar to that observed in FQ1, 2 and 3 of flint mine complex 1. A quantity of plain bowl pottery was located within a gallery that opened from the quarry pit in its south west sector.

Discussion

As the ubiquitous surface erosion on the site led to the loss of about a metre of chalk surface from the hilltop, the quarry pits at present about 70 cm deep such as FQ1 (see Fig. 19.2) would originally have been *c.* 1.70 m deep. The galleries in almost every instance are now covered by 'roofs' only 10–20 cm thick, which during excavation could not be trodden on for fear of bringing about their collapse – a clear demonstration of the degree of erosion on the site. If

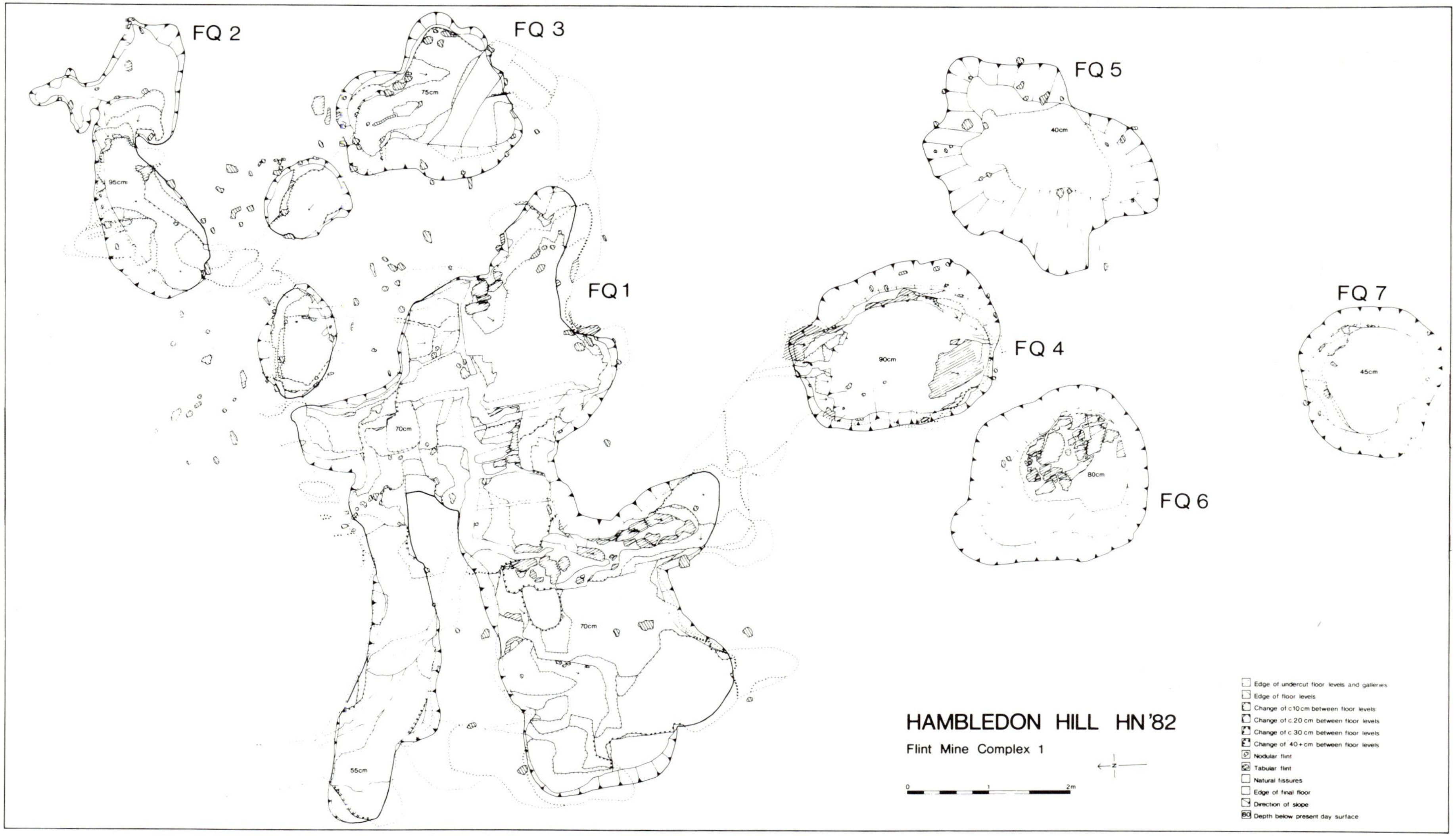

Fig. 19.2. Plan of flint mine complex 1 at Hambledon.

the postulated, and necessarily approximate, figure of 1 m for the depth of eroded, and now absent, sub-soil is accepted, then the volume of sub-soil and chalk removed when flint mine complex 1 (FQ's 1–7) was dug out would have amounted to approximately 50 m^3. Startin's (1982) calculations indicate that this degree of block chalk digging would take approximately 130 man hours. This effort was invested to obtain flint which seems to the writer to be of relatively poor quality and which, he would suggest, could not have been expected to produce large edge tools. Its use must have been for small 'domestic' tools (scrapers, borers, knives, etc.) for general constructional and domestic activities possibly associated with the construction of the outwork system on the site. It would seem likely that these workings are essentially *ad hoc* and of short duration.

Inspection of the very extensive aerial photographic cover of the Hambledon complex revealed a series of indistinct anomalies that appear to coincide with the disposition of these flint quarries on the ground. A series of other, somewhat similar, anomalies of equally indistinct form occur outside the excavation area. These anomalies may possibly mark the position of a further extension of the mine complex on the southern spur of Hambledon Hill.

Parallels for the flint quarries at Hambledon are hard to find in Britain, although one expects that *ad hoc* quarrying in direct association with occupation sites and heavy engineering work must have been very common indeed. Two examples are known to the writer.

(1) *'Trial flint mine pits' 200 m east of Stonehenge Inn, Durrington, Wilts*

This complex, 400 m north east of the great 'henge' earthwork enclosure at Durrington Walls, was discovered during the course of pipeline laying in 1952 (Booth & Stone, 1952). Six pits ranging from some 60 cm–1 m in depth had been dug apparently to gain access to a seam of relatively poor quality flint. One of the deeper pits had undercuts and galleries extending from its base and another pit also had undercutting. Antler picks were found in the spoil dump located beside the pits and the floor of Pit 5 produced antler picks and a rake. An apparently broken transverse arrowhead also lay on the floor of Pit 5, suggesting that this quarry may have been contemporary with Durrington Walls itself. If this observation is accepted, then these pits would appear to be chronologically quite distinct from those at Hambledon Hill, as they belong to the beginning of the second millennium bc. Nevertheless, the possible use of flint mining as an adjunct to a great engineering project like Durrington is an interesting parallel, which may lead us to expect such activity in the vicinity of many sites of similar size.

(2) *Ballygalley Hill, County Antrim, Northern Ireland*

This site was excavated in 1957 by A. E. P. Collins (Collins, 1978).* A neolithic occupation site, set on a prominent hilltop, was sealed beneath a later 'rath'. Only a small was excavated. It produced a very large quantity of fl knapping débris and sherds of plain carinated bowls Case's Ballymarlagh type (Case, 1961, p. 223). An excavation cutting laid out approximately 100 m to the south of the site revealed a series of shallow 'open cast' cuts through a capping of Boulder Clay, ranging from 1–1.5 m in depth, revealing flint seams obtruding from the underlying chalk matrix. Basalt blocks had been obtained from the overlying Boulder Clay and these were used as anvils for the intensive flint knapping that was undertaken on the site. Again the discovery of small-scale flint mining in direct association with a nearby settlement site is an interesting parallel for the situation at Hambledon Hill and one that must be approximately contemporary.

The retrieval of the diminutive flint quarry at Hambledon Hill and its context there, together with the evidence from the other two sites (both totally unsuspected prior to excavation), suggests the existence of a practice that almost certainly was extremely widespread. This must have important implications in terms of the practice of flint extraction, the often devolved nature of its organisation, and may lead to further developments in flint mine studies in the future, for example in the pursuit of scientific locational techniques for determining the origin of flint implements.

* The writer is most grateful to Mr Trevor Cowie of the National Museum of Antiquities of Scotland for drawing it to his attention.

Bibliography

Bamford, H. (1979). 'Briar Hill neolithic causewayed enclosure – second interim report.' *Northamptonshire Archaeol.*, **XIV**, 3–9.

Barker, G. & Webley, D. (1978). 'Causewayed camps and early neolithic economies in central southern England.' *Proc. Prehist. Soc.*, **XLIV**, 161–86.

Booth, A. St J & Stone, J. F. S. (1952). 'A trial flint mine at Durrington, Wiltshire'. *Wilts. Arch. Natur. Hist. Mag.*, **LIV**, 381–8.

Case, H. J. (1961). 'Irish neolithic pottery: distribution and sequence.' *Proc. Prehist. Soc.*, **XXVII**, 174–233.

Collins, A. E. P. (1978). 'Excavations on Ballygalley Hill, County Antrim.' *Ulster Journ. Archaeol.*, **XLI**, 15–32.

Dixon, P. (forthcoming). *Crickley Hill*, Edinburgh University Press.

Field, N. H., Matthews, C. L. & Smith, I. F. (1964). 'New Neolithic Sites in Dorset and Bedfordshire, with a note on the distribution of Neolithic storage pits in Britain.' *Proc. Prehist. Soc.*, **XXX**, 352–81.

Liddell, D. (1930–35). Successive reports entitled: 'Excavations at Hembury Fort.' *Trans. Devon Archaeol. Explor. Soc.*, **1, 2** (1930), 40–63; **1, 3** (1931), 90–113; **1, 4** (1932), 162–90; **11, 3** (1935), 135–75.

Mercer, R. J. (1980). *Hambledon Hill – A Neolithic Landscape* (1980), Edinburgh University Press.

Mercer, R. J. (1981). 'Excavations at Carn Brea, Illogan, Cornwall 1970–73 – a neolithic fortified complex of the third millenium bc'. *Cornish Archaeology*, **XX**, 1–204.

Startin, D. W A. (1982). 'Prehistoric earthmoving.' In: Case, H. J. & Whittle, A. W. R., eds. *Settlement Patterns in the Oxford Region: Excavations at the Abingdon Causewayed Enclosure and Other Sites*, CBA Res. Rep. No. 44, pp. 153–6.

20

The ancient chert mines at Wadi el-Sheikh (Egypt)

G. WEISGERBER

Introduction

In 1896 Heywood Walter Seton-Karr discovered the 'lost flint mines' of Wadi el-Sheikh in Egypt. However, the mines were later 'lost', as researchers have almost completely ignored these magnificent ruins. Few reports were published about the chert mines, preserved in the arid eastern desert as an open-air museum.

In April 1981 the author, accompanied by Mr J. Kunkel,* visited one part of the mine district; a brief report on this trip was published in 1982 (Weisgerber, 1982). Shortly before the field-trip, local Bedouin had dug out some of the old shafts, so that for the first time it was possible to observe and describe underground mining works in this area.

Wadi el-Sheikh emerges from the east into the Nile Valley some 160 km south of Cairo, south of Beni Suef between El-Fashn and Maghagha opposite the locality of El-Fant. Sparse vegetation runs along the dry wadi bed; the higher-lying limestone terraces are void of any growth. Along these terraces, rising some 40 m above the wadi bed, on the slopes and at the foot above the Lower Terrace are located the ancient chert mines.

Historical survey of discovery

Heywood Walter Seton-Karr (1859–1938)

In 1896 H. Seton-Karr discovered the chert mines of Wadi el-Sheikh and Wadi Sojoor (Fig. 20.1), located east of the Nile in the Maghaghah region (Seton-Karr, 1898). He was known as a big-game hunter, and combined this with a great interest in Archaeology and in Palaeolithic cultures. During his travels through Egypt, on the way to Somalia, traces of mines had been shown him by the Bedouin, who described the immense ruins as the remains of ancient gold mines. Seton-Karr was actually the first person to realise that the traces were those of ancient chert mines (Forbes, 1900).

Seton-Karr set up 15 'collecting camps' and at the site carried out fieldwork for a period of three weeks in 1896 and 1897. He had the area covered systematically and all finds documented. It became apparent that different frequency concentrations obtained for the various semi-finished products and implements and that knives, axes, sickles, rings, etc., were deposited in different areas, showing that specialist workshops existed near the mines.

Many thousands of objects must have been collected during these campaigns. Seton-Karr donated the finds to numerous European and foreign museums which has made it difficult to make a complete inventory of all the finds today. The major part of the collection was sold to the Mayer Museum in Liverpool, known today as the Merseyside County Museum. Additionally, photographs and a map of the valley were turned over to the Museum.

Henry O. Forbes

Personal contacts between Seton-Karr and Henry O. Forbes, then Director of the Mayer Museum, probably influenced the sale of most of the collected finds to Liverpool. Forbes was the only one to report which type of implements were found near each of Seton-Karr's camps. The original plane-table map has been preserved at Liverpool and camp sites can be localised on it (Forbes, 1900, map).

Max Blanckenhorn

Two years after their discovery, the German geologist Dr Max Blanckenhorn visited the mines, during his period of field studies in Egypt. He published a report on the results in 1901 (Blanckenhorn, 1901). In 1898 he remained in the field for a period of six weeks and was convinced that he had studied the 'most extensive and interesting flint mine-workings in Egypt'.

In addition to flint deposits in the Eocene Lower Libyan layer between Thebes and Akhmin, Blanckenhorn characterised the Lower Mokattam of northern Egypt as having good flint material, specifically on the eastern side of the Nile. Blanckenhorn's description of the chert mines is as follows:

> The mines can be found close to the Nile, near the village of Der al-Hadid; then upstream along the right bank of Wadi el-Sheikh, some twenty-four kilometres (as the crow flies) from the mouth in southeast direction; then south of the wadi on the southern arm – Wadi Agerum . . . The old stone-quarries and dumps are in such fresh condition . . . that I was inclined to think they were modern

* I would like to express my thanks for Mr Kunkel's accompanying me on this strenuous trip. As an archaeological conservator in the German Mining Museum, Mr Kunkel is well acquainted with fieldwork in arid regions.

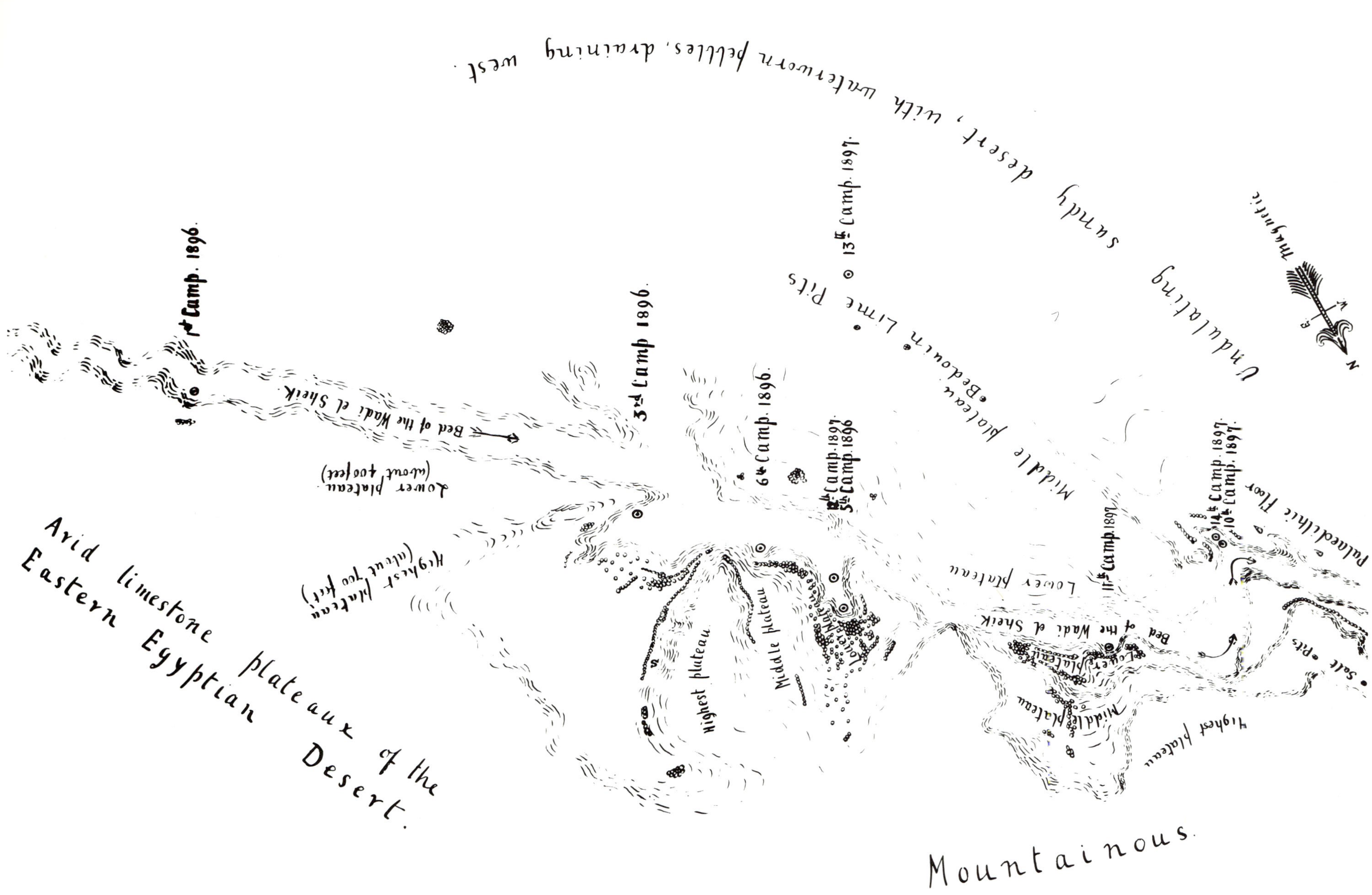

Arid limestone plateaux of the Eastern Egyptian Desert.
Mountainous.
Undulating sandy desert, with waterworn pebbles, draining west.
1st Camp. 1896.
3rd Camp 1896.
6th Camp. 1896.
12th Camp. 1897.
5th Camp. 1896
13th Camp. 1897.
11th Camp. 1897
14th Camp. 1897.
10th Camp. 1897.
Bed of the Wadi el Sheik
Lower plateau (about 400 feet)
Highest plateau (about 700 feet)
Highest plateau
Middle plateau
Lower plateau
Bedouin Lime Pits
Middle plateau
Lower plateau
Bed of the Wadi el Sheik
Highest plateau
Salt Pits
Palaeolithic Floor
Magnetic
N
E
W

gypsum and salt pits, which had recently been abandoned. The only sign of age is the brownish to blackish crust or patina on the exposed sides of the artifacts.

Chert is contained in outcrops only in the Eocene terrace; on the 40 metres lower-lying 'Lower Plateau' (Seton-Karr) – the terrace above the wadi bed – chert is found only as detritus in secondary bedding, most often at the base of high cliffs.

Blanckenhorn therefore distinguished two types of mining: open-diggings in the secondary deposits on the 'Lower Plateau' and open-pit mining on the rim of the Upper Terrace: 'On the actual flint terrace, where the respective beds are naturally most open along the terrace edge, mining took place almost invariably from the rim inwards . . . The rim is covered with tip-heads and open-diggings, mostly horse-shoe shape with convex rims having walls of $\frac{1}{3}$ to $1\frac{1}{3}$ metres in height, these resemble entrenchments (Blanckenhorn, 1901, p. 488).' As the description illustrates the situation prevailing on the terrace edges, it is obvious that Blanckenhorn did not notice the mine shafts on the plateau surface itself, or did not realise what they were. He refers therefore only to flint quarries in the publication and in the two sketches he made of the area (Blanckenhorn, 1901, p. 34f).

Blanckenhorn described the chert found in Wadi el-Sheikh as light grey, yellow-brown or blackish and non-opaque (Blanckenhorn, 1901, p. 486). It can be added that the material feels slightly grainy to the touch and when exposed to weathering the chert develops a dark brown patina. These characteristics allow the Sheikh chert to be easily distinguished from other varieties of Egyptian flint.

Jacques de Morgan

The finds which had been donated to the Musée des Antiquités Nationales at Saint-Germain were illustrated, along with maps constructed by Jacques de Morgan in two handbooks (de Morgan, 1924, 1926) and reprinted in other reference works (Vandier, 1952; Jahn, 1960).

Elise J. Baumgärtel

The third visitor to Wadi el-Sheikh was Elise J. Baumgärtel, who has done much research on Egypt's prehistory (Baumgärtel, 1960, 1971). She completed a report on her researches in Wadi el-Sheikh (Baumgärtel, 1930). Originally Baumgärtel became acquainted with the flint finds of Wadi el-Sheikh during the course of research on stone age Mediterranean cultures.

Before proceeding to Egypt she visited the ancient flint mines in the Netherlands and Belgium. In her publication with Brotzen (1927), the Wadi el-Sheikh chert material was ordered typologically and assigned to the Campignian Stage. This is a fate often shared by blanks and semi-finished artifacts found at sites of prehistoric mines, in cases where the implements are not recognised as being half finished (Reisch, 1974).

The 1930 expedition showed the true extent of mining activities on the terrace. No remains of miners' huts were found. The pits were described as round, with a surrounding wall of chalk tip and centres filled with sand. The layer of flint must have been thin, as shown by numbers of broken artifacts lying on the sites and indicated by a thin cortex on the Berlin finds. Baumgärtel, however, recognised the shaft mouths ringed with waste-heaps as being mine-pits, considering them to be flat depressions. On the other hand, she did not notice that there were a large number of open-air quarries. Baumgärtel was impressed by the enormous extent of flint production at the site. As spectacular finds were lacking, she reached the conclusion that most of the impressive flints had been collected during Seton-Karr's campaigns. Hand-made and wheel-turned pottery on the site indicated how long flint mining had been carried out.

Martin Jahn

Martin Jahn mentioned the previously ignored pits of Wadi el-Sheikh in his book *Der Älteste Bergbau in Europa* (Jahn, 1960, p. 17f, fig. 2(*f*)). Jahn felt that the pits could be compared with the open-diggings of Grand Pressigny in France, referring to both as 'mardell'-type mines. This term is still in use in mining archaeology, although it cannot be considered a correct technical term for open-cast quarrying.

Mining and production of raw materials

Surface quarrying

During the author's trip in 1981, depressions on the Middle Plateau, especially near the Upper Plateau Cliff, were visible as numerous round areas (about 0.6 m diameter) filled with deposits of sand and dust. These appeared to be filled-up small depressions, typical of surface quarrying. This is an indication of chert layers lying very near the surface, as prospection pits are not usually so close to each other. These features are similar to the 'dish' pits found at Timna on the Red Sea, which were actually filled-up shafts. However, the diameter of these 'dish' pits was larger than those found in Wadi el-Sheikh (Conrad & Rothenberg, 1980, p. 72f, fig. 49). In Wadi el-Sheikh's arid climate the tip-heaps would also be visible. These particular round areas are, however, completely flat.

Slope and trench open workings

The district examined in the 1981 field season showed traces of extensive earth movement along the slope of a fossil wadi bank. Trenches run along the sloping banks of the wadi as far as the eye can see; presumably further remains are

Fig. 20.1. Original plane-table map of Wadi el-Sheikh by Seton-Karr, showing locations of the mines and collecting camps. (Copy courtesy of the Merseyside County Museum, Liverpool.)

buried under drifted material. Dug-out earth has been heaped up on the wadi side. Here trenches were dug in order to lay bare the beds containing chert and to simplify mining the material. The workings do not seem to have been very extensive, as most of the fossil wadi bank remained intact.

Raw material extraction was more extensive on the Middle Plateau and also in the area of the cliff, where a series of pit open-works exists. Waste heaps, more than 10 m in length run parallel to one another, indicating that mining was carried out systematically, although many of the trenches have been covered by blown-over material. The tip-heaps consist of weathered chalk fragments, with heights up to 1 m and widths of 2 m and more, remaining at the present day. The function of such trenches and pits could be easily determined by use of aerial photography.

Between the tip-heaps are waste dumps of the tool-makers. The various semi-finished flint products were flaked both here and near the mineshafts. There are remains not only of mounds of waste flaking debris, many cubic metres in extent near each pit, but also remains of finely worked artifacts broken during processing.

Slope diggings, which seem to have been Blanckenhorn's chief interest (Blanckenhorn, 1901, p. 487f, fig. 34(*f*)), also occur on the edge of the Middle Plateau in the direction of Wadi el-Sheikh. On the plateau rim, on the slope, and on the plateau plain can be seen traces of minor workings. On the plateau, the remains are evidently small tip-heaps; on the steep slope the traces are more in the nature of dug-out holes.

Underground mining

Adit-mining

It seems possible that adit-mining was carried out on the slopes of Wadi el-Sheikh or in the small valleys of the Eocene Terrace. When mining chert in slope workings, the hanging wall would be undermined to some extent and tunnels could be dug and raw material exploited from the adit itself.

During the 1981 field season several dark openings in the terrace rim were located that may be remains of adit mouths. Unfortunately, there was not enough time during the expedition to explore these tunnels.

It is presumed that large-scale tectonic movements or extensive erosion processes seen west of Wadi Arabah (Timna), are not likely to be repeated in this area. In Wadi Arabah, shafts and galleries were badly eroded, so that traces of former underground galleries are nowadays apparent as tunnels along the valley walls (Conrad & Rothenberg, 1980, 65, figs. 152, 164).

Shaft mining

The remains of the Wadi el-Sheikh mines are characterised by high mining dumps and create an imposing picture; it is as if they have only recently been abandoned. The steep tip-heaps appear freshly dumped and the close rows of dumps indicate how actively mining was carried out. A closer view shows, however, that dust has collected at the foot of the tip-heaps; the chalk fragments and blocks are weathered and the shaft mouths are closed by sand. Semi-arid conditions have contributed to preservation of the prehistoric mining district in a state known at no other site. Other flint mining-districts, such as Spiennes (Belgium B 1), Ryckholt–St Geertruid (Netherlands NL 1) and Grimes Graves (Great Britain GB 13), must have presented the same imposing picture in prehistoric times before they were flattened (Weisgerber *et al.*, 1981).

In relation to other prehistoric sites, the tip-heaps are enormous: some are piled up to 3 m high above the plateau surface. Heap forms differ according to the amount of debris accumulated, some are sickle-shaped, some horse-shoe shaped and some ring-shaped. The forms are due to the fact that dumps were thrown up in such a way as to keep the pit-mouth clear on the side of access. Even ring-shaped dumps have a pathway over the top. The diameter of ring-heaps may reach dimensions of 20 m, and a breadth of up to 8 m. Central areas within the rings are filled to varying heights with drifted sand.

One of the shafts must originally have had a diameter of 1.2–1.8 m; it had recently been partly emptied by the Bedouin in an attempt to mine salt. But though they dug through some 3 m depth of sand deposits they had not reached the sides and bottom of the original shaft.

Another shaft, examined was 4 m deep, with a diameter of only 1 m. Its sides have only been roughly cut and the various degrees of hardness of the different chalk–limestones layers can still be seen. The irregularities these cause permit easy access to the shaft. The original shaft was partly filled by recently deposited sand. A shaft landing was found at the bottom of the shaft from which two galleries branch off. The first gallery opening is 1 m wide and can only be seen for a few metres as it is filled with detritus; the second opening is 2.50 m wide and free for a length of approximately 20 m. The central shaft opening is filled on both sides with piles of chalk, leaving a narrow track in the middle. The lower layer of shaft fill is composed of fine, small-sized fragments while on top are deposits of limestone slabs and blocks. A third landing, near the two described, was filled almost to the ceiling with finely sorted debris. In both the open galleries remains of chert blocks hanging on the roofs could be seen which were obviously mined. Presumably another flint bed was mined from the tunnel sole.

Tool marks on the roof differ completely from those found in European flint mines: they appeared to be marks made by metal tools. This strengthens the impression that the mine might have been constructed and used in two different stages – a stone-using and a metal-using period.

Miners' tools

Miners' tools belonging to one phase were found mixed up in the detritus tip-heap. They are roughly hewn grooved hammers, distinguished from known European counterparts found in prehistoric copper or flint mines in that the groove is closer to the end rather than the middle of the hammer (Jovanovic, 1982; Schmid, 1980). This type of hafting is seen on Egyptian wall-paintings (Weisgerber *et al.*, 1980, p. 46, fig. 22). A find in the Cairo Museum of this type of tool has a thin wood shaft, fastened with rawhide (Zuber, 1956, p. 168, fig. 4).

Chert picks, which Seton-Karr had collected in Wadi el-Sheikh, seemed to have been hafted in the same fashion, as far as can be determined from the finds now in Merseyside County Museum. These are heavy core-tools, with rhombic or triangular cross-sections, and notches on the end.

Infrastructure of the mining district

Since the discovery of the Wadi el-Sheikh chert mines, researchers have concerned themselves with various questions.

Where did the miners live?
Was work carried out seasonally?
Where were rough blanks finished?
To which regions in Egypt were products distributed?

Those who previously inspected the mining district found no traces of anything that could be considered miners' houses or huts but during the 1981 fieldwork remains of a hut were found next to a shaft mouth.

If we assume that the miners lived locally, the next question is: who provided them with water and food? A watering site has not yet been found in Wadi el-Sheikh; the Nile is many donkey's hours distant from the mines. A large group of miners could not survive for longer periods in the district without logistic support. An interdisciplinary survey including a hydrologist is needed to provide the answers to this question.

The evidence of finds from the enormous flake-heaps shows that raw material was worked to the stage of semi-finished products in the mine area. Artifacts such as axes, blades, rings, etc., were produced. In many cases the flake-heaps were overlaid by later tip-heaps.

In the course of time, large heaps – flakes, old cores and broken artifacts – grew up next to the larger mines. The district studied in the field in 1981 was obviously devoted to production of blades and specially prepared cores. Other types of artifacts were found only in minor quantities. A huge heap of flaking debris was discovered at Qarara near the mouth of Wadi el-Sheikh (Kaiser, 1961). What relationship existed between this manufacturing centre some distance away and the mines themselves and during what periods did this operate?

Dating

The remains of mining in Wadi el-Sheikh cover an area of many square kilometres; and therefore it can be presumed that flint mining also took place over a long period of time. We may also presume that some surface finds collected during field-trips belong to a more recent period of mining activities.

Large masses of materials are moved by mining processes destroying or covering over older workings on the same site. In relation to the Wadi el-Sheikh chert mines, this suggests that the numerous blanks and ornaments collected most probably belong to the latest period of mining operations. These artefacts appear to belong to a single chronological period, particularly the large knives which are typical of the Middle Kingdom and illustrated in tomb drawings (Beni Hasan tombs 2 and 15) of the period. So it looks as if flint mining was carried out here on a large scale during the Bronze Age. Perhaps this was due to the specific characteristics of Wadi el-Sheikh flint, which occurs in big plates suitable for the production of large, flat artifacts, such as knives.

It is also possible that older mining works can be found on the site: Baumgärtel found indications of earlier workings in the presence of hand-thrown pottery. Another indicator for older workings was the discovery in 1981 of two different techniques in the construction of a mine gallery. Such exploitation may have started in the Old Kingdom.

Further research will have to be done, to clear up these problems. An archaeological survey and excavations in the mining area will be helpful as will archaeometric methods of dating the finds at this site which may include organic materials, excellently preserved in the arid climate. More knowledge of the varieties of Egyptian chert and their origins will enable us to determine the provenance of finds. Cherts from Wadi el-Sheikh found in the Nile Valley settlements and dated by archaeological methods will allow us to determine the periods in which mining operations were carried out in the area.

Acknowledgement

The author is grateful to Mrs A. M. Martin, Bochum, for translating this paper.

Appendix 1: remarks concerning museums with collections of material from Wadi el-Sheikh

Australia

National Museum of Victoria, Melbourne

Seton-Karr donated a collection to the Ethnological, Industrial and Trade Museum, in Melbourne presumably in 1897 (Seton-Karr, 1898). The Anthropological Museum has 139 artifacts, which were those previously in the Ethnological Museum. At present, no information is available on these finds.

Belgium

Musée du Cinquantenaire, Brussel

In an obituary on Seton-Karr, mention was made of a collection of Wadi el-Sheikh material donated to the museum (*Chronique d'Egypte*, 1938, **25**, 327).

Egypt
Egyptian Museum, Cairo
A collection of 125 objects is stored at the Museum. The objects are listed under the 'Catalogue Général' as special types of Neolithic artifacts (Currelley, 1913).

France
Musée de Saint-Germain, Paris
Several cores and picks are described by de Morgan (1924, fig. 75, 1926, fig. 204f.) and he mentions that several generous donations of material from Wadi el-Sheikh were made to the Museum by Seton-Karr.

Germany
Museum für Völkerkunde, Prähistorische Abteilung, Berlin
Objects from Wadi el-Sheikh were donated as a collection by Seton-Karr; the material has been published by Baumgärtel & Brotzen (1927). The material was destroyed during World War II but was a representative collection of artifacts similar to those in other museums.
Staatliche Sammlung Agyptischer Kunst, Munich
Twenty artifacts were donated by Seton-Karr to the Museum in 1903.

Great Britain
Merseyside County Museum, Liverpool
In 1897 a large collection of finds from Somalia and Wadi el-Sheikh was bought by the Corporation of Liverpool for the Mayer Museum from Seton-Karr. In 1899 a further grant was given to obtain finds from Wadi Sojoor and the photographs of the Wadi el-Sheikh mines. The finds were published by Forbes (1900) who mentions some 2000 objects. During the trip to the Museum in 1981, only the finds described by Forbes could be checked, and these included arm rings, axes, chisels, leaf-shaped and blade-shaped artifacts, hoes, agricultural implements (picks and cores?), chipping-stones, scrapers, cores and flakes.
British Museum, London
A small collection of 10 artifacts was donated to the Museum in 1901 by Seton-Karr. The finds were photographed in 1981.
Department of Recreation and Amenities, Borough of Ipswich, Ipswich
An indefinite number of artifacts were donated to the Museum by Seton-Karr. Not all of these have been catalogued. Three cores have been described in a publication which erroneously identifies them as Upper Palaeolithic artifacts (Moir 1918).

Italy
Museo Civico Archeologico, Bologna
Seton-Karr donated a collection of artifacts from Wadi el-Sheikh, together with 33 artifacts from Fayum, to the Museum in 1901. Seven cases with material from Wadi el-Sheikh are exhibited.
Museo e Istituto Fiorentino di Preistoria, Florence
A collection of 130 flints was presented to the Museo di Antroplogia e Etnologia, Florence by Seton-Karr in 1902. The collection was later integrated into the Prehistoric Museum, where the finds are both on display and in the depot.
Soprintendenza Archeologica per la Toscana, Florence
A collection of some 50 flint artifacts was donated to the Museum in 1901. A forthcoming publication will give a complete description of the implements.
Museo Egizio, Soprintendenza per la Antichita Egizie, Turin
Seton-Karr donated a large number of artifacts to the Museum in 1904; other artifacts were transferred from the Museum in Bologna to Turin. According to a description of the finds (Curto, 1978), this material may have come from Wadi el-Sheikh; however this still has to be checked.

Spain
Museo Arqueologico Nacional, Madrid
In the ancient Egypt section there are some objects from Wadi el-Sheikh on display including the typical picks and some semi-finished knives and flakes.

Malta
National Museum of Archaeology, Valetta
De Morgan (1924) mentions that he had seen many objects from Wadi el-Sheikh in Valetta, which had been donated to the Museum by Seton-Karr. The collection was destroyed during World War II.

References

Baumgärtel, E. J. (1930). 'The flint quarries of Wadi el-Sheikh.' In: *Ancient Egypt*, pp. 103–8, 6 figs., 1 sketch-map.

Baumgärtel, E. J. (1960). *The Cultures of Prehistoric Egypt*. Vol. 1, London, New York, Toronto.

Baumgärtel, E. J. (1971). *The Cultures of Prehistoric Egypt*. Vol. 1, 3rd ed., Michigan.

Baumgärtel, E. J. & Brotzen, F. (1927). 'Steinzeitliches Material aus den südlichen Mittelmeerländern im Museum für Völkerkunde, Berlin.' In: *Prähistorische Zeitschrift*. Vol. 18, pp. 100–3, fig. 1.

Blanckenhorn, M. (1901). 'Geologie Ägyptens.' *Zeitschrift der Deutschen Geologischen Gesellschaft*, **52** (3), Berlin.

Conrad, H. G. & Rothenberg, B. (1980). *Antikes Kupfer im Timna-Tal* (=Der Anschnitt, Beiheft 1; Veröffentlichungen aus dem Deutschen Bergbau-Museum, Bochum, No. 20), Bochum.

Currelley, C. (1913). 'Stone implements.' *Catalogue Général des Antiquités Egyptiennes du Musée du Cairo Nos. 63001–64906*. Cairo.

Curto, S. (1978). 'Storio del Museo Egizio di Torino.' *Catalogo del Museo Egizio di Torino, Vol. 1*. Turin, pp. 53, 108f.

de Morgan, J. (1924). *L'humanité préhistorique*. Vol. 2, Paris, pp. 156–9.

de Morgan, J. (1926). *La Préhistoire Orientale*. Vol. 1, Paris, pp. 156–62, Figs. 201–6.

Forbes, H. O. (1900). 'On a collection of Stone Implements in the Mayer Museum made by Mr H. W. Seton-Karr, in Mines of the Ancient Egyptians discovered by him on the Plateau of the Nile Valley.' In: *Bull. of the Liverpool Museums*, **2**, 76–106, plates, map.

Jahn, M. (1960). 'Der älteste Bergbau in Europa. Berlin' (=*Abhandlungen der Sächsischen Akademie der Wissenschaften zu Leipzig, Phil.-hist. Klasse*, Vol. 52, No. 2), Ref. p. 17f, Fig. 2f.

Jovanovic, B. (1982). 'Rudna Glava. Der älteste Kupferbergbau im Zentralbalkan,' Bor/Beograd, (=*Museum für Bergbau und*

Hüttenswesen, Archäologisches Institut. Besondere Ausgaben, No. 17).

Kaiser, W. (1961). 'Bericht über eine archäologisch-geologische Felduntersuchung in Ober-und Mittelägypten.' In: *Mitt. des Deutschen Archäologischen Instituts Kairo*, Vol. 17, S. 4–68.

Moir, J. R. (1918). 'Some flint implements of rostro-carinate form from Egypt.' In: *Man*, Jan. 1918, pp. 1–5.

Reisch, L. (1974). 'Der vorgeschichtliche Hornsteinabbau bei Lengfeld, Ldkr. Kelheim' (=*Materialheft zur Bayerischen Vorgeschichte, No. 29*). Kallmünz.

Schmid, E. (1974). *Der jungsteinzeitliche Abbau auf Silex bei Kleinkems* (Baden-Württemberg) (D 1).

Schweinfurth, G.-A. (1885). 'Kiesel-Nuclei aus der arabischen Wüste. In: *Verhandlungen der Berliner Gesellschaft für Anthropologie, Ethnologie und Urgeschichte*, pp. 128f.

Seton-Karr, H. W. (1898). 'Discovery of the lost flint mines of Egypt.' In: *Journal of the Royal Anthropological Institute of Great Britain and Ireland*, **27**, 90–92, Pl. x.

Vandier, J. (1952). *Manuel d'Archéologie Egyptienne.* Vol. 1, Paris, Ref. pp. 62f.

Virchow, R. (1885). 'Steingeräte von Helwan und aus der arabischen Wüste.' In: V.d. Berl. *Gesellschaft für Anthropologie, Ethnologie und Urgeschichte*, pp. 128f.

Weisgerber, G. (1982). 'Altägyptischer Hornsteinbergbau im Wadi el-Sheikh.' In: *Der Anschnitt*, **34**, H. 5–6, S. 186–210.

Weisgerber, G., Slotta, R. & Weiner, J. (eds.) (1980). '5000 Jahre Feuersteinbergbau – Die Suche nach dem Stahl der Steinzeit. Deutsches Bergbau-Museum, Bochum' (=*Veröffentlichungen aus dem Deutschen Bergbau-Museum, No. 22*).

Zuber, A. (1956). 'Techniques du travailles des pierres dures dans l'ancienne Egypte.' In: *Techniques et Civilisations*, No. 5.

21

A sociocultural perspective on the spatial analysis of commodity and stylistic distributions

ROBERT N. ZEITLIN

Introduction

The spatial patterning of material culture has been recognized as a readily accessible source of interpretive insight almost from the very inception of scientific archaeology. On the scale of inter-site analysis, spatial distributions of material goods and cultural traits have served such ends as demonstrating the existence of long-distance exchange, measuring the intensity of extra-local sociocultural relationships, delineating the extent of diffusion, and tracing the paths of migration.

Over the years, as method and theory have grown less naive, the pitfalls in some of these early attempts at archaeological reconstruction have become well known. No longer are tribal migrations or foreign invasions postulated solely on the observed intrusion of artifact styles into a new geographic area, nor are ethnic boundaries drawn on the basis of single or even multiple trait distributions. Rather, there has been a developing awareness of the complex, shifting, and uneven relationship between the spatial occurrence of different classes of material culture and an attempt to better understand the sociocultural implications of these occurrences. Just as the unequal quantitative and qualitative distribution of artifactual material within a single archaeological site typically reflects social and cultural differentiation so, on a multi-site or multi-regional level of analysis, differences in distributional patterning are recognized as having structural and functional referents.

In short, it cannot simply be assumed that spatial distributions of cultural traits tend to overlap and move together in space, even though in many cases they appear to do so. Where coextensive distributions occur, just as where they do not, there are invariably underlying causes and it is a central task for the archaeologist to ascertain the reasons for them. Most immediately, there is the requirement of isolating the different sets of variables acting upon the distribution of each class of traits, whether it be of economically valuable commodities, technological innovations, or art styles.

Models of diffusion, interaction, exchange, and communication, some developed within anthropology, some adopted from geography, economics, and other disciplines, have served to describe these spatial processes and to elucidate their sociocultural concomitants (e.g., Ammerman & Cavalli-Sforza, 1979; Clarke, 1977; Davis, 1983; Deetz & Dethlefsen, 1965; Hodder, 1978; Hodder & Orton, 1976; Plog, 1980; Renfrew, 1972, 1977). In the pages that follow I draw upon a number of these ideas as two classes of prehistoric spatial distributions are examined: that of an economically valuable commodity, obsidian, and that of stylistic traits displayed on pottery. My strategy is to analyse the two sets of distributional patterns in the context of the social and political environment that I believe must at various times have stimulated or constrained their spatial limits. The effort is aimed at advancing an argument that sociopolitical variables often had an archaeologically measurable influence on the processes by which these spatial distributions were created. A theoretical rationale for this position is also suggested and, in a tentative attempt at operationalizing the analysis, I outline an elementary procedure for measuring correspondences in the range and intensity of inter-regional commodity procurement and stylistic sharing. Applying the methodology to data from Precolumbian Mesoamerica's southern Isthmus of Tehuantepec, I provide a case study not only of how a pattern of linked changes in the spatial distribution of material goods and style was discerned archaeologically, but of how these parallel changes can serve as a source of inference about temporal variations in inter-regional social and political relationships.

An historical overview

A recognition of sociocultural variables reflected in the spatial patterning of material goods and cultural traits is, as noted above, not new to prehistorians. In the European tradition, for example, it provided the methodological basis for the early twentieth century culture–historical school, as represented in the writings of Graebner (1911) and Schmidt (1939). Envisioning the ethnographically observable distribution of humanity as the consequence of outward migrations from a small number of primordial tribal abodes, each with its distinctive culture complex, the task for prehistorians, as they saw it, was to trace the development and spread of these culture circles (*kulturkreise*) by searching for coexisting complexes of shared traits among spatially separated societies.

At about the same time the culture–historical school was holding forth, a similar involvement in the dynamics and sociocultural significance of spatial distributions was taking place in American anthropology under the rubric of the culture–area concept, as championed by Clark Wissler (1917, 1923). Having observed the way cultural traits tended to cluster geographically in his classificatory study of indigenous New World peoples, Wissler saw a way to delimit the geographic boundaries of culturally related societies. He further posited that within each culture–area there would be a definable centre of dispersal, radiating outward from which one would observe an attenuated spatial distribution of shared cultural traits, presumably as a result of diminishing sociocultural interaction with increasing distance from the centre.

Proponants of the culture–historical school were some of the first to criticize the mechanistic culture–area model of concentric zoned diffusion described by Wissler. Cognizant of the potential effects of social and cultural variables on the spread of culture, Schmidt, for one, pointed out that: 'An equal radiating diffusion from one centre would . . . only be possible . . . and merely possible then . . . if the external conditions of communication and the inner conditions of receptivity and inclination of men were the same in all these radiating circles, which of course is never the case (1939, p. 49).' To the contrary, he claimed the spread of culture would be characterized by the movement of organically integrated and, consequently, more or less spatially unattenuated complexes of social and cultural forms. Hedging, however, Schmidt did allow that the greater the distance separating two related groups, the greater the possibility of differences between them because of the intrusive influence of intervening alien cultures.

Curiously, the American advocates of concentric zoned diffusion never, as far as I have been able to determine, explicitly acknowledged or sought corroboration from a tantalizingly analogous conceptualization which had begun to take form in economic geography during the mid-nineteenth century. Inspired by the then prevailing view that natural laws could be applied to both social and physical phenomena, the so-called gravity concept modelled human interaction on the force of terrestrial attraction. It predicted that the intensity of interaction between two geographically separate groups of people would vary in direct proportion to the population mass of each and inversely with the square of the distance separating them (Carey, 1858).

Although the application of the gravity model to questions in economic geography about the acquisition and subsequent spatial distribution of material goods was first attempted at a time when both the culture–circle and culture–area concepts were still in vogue (Reilly, 1931), not until quite recently did it find its way into archaeology, where it was first employed to account for spatial distributions such as those of Romano–British pottery and Neolithic stone axes in England (Hodder, 1974*a* and *b*). While it proved relevant to archaeological data, a major reason the gravity model has failed to achieve wider usage may have to do with the fact that reliable information regarding absolute population size and total amounts of commodity production is difficult or impossible to retrieve for much of prehistory. Estimates of population based on settlement area, house counts, artifact discard, etc., and relative measures of productivity calculated from ratios of one recovered commodity to another have, in some instances, provided an avenue around these deficiencies, but they are all fraught with serious problems.

Instead, population factors are commonly ignored and relative variations in the spatial distribution of goods are ascribed exclusively to the frictional effects of distance. Quite simply, the greater the distance from its source, the less frequently a commodity or culture trait is expected to show up in the archaeological record. Alternatively, one could reformulate the concept to say that sources of commodities or centres of cultural innovation increasingly distant from a particular archaeological site should be less frequently represented in the assemblage from that site. Here the proximity of non-local sources, where there is more than one available, is used as a measure of the expected contribution from each source.

Distance–decay models, as these simplified gravitational analogues are known, were first introduced into the archaeological literature in the late 1960s to account for the distribution of obsidian in the Near East (Dixon, Cann & Renfrew, 1968; Renfrew, Dixon & Cann, 1968). Eventually use was made of statistical procedures such as regression analysis and trend surface analysis to demonstrate the covariation between commodity abundance and proximity to source. Further efforts have also been directed towards relating the slope and shape of regression lines or the form and interval of density contours as a way of discriminating the spatial effects of different exchange and distribution mechanisms, moving different classes of goods, and other variables. A steeper gradient of commodity fall-off is predicted where transportation is difficult or where commodities are bulky; where distribution is implemented by a large number of exchanges, a more convex regression line is anticipated; where regional distribution derives from or is restricted to major centres or marketplaces, secondary peaks indicating high commodity frequency are often superimposed on the regression line at points representing these distant loci of supply, etc. (Findlow & Bolognese, 1982; Hodder & Orton, 1976; Renfrew, 1977).

As with conventional gravity models, the principles of distance–decay have proven most useful to archaeologists when applied to relatively simple or localized commodity distributions among socially undifferentiated populations. In larger scale or more complex applications, they often produce results other than the predicted fall-off. But these seeming anomalies merely point out what is intuitively obvious and what geographers (e.g., Stoufer, 1940), among others, have made explicit for years: that any realistic model of human interaction must take into consideration variables over and above distance and population size. So, the principal but still significant role in archaeology of the distance–decay distributional concept reduces to that of a baseline formulation. When things don't work out as predicted, it simply signals the need to consider other variables to account for any observed lack of conformity with the ideal model.

As an example of this usage, my own research into prehistoric commodity procurement on the southern Isthmus of Tehuantepec will be called forth later in the paper to show how obsidian, highly prized for making stone tools and ornaments, was not always obtained from the closest quarry source, as would be expected were distributions to conform to the rules of distance–decay. Moreover, spectrochemical analysis of obsidian specimens reveals that supplies procured from at least 15 identified non-local sources lying to the northwest and southeast at distances ranging from about 400 to 1000 km varied widely over the three millennia between 1500 BC and AD 1500, producing another disparity with the model (Zeitlin & Heimbuch, 1978). After weighing a number of possible variables (e.g., material quality variation between sources; new source discoveries or source exhaustion; changes in mode of procurement; route shifts), I came to the tentative conclusion that sociopolitical relationships among participants in the procurement network through which materials were conveyed played heavily in determining the ultimate distribution of obsidian in Precolumbian Mesoamerica. Alterations in those relationships were seen as accounting in large measure for diachronic changes in source utilization.

In arriving at this conclusion I was guided not only by an awareness of how political events and social circumstances affect non-local commodity procurement and exchange in modern times, but by an understanding of prehistoric inter-regional sociopolitical relationships derived from what is currently known about those ancient Mesoamerican localities through which obsidian would have passed *en route* to the southern Isthmus. I saw in the waxing and waning of relationships between the southern Isthmus and neighbouring regions a remarkable correspondence with the use and disuse of various obsidian sources (Zeitlin, 1982).

If this example is as typical as I believe it to be, then it should serve to corroborate the argument that studies of non-local commodity procurement and exchange cannot proceed beyond the simple description of spatial distributions and attempts to explain them in terms of mechanistic formulations unless they introduce other variables into the picture, and here I refer specifically to social and political ones. Following that lead, it seemed to me that a source of insight into some of the ways that these social and political variables affect commodity distributions might be explored through the analysis of stylistic distributions.

Stylistic analysis in archaeology

The potential value of stylistic analysis for better comprehending the process of commodity distribution lies in the perceived communicative role of stylistic traits as a marker of social and political affiliation (Conkey, 1978; Wobst, 1977). If style is employed by people to signal social, political, and even ideological affiliation then, hypothetically, the maximal spatial extent of similar styles would predict the geographic range of sociocultural relationships. Insofar as the ebb and flow of inter-regional social and political relationships, as inferred from stylistic distributions, alternately limits and stimulates the acquisition of non-local goods, spatial analysis of inter-regional homologies in style could conceivably provide a means for delineating the changing structure of procurement networks.

It could be claimed that all material goods communicate cultural information about their owners (Douglas & Isherwood, 1979), but for the purposes of analysis there is a need to distinguish goods that communicated some form of social identity through their style from those which display little evidence of function beyond utilitarianism (Binford, 1962). This can be a difficult task in practice since there is often considerable overlap, between utilitarianism and style. Nevertheless, we might infer that an object served in the cognitive realm if at least one, and ideally more than one, of the following formal or contextual criteria are met:

1 The form, decoration, and/or material employed compromise or are superfluous to utilitarian function. Highly decorated ceremonial pottery such as the fragile, polychrome painted vessels of the Classic Maya or the chalk versions of stone axes in Neolithic England are examples.
2 The effort and care involved in producing the object are measurably greater than that necessary for any utilitarian function. As an example there are the delicately engraved Magdelanian bone and antler 'implements' found in Upper Paleolithic sites of Contabrian Spain (Conkey, 1980).
3 The effort involved in procuring the object or material, when it is of non-local origin, is significantly greater than that which would have been required to employ a locally available equivalent. The use of exotic cherts from distant sources by Australian Aborigines for making adzes and knives when better quality lithic material was locally available (Gould, 1980) provides a case in point.
4 The object or material is found predominantly in a ritual or ceremonial setting. Typical contexts are shrines, public buildings, offerings, etc.

Where the fulfillment of these criteria points to a cognitive function for a class of goods, it seems reasonable to predict that its spatial distribution would normally be circumscribed by the geographic sphere within which communication prevailed. Some communication is normally delimited by a household, other kinds by a community, and still others extend throughout an entire region and beyond. At any scale, one would not ordinarily expect the archaeological distribution of a particular style to exceed the spatial boundaries of social or political interaction, whether the style was disseminated by the physical transport of style-bearing commodities or by the diffusion of stylistic concepts. Consequently, temporal changes in spatial distributions of style, be they displayed on Paleolithic projectile points or Neolithic pottery, might serve to point out alterations in inter-regional social or political relationships.

By informing on sociopolitical relationships, stylistic distributions might then be used to delimit the range of commodity procurement and exchange. The underlying premise here is that the movement of goods would tend to be inhibited between neighbouring societies which lacked at least a degree of positive interaction, as reflected in stylistic homology, or worse, whose sociopolitical interactions were hostile. There are, of course, exceptions to all this. For one, *ideas* of utilitarian value often diffuse across even the most sharply divided social or political boundaries. It is thus important to differentiate technological goods which were actually procured from another region, suggesting some kind of direct or indirect inter-regional interaction, from locally manufactured copies of a technological innovation where the concepts alone were acquired by diffusion, in which case sustained inter-regional relationships are not necessarily implied.

Having stuck my neck out in making what may appear to be some rather simplistic generalizations, let me hasten to acknowledge some of the other circumstances under which the spatial distribution of material commodities might not correspond to the spread of stylistic elements. The procurement of commodities through multiple down-the-line exchanges where only adjacent trade partners are necessarily in contact; acquisition through silent trade, booty, tribute, contraband, or as occasional collector's oddities; direct procurement of goods by the end-user; the mediation of procurement from an unfriendly society by neutral middlemen; these are but a few conceivable instances where commodity distribution could extend beyond the boundaries of inter-regional social or political relationships and therefore not correspond to the distribution of stylistic attributes. Conversely, two societies occupying separate but equally bountiful habitats might share a number of stylistic traits by virtue of their amicable relationships yet have had little or no call for the exchange of utilitarian commodities.

Indeed, there may well be any number of circumstances where commodity distribution does not seem to relate at all or in the anticipated manner to stylistic sharing. Where this discouraging situation occurs, one might profitably first attempt to exclude the possibility that the stylistic attributes measured were just not the ones used by communicating societies to manifest their relationships. From a practical standpoint, the problem does not so much entail excluding attributes that may reflect idiosyncratic behaviour or other irrelevant variables, but is one of finding a sufficiently large corpus of stylistic traits which can be shown to be distributed inter-regionally. Beyond this problem, a particular class of goods, such as pottery, or lithic implements, might have conveyed stylistic information at some time periods and not others. In this case, a sudden loss of inter-regional relationships in style could give the erroneous impression of broken social or political ties. Finally, there is the potential that circumstances such as those of economic stress and competition, as cited in a recent study by Hodder (1979), might result in a stylistically observable 'cultural fragmentation' concomitant with an intensification of inter-societal commodity exchange.

With all these possible exceptions, perhaps it might be best to think of correspondences between the spatial distributions of utilitarian commodities and stylistic elements as an imperfect, not always consistent reflection of positive social or political relationships. Where significantly numerous and complex correspondences in the spatial distribution of goods and style can be demonstrated archaeologically and, better yet, where parallel changes in the two kinds of distribution occur through time, important insights into the nature of inter-regional relationships can often be gleaned. On the other hand, where inter-regional interaction on the level of commodity procurement is not matched by stylistic relationships, this need not necessarily be written off to defects in the scheme of analysis. Rather, the very inconformities can stimulate the formulation of testable hypotheses about the specific characteristics of inter-regional relationships and sociocultural differentiation.

A trial procedure for delineating the spatial distribution of material goods and style

In order to put all this into practice there must be a way of operationalizing the distributional analysis of material commodities and styles. A simple and relatively easy to implement procedure can be designed from the position of an egocentric network where only linkages between a central site or region and a specified number of surrounding localities need be considered. Starting with commodity procurement, absolute or relative amounts of a class of goods procured from identified outlying sources are calculated at the central site or region for each time period. The percentages of the total attributable to each non-local source serve as an indication of interaction, in terms of the particular commodity, between the central site or region, the source locale, and intervening regions. Determination in this manner that the most proximate source was not always the one most favoured by the central site or region implies that factors beyond distance–decay were influencing procurement. Perhaps more significantly, changes over time in the relative contribution of different sources to local supply provide further evidence that complicating variables have affected interaction with the source or possibly with inhabitants of regions in between.

As for the delineation of inter-regional relationships through stylistic sharing, similarity coefficients can be calculated which quantify the degree to which related styles occur in paired regions. A coefficient of similarity, modified from Jaccard's formula (Sneath & Sokal, 1973; Zeitlin, 1983), was used in my study to estimate the association between ceramic styles of a central region, the southern Isthmus in this example, and its neighbours, the highland Valley of Oaxaca, the Central Depression of highland Chiapas, coastal lowland Soconusco, and the lowland northern Isthmus (Fig. 21.1).

The objective of the analytical procedure was to determine how many classes of stylistic attributes from the southern Isthmus were found elsewhere. This was accomplished by calculating the number of positive matches

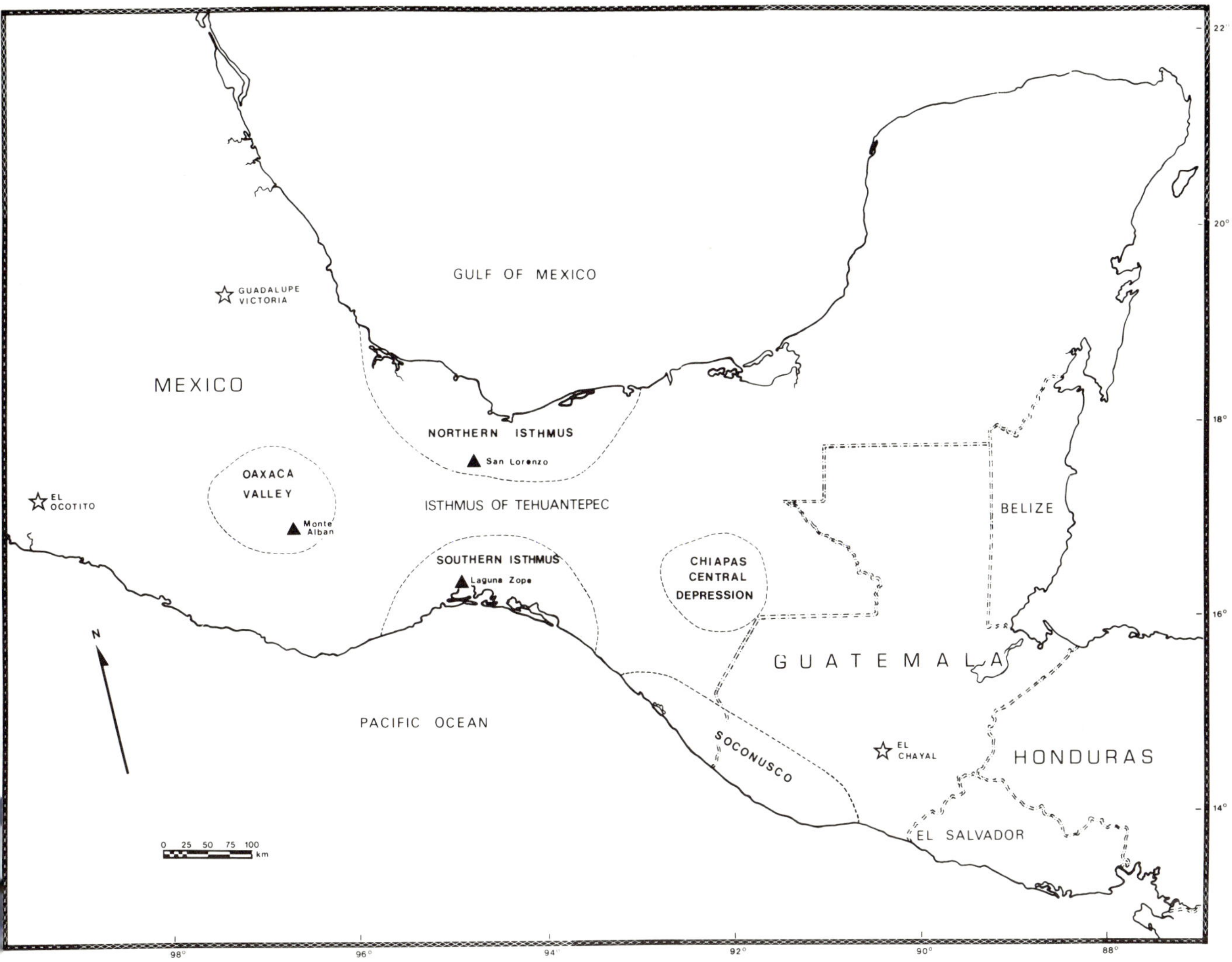

Fig. 21.1. Map of Mesoamerica indicating Preclassic obsidian sources supplying the southern Isthmus of Tehuantepec and locations of regions discussed in text. Obsidian sources: (1) El Ocotito; (2) Guadelupe Victoria; (3) El Chayal. Regions: (A) northern Isthmus of Tehuantepec; (B) southern Isthmus; (C) Oaxaca Valley; (D) Chiapas Central Depression; (E) Soconusco.

where attributes seen on pottery of the southern Isthmus were also noted as occurring in surrounding regions. For the southern Isthmus and one of the other regions, the sum of positive matches divided by the total of positive matches and unmatched pairs produced the desired pairwise coefficient of similarity. A coefficient equaling one signified complete similarity between the paired regions with respect to the selected styles; any coefficient less than one was taken to indicate diminished similarity.

There may be instances, as was my experience, where data from one region are unavailable or unclear for a particular stylistic attribute. When this occurs, the attribute may be excluded from the similarity calculation for that particular pair of regions. The overall procedure can be made more rigorous if one has a large number of systematically or randomly selected attributes to compare and made more sensitive if there are percentage figures for attribute occurrences in each region rather than mere presence–absence values. Despite any such deficiencies, I recommend the resultant coefficients as a meaningful and easily obtained measure of inter-regional stylistic similarities and, interpreted with caution, as an indicator of the range and intensity of social and political relationships.

A synopsis of stylistic inter-relationships from a south Isthmian perspective

To illustrate the potential usefulness of calculating coefficients of inter-regional stylistic similarity and relating them to patterns of commodity procurement, let me briefly summarize some results of a trial analysis utilizing ceramic remains from the southern Isthmus of Tehuantepec (Zeitlin, 1983). The pottery comes from excavations at the region's most prominent Preclassic settlement, Laguna Zope, where a continuous span of occupation extended from about 1500 BC to AD 300. Previous analysis had enabled the period of habitation to be divided into six cultural phases, distinguished ceramically by alterations in the modes of pottery form, finish, and decoration favoured by the ancient inhabitants of the site (Zeitlin, 1978).

Inter-regional stylistic relationships for the earliest cultural phase, dated from about 1500 to 1100 BC, were analyzed on the basis of 20 ceramic attributes. Among these were use of white slip, red or specular-red painted vessel rims, brushed or striated surface treatment, black-and-white designs produced by differential-firing, incised exterior rim lines, vertical fluting or gadrooning, lunate-, dentate-, and rocker-stamping, indented-line lattice designs, and incised parallel triple-vertical or -diagonal lines. Coefficients of similarity between stylistic attributes of the southern Isthmus and its northern Isthmus, Valley of Oaxaca, Chiapas Central Depression, and Soconusco comparison regions indicate a marked degree of commonality (Table 21.1). From the perspective of style, it is apparent that the southern Isthmus, by the Early Preclassic period, was already participating in a network of inter-regional communication and shared values.

There is no archaeological evidence suggesting any social or political impediment to interaction between the southern Isthmus and other regions during the Early Preclassic. Lacking indications to the contrary, I would charge the somewhat lower coefficients of stylistic similarity with Chiapas and Oaxaca to the attenuating effects of more difficult access to these highland regions in comparison to the relative ease of communication between the southern Isthmus and its contiguous lowland neighbours on the northern Isthmus and in Soconusco.

The little we know about the procurement of obsidian by Early Preclassic south Isthmian peoples tends to accord with this modified distance–decay model of interaction. Of the obsidian from three geological sources which have been determined to have served the region at this early date, that from Guadalupe Victoria, located about 400 km to the northwest in highland central Mexico, would have been transported either through the northern Isthmus or the Oaxaca Valley (Fig. 21.1). El Ocotito obsidian, also from the Mexican highlands about 500 km to the west, would also most probably have passed through the Valley of Oaxaca. Almost surely, El Chayal obsidian, at somewhat over 500 km to the southeast in highland Guatemala, would have had to reach the southern Isthmus via Soconusco. For all these intermediate regions, the analysis of ceramic style suggests that a network of sociopolitical relationships existed to facilitate commodity procurement.

What I am defining as the Middle Preclassic period can be divided into two ceramically distinctive cultural phases on the southern Isthmus, running from 1100 to 800 BC and from 800 to 400 BC. In addition to a number of traits which continue from the Early Preclassic, the initial Middle Preclassic phase is characterized, in my preliminary analysis, by eight new stylistic attributes. Some of these, like the 'jaguar-dragon-paw-wing', 'flame-eyebrow', and 'crossed-bands' motifs, are clearly inspired by the spread of Olmec influence from its heartland on the northern Isthmus; others, such as the application of circumferential punctations below an incised rim line, interior finger-punched bosses, and exterior-bolstered rims, are of more obscure origin. It is notable that the terminal phase of the Middle Preclassic, which witnesses a cataclysmic end to the major Olmec centre at San Lorenzo, is marked by a suite of nine new ceramic attributes, the best known of which are a fine white pottery, sgraffito-incised, circumferential double lines or 'double-line-breaks' on the inner rim of outslanted or outflaring wall, flat-bottom bowls, incised vertical scal-

Table 21.1. *Coefficients of similarity in ceramic style between the southern Isthmus of Tehuantepec and four surrounding regions*

Approximate dates	Cultural phase	Comparison regions: Northern Isthmus	Soconusco	Chiapes Central Depression	Oaxaca Valley
AD/BC					
	Terminal Late Preclassic	0.50	0.18	0.64	0.92
200					
	Initial Late Preclassic	0.93	0.87	1.0	0.60
400					
	Terminal Middle Preclassic	0.78	0.89	0.89	0.78
800					
	Initial Middle Preclassic	0.86	1.0	0.75	0.75
1100					
	Early Preclassic	0.95	0.90	0.63	0.44
1500 BC					

loped and straight lines, incised and excised scrolls, and jabs zoned by incised scalloped lines.

Coefficients of similarity between the southern Isthmus and the northern Isthmus, Valley of Oaxaca, Chiapas Central Depression, and Soconusco for the Middle Preclassic period indicate continued inter-regional communication (Table 21.1). Higher coefficients than previous for the Valley of Oaxaca and Chiapas Central Depression may signify that ties between the lowland southern Isthmus and these highland regions had strengthened since Early Preclassic times. This apparently intensified highland–lowland stylistic relationship might reflect the growing sociopolitical complexity of Middle Preclassic Mesoamerican societies, a complexity which translated into a greater demand for exotic status-marking goods as well as for utilitarian commodities from other regions (Flannery, 1968).

The analysis of obsidian distribution appears to confirm the impression of intensifying inter-regional interaction beginning in the Middle Preclassic. The two principal early sources, Guadalupe Victoria in Mexico and El Chayal in Guatemala, show evidence of increasingly heavy utilization by initial Middle Preclassic peoples of the southern Isthmus and neighbouring regions (Cobean *et al.*, 1971; Hester, Heizer & Jack, 1971; Pires-Ferreira, 1975; Zeitlin, 1978), a development probably attributable to the establishment of more regularized social, political, and economic relationships between regions. During the terminal Middle Preclassic, however, we note what appears to be a 400 year hiatus on the southern Isthmus in the importation of Guadalupe Victoria obsidian. This discontinuity in the use of obsidian from Guadalupe Victoria corresponds in time with the demise of Olmec San Lorenzo. Since the north Isthmian region of San Lorenzo was a likely route for Guadalupe Victoria obsidian travelling to the southern Isthmus and since San Lorenzo itself was a heavy user of Guadalupe Victoria obsidian, possibly even controlling its distribution, I infer that the disrupted obsidian supply to the southern Isthmus can be attributed to the same Olmec sociopolitical upheaval held responsible for the terminal Middle Preclassic changes in ceramic styles.

From about 400 to 200 BC, the initial Late Preclassic phase on the southern Isthmus is characterized by a strikingly new corpus of ceramic styles, many of which are closely related to those found as far to the east as the Maya area and southeast into what is now El Salvador. Prominent among the 15 stylistic attributes isolated for the present analysis are highly burnished, glossy–waxy finishes in combinations or monochromes of red, orange, or brown slip, a distinctive 'Usulután' mode of resist-fired decoration, bowl forms with composite or inflected profiles, some with notched midline breaks, and complex patterns of deeply grooved, multiple parallel curvilinear, straight, or zig-zag lines. Similarity coefficients point up a cultural schism in which regions from the Isthmus of Tehuantepec eastward – both the south and north Isthmus, the Chiapas Central Depression, and Soconusco in my sample – share in this new stylistic horizon (Table 21.1). To the west, however, the Valley of Oaxaca was beginning to go its own way as the nascent centre of Zapotec civilization at Monte Albán. From a ceramic perspective, the Oaxaca Valley's growing independence is manifested in a significantly lessened intensity of stylistic relationships with its south Isthmian neighbour.

By the terminal phase of the Late Preclassic on the southern Isthmus, however, a further alteration of inter-regional affiliations is highlighted by stylistic analysis. The previous eastward-looking glossy–waxy slipped ceramic tradition is abandoned there in favour of a highly burnished, unslipped monochrome grayware, unmistakably related to pottery typical of Monte Albán in the Valley of Oaxaca. I hesitate at this point to be very explicit about what kind of sociopolitical relationship between the southern Isthmus and Monte Albán civilization might be inferred from this strong ceramic relationship. Let me just note that by this time Monte Albán had grown into what has been described as an aggressive, territorially acquisitive state (Blanton *et al.*, 1981; Spencer, 1982). Of the other 12 stylistic attributes employed in the comparative analysis of this phase, among which are the use of a stucco polychrome surface finish, incision of concentric circles or spirals on vessel floors, bowls or plates with effigy rims, animal effigy vessels, hollow mammiform or conical vessel supports, and trough-handle ladles, almost all are characteristic of Late Preclassic pottery in the Valley of Oaxaca.

Expressed as inter-regional similarity coefficients, the greatly intensified terminal Late Preclassic relationships between the southern Isthmus and highland Oaxaca are more than matched by a considerable weakening of connections with Soconusco (Table 21.1). For some reason, Soconusco, more than any of the regions considered in the study, maintains a stylistic, and probably a strong social and political, affiliation with the developing Maya civilization in the Yucatan lowlands.

In general then, inter-regional stylistic relationships depict the terminal Late Preclassic as a time of great sociopolitical restructuring in southern Mesoamerica. Not unexpectedly, from the point of view of my thesis, these changes have repercussions on obsidian distribution. El Chayal, located in highland Guatemala, just upland from Soconusco, begins to be a source of supply reserved increasingly for Maya consumption, so that by the beginning of the Classic period, around AD 300, virtually none of its obsidian was reaching the southern Isthmian region. Paralleling the shift in stylistic relationships away from Soconusco and, by extension away from the Maya area, the southern Isthmus turns to regions to its west and north for obsidian supplies and perhaps for other commodities and sociopolitical affiliations as well.

Conclusion

Let me end by returning to my original premise. I had suggested that insofar as social and political variables are at times responsible for limiting and at other times stimulating the long-distance movement of commodities, and if these same social and political variables tend to limit or stimulate the diffusion of style, then an analysis of stylistic distribution can be informative in accounting for what otherwise might seem like the capricious procurement of commodities. The reasoning I adopted for this argument was that certain stylistic traits served, consciously or unconsciously, directly or indirectly, as emblems of inter-regional social and political relationships. It was the nature of these relationships that I held as responsible, at least in part, for facilitating or disrupting the long-distance movements of commodities such as obsidian.

If this is acceptable, then the archaeological delineation of covarying shifts in inter-regional stylistic similarities on the one hand and commodity procurement patterns on the other may be indicative of the effects of changing social and political relationships. By reconstructing these relationships from spatial distributions as well as from other supporting archaeological evidence, we may be better able to account for the patterns of long-distance commodity procurement which we have grown increasingly successful at identifying through chemical trace element characterization and other analytical techniques.

Bibliography

Ammerman, A. J. & Cavalli-Sforza, L. L. (1979). 'The wave of advance model for the spread of agriculture in Europe.' In: *Transformations: Mathematical Approaches to Culture Change*, edited by C. Renfrew & K. L. Cooke, pp. 279–93. New York: Academic Press.

Binford, L. R. (1962). 'Archaeology as anthropology.' *American Antiquity*, **28**, 217–25.

Blanton, R. E., Kowalewski, S. A., Feinman G. A. & Appel, J. (1981). *Ancient Mesoamerica: A Comparison of Change in Three Regions.* Cambridge University Press.

Carey, H. C. (1858). *Principles of Social Science.* Philadelphia: J. B. Lippincott and Co.

Clarke, D. I. (ed.) (1977). *Spatial archaeology.* London: Academic Press.

Cobean, R. H., Coe, M. D., Perry, E. A., Jr, Turekian, K. K. & Kharkar, D. P. (1971). 'Obsidian trade at San Lorenzo Tenochtitlán, Mexico.' *Science*, **174**, 666–71.

Conkey, M. W. (1978). 'Style and information in cultural evolution: toward a predictive model for the Paleolithic.' In: *Social Archaeology: Beyond Subsistence and Dating*, edited by C. L. Redman *et al.*, pp. 61–85. New York: Academic Press.

Conkey, M. W. (1980). 'The identification of prehistoric hunter–gatherer aggregation sites: the case of Altamira.' *Current Anthropology*, **21**, 609–30.

Davis, D. D. (1983). 'Investigating the diffusion of stylistic innovations.' *Advances in archaeological method and theory*, **6**, 53–89. New York: Academic Press.

Deetz, J. A. & Dethlefsen, E. (1965). 'The Doppler effect and archaeology: a consideration of the spatial effects of seriation.' *Southwestern Journal of Anthropology*, **21**, 196–206.

Dixon, J. E., Cann, J. R. & Renfrew, C. (1968). 'Obsidian and the origins of trade.' *Scientific American*, **218**, 38–46.

Douglas, M. & Isherwood, B. (1979). *The World of Goods.* New York: Basic Books.

Findlow, F. F. & Bolognese, M. (1982). 'Regional modeling of obsidian procurement in the American southwest.' In: *Contexts for prehistoric exchange*, edited by J. E. Ericson & T. K. Earle, pp. 53–81. New York: Academic Press.

Flannery, K. V. (1968). 'The Olmec and the Valley of Oaxaca: a model for interregional interaction in formative times.' In: *Dumbarton Oaks Conference on the Olmec*, edited by E. P. Benson, pp. 79–110. Washington, DC: Dumbarton Oaks Research Library and Collection.

Gould, R. A. (1980). *Living archaeology.* Cambridge University Press.

Graebner, F. (1911). *Die Methode der Ethnologie.* Heidelberg: Carl Winter's Universitäts Buchhandlung.

Hester, T. R., Heizer, R. F. & Jack, R. N. (1971). 'Technology and geologic sources of obsidian artifacts from Cerro de las Mesas, Veracruz, Mexico with observations on Olmec trade.' *Contributions to the University of California Archaeological Research Facility*, **13**, 133–41.

Hodder, I. R. (1974*a*). 'Some marketing models for Romano–British coarse pottery.' *Britannia*, **5**, 340–59.

Hodder, I. R. (1974*b*). 'A regression analysis of some trade and marketing patterns.' *World Archaeology*, **6**, 172–89.

Hodder, I. R. (1979). 'Economic and social stress and material culture patterning.' *American Antiquity*, **44**, 446–54.

Hodder, I. R. (1982). 'Towards a contextual approach to prehistoric exchange.' In: *Contexts for prehistoric exchange*, edited by J. E. Ericson & T. K. Earle, pp. 199–211. New York: Academic Press.

Hodder, I. (ed.) (1978). *The Spatial Organisation of Culture.* Pittsburgh: University of Pittsburgh Press.

Hodder, I. & Orton, C. (1976). *Spatial Analysis in Archaeology.* Cambridge University Press.

Pires-Ferreira, J. W. (1975). 'Formative Mesoamerican exchange networks with special reference to the Valley of Oaxaca.' Ann Arbor: *Memoirs of the Museum of Anthropology, University of Michigan*, No. 7.

Plog, S. (1980). *Stylistic Variation in Prehistoric Ceramics. Design Analysis in the American Southwest.* Cambridge University Press.

Reilly, W. J. (1931). *The Law of Retail Gravitation.* New York: W. J. Reilly Co.

Renfrew, C. (1972). *The Emergence of Civilization.* London: Methuen.

Renfrew, C. (1977). 'Alternative models for exchange and spatial distribution.' In: *Exchange Systems in Prehistory*, edited by T. K. Earle & J. E. Ericson, pp. 71–90. New York: Academic Press.

Renfrew, C., Dixon, J. E. & Cann, J. R. (1968). 'Patterns of trade: Further analysis of Near Eastern obsidians.' *Proceedings of the Prehistoric Society*, **34**, 319–31.

Schmidt, W. (1939). *The Culture Historical Method of Ethnology.* New York: Fortuny's.

Sneath, P. H. A. & Sokal, R. R. (1973). *Numerical Taxonomy.* San Francisco: W. H. Freeman and Co.

Spencer, C. S. (1982). *The Cuicatlán Cañada and Monte Albán. A Study of Primary State Formation.* New York: Academic Press.

Stoufer, S. A. (1940). 'Intervening opportunities: theory of mobility and distance.' *American Sociological Review*, **5**, 845–67.

Wissler, C. (1917). *The American Indian.* New York: Douglas C. McMurtrie.

Wissler, C. (1923). *Man and culture.* New York: Thomas Y. Crowell.

Wobst, H. M. (1977). 'Stylistic behavior and information exchange.' In: 'Papers for the director: research essays in honor of James B. Griffin', edited by C. E. Cleland. *Anthropology Papers, Museum of Anthropology, University of Michigan*, **61**, 317–42.

Zeitlin, R. N. (1978). 'Long-distance exchange and the growth of a regional center: an example for the southern Isthmus of Tehuantepec, Mexico.' In: *Prehistoric Coastal Adaptations. The Economy and Ecology of Maritime Middle America*, edited by B. L. Stark & B. Voorhies, pp. 183–210. New York: Academic Press.

Zeitlin, R. N. (1982). 'Toward a more comprehensive model of interregional commodity distribution: political variables and prehistoric obsidian procurement in Mesoamerica.' *American Antiquity*, **47**, 260–75.

Zeitlin, R. N. (1983). Adding style to the study of prehistoric commodity exchange and sociopolitical interaction in southern Mexico. Manuscript.

Zeitlin, R. N. & Heimbuch, R. C. (1978). 'Trace element analysis and the archaeological study of obsidian procurement in Precolumbian Mesoamerica.' In: 'Lithics and subsistence: the analysis of stone tool use in prehistoric economies', edited by D. D. Davis. *Vanderbilt University Publications in Anthropology*, **20**, 117–59.

22

Exchange pathways at a stone axe factory in Papua New Guinea

JOHN BURTON

At the time of first contact with Europeans in 1933 (Leahy & Crain, 1937), the Tungei tribe of the Tuman River area of Papua New Guinea owned and exploited a group of stone axe quarries (Chappell, 1966, pp. 102–3) whose products were traded widely in the highlands wealth economy, reaching places as far away as Porgera, in western Enga, Mendi, in Southern Highlands Province, Karimui, in southern Simbu, and the Simbai area to the north of the Jimi River in Western Highlands Province (Fig. 22.1).

Among the Tuman tribes, political and military alliances of related clans (Brown, 1960), the primary pathways of trade and exchange were, as they are today, those established between clanspeople by marriage. Before the Pacific War, stone axes had an important role to play in the payment of brideprice; today their place has been taken by herds of pigs, cash and items of the modern economy. In this paper, I will show that the analysis of Tungei marriage patterns over the earlier part of the century provides an unusual but effective window on the trade in stone axes in the vicinity of the Tuman axe factory.

Brideprice and the axe trade: a special relationship

The ethnography of brideprice payments prior to 1933 is not entirely satisfactory, but the general pattern is clear. Stone axes were regionally almost always a major part of

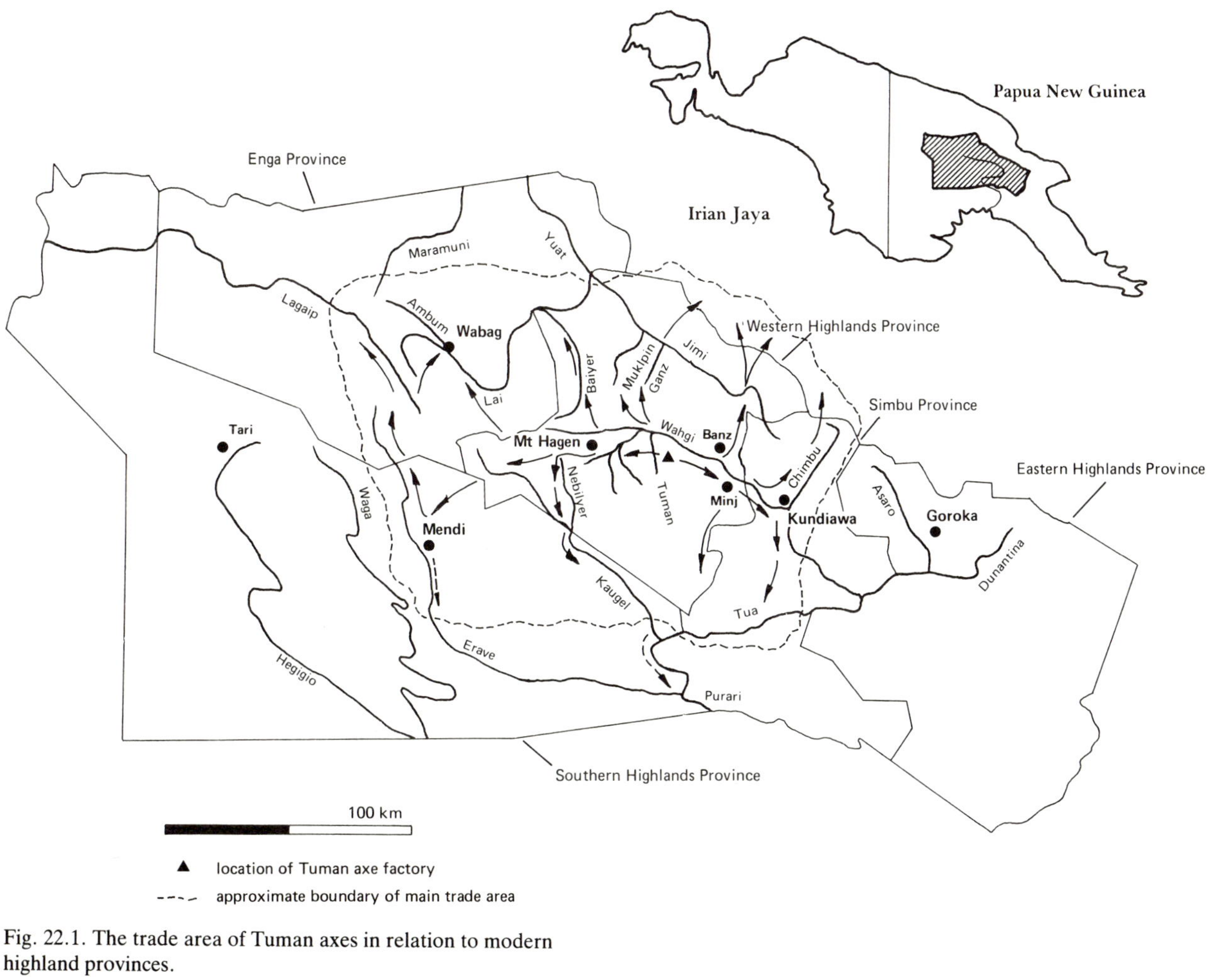

Fig. 22.1. The trade area of Tuman axes in relation to modern highland provinces.

brideprice and locally they were given in very large numbers indeed. Why axes should have dominated brideprices to such an extent – rather than pearlshells or pigs – is beyond the scope of this paper. It is likely that they did so because of an adequate supply and a high – if not insatiable – demand across the highlands as a whole.

The phenomenon was a regional one; in North Melpa, the Kawelka big-man Ongka (A. J. Strathern, 1979, p. 26) was quite blunt on the subject:

> . . . pigs might be given too, but the axes were the essential thing; if axes were not given they would refuse the marriage, saying 'He's a rubbish, useless man, what shall we do? Don't let's give him the girl.'

A sample of Tungei brideprices illustrating the traditional importance of axes is set out in Table 22.1. In the 14 brideprices given before the Pacific War an average of 5.4 finished axes (*tui*) were given, together with 8.1 roughouts (*tui enjim pei*). As the table shows, Tungei men tended to answer in large round numbers of axes. At other times they did not mention giving unsharpened roughouts, or they did not specify what state of completion the axes were in, so that the items included are variable. However, in the eight brideprices where hafted axes were mentioned, an average of 10.7 were given; in the nine brideprices where roughouts were mentioned, an average of 14.1 were given.

From ethnographic work in the Banz area, Heaney has provided data on 11 marriages up to 1934 (1982, table 24.1). The brideprices consisted of an average of 3.2 stone axes, 1.3 pearlshells, 2.6 pigs, 0.4 baler shells, 0.1 *Nassa* shell headbands, 0.1 salt packs, and 4.1 bird of paradise plumes. Presumably the pigs included in this list were given in the form of cooked pork as well as live animals. The axes and pearlshells were the main wealth objects in this list; axes outnumber pearlshells by about 2.5 : 1.

For the part of Simbu known as Sinasina, Hide (1981, table A3.1) has presented data on eight brideprices given before 1933, and a further 23 given in the period 1933–38. An average of 1.2±1.7 stone axes were given in the earlier period, and 1.7±1.4 in the second. This compares with 0.5±0.5 and 0.9±1.1 pearlshells given in the same periods, among other things (1981, table A3.2).

In central Simbu, the Patrol Officer L. G. Vial saw some particularly impressive brideprice ceremonies in the 1930s in which 20–60 axes were given (1940, p. 162). The missionary W. Bergmann also witnessed gifts of '30–40 stone axes, beginning with the very big axes only used for the purpose of bride price, down to the small working axes'; axes far outnumbered pearlshells, *Nassa* shell headbands, and pigs (2–3 per payment each), plumes (2–4 per payment), and miscellaneous other items (1971, p. 174).

Steel axes replaced stone ones after the Pacific War in

Table 22.1. *Brideprices paid by the Tungei in the 1920s and 1930s*

	Finished axes with handles	Unmounted roughouts	Cooked pigs	Pearlshells	*Nassa* head-bands	Cowrie ropes	Baler head-pieces
Marriages made before 1933:							
Gełu		3	4	1			
Pok	10	10		2	1	5	
Malimbe	20	20		+		+	+
Warke			2–3			1	
Warke	—	—	—	—	—	—	—
Aip	10	20	5	3	—	—	—
Kandeł		20	2				
Kandeł		20	3				
Kandeł	—	—	—	—	—	—	—
Numndi		20	4				
Pam Wu	10			1			
Totals:	50	103	mean number of axes:				4.5±6.9
			Mean number of roughouts:				9.4±9.2
Marriages made 1933–43:							
Gełu		3	4	1			
Pok	10	10		2	1	5	
Aus	10	10	10	5		5	
Aus	5		5	3			
Gełu	10		4–5		1		1
Grand totals:	75	113	grand mean number of axes:				5.4±6.3
			grand mean number of roughouts:				8.1±8.8

\+ indicates presence of item
– indicates absence of item
(blank) indicates not mentioned at interview

mean number of axes in payments mentioning axes: 10.7±4.5
mean number of roughouts in payments mentioning roughouts: 14.1±6.7

everyday contexts, and pearlshells replaced them in ceremonial exchanges. In the 1930s, steel axes changed hands for extravagantly sized pigs and even in the 1940s and early 1950s, steel axes and spades were featured in bride-prices to a certain extent. But, by this time, pearlshells had already taken over the role of stone axes as wealth items (Reay, 1959, p. 99; Brown, 1969, p. 88; A. J. Strathern, 1971; Hughes, 1978). Today, the latter are regarded merely as curios in the Wahgi Valley.

As an aside to this discussion, I can only provide anecdotal evidence for the kinds of payment made to the axe-makers by men of other tribes. In return for their daughters and sisters Tungei men did not seek to receive stone axes. They say they wanted pearlshells (*mong kine*), cowrie ropes (*tun mong*), *Nassa* shell mats (*geram*), and other valuable things like pigs. This was confirmed in interviews with men of Andakelka tribe, who are western neighbours of the Tungei, and with men of the Kuma clan Kondika at Minj. There was an asymmetry in material transactions with the Tungei: shells, plumes and pigs flowed in and axes flowed out.

Wife-giving and axe-taking: a demographic approach

As in most other highlands societies, virilocal residence predominates in the Tuman area and men are polygamous to a certain extent. The factors which limit their willingness or ability to marry more women than they do are only in part economic; personal and social choice plays a strong role. During fieldwork in 1980–81, I was unable to discover whether the Tungei were more, or less, polygamous in axe-making times than they are today. However, I found good reason to believe that Tungei marriage patterns have been distorted by unusual economic forces in the vicinity of Tungei territory. The evidence for this can be seen in relation to the history of Australian patrol activities in the Tuman area.

In the postwar period, census-taking was a frequent occurrence in the Western Highlands and, despite many shortcomings (van de Kaa, 1973), the records which survive provide a framework upon which to build more elaborate and detailed sub-clan genealogies. In the case of the Tungei, the 1958 and 1968 Village Book compilations survive and provide a reasonably complete roll of resident Tungei clansmen of the time, together with their then current wives and children. However, the years of birth attributed to many of the people listed there are extremely inaccurate.

I was able to extend these records to build up a picture of the marriages contracted by the Tungei between about 1900 and the present day, and it proved possible to divide this span of time into six periods of equivalent length. The two Village Books readily enabled the marriages of the younger generation of men to be attributed to the periods 1958–68 and 1968–81 respectively. In some cases the actual year of marriage was recorded by the Patrol Officer when he called the roll (recently this task has been taken over by the Bureau of Statistics). In the Wahgi Valley, census was taken at each Patrol Post about every two years during the 1950s and 1960s.

There was no difficulty in establishing whether men of the older generation were married before or after 1933, the date of the first Australian expedition to Mt Hagen, an event well remembered by elderly Tungei men and women. Indeed, two periods could be distinguished before this time. Living men who married before white men arrived were mostly married between about 1920 and 1933. The *fathers* of the next oldest set of men were married earlier than this, in a period that I set nominally at 1900–20. This classification principally refers to first marriages.

So far I have defined Periods 1 and 2 in my scheme, and Periods 5 and 6. The remaining men, aged between 45 and 70 years at the time of fieldwork, must have been married in the period between 1933 and the 1958 census. By considering the age of other members of their families, I was able to class the marriages of these men in two periods, divided at the nomimal date of 1943; the mid-point of the Pacific War. This was also the year of a well-remembered event, the bombing of Mt Hagen by Japanese warplanes. These periods, labelled Period 1–Period 6, are summarised in Table 22.2.

Some 28 clans and tribes were listed by informants as the natal groups of their wives and those of their kinsmen in traditional times, that is, until the end of Period 3. There

Table 22.2. *Summary of the six periods of Tungei marriages in the study*

Period	Dating	Cohorts married	Number of marriages (a)	(b)	(c)
1	1900–20	fathers of senior men	29	102	—
2	1920–33	oldest living men (70+ years)	22	61	—
3	1933–43	senior men (60–70 years)	32	84	—
4	1943–58	older middle-aged men (45–60 years)	43	57	2
5	1958–68	younger middle-aged men (35–45 years)	40	52	24
6	1968–81	young men (22–35 years)	34	104	67
		Total	200	460	93

(a) Women of clans within Tungei
(b) Women of outside clans traditionally in contact with Tungei
(c) Women of clans only in contact with Tungei in modern period

were some important changes during the first three periods, a span of some 40 years, which can be put down partly to the effect of tribal wars. Nevertheless, no new clans or tribes were added to the roll until after the Pacific War. In the 1950s and 1960s, the social horizons of the Tungei widened rapidly as road transport became more widely available and the region emerged as a newly developing part of the Pacific economy. Marriages began to be contracted with women from quite distant groups, indeed some men brought home brides from as far away as Hanuabada (Port Moresby) and Madang (north coast of Papua New Guinea).

The geographical relationship between the territories of the groups sending women to the Tungei and Tungei territory can be seen in Fig. 22.2 based on maps prepared with the computer graphics system SYMAP. A total of 753 marriages are represented over the six periods (data in Table 22.2), reflecting perhaps 90% of the actual number of marriages contracted during the periods concerned.

The maps, which display the percentages of wives drawn from different areas over the six periods, illustrate the diminishing strength of affinal ties with distance from Tungei territory. Traditionally (i.e. in Periods 1–3) women had about a 1 in 3 chance of marrying within a 2.5 km radius of their homes. Typical examples were Tungei girls who married men from those Tungei clans where marriage rules permitted them to find husbands. More than 2 in 3 married within a 5 km radius, while almost all the remainder were married within 15 km of their homes.

The significance of this finding for an understanding of the axe trade is plain. Brideprices were handed over by Tungei clansmen to the close kin of girls whom they married. Moreover, the lifelong obligations of husbands to compensate and 'do right by' their in-laws drew the axe-makers into consistently parting with their production along well-worn exchange pathways (*ondłam*, 'roads'). 'Free' trade played a lesser role in articulating exchange in the Wahgi Valley Wealthy economy, so that the maps provide a reasonable snapshot of the first steps taken by stone axes in a 'down-the-line' exchange system (Renfrew, 1975, pp. 41–4). Subsequent movements of axes took place in a similar way; the first recipients of axes kept some of them and passed on others along *their* exchange paths.

As a secondary point, it is noteworthy that very few women from north of the Wahgi River married into the Tungei, despite the absence of a language barrier and the apparently friendly relations between northern and southern groups. Cane bridges (*noł poł*) spanned the Wahgi and were used by the Taylor–Leahy patrol (Taylor, 1933, diary entry for 11 April). All the same, several kilometres of empty swamplands separated the northern and southern groups and seem to have acted as the real barrier to social interaction. As among the Mae Enga, the pattern of marriages made by the Tungei can be seen to have been subject to 'topographical constraint' (Meggitt, 1974, p. 80).

Changes over time

A. J. Strathern (1972, table 11) shows that the wives of Kawelka men came, in the postwar period, from tribes further afield than traditionally. This was noticeable rather than marked by the time of fieldwork in the 1960s. Changes in Tungei marriage patterns have been rather greater. One measure which is particularly interesting is the proportion of wives taken from inside the tribe as opposed to those from outside groups. Fig. 22.3 contrasts for both Tungei and Kawelka, over the six time periods already distinguished in the former case, and for Strathern's 'senior' and 'junior' (1972, pp. 102–3) generations of Kawelka men.

Kawelka men are able to marry women from other Kawelka clans. Women from these clans – Mandembo, Membo and Kundmbo – comprised 16% of Kawelka wives of the senior generation. In the junior generation 14% were from Kawelka clans. Equally it could be said that the proportion of wives from beyond the tribe rose from 84% to 86%. This modest rise is, however, not enough to be significant in a chi-square test.

Among the Tungei, outside wives comprised 78%, 73% and 72% of the totals in Periods 1–3, but this dropped to 58% in Period 4 (1943–58). In Period 5 (1958–68) the percentage climbed again to 66%, and in Period 6 (1968–81) to 83%. Outside groups were an important, and constant, source of wives in traditional times, as they are today. Nevertheless, between the 1930s and the 1950s there was a drop in the level of out-marriage of 15% in round terms. This change coincided, for the Tungei, with the loss of the axe trade. Is it therefore possible to say that the 15% change is a significant one? Could it be taken to show that Tungei men could no longer pay large brideprices to outsiders?

Firstly, marriages contracted by men of the Tungei clans with women from elsewhere are all 'outside' marriages, but today they can be classed into two categories: local marriages with women of traditionally affinal groups, and those with groups from new areas. Patterns of marriage in the postwar periods are therefore not strictly comparable with that for Periods 1–3, when only local marriages were made. In Fig. 22.3 the two rates are separately graphed. If only local marriages are considered, the proportion of outside marriages to those within the Tungei was the same in Period 5 as in Period 4 (both 57%). In Period 6 the percentage finally rose, to 75%, which is about the average of what it had been in Periods 1–3.

An appropriate test of significance for these data is the 2×2 chi-square test. The cell frequencies are set to be the numbers of wives obtained from inside and outside the tribe, before and after the supposed change. If the change is given as around the time of the Pacific War, then the two periods to test are Period 3 and Period 4. The test result is $\chi^2_{[1]} = 5.11$, or a probability of the null hypothesis of <0.025 (significant at better than the 5% level).

The scope of the test can be widened by including an earlier period and a later one, so that the numbers in Periods 2 and 3 and Periods 4 and 5 are added together. A problem here is that brides began to be found among 'new'

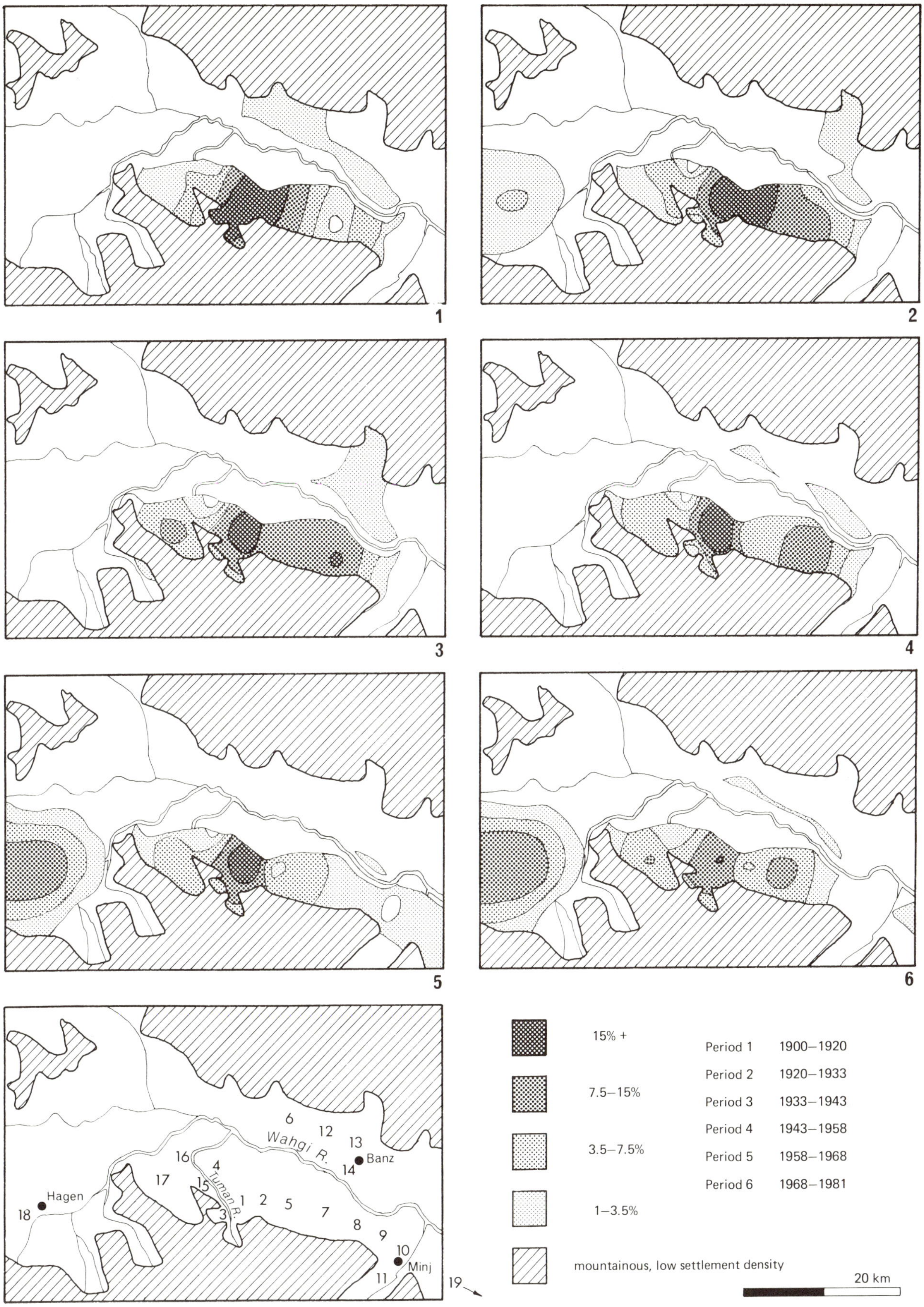

Fig. 22.2. Sources of Tungei wives; percentages of wives born in various parts of the Wahgi Valley, in six periods. *Key to tribes and clans* – Ek Nii speaking groups: (1) Tungei, (2) Wałpi, (3) Mengka, (4) Mamelka, (5) Sikeing, (6) Kelaka. Middle Wahgi speaking groups: (7) Kurupka, Miamka and Ngeneka, (8) Kisu and Deimanka, (9) Berepka, (10) clans at Minj township, (11) clans in headwaters area. Middle Wahgi, Banz dialect groups: (12) Komblo, (13) Bamblinge, (14) Andpang. Melpa speaking groups: (15) Andakelka, (16) Onembe, (17) Kuli, (18) Central Melpa tribes. Other groups in Wahgi Valley; (19) unidentified Simbu tribes.

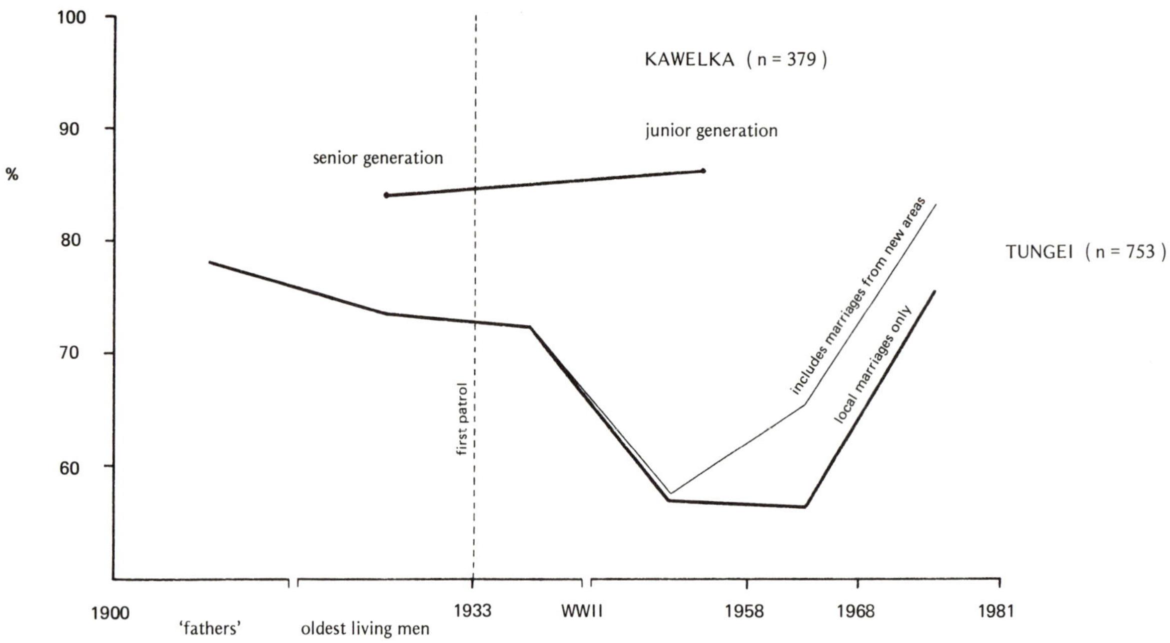

Fig. 22.3. Percentages of Tungei wives originating inside and outside the tribe, compared with Kawelka.

clans and tribes during Period 5 (see Table 22.2). The growth in importance of this class of wives – for reasons that are independent of the ending of the axe trade – acts to dilute the effect being tested for, namely a decrease in the number of outside women being married.

In the event, the test result is significant whether or not the non-local marriages are included or excluded. If the non-local marriages are excluded, the test result is $\chi^2_{[1]}$ =11.12, with a probability of <0.001 (significant at the 0.1% level). If the non-local marriages of Period 5 are included, $\chi^2_{[1]}$=5.64, with a probability of <0.025 (significant at better than the 5% level).

The change in marriage patterns is thus a statistically significant one. Is it now safe to assume that the axe trade was so important to the Tungei that its loss caused a major social upheaval? This is not a question of seeing the change as the collapse of a traditional way of life, but of a 'reduction in circumstances', as it were. The Tungei were initially an unusual group, in that they had exclusive access to a major source of wealth. When this became valueless in the late 1930s and early 1940s their level of means fell to that of the groups around them. The obvious interpretation is that this caused the change in marriage patterns. But, as many ethnographers have shown, a large range of social, demographic and political factors must be taken into account when discussing choices in inter-personal and inter-group relations. Before a firm conclusion can be drawn it is necessary to consider whether any of them complicate the account I have given so far.

Demographic change and clan fission

Sustained population growth is a clear determinant of change in the Middle Wahgi. According to Reay (1959, pp. 29–30) clans pass through several phases of segmentation, in step with an increase in size from 100–300 persons (Phase A), to 200–900 (Phase B), and to 700–1700 (Phase C):

> . . . in accounting for the development of phratries (the descent groups of widest span) there is always a tradition that there was originally a single clan that underwent progressive segmentation through phases A and B until, at C, the main segments of the clan began to intermarry. Separate clanship was established as the groups became exogamous.

Perhaps clan fission of this kind was a factor after the Pacific War, adding to the permissible sources of brides within the tribe. Subsequently, restrictions may have been placed on internal marriages once these new sources had been 'exhausted'. The effect of Tungei marriage rules may have been to require young men to look outside the tribe again for wives during the 1960s and 1970s, when the trend of the 1950s was reversed and tribe exogamy became more important again.

In the case of some clans steadily growing in numbers, low level segmentation has occurred fairly recently. Malimbe of Komnemb clan told me that the smallest Komnemb sub-divisions had been named in his adult lifetime, perhaps in the 1930s. On the other hand, another clan, Ekiemb, has declined in numbers – its sub-clan genealogies show many more men living in 1933 than now – and two other clans, Menjpi and Kupaka, have a number of relict sub-clans with few or no members today.

But even by 1981, no great structural changes had occurred that might have affected clan exogamy. Many years of population growth at over 2.50% p.a. must, in round terms, have caused an increase in numbers of about 200% since the last sub-clan divisions were made, but most of this has been seen in the unusually large family sizes since

1950. This has simply swollen the ranks of existing clans. While major changes can be expected eventually, fission should not occur until the present generation of young clansmen are mature, family men (*wu gał*), or even the elders (*wu mam*) of the Tungei. The lineages which do become sub-clans will take the family heads of the 1930s as their apical ancestors.

Political change in the Tuman River area

The relationships of the Tungei with their neighbours have often been dictated by the shifting military alliances of the area. The Tungei fought two major wars before 1933, one around 1920 with the Mengka, and one around 1930 with the Mamelka. As far as marriage with Mamelka is concerned, the conflict with them effected little change, since few brides had come from them beforehand. But the Mengka traditionally supplied a large fraction of Tungei brides. One-third of the clan ancestresses (women raising families before Period 1) were Mengka by birth, and in Period 1 22% of Tungei brides were Mengka. After the Tungei–Mengka war this proportion dropped to 2%. Clearly the viciousness of the hostilities, in which several men were killed on each side, prevented any early resumption of friendly social intercourse.

The Tungei also fought, around 1934, with the Wałpi, a sub-tribe who have now become incorporated into a larger Tungei polity. In Period 2, 20% of Tungei brides were from the Wałpi; after the fighting, in Period 3, the corresponding figure dropped to 9%.

The removal of any of these exterior sources of wives over a 15-year period might well have been expected to increase the numbers of men finding brides among other Tungei clans. But this did not happen and until the end of Period 3 no significant change occurred. In fact, other sources of extra-tribal women were found to replace Mengka and Wałpi, a phenomenon seen in the rise in importance as sources of wives of Kurupka and Miamka, Kuma clans to the east of the Tungei at Kudjip, and to a lesser degree the increasing importance of Kuli, a large Melpa tribe to the west of the Tungei.

Conclusion

Elderly Tungei men say that stone axe production brought them fame and great wealth in traditional times. They can recite songs telling of the hard work put into axe-making and the fabulous riches which came to them at the conclusion of a successful quarrying expedition. Archaeologists have also stated that Tuman blades were the most widely traded of all the highlands axe types (Chappell, 1966; Lampert, 1972; Hughes, 1977). Neither form of evidence leads to a real understanding of how the axe-making economy worked, however. Both give an impression of scale, but little more than this.

To put this in perspective, other axe-makers I interviewed said very similar things about their importance in axe-making times. The Kawelka, for example, obtained axe stone from their quarry at Mbukl, a minor quarry (A. M. Strathern, 1965), and were presumably able to trade their production into the wealth economy just as the Tungei did. The difference is that their axe-making activities do not seem to have affected their relationships with neighbouring tribes. Nor were their marriage choices distorted by a desire to secure axes from the nearby Jimi Valley sources (A. M. Strathern, 1965, p. 191). Matched against this experience, the distortions to Tungei marriage patterns that I have documented here suggest therefore, that the Tungei experience was something out of the ordinary.

Various scenarios can be put forward to illustrate what may have been uppermost in men's minds in their relationships with the Tungei. It seems likely that the kinsmen of a girl being courted by both a Tungei youth and a youth from an 'ordinary' tribe would have preferred her to choose the partner whose kinsmen could give more axes of a higher quality. Equally, and for the same reason, a man who was thinking of finding a suitable husband for his daughter, would have been inclined to test out his Tungei contacts first. Once his daughter was married to a Tungei, moreover, he would have established more than a lifetime's access to one of the valley's more important sources of wealth. All things being equal, a girl's kinsmen prefer affines who can pay a larger and better brideprice than those who cannot.

The advantages of having a young girl marry into the Tungei were thus considerable, and it is reasonable to think that the Tungei were continually visited by exchange partners and distant relatives hoping to forge stronger links with them by giving them a daughter or a sister in marriage. The 'pressure' on Tungei men to accept women from far-flung groups must have been considerable in axe-making times.

All this changed when stone axes dropped out of the wealth economy. Other valuables, principally pearlshells, took their place as brideprice and by the time anthropologists settled among highlands societies in the 1950s and 1960s, pearlshells completely dominated wealth transactions (e.g. A. J. Strathern, 1971; Hide, 1981, fig. 4.3; Feil, 1982). The new pearlshells, moreover, came from new places. Groups who were formerly rich, by virtue of the fact that they had good access to the shell trade from the coast, became poor; those who lived near the airstrips of patrol stations and missions became rich (see Hughes, 1978). As soon as the new sources of wealth were perceived, men turned towards them. The Tungei then became the owners of a valueless resource and they no longer 'pulled women' as before.

For a time the strong pressures upon the Tungei to marry girls from everywhere but among their own clans were withdrawn. Later still, from 1958 on in my classification, men began to look beyond traditional horizons to marry women of distant groups. In time the coffee-based cash economy reached Aviamp, and Tungei business concerns widened their sphere of interest in a permanent way. By the 1970s a high rate of out-marriage was regained. Over 80% of brides were from other tribes.

In a number of other societies, anthropologists have seen

evidence of differential marriage patterns. Salisbury (1962, p. 104) states that there was a statistically significant drift of women toward those Siane groups in the northeast of his region that were nearer to the ultimate source of valued shells (the north coast of New Guinea). The drift was intensified by European settlement at Goroka after 1933, an added source of wealth objects. Other authors have investigated this kind of directional effect, notably Hide (1981, pp. 181–204) and Nurse (1981). To my knowledge, however, only Podolefsky (1984) has drawn attention to the kind of outward–inward effect seen among the Tungei.

The Tungei stand out as a tribal community that was able to compete in the regional wealth economy on a higher level altogether than their neighbours. Mastery of the techniques of quarrying and axe-making (Burton, 1984*a*, *b*) were only part contributors to this success. Other factors include geological good fortune – not simply a high quality axe stone but one which almost deterministically fostered a particular production system – and a lack of competition from other wealth objects before 1933.

From the archaeologist's viewpoint several lessons may be learned. It emerges that an economy's destiny can be bound up in its 'relations of distribution' as much as in its relations of production. Indeed it is certain that the limits placed on the ultimate wealth-generating ability of the Tungei were set by the amount of stress that could be sustained by their social system and that of their neighbours. Any increases of production that the Tungei might have been able to make would not, in these circumstances, have made them wealthier or more powerful; without changing the social arrangements for receiving payment for their labour, they would simply have had to give away more axes in each brideprice.

A second point is that the immediate threat to stone axe production and exchange may have come from the wealth economy – from the introduction of new sources of pearlshells after 1933 – rather than from the technical change from stone to steel. In Europe, the demise of the great stone axe factories of the neolithic period coincided not with a comparable upsurge in the production of hard-edged metal tools, but with the appearance of many markers of social change and the introduction of the new wealth objects of the succeeding early Bronze Age. In Papua New Guinea, the transition from a stone-using economy to a metal using one was accomplished almost too swiftly for the actual sequence of events to be seen with clarity. But with hindsight, it can be seen that apparently technologically-led changes can be a lot more complex than they seem at first sight.

Note on orthography

In line with current linguistic and educational practice (Ramsey, 1975) the 'barred l' symbol 'ł' is used in this paper to represent the /gl/, /kl/ and /l/ allophones in the Tuman language Ek Nii.

Acknowledgements

The fieldwork on which this paper is based was carried out with the financial support of the Research School of Pacific Studies, Australian National University. I would also like to thank Jack Golson, Linda O'Hanlon and Michael O'Hanlon for a variety of fruitful discussions and assistance.

Bibliography

Bergmann, W. (1971). *The Kamanuku.* Mutdapilly, Queensland: privately circulated.

Brown, P. (1960). 'Chimbu tribes: political organisation in the eastern highlands of New Guinea.' *Southwestern Journal of Anthropology*, **16**, 22–35.

Brown, P. (1969). 'Marriage in Chimbu.' In: R. Glasse & M. J. Meggitt (eds.), *Pigs, Pearlshells and Women*, pp. 77–95. Englewood Cliffs: Prentice-Hall.

Burton, J. (1984*a*). 'Quarrying in a tribal society.' *World Archaeology*, **16**, 234–47.

Burton, J. (1984*b*). Axe makers of the Wahgi. Unpublished PhD thesis, Australian National University.

Chappell, J. M. A. (1966). 'Stone axe factories in the highlands of East New Guinea. *Proceedings of the Prehistoric Society*, **32**, 96–121.

Feil, D. K. (1982). 'From pigs to pearlshells: the transformation of a New Guinea highlands exchange economy.' *American Ethnologist*, **9**, 291–306.

Heaney, W. (1982). 'The changing role of bird of paradise plumes in bridewealth in the Wahgi Valley.' In: L. Morauta, J. Pernetta & W. Heaney (eds.), *Traditional Conservation in Papua New Guinea: Implications for Today*, pp. 227–31. Monograph 16, Institute of Applied Social and Economic Research, Port Moresby.

Hide, R. L. (1981). Aspects of pig production and use in colonial Sinasina, Papua New Guinea. Unpublished PhD thesis, Columbia University.

Hughes, I. (1977). *New Guinea Stone Age Trade.* Terra Australis 3. Canberra: Department of Prehistory, Research School of Pacific Studies, Australian National University.

Hughes, I. (1978). 'Good money and bad: inflation and devaluation in the colonial process.' *Mankind*, **11**, 308–18.

van de Kaa, D. J. (1971). The demography of Papua New Guinea's indigenous population. Unpublished PhD thesis, Australian National University.

Lampert, R. (1972). Hagen axes. Unpublished manuscript, Department of Prehistory, Research School of Pacific Studies, Australian National University.

Leahy, M. J. & Crain, M. (1973). *The Land that Time Forgot.* Sydney: Hurst and Blackett.

Meggitt, M. J. (1974). ' "Pigs are our hearts!", the Te exchange cycle among the Mae Enga of New Guinea.' In: M. J. Meggitt (ed.), *Studies in Enga History*, pp. 57–95. Oceania Monograph No. 20.

Nurse, G. T. (1981). 'Marital migration and gene flow on the mainland of New Guinea.' *Journal of Biological Science*, **13**, 357–62.

Podolefsky, A. (1984). 'Contemporary warfare in the New Guinea highlands.' *Ethnology*, **23**, 73–87.

Ramsey, E. M. (1975). *Middle Wahgi Dictionary.* Mt Hagen: Church of the Nazarene.

Reay, M. (1959). *The Kuma. Freedom and conformity in the New Guinea highlands.* Melbourne: Australian National University Press.

Renfrew, A. C. (1975). 'Trade as action at a distance: questions of integration and communication.' In: J. A. Sabloff & C. C. Lamberg-Karlovsky (eds.), *Ancient Civilization and Trade*, pp. 3–59. Albuquerque: University of New Mexico Press.

Salisbury, R. F. (1962). *From Stone to Steel. Economic Consequences of Technological Change in New Guinea.* Melbourne: Australian National University Press.

Strathern, A. J. (1971). *The Rope of Moka. Big-men and Ceremonial Change in Mount Hagen, New Guinea.* Cambridge University Press.

Strathern, A. J. (1972). *One Father, One Blood. Descent and Group Structure Among the Melpa People.* Canberra: Australian National University Press.

Strathern, A. J. (1979). *Ongka. A Self-account By a New Guinea Big-man.* London: Duckworth.

Strathern, A. M. (1965). 'Axe types and quarries.' *Journal of the Polynesian Society*, **74**, 182–91.

Taylor, J. L. (1933). Patrol Report of 1933, Purari River Headwaters Area, Territory of New Guinea. *Archived Government Report*, Port Moresby and Mitchell, A. C. T., Australia.

Vial, L. G. (1940). 'Stone axes of Mount Hagen, New Guinea.' *Oceania*, **11**, 158–63.

23

Neolithic exchange systems in central Europe 5000–3000 bc*

ANDREW SHERRATT

The results of identification programmes dealing with flint, hard rocks and obsidian have long since exploded the idea that the Neolithic period was characterised by self-sufficiency and an absence of trade (Sherratt, 1976; Willms, 1982). Exploration of the range of items traded by Neolithic communities has revealed a bewildering diversity of exchange systems covering impressive distances and often involving large volumes of material. At the same time, the application of radiocarbon chronologies and acceptance of the implications of re-calibration have emphasised just how large a span of time we are dealing with – something like 4000 years between the introduction of agriculture and the regular use of tin–bronze in central Europe. It would thus be quite false to try and force the evidence for traded materials into a single model of 'Neolithic trade': the diversity of evidence truly reflects the major contrasts that must have existed in social and economic organisation during these millennia.

Equally, however, we must resist the temptation to apply a simplistic evolutionary model, in which trade is always expanding, the mines getting deeper and deeper, and technological change constantly producing new novelties. What we see in the record of material distributions is a series of responses to opportunities and problems – phenomena which may have a much more kaleidoscopic pattern when viewed with a finer chronological resolution. One reason for this is the constantly shifting configuration of settlement, and the increasing scale of contact made possible by the clearance and colonisation of new parts of the landscape. From an initial pattern of scattered nodes of pioneer settlement, the exploration of surrounding resources led to the emergence of lively local exchange systems: but these localised exchange cycles were themselves superseded by interaction spheres of increasing scale so that by the third millennium there are complexes like Corded Ware or Bell Beakers that encompass whole segments of the European continent. The motives for these wider types of exchange were clearly different from those of earlier systems, and this is reflected in the items which were carried. The demands made on a particular source of raw materials may therefore change in character with the changing nature of trade.

What, then, are the implications of these arguments for the way in which we approach Neolithic exchange systems? In the first place, we cannot restrict ourselves to single materials: flint or chert was only one commodity among many, and moreover its significance changed as the items which were produced from it circulated in different ways. In central Europe, for instance, it is important that we look also at copper, which in certain areas circulated alongside (and perhaps in exchange for) stone resources. Yet even so, we run the risk of studying only one side of the trading equation, for there are many instances where we can detect a variety of stone sources moving in one direction, and nothing in the other. Indeed, it is possible to argue that in much of prehistory we are in fact looking at the visible spinoff from the circulation of other commodities now invisible, of which livestock and its products was probably the most important. Thus in the Neolithic we may be seeing, as it were, a negative impression of the movement of cattle; in the Bronze Age, the residual from the movement of woollen textiles; and in the Iron Age, the salt mines and stock enclosures for the production of salt beef, whose export balanced the importation of Mediterranean wine amphorae and drinking vessels. Even if we cannot substantiate these possibilities, we cannot ignore them; for if we do, our reconstructions and explanations may be simply wrong.

With these points in mind, therefore, let us turn to a specific set of examples in order to follow through a succession of exchange systems over a period of two thousand years, as they are reflected in the lithic assemblages of one part of central Europe that was a major centre of Neolithic population: the Carpathian Basin. The area to be examined in detail here (Sherratt, 1982–83)* is the central part of the Great Hungarian Plain (Fig. 23.1) in the time from the first agricultural settlement (*c.* 5000 bc) down to the end of the early Copper Age (*c.* 3000 bc): a period which precedes the major development of flint-mining in the North European Plain (Lech, Schild, this volume) and the large-scale distribution systems associated with it. The large, relatively uniform alluvial plain of eastern Hungary provides a natural laboratory for the study of lithic exchange systems,

* All dates quoted are uncalibrated radiocarbon dates.

* Study of the lithic materials from this area is part of a programme of fieldwork on its settlement–history undertaken jointly by the Archaeological Institute of the Hungarian Academy of Sciences and the Ashmolean Museum. The sites summarised in Fig. 24.2 were investigated by Dr István Torma, György Goldman, and the author. The identifications of raw materials were made by Mrs C. Takács-Biro of the Hungarian Geological Institute, Budapest. I would like to thank my Hungarian colleagues for their advice and help.

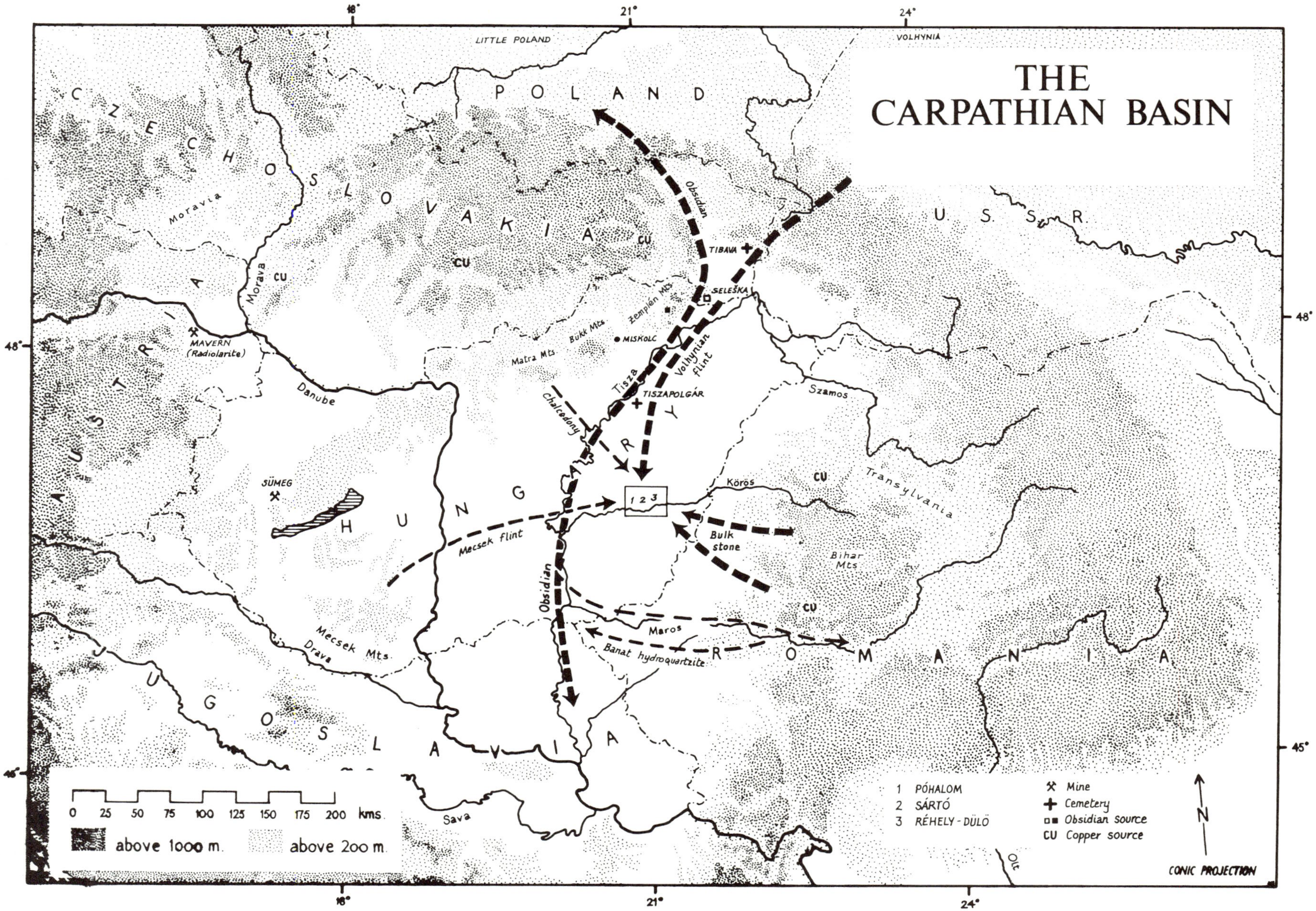

Fig. 23.1. Map of Carpathian Basin, showing identifiable sources and distribution routes of traded materials in the Neolithic and Copper Ages, 5000–3000 bc.

being practically devoid of stone resources but surrounded by mountain and piedmont areas with a wealth of stones (including obsidian) and metal ores. The flint and chert sources of Hungary are described elsewhere (Takács–Biro, volume I of the Symposium Proceedings): in general they are of rather poor quality by comparison with siliceous materials from Jurassic and Cretaceous deposits north of the Carpathians. The acquisition of cutting materials of any kind thus required special procurement, which the acquisition of high-quality materials for specific purposes involved particular effort. The sequence of stone sources used in this area is thus an effective index of changing needs and the scale of the sustaining area upon which local communities were able to draw. That this area was able to acquire stone at all is significant; that it did so in certain periods in enormous quantities or from great distances is particularly interesting. However, these patterns only make sense when seen in the context of complementary flows of other materials and the social setting in which they took place. (For discussion see Sherratt, 1982).

The proportions of various types of flaked stone materials at three neighbouring sites of the period under consideration in Co. Békés, eastern Hungary is given in Fig. 23.2. These assemblages are characteristic of their respective periods, and similar observations (though not cast in quantitative form) have been made on other contemporary sites in this region. The assemblages represent three distinctive systems: an early, pioneer phase following the first introduction of agriculture; a phase of regional interaction in the Late Neolithic; and a phase of interregional exchange in the Early Copper Age.

The pioneer phase

The quantities of chipped stone material found on early Neolithic sites in the centre of the Plain are not large, and the pieces themselves are generally small. Partly worked cores were imported, but these and their debitage are rare and good flint was carefully conserved – in one case the flakes and debris from a knapping operation were carefully collected and stored in a vessel (Kaczanowska, Kozłowski & Makkay, 1981). At first, stone supplies came mainly from the south; but as agricultural communities reached the obsidian sources of the Zemplén mountains in the north, this material came into widespread use. At Méhtelek, a site of the Szatmár group near Szatmár (Satu Mare) on the Hungarian–Romanian border, it formed up to 80% of the chipped stone, the remainder being a great variety of other stones. It was traded both to surrounding Mesolithic groups in Moravia and Little Poland (Kalicz & Makkay, 1977b, p. 17) and southwards to the agricultural communities of the Plain. At one of these, the site of Réhely–Dülö (Körös culture), some 170 km from the sources, it was the dominant material (Fig. 23.2). The rest of the assemblage was made of small quantities of less useful jasper, probably from river pebbles. Supplies of obsidian were limited and perhaps sporadic. There was great variability in the size of blades, and extensive re-working shows an intensive utilisation of the material available.

Regional interaction

Within a few hundred years, however, this situation had been transformed: by *c.* 4300 bc large, prosperous lowland sites had established intensive relations with an expanding periphery that was actively supplying quantities of desirable rocks from the highland fringe. The amounts of material recovered from settlements of this period are impressive: surface investigations at the Late Neolithic site of Sártó (Szálkahát and Tisza cultures), a settlement some 500 m in diameter, have yielded over 50 greenstone axes and more than 45 kg of grinding stones, for instance (Fig. 23.3(*a*), (*b*)). The flaked stone assemblage is dominated by large quantities of limnic hydroquartzite, imported in the form of tabular pre-cores weighing up to 1.35 kg (Fig. 23.3(*c*)). It is not a high-quality material, but its usage was prodigal – large cores and quantities of debitage are found in abundance, as well as tools such as knives, scrapers and sickles (Fig. 24.3(*d*)). Supplies were clearly plentiful and reliable. Jasper was also used in small amounts, but more significant is the sporadic appearance of small quantities of longer-distance imports such as the type of flint found in the Mecsek mountains of south-west Hungary, and of the radiolarite which occurs at a number of sources on the southern fringes of the Carpathians and which was exploited by Lengyel groups in western Hungary, Austria and Slovakia (see Sherratt, 1976, p. 566). Obsidian, although proportionally far less than the abundant hydroquartzite, was nevertheless available in large quantities: it occurs as large cores, little smaller than the 'bombs' (ejecta) of the original raw material, and as regular blades, implements and debitage (Fig. 23.3(*e*)). With the obsidian at this site should also be mentioned a number of sherds of the fine, decorated Bükk pottery which was made in the areas around the obsidian sources to the north (Fig. 23.3(*f*)).

Sites such as Sártó, which are conspicuous for their size and wealth, probably represent the upper tier of a settlement hierarchy, being surrounded by smaller settlements (Makkay, 1982; Sherratt, 1982–83). The major sites occur at something like 30 km intervals, and are located at points along the major rivers (notably the Körös and lower Tisza), which must have played a major role in the distribution of such items. However, within these sites there is remarkably little evidence for social differentiation in the form of rich households or well-equipped graves, or of material items distinctive of an elite. Nevertheless, imported materials came in regular flows: their movement was well organised and not just a sporadic windfall.

Where were these materials coming from? Figure 23.4 represents a first attempt to define the supply radii for some of them. Two areas in particular stand out: first the edges of the Bihar Massif, to the east, along the upper reaches of the various branches of the Körös River in Romania; and

Fig. 23.2. Changing proportions of imported materials on three sites in the centre of the Great Hungarian Plain from 5000 to 3000 bc. The histograms give the percentages by weight of different raw materials in the flaked stone assemblages from three adjacent sites. Note that this diagram shows only relative proportions: the absolute quantity of obsidian, for instance, greatly expands in the Late Neolithic.

secondly the north Hungarian mountains, linked either along the Tisza and lower Körös or more directly along the flood-course of the Hortobágy. Both areas are known as sources of cherts. Comşa (1967, 1976) has described sources for cherts widely used in Neolithic times in the region of Crişana (Ér Valley), at Iozaşel (on the White Körös) and in the Banat (Poiana Ruscăi mountains near Bega), while small shafts for limnic hydroquartzite are known near Miskolc in north-east Hungary (Vértes, 1964, p. 210). Axes of volcanic and metamorphic rocks could have either a northern or an eastern origin, but in the days before wheeled vehicles and pack animals the bulky products like grinding-stones are likely to have been transported directly by boat from the nearest eastern source. Direct access at certain seasons is a possibility here, though the distances are impressive: 100 km or more.

Higher-value items – obsidian and decorated pottery – came from the north over a slightly longer distance. Their bulk was smaller, but in the case of pottery more fragile. The northern mountains form an interesting study of the relationship between settlement and the exploitation of raw materials (Fig. 23.5). The mountains fall into three blocks: Matra, Bükk and Zemplén. The Bükk, in the middle, is the exception – it is limestone, and has plentiful caves. Zemplén, on the other hand, is volcanic and metamorphic, supplying a variety of rocks of which the most important was obsidian. Other siliceous rocks also occur there. Two main sources of obsidian have been distinguished analytically (Williams-Thorpe, 1978; Williams-Throrpe *et al.*, 1984): one, which was the major supplier throughout prehistory, and the sole Carpathian source for most of it, is just over the Czech border at Seleška (Szöllöske), in a fairly low-lying position. But just at the period under discussion, smaller secondary sources within the Zemplén mountains were opened up and specialised workshop settlements with hoards and production–debris are known from the valleys leading down from them (Fig. 23.5, bottom). What is even more remarkable is the extent of settlement in the Bükk mountains themselves. Occupation traces occur up to 900 m above sea level in the limestone caves that are famous for their Palaeolithic assemblages. This upland occupation lasted only for a short time from *c.* 4300–4000 bc, and much of the area was then abandoned. The sites seem to be tied in with the extensive trading opportunities of the period. Remains from the caves show traces of hunting, sheep herding, and evidence for pottery-making, as well as hoards both of axes and blades (Lichardus, 1974). It seems that, as well as being concerned with the direct exploitation of sources in the Zemplén and nearby, these groups were also acting as middlemen in the dispersal of this range of materials and were also perhaps engaged in forms of secondary specialisation such as the manufacture of specially fine pottery or the transhumant movement of sheep. The most easily recognisable product, obsidian,

(*a*)

Fig. 23.3. Types of imported materials at Sártó (Dévaványa Parish, County Békés) in the Late Neolithic (Szakálhát and Tisza periods): (*a*) axes of volcanic and metamorphic rocks, including trapezoid (left) and shoelast (right) forms; (*b*) saddle quern (weight: 8.73 kg); (*c*) tabular pre-core of hydroquarzite (weight: 1.35 kg); (*d*) hydroquarzite cores and core-trimming flakes (top), blades and scrapers (bottom) – one blade with sickle-gloss; (*e*) obsidian cores and core-trimming flakes (top), blades and scrapers (bottom); (*f*) sherds of imported fine pottery of the Bükk culture, incised and incrusted with red paint.

(c)

Fig. 23.3. (*continued*)

(d)

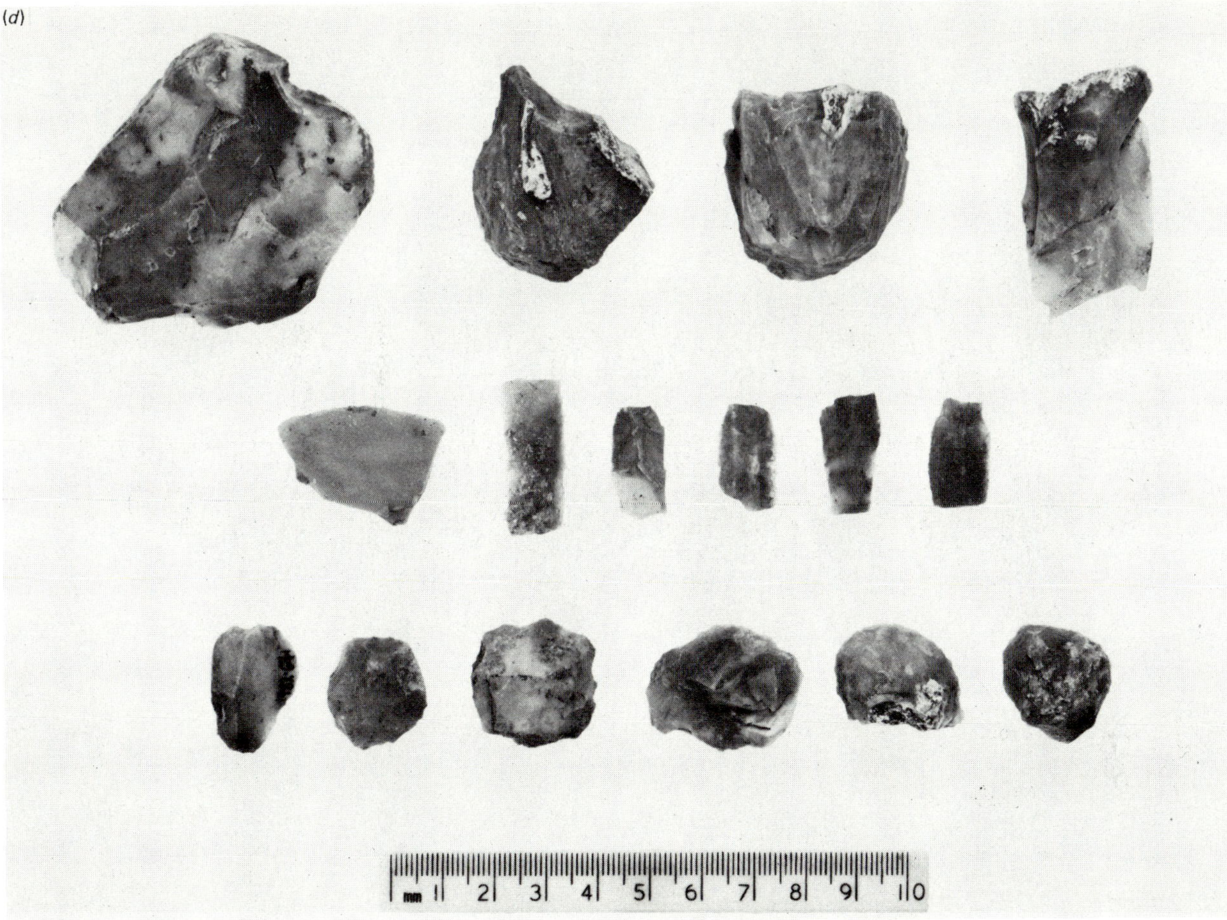

Fig. 23.3. (*continued*)

found its way 400 km down river to Transylvania and Jugoslavia, and also (along with some pottery) across the Carpathians for 200 km into Little Poland.

The pattern of trade, then, was one of intensive localised movement of highland products to adjacent lowlands, with one or two more special items moving from one side of the Basin to the other, and occasionally just into the area over the mountains from important sources. The assymetry of this reconstruction, in which durable items of many kinds were moving into the Plain, strongly suggests the counter-movement of some balancing exportable commodity, of which cattle is the most probable (Sherratt, 1982). The cultural pattern at this period is one of fragmentation, with a multitude of small groupings defined by their fine pottery, suggesting a rather localised scale of interaction in which middleman positions could be exploited, since there was no elite participation in longer-distance trade by which intervening areas could be bypassed. This then is what I have defined as a regional network, and marks the first climax of trade activity within the Carpathian Basin.

Inter-regional exchange

We now move to a very different situation. This is defined in the Hungarian sequence as the beginning of the Copper Age (*c*. 3500 bc), although populations in surrounding areas were still stone-using and indeed in Poland this period saw the development of open shaft-mining and specialist flint-workshops. Technological labels – Neolithic, Eneolithic or Copper Age – are less important than the recognition that a major change had occurred in social organisation and the character of trade.

In the examples presented here, this period is represented by the Early Copper Age site of Póhalom (Tiszapolgár culture). The absolute quantities are no longer as impressive as those of the preceding period, and the sites have no obvious hierarchy; but the component of long-distance imports is substantial. Hydroquartzite continues as an everyday material for small cutting tools, and obsidian is also present; there are also significant numbers of pieces of Mecsek flint and radiolarite. But almost half of the chipped stone industry, including the larger and longer blades, is made of a Cretaceous flint which has been identified as coming from deposits in Volhynina (the region east of L'vov on the upper Driestr: Kaczanowska, 1981). This was capable of being made into the longer blades which were thought particularly desirable (see below). Such supplies were obtained on a regular basis for some hundreds of years: Volhynian flint makes up to 60% of assemblages of the succeeding Bodrogkeresztúr period (*c*. 3000 bc).

(e)

Fig. 23.3. (*continued*)

This change in the nature of exchange networks was part of a radical re-ordering of settlement patterns and social structures. In the central part of the Hungarian Plain there was a complete change in settlement pattern: the very large sites disappeared and were replaced by a much more dispersed pattern of settlement in small units (Sherratt, 1982). The graves, however, took on a new importance; for instead of burials within the household unit, larger cemeteries of up to 200 graves appeared at this time. These are regularly laid out and have carefully differentiated grave inventories, marking gradations of wealth and social role (Bognár-Kutzián, 1963). The graves give the clue to some of the motives behind new patterns of trade, for the items interred in them carry a specific social message. Besides work objects, a special prominence is given to long blades and to shafthole axes. There is some suggestion that the former are associated with older males, the latter with younger adult males and these items occur for the first time in copper; gold ornaments also appear. The very large numbers of stone objects on settlement sites tend to diminish, and the scale of obsidian distribution declines, except for specific items like bifacial arrowheads; moreover the smaller or more inaccessible sources went out of use. The emphasis seems to have shifted from an abundance of fine materials, used to make everyday items, to a specific set of prestige objects made from appropriate materials.

Culturally, too, a new pattern is observable: instead of the multitude of regionally restricted ceramic groups, a uniform culture appears over the whole of the eastern part of the Carpathian Basin. Nor is this restricted to the Basin itself: the cultural group shows a tendency to mix at the edges, as indicated by the Polish term Lengyel–Polgar culture to define this phase north of the Carpathians. Certain inter-regional ceramic types occur in both these

Fig. 23.3. (*continued*)

regions, and indeed in south-western direction also, where a similar cultural mixing occurs with the Salcutsa culture of western Romania and Bulgaria.

The wealthiest area in this period seems to have been on the edge of the Plain rather than in the middle, particularly at points articulating with routes leading out of the Carpathian Basin up in the north east. Here the richer graves are distinguished by the presence of copper, gold and imported flint objects, such as the core of Volhynian flint in grave 10 at Tibava (Šiška, 1964).

One item that was leaving the Carpathian Basin was copper. Shafthole axes appear in small numbers in adjacent areas, and were broken down and re-used as sources of copper away from the area of primary supplies. These copper items entered local exchange networks reaching northwards as far as Scandinavia, and eastwards to the forest–steppe area of the Ukraine (Chernikh, 1966; Randsborg, 1970). Interestingly, one of the items which may have come in return was the horse: for the first occurrence of horse bones in cemeteries of the Hungarian Copper Age demonstrates a link with the contemporary centre of horse-domestication out on the steppes south of Kiev (Bökönyi, 1978). This wider movement of materials and other commodities, particularly high-value goods, marks the emergence of an inter-regional system which had extensive links outside the Carpathian Basin.

Discussion

The three types of exchange systems characterised above form only a part of the history of trade in lithic materials, which continued to circulate alongside copper for a further 1000 years. This latter period saw the development of the Sümeg flint mines in western Hungary and the major development of bulk production and distribution in the flint-mining areas of the North European Plain (Lech; Schild, this volume). The enlarged scale of flint production must be related to the increasing population and

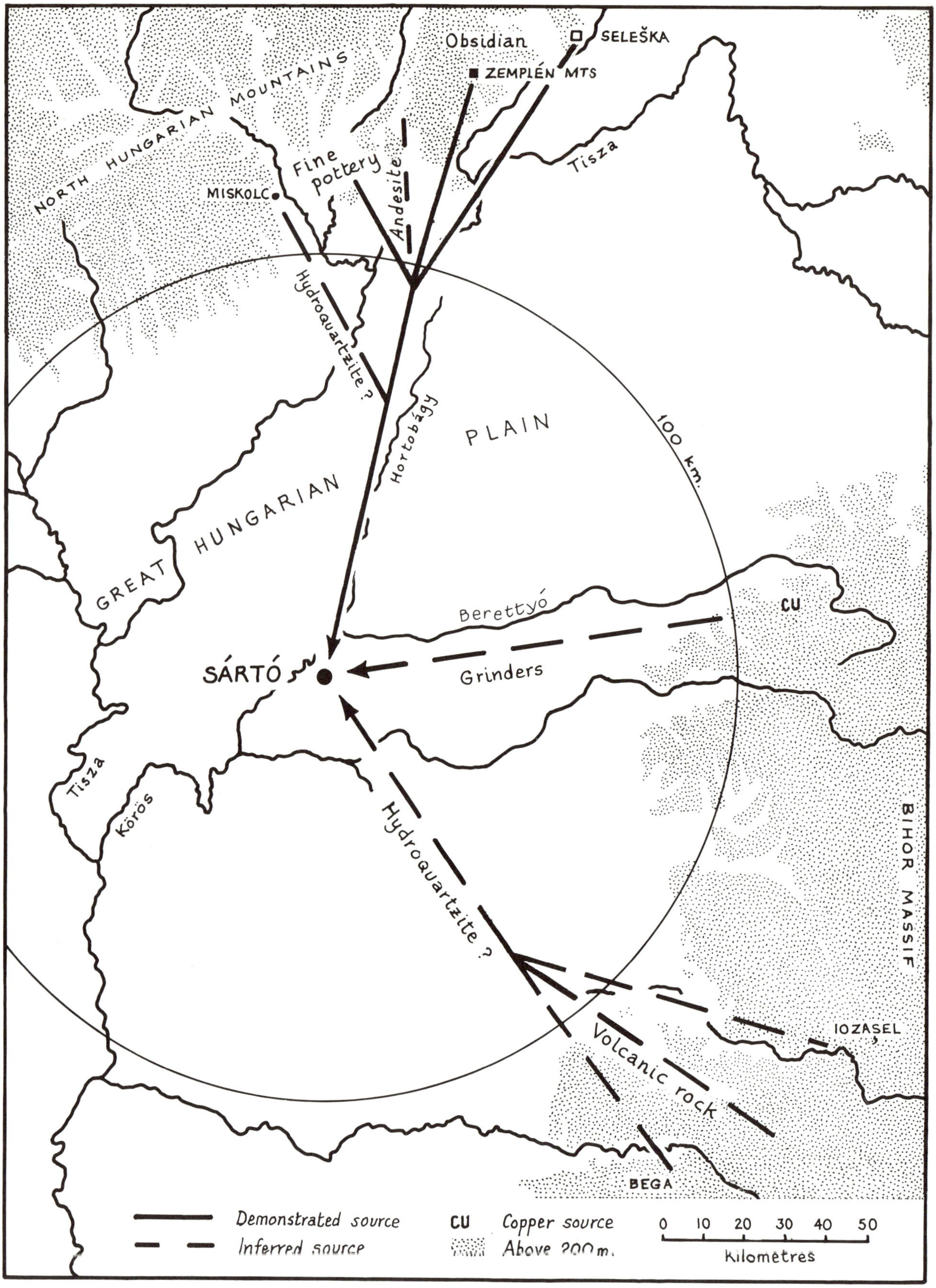

Fig. 23.4. Supply distances for the major classes of lithic materials at Sártó in the Late Neolithic.

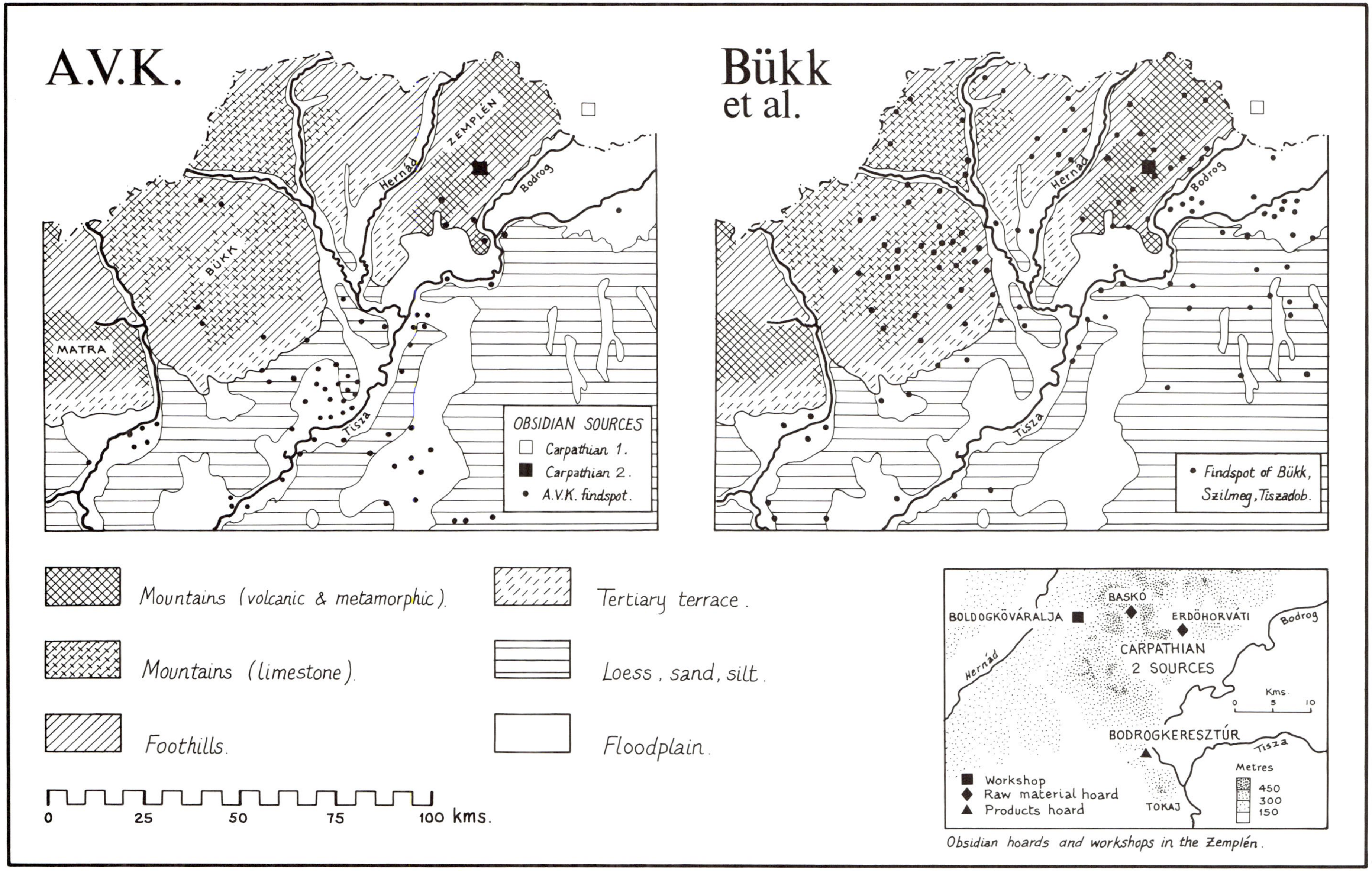

Fig. 23.5. Colonisation of the North Hungarian Mountains 4500–4000 bc in the AVK (Bandkeramik) and Bükk periods, in relation to the exploitation of lithic resources. Obsidian sources in the Zemplén (Carpathian 2) became important in the Bükk period (detail map, bottom right). (Distributions after Kalicz & Makkay, 1977.)

accelerated rate of forest clearance linked to the introduction of plough agriculture and more extensive types of farming (Kruk, 1980; Sherratt, 1981).

In northern Europe the need for stone axes in forest clearance was marked by the development of effective upright axes with a long cutting edge and rectangular cross-section (Montelius' 'thin-butted' and 'thick-butted' types) that have no counterparts further south. These were produced in very large quantities from mined flint. However, towards the end of the third millennium there was a shift in north European flint production away from axes and towards large cutting implements – especially daggers but also knives and sickles. This is associated with the very widespread distribution of specialised products such as the daggers of Grand Pressigny flint, whose use as equipment for warrior males shows continuity with the first copper daggers in north-west Europe, as demonstrated by Beaker grave-goods in the Netherlands (Lanting, Mook & van der Waals, 1973, p. 45). In some respects, therefore, this change in northern Europe recapitulates the type of transition described here for central Europe about 1000 years earlier. The mass circulation of everyday items gave way to a concentration on specific exotic materials and types with a more explicit symbolism of social status. It is significant that the regular use of copper occurs in this context in both areas. Materials such as obsidian seem to have been incapable of carrying such a message: as with *Spondylus* shell, its main period of use occurs in earlier Neolithic contexts at a lower level of social differentiation.

The interpretation of studies of lithic material sources thus inevitably leads not only to a consideration of the other items circulating in exchange networks, but also to attempts at evaluating the social significance of the materials in question and how they were manipulated by the individuals concerned (Douglas & Isherwood, 1980). The complementary evidence from settlements and graves provides a starting-point for the analysis of how such commodities were used by Neolithic populations, and wider comparative studies should clarify the kinds of contrasts identified here.

The study of lithic artefacts thus remains a challenging area of archaeological research, but one which cannot be isolated from more general consideration of the archaeological record, and indeed of ethnography as well. If this account has indicated the relevance of flint characterisation to a wider body of problems, then it will have succeeded in its aims.

Bibliography

Bognár-Kutzián, I. (1963). *The Copper Age Cemetery of Tiszapolgár Basatanya*. Budapest, Akadémiai Kiadó.

Bökönyi, S. (1978). 'The earliest waves of domestic horses in east Europe.' *Journal of Indo-European Studies*, **6**, 17–76.

Chernikh, E. (1966). *Istoriya Drevneishei Metallurgiya Vostochnoi Evropi* Moscow, Nauka.

Comşa, E. (1967). 'Ueber die Verbreitung und Herkunft einiger von den jungsteinzeitliche Menschen auf dem Gebiete Rumäniens verwendeten Werkstoffe.' *A Móra Ferenc Múzeum Évkönyve*, Szeged 1967, pp. 26–8.

Comşa, E. (1976). 'Les matières premières en usage chez les hommes neolithiques de l'actuel territoire Roumain.' *Acta Archaeologica Carpathica*, **16**, 239–49.

Douglas, M. & Isherwood, B. (1980). *The World of Goods*. Harmondsworth, Penguin.

Kaczanowska, M. (1980). 'Uwagi o surowcach, technice i typologii przemysłu krzemiennego kultury Bodrogkereszturskiej i grupy Laznany.' *Acta Archaeologica Carpathica*, **20**, 19–56.

Kaczanowska, M., Kozłowski, J. & Makkay, J. (1981). 'Flint hoard from Endröd Site 39, Hungary (Körös culture).' *Acta Archaeologica Carpathica*, **21**, 105–17.

Kalicz, N. & Makkay, J. (1977*a*). 'Frühneolitische Siedlung in Méhtelek-Nadas.' *Mitteilungen des Archaeologischen Instituts der Ungarischen Aademie der Wissenschaften*, **6**, 13–34.

Kalicz, N. & Makkay, J. (1977*b*). *Die Linienbandkeramik in der Grossen Ungarischen Tiefebene*. Budapest, Akadémiai Kiadó.

Kruk, J. (1980). *The Neolithic Settlement of southern Poland* (edited by J. M. Howell & N. J. Starling). British Archaeological Reports, Int. Ser. 96, Oxford.

Lanting, J. N., Mook, W. G. & van der Waals, J. D. (1973). 'C14 Chronology and the Beaker problem.' *Helinium*, 13, 38–58.

Lichardus, J. (1974). *Studien zur Bükker Kulture*. Bonn, Habelt.

Makkay, J. (1982). *A Magyarországi Neolitikum Kutatásának Uj Eredményei*, Budapest, Akadémiai Kiadó.

Randsborg, K. (1970). 'Eine kupferne Smuckscheibe aus einem Dolmen in Jütland.' *Acta Archaeologica*, **41**, 181–90.

Sherratt, A. G. (1976). 'Resources, technology and trade: an essay in early European copper metallurgy.' In: Sieveking, G., Longworth, I. & Wilson, K. (eds.), *Problems in Economic and Social Archaeology*. London, Duckworth, pp. 557–81.

Sherratt, A. G. (1981). Plough and pastoralism: aspects of the Secondary Products Revolution.' In: Hodder, I., Isaac, G. & Hammond, N. (eds.), *Pattern of the Past: Studies in Honour of David Clark*. Cambridge University Press.

Sherratt, A. G. (1982). 'Mobile resources; settlement and exchange in early agricultural Europe.' In: Renfrew, A. C. & Shennan, S. J. (eds.), *Ranking Resources and Exchange*. Cambridge University Press.

Sherratt, A. G. (1982–83). 'The development of Neolithic and Copper Age settlement in the Great Hungarian Plain *Oxford Journal of Archaeology*' (Part 1) **1**, 287–316; (Part 2) **2**, 13–41.

Šiška, S. (1964). Gräberfeld der Tiszapolgár Kultur in Tibava *Slovenská Archaeologia*, **12**, 352.

Williams-Thorpe, O. (1978). The Distribution and Provenance of Archaeological Obsidian in Central and Eastern Europe PhD dissertation, University of Bradford.

Williams-Thorpe, O., Warren, S. E. & Nandris, J. G. (1984). 'The distribution and provenance of archaeological obsidian in central and eastern Europe.' *Journal of Archaeological Science*, **11**, 183–212.

Willms, C. (1982). *Zwei Fundplätze der Michelsberger Kultur aus dem westlichen Münsterland, gleichzeitig ein Beitrag zum Neolitischen Silexhandel in Mitteleuropa*. Hildesheim, August Lax.

Vértes, L. (1964). 'Eine prähistorische Silexgrube am Mogyorósdomb bei Sümeg.' *Acta Archaeologica Hungarica*, **16**, 187–215.

24

Sources and specialists: three ancient Near Eastern urban flint industries

ROBERT MILLER

Neither the techniques of production nor the end-products of specialist stone working are necessarily unique, as technological convergence can account for many resemblances between artefacts produced by specialists and non-specialists using the same or similar raw materials. Recognition of specialist stoneworking requires a number of technological, economic and demographic factors to be taken into account. In archaeological contexts the presence of a specialised industry may be shown by the concentration or separation of stages of production in well-defined areas, as well as by the scale of production. Output may be limited to a few, well-defined forms such as arrowheads, borers or roughed-out cores. Specialisation does not have to be full-time; intense seasonal work can provide extra sources of income from the sale of finished or semi-finished goods.

However, one important feature of craft specialisation which needs to be stressed is that craft specialisation is related to population density. In Durkheim's formula, division of labour is the result of increasing population density which does not merely permit craft specialisation; it causes division of labour to develop (Durkheim, 1926, pp. 237–45).

The relatively dense demographic context of craft specialisation does not have to be urban, although the industries dealt with in this study are from urban or urban-related contexts usually centres for considerable artisanal activity (Adams, 1960; Jacobs, 1972). At Beidha, an early Neolithic village in southern Jordan, there is already evidence of specialist activity and craft specialisation is an important component of rural economies. It can also occur in hunting and gathering economies, as at Mount William, Australia, which was the source of raw material for a group of part-time stone-workers who met demand for greenstone axes in areas of Victoria already densely populated before European contact and immigration (McBryde, 1978). By the time of city formation in the Near East such craft specialisation had a respectable ancestry.

What are the characteristic relations between sources of raw material and the technology and organisation of production in urban-based flaked stone industries? Systematic research into urban-period stone industries is just beginning, and each site or set of sites has its own unique characteristics. Nevertheless, by looking at a number of different sites together it is possible to observe the craft of flintworking in a context which is too often neglected, as lithic specialists generally focus on prehistoric stone tool technology and overlook the evidence from later periods. It is also possible to observe how different areas with different concentrations of human and natural resources tend to specialise in the production of different goods.

To illustrate the characteristics of specialist flintworking in relation to the use of raw material, I would like to consider case studies from three early urban centres in Syria, Saudi Arabia and Egypt. These three flaked stone industries developed at the central points of well-developed early exchange networks and share similar features in the organisation of production which transformed local flint sources into articles for use and for export elsewhere.

The flaked stone industries from Thebes in upper Egypt and Teima in northwest Arabia were produced by urban artisans, working in centres of population which were the capitals and administrative centres for large parts of the ancient Near East in the middle of the second and first millennia respectively. Another flintworking industry was found in the upper Euphrates valley, Syria, at Hadidi, Halawa and Sweyhat, where trade and contacts from Anatolia and the Levant converged on northwest Mesopotamia in the late third and early second millennia BC.

The basic sources of information are the sites themselves, the presence of knapping waste, or the absence of certain stages of knapping waste. Sometimes preliminary stages of tool manufacture and raw material reduction can be inferred from relict features of flake and core topography, which provide information on knapping technology. Cores, flakes and location of working areas can also provide valuable data on some features of quarrying and roughing-out, as well as on the division of labour between sources of raw material, workshops where the tools were produced, and the areas where tools were used and discarded.

This can be seen clearly in the New Kingdom industry of Thebes. Recent excavations in East Karnak produced beautifully worked bifacial knives, as well as burins, borers and other tools. But these flint tools represent only the final stages of tool life: use, edge renewal and eventual discard. Although many of these tools are made on bifacial or blade blanks, no New Kingdom bifacial rough-outs or blade cores were found in several seasons of excavation around the Akhenaten Temple. Early stages of production were carried out elsewhere, perhaps in workshops located in other parts of the city or on its outskirts, as at Tell Amarna (Spurrell, 1894, p. 37). Some flintworking also took place across the river where the flint sources used by New Kingdom knappers were located.

Large flat nodules are found in dry valley beds of former western tributaries of the Nile, and these provide a very high quality of knapping flint without frost fractures. These nodules were used to make bifacial knives and other tools. Another source of raw material used by the New Kingdom toolmakers was the supply of flint nodules produced as a by-product of tomb cutting. The tombs in the Valley of the Kings are cut into limestone with flints, and it is sometimes possible to see where slabs of flint have been removed from the shafts of tombs and their entrances (Fig. 24.1). Some of the flint was immediately worked up into picks and mauls used for cutting the tombs themselves (Seton-Karr, 1905; Mackay, 1921, pp. 155–6). But more flint was produced than would have been needed to make flint picks to cut the tombs. This leftover flint was not wasted. Howard Carter noted that the concentration of flint boulders around the entrances to tombs in the Valley of the Kings was one way tombs could be detected (1972, pp. 29–30). In the photograph he published of the early stages of the excavation of King Tutankhamen's tomb, cortical flakes and artefacts which resemble flint cores are visible in section above the workmen's huts built over the fill of the tomb shaft to conceal the burial (*ibid.*, p. 29). Was flintknapping a part-time activity for some of the guards watching the tomb? It was still possible in 1979 to observe on top of tomb shafts in the Valley of the Kings concentrations of blade cores, together with long cortical flakes struck in order to remove the outer surface of the nodules, as well as crested flakes and plunging flakes shooting over the base of the core (Fig. 24.2). Further study of these concentrations of knapping waste could reveal technologically significant series of refitted flakes. It might also be possible to run tests for patterns of change or consistency in knapping habits through time, by comparing knapping waste from earlier and later tombs in a series whose individual members are clearly dated.

What was the volume of flint which could have been produced from the New Kingdom tombs in Thebes? How much of Thebes' output would have been needed locally for a population of about half a million? Was some exported to those contemporary eastern Mediterranean sites clearly dated to the second millennium BC, which have finished flint blade tools but no evidence of the earlier stages of production. Further study of these well-dated New Kingdom industries and their Middle Kingdom ancestors, as well as their successors in late dynastic times is urgently needed.

A second major historic period flaked stone industry comes from Teima, Saudi Arabia, which was the site of major urban development in the mid-sixth century BC.

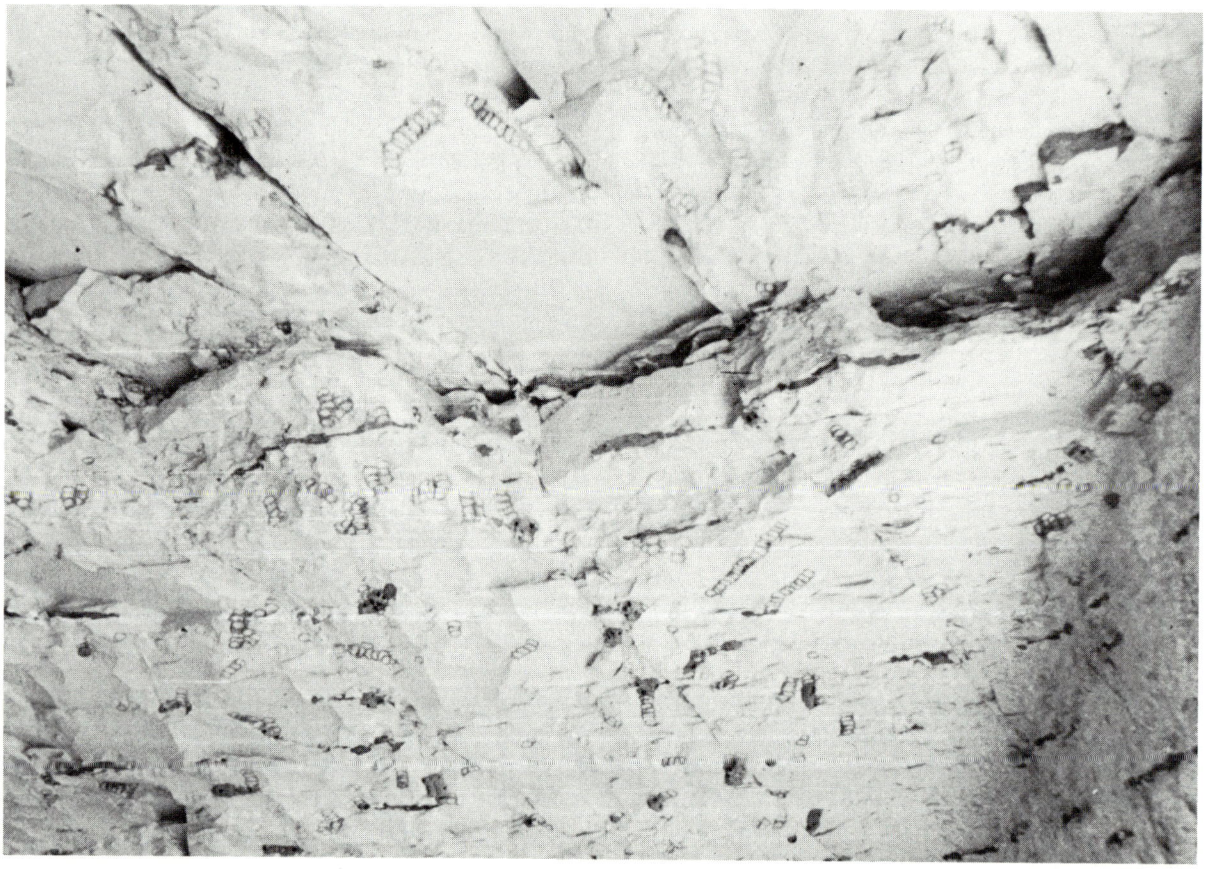

Fig. 24.1. Entrance to tomb shaft, Valley of the Kings, showing removal of flint slabs.

Fig. 24.2. Knapping waste over entrance to tomb in Valley of the Kings.

Excavation and survey show that a lapidary and beadmaking industry which began at this time used large numbers of drill bits made from shatter chunks of coarse-textured beige flint (Miller, 1980). Although the flint used to make the borers would have been partly derived from alluvium eroded from the sandstone with flints found west of the site, no pebbles from which such borers could be made were found in surface survey around the site, even though the wadis which drained the sandstone with flints led to the Teima basin. This suggested that flint nodules had either been brought in from outside, or intensively collected from the surface around Teima at the time of the lapidary industry. In either case the quantities were immense; an estimated 200 metric tons of knapping waste, mostly pyramidal cores, flakes used as cores and drills with bits rounded by abrasion.

Blanks for cores and tools at Teima were often made by shattering nodules. I did some experiments with shattering which showed that complex infolded and sub-conical fracture surfaces were typical of pieces produced by shattering. Lower Palaeolithic stoneknappers, Brandon gunflint makers and aboriginal Australian toolmakers are among the groups which resorted to shattering to break up material from the quarry or source, especially when nodules or large blocks of material were used. Shattering is more wasteful at later stages of production, so consistent use of shattering to produce tool and core preforms indicates a primary stage of raw material reduction just as much as the presence of cortical flakes. By throwing semi-spherical and ellipsoid nodules against a fixed anvil or placing them on an anvil and hammering them cleanly in the middle to break them open, it was possible to judge the quality of the raw material and select plano–convex pieces with workable flaking angles for further reduction. Cores excavated from two of the upper Euphrates valley flint-producing sites, Hadidi and Sweyhat also show that core rough-outs were produced by shattering nodules or slabs of flint.

Another way of economising on the transport of raw material at Teima was the use of flakes as cores. These were also used at Hadidi and other upper Euphrates valley sites (Miller, 1985). Such cores, *nucléus sur éclat* in the terminology of Newcomer & Hivernel-Guerre (1974) are a useful way of economising on the transport of material from source to working area. By detaching flakes to use as cores from blocks of raw material found at source, the knappers or their suppliers cut down on the time and energy needed to carry the flaking material to their camp, home or workshop. It also made it possible to get a good idea of the working quality of the flake to be used as a core. When a bulb of percussion is found on the striking platform of the core, or on the dorsal surface of a flake, it is clear that a flake has been used as a core.

A similar desire to economise on unnecessary expenditure of energy and to provide knappers with a supply of flaking stone of proven quality seems to have governed the procurement of raw material which could be inferred from the workshop excavated at Hadidi. This workshop, dated to the late third millennium BC, specialised in the manufacture of arrowheads. Roughed-out, resharpened and almost finished arrowheads were found, together with discoidal and Levallois cores (Fig. 24.3), flakes used as cores and other flint artefacts. Cores for the production of the flakes which were used as cores were missing from the workshop assemblage; cortical flakes from roughing-out the bifacial blanks were also lacking. The flint raw material had been brought to the workshop in a semi-finished state, perhaps after an intermediate stage of production, when the plano–convex blanks were subjected to heat treatment before being brought to the workshop to apply the finishing touches.

Carrying out some preliminary roughing-out elsewhere was a way to reduce the energy needed to modify and handle the flint in the workshop. The knappers could thus work with blanks, flakes used as cores, bifacial rough-outs and plano–convex shatter pieces. All of these would be smaller pieces closer to the size of the finished pieces they were producing.

Hadidi provides another example of the knappers' tendency to economise on the amount of raw material processed in a workshop by dividing labour between source and workshop.

At least two sources of flint were used to supply the Hadidi workshop: patinated pieces from earlier, Palaeolithic, scatters and fresh flint brought from outcrops exposed along the river or in mountains near the site. Chunks of tabular flint were shattered or flaked at the outcrop to provide the plano–convex pieces used in the workshop.

The artisans working at Hadidi may have had batches of flint brought to them by intermediate suppliers, such as shepherds exchanging an occasional bag of flint for a few arrows or arrowheads. Rough-outs could also have been produced by people attached to the workshop, perhaps by the apprentice knappers, whose presence could be inferred from the level of skill shown by some of the attempts at knapping found in the workshop. There may thus have been a division of labour between rough flaking and final knapping analogous to that observed at Brandon a century ago by Skertchly (1879), where different groups of workmen were involved in quarrying, flaking and knapping.

Skertchly's work also suggests that about 10% by weight of raw material could be made into tools. The combined weight of the flint cores and blanks found in the largest group of knapping debris on the floor of the Hadidi workshop was about 2.5 kg, which would suggest the knappers were working with a load of what was initially about 25 kg of raw material. This is close to the range of 20–30 kg which miners could carry for short distances from raw material sources (Fober & Weisgerber, 1980, p. 41). It is also significant that threshing sledge flints were sold in 25 kg lots in recent years by knappers in Cakmak in northwest Turkey (Weiner, 1980, p. 388).

If the Hadidi knappers were working with batches of about 25 kg at a time, how was the flint carried? It is possible that leather bags such as those Shepherd believes were used in prehistoric European flint mines (1980: p. 20) were used to bring loads of flint into Hadidi. Leather bags were used by workers carrying things to and from cities in the late third millennium BC Mesopotamia, contemporary with the Hadidi workshop (Sollberger, 1966, nos. 95 and 173, pp. 31, 49, 98). However, leather bags were generally used in the city-states of Mesopotamia to carry either water or dry goods, such as corn which it would not have been convenient to carry in baskets. It seems more likely that the latter were used for flint transport. Stones, like bricks, were certainly carried in baskets as is shown by the Ur III figures representing building material being carried (Moortgat, 1969: 56).

Evidence of knappers working another kind of flint was found across the river from Hadidi at Halawa and Sweyhat. Instead of working the translucent brown outcrop flint used for arrowheads, knappers used nodules of a coarser-textured pinkish grey flint present in Euphrates river gravels. Some of these nodules could have been a by-product of tomb digging, as large cemeteries of the mid third–early second millennium BC, contemporary with the flintworking

Fig. 24.3. Cores from knapping workshop, Hadidi, Syria.

industry, are found near the sites. Flint from some unflaked nodules collected from the surface in 1979 appeared to have been subjected to heat from fire, according to preliminary ESR spectroscopy (see Robins *et al.*, 1981). However, this could merely have reflected the fact that a stone lying on the surface is likely to have been affected by fire at some time.

In contrast to the workshop at Hadidi, where only discoidal and Levallois cores were found, another type of core was also produced at Halawa and Sweyhat and at Hadidi outside the workshop. This other core type is a distinctive type of large single-platform core, made by striking a flake off one end of a nodule or a suitable shatter-block from a nodule, creating a striking platform which can be used for the removal of a series of parallel cortical flakes. Most of the flakes found with these cores still have cortex on the dorsal surface. Either raw material was so abundant at the site that knappers did not bother to develop complex reduction sequences, or something else was happening.

Until the second half of the third millennium BC, flint sickle blades were commonly used as harvesting tools, and although copper seems to have temporarily replaced flint for harvesting tool edges in the late third millennium BC, blade blanks would still have been used for other tools. There are also sickle blades of opaque pinkish grey flint found at Halawa in the early second millennium BC, part of a final reversion to lithic technology due to either fashion, reduced access to copper metallurgy, or to continued circulation of lithic artefacts in some spheres of activity.

The use of blades without cortex as tool blanks suggests that excavation and survey results available for this study in the upper Euphrates valley area generally record primary reduction areas for blade core production. The single platform cores of the upper Euphrates valley sites may be roughed-out blade cores intended for flaking elsewhere, like the roughed-out cores (precores) of chocolate-brown flint from the Polish Neolithic which Schild discusses in his contribution to this symposium.

Considerable economies of raw material are possible if knappers use such roughed-out, preformed cores for blade production (Sheets & Muto, 1972). In addition to reducing the amount of core and flake waste which needs to be disposed of, knappers able to select roughed-out cores will be sure of the quality of the flaking material.

Other economic considerations related to export and exchange systems should also be taken into account when large numbers of roughed-out cores are being produced. It has been suggested that the way stone is quarried and worked up at or near the source depends on whether or not it is intended for immediate or local use, or for distribution and export through exchange or market systems (Torrence, 1981, 1982). Torrence has described an example of this process of adapting stages of production to market specialisation. A type of obsidian core was produced for export on Melos which was only roughed-out, as it was intended for further reduction and blade production elsewhere (1981, p. 72). Coincidentally, the way these obsidian export cores were made is almost identical to the way knappers in the upper Euphrates valley made single-platform flint cores in the mid third–early second millennia BC. Market specialisation may thus lead to technological convergence.

The knappers who made these cores did not have to be specialists. The single-platform cores from the upper Euphrates valley could have been destined for sites in the central and southern Mesopotamian plain which lacked local sources of raw materials, including flint, and depended on participation in exchange and distribution networks to secure them. The core rough-outs were ready for use. The ridges between blade or flake scars could be used as guide ridges for striking off further blades, perhaps by pressure flaking. This stage of manufacture would have occurred, either in workshops in the cities of the upper Euphrates valley, or in workshops downriver. Workshop deposits from southern Mesopotamia where blade cores were used to produce blanks for sickle blades and borers have been published by Mrs Payne (1978 and 1980). Was the upper Euphrates one of the sources of supply for these workshops?

The hypothesis that the knappers of the upper Euphrates valley were producing semi-finished cores for export helps to explain some of the features of the single-platform cores found in survey and excavation. There was usually a place on the striking platform and ridges of the core flaking face for further flake removals as good as any of the ones struck off so far. Why hadn't the knappers continued? The majority of the single-platform cores at the Halawa and Sweyhat sites showed signs of knapping accidents such as hinge fractures on the core face or of incipient cones of percussion on the striking platform or core face. Both types of accident would have made further flaking difficult. The cores which were left behind may thus be analogous to factory seconds, rejected after quality control checks. Semi-finished cores which passed inspection would probably have been loaded on river transport and shipped downriver to lower Mesopotamia, as this would cause less damage to the working material than if pack animals had been used to bring flint overland from sources closer to the towns of southern Mesopotamia as the crow flies.

It has already been suggested that as early as the Uruk period, in the mid–late fourth millennium BC, building stone could have been quarried in the upper Euphrates and shipped to Uruk (Deshayes, 1980, p. 404). A cuneiform text of about 2200 BC provides evidence that, at a time contemporary with one phase of the Sweyhat and Halawa stone flaking industry, large stones were shipped from the 'Mountain of the Amorites' on the middle Euphrates to build a temple in Lagash, in southern Mesopotamia (Kupper, 1957, p. 150). Another related trade in partly finished stone for toolmaking as well as building could have linked the towns on the upper and lower Euphrates.

What was the scale of this rough-out core production? The same sorts of cores and cortical flakes from the pinkish–grey terrace nodules are found on every late third–early second millennium BC site in the upper Euphrates valley that I have visited. This was a time of high settlement density; local demand for flint goods certainly existed, and as well as flaked stone quarrying tools, weapons and farming tools were needed and produced locally. But it is also likely that it was the favourable position of the sites east of Aleppo, where the Euphrates emerges from the mountains at the intersection of land and water trade routes, that led to the development of a local industry for the manufacture and export of flint arrowheads and roughed-out blade cores.

In conclusion, we can see that knapping technology can provide us with categories for the classification and study of the relation between sources, workshops and markets or other centres of exchange and distribution. Although it is possible to draw inferences from observing different stages of roughing-out, finishing and using tools, it remains difficult to correlate a particular workshop with a particular quarry. But we can document patterns of raw material use and transformation which enable us to reconstruct the outline of complex industries where different stages of production were concentrated in different areas.

Acknowledgements

A grant from the University of London's Central Research Fund supported fieldwork during which part of the data on which this study is based was collected.

Bibliography

Adams, R. McC. (1960). 'The origin of cities'. *Scientific American,* **203**(3), 153–68.

Carter, H. (1972). *The Tomb of Tutankhamen* (1st edn, 1923–33). London: Sphere.

Deshayes, J. (1980). 'Conclusions'. In: *Le moyen Euphrate, Zone de Contacts et D'échanges*. J. C. Margueron (ed.). Leiden: Brill, pp. 403–6.

Durkheim, E. (1926). *De la Division du Travail Social*, 2nd edn. Paris: Alcan.

Fober, L. & Weisgerber, G. (1980). 'Feuersteinbergbau – Typen und Techniken. In: *5000 Jahre Feuersteinbergbau*. G. Weisgerber (ed.). Bochum: Deutschen Bergbau-Museum, pp. 32–47.

Jacobs, J. (1972). *The Economy of Cities*. Harmondsworth: Penguin.

Kirkbride, D. (1967). 'Beidha 1965: an interim report'. *Palestine Exploration Quarterly,* **99**, 5–13.

Kupper, J. R. (1957). *Les nomades en Mésopotamie au Temps des Rois de Mari*. Paris: Belles Lettres.

Mackay, E. (1921). 'The cutting and preparation of tomb-chapels in the Theban necropolis'. *Journal of Egyptian Archaeology,* **7**, 154–68.

McBryde, I. (1978). '*Wil-im-ee Moor-ring*: Or, where do axes come from?' *Mankind,* **11**, 354–82.

Miller, R. (1980). 'Appendix B: Chipped Stone'. In: G. Bawden, C. Edens & R. Miller. 'Preliminary archaeological investigations at Tayma. *Atlal,* **4**, 99–103, pl. 68.

Miller, R. (1985). *Flint Knapping and Arrowhead Manufacture at Tell Hadidi, Syria*. Milwaukee: Milwaukee Public Museum.

Moortgat, A. (1969). *The Art of Ancient Mesopotamia*. London: Phaidon.

Newcomer, M. H. & Hivernel-Guerre, F. (1974). 'Nucléus sur eclat: technologie et utilisation par différentes cultures préhistoriques'. *Bulletin de la Société Préhistorique Française,* **71**, 119–28.

Payne, J. C. (1978). 'Flint and obsidian industries'. In: P.R.S. Moorey. *Kish Excavations, 1923–1933*. Oxford: Clarendon, Microfiche D09–E09.

Payne, J. C. (1980). 'An Early Dynastic III flint industry from Abu Salabikh'. *Iraq,* **42**, 105–19.

Robins, G. V., Seeley, N. J., Symons, M. C. R. & MacNeil, D. A. C. (1981). 'Manganese (II) as an indicator of ancient heat treatment in flint'. *Archaeometry,* **23**, 103–7.

Seton-Karr, W. H. (1905). 'How the tomb galleries at Thebes were cut and the limestone quarried at the prehistoric flint-mines of the eastern desert'. *Annales du Service des Antiquités de l'Égypte,* **6**, 176–84.

Sheets, P. D. & Muto, G. R. (1972). 'Pressure blades and total cutting edge: an experiment in lithic technology'. *Science,* **175**, 632–4.

Shepherd, R. (1980). *Prehistoric Mining and Allied Industries*. London: Academic.

Skertchly, S. B. J. (1879). *The Manufacture of Gunflints*. London: Geological Survey.

Sollberger, E. (1966). *The Business and Administrative Correspondence under the Kings of Ur*. Locust Valley: Augustin.

Spurrell, F. C. J. (1894). 'Flint tools from Tell el Amarna'. In: *Tell el Amarna*. W. M. F. Petrie (ed.). London: Methuen, pp. 37–8.

Torrence, R. (1981). 'The Melos obsidian quarries; a case study in lithic technology'. *Staringia,* **6**, 70–3.

Torrence, R. (1982). 'The obsidian quarries and their use'. In: *An Island Policy*. C. Renfrew and M. Wagstaff (eds.) Cambridge: CUP, pp. 193–221.

Weiner, J. (1980). Die Flintminen von Cakmak – Eine im Aussterben begriffene heute noch produzierende Feuerstein-industrie in Nordwestanatolien. In: *5000 Jahre Feuersteinbergbau*. G. Weisgerber (ed.). Bochum: Deutschen Bergbau-Museum, 383–95.

25

Recent developments in the application of the *k*-means approach to spatial analysis

ALBERT J. AMMERMAN, KEITH W. KINTIGH AND JAN F. SIMEK

Embarking upon spatial analysis in archaeology is in some respects like sailing through the Straits of Messina in Homer's time. One gets to choose between the celebrated perils of Scylla on one hand and those of Charybdis on the other. In one case, we can search informally by means of visual inspection for meaningful patterns in our distribution maps. If the maps that we are working on have many points, we may find, much as in the case of Scylla's insatiable appetite, that our time is endlessly consumed by the business of pattern matching. Clearly, there must be a better way. Hence the promise held out by quantitative methods on the opposite shore. But as we watch one ship after another attempt to sail through the statistical rigors of Charybdis, few results of interest appear to emerge from its reductive whirlpool. Obviously, the reasonable thing to do, although this may be much easier to say than to accomplish, is to steer a middle course.

The aim of this paper is to describe some of the efforts that we have made in this direction during the last few years. The argument can be made that there is a need to develop new methods of spatial analysis in archaeology which are tailored more closely to archaeological data and objectives (Kintigh & Ammerman, 1982). It will be useful to begin with a brief review of the motivation behind the alternative approach that we advocate. The discussion will then turn to *k*-means cluster analysis as a means of implementing a more heuristic approach to the study of spatial patterns in archaeology. The method has been applied to two Upper Palaeolithic sites in France, Pincevent and Le Flageolet, and selected results from these two studies will be presented. The two sites differ considerably as far as their contexts are concerned and thus help to underscore the larger role that context should probably play in spatial analysis. Finally, some aspects of the development and implementation of the approach that we are currently exploring will be discussed.

Traditional approaches to spatial analysis and newer quantitative ones are complementary to one another in terms of some of their main strengths and weaknesses. While intuitive visual approaches often display a richer substantive problem formulation, they are hindered by a limited capacity for information processing. Quantitative methods generally have such a capacity but tend to suffer from the naiveté of their assumptions. The alternative that we have proposed attempts to combine the problem-oriented intelligence of less formal approaches with the information processing capacity and systematic benefits of quantitative treatments. The approach is heuristic in the sense that spatial analysis is seen as involving a *search* for patterns in distribution maps and also in the wider sense that one is at the same time trying to *discover* those principles of organization that may underlie patterning that is recognized. What is needed in order to make a start are methods of directing a search through a spatial distribution. One of the challenges here is that the unordered components of a distribution may serve to obfuscate patterns. Moreover, if a distribution contains many points, it may be necessary to search through a large number of relationships. The set of potential relationships will probably be much too large in most cases to permit a simple exhaustive check of all the combinations against a list of possible patterns. One option offering a way out of this situation is to conduct a more informed search guided by heuristic rules. 'Heuristic rules may be thought of as rules of thumb, rules which facilitate discovery but which are not absolute rules leading directly to the solution of a problem. In this kind of heuristic search, context plays a basic role in the evaluation of the relationships between distributions, patterns and structural models (Kintigh & Ammerman, 1982, p. 23).' The strategy suggested here has clear parallels with those adopted in cognitive science and related fields, where the intractability of many complex problems (at least as far as simple analytical solutions are concerned) led to a shift toward more heuristic approaches (e.g. Simon, 1969, 1980).

In our case, we turned to *k*-means cluster analysis as a means of conducting the search for patterns in a distribution map. Before considering the method itself, it is perhaps useful to provide some background on how it came to be chosen. We did not begin with an interest in the *k*-means technique and a wish to see it applied in particular to the problem of spatial analysis. Rather, our initial efforts were concerned with identifying some of the limitations of previous methods and some new directions that might help in getting around them. The appearance of *Spatial Analysis in Archaeology* (Hodder & Orton, 1976) stimulated a broad interest in the subject. Yet when students set out to use many of the methods presented in the book, one of the common results was frustration. It was not that the methods did not work but that the 'answers' obtained did not speak to the larger and often more interesting spatial problem that the student had in mind at the outset. The methods were too reductive: a map was often characterized in terms of only one or a few summary statistics. Location and configurational patterning were in effect often lost during the course of analysis. There was a need to put location back into spatial analysis. There was the further possibility to consider that patterning might be present not just on one level but on multiple levels of spatial refinement within a distribution map. This possibility should be taken into account in the search strategy. One of the ghosts in the machine when it comes to making 'form to process' arguments along statistical lines (e.g. Cliff & Ord, 1975) is that a single process is presumed to be mainly responsible for generating the distribution of points on a map. Perhaps a more realistic point of departure for most archaeological studies would be that a range of different processes contributed to the generation of a map. The task of sorting out different processes or factors would not necessarily be an easy one. As mentioned above, one resource that could probably be used more effectively toward this end was contextual information. Another line of thought was that more credit should be given to the human eye, when it comes to pattern recognition and especially pattern matching between maps. Machine analogues were to be encouraged but it was likely that much time would pass before such analogues would be on a par with the experienced eye. One implication here was that the search strategy should be, at least in part, display-oriented. These were some of the considerations that we had in mind when initial efforts were made to sketch out a computer assisted search procedure. The actual programming of the procedure outlined would involve a major piece of work. There was the realization at this point that *k*-means cluster analysis could do many of the things that were wanted and that it would be much easier to adapt existing *k*-means programs to the task of spatial analysis than to begin writing a new program from scratch.

The *k*-means procedure is a form of nonhierarchical divisive cluster analysis in which an attempt is made to minimize the intracluster variances while maximizing the intercluster distances. In archaeology, it had previously been used mainly for purposes of artifact classification (e.g. Hodson, 1971; Doran & Hodson, 1975; Johnson & Johnson, 1975). A basic description of the *k*-means procedure is given by Doran & Hodson (1975). We have found that, modified to compute some additional statistics, it provides a useful tool for examining spatial patterns (Kintigh & Ammerman, 1982). The method is divisive since it starts with a single group or cluster of points and then proceeds to subdivide that group to obtain more clusters. It is nonhierarchical in that points can be reassigned to different groups at different clustering stages. It differs in this respect from hierarchical agglomerative techniques such as single link and average link cluster analysis where reallocation is not possible. Another advantage of the *k*-means procedure is that it can readily handle large data sets: one of the distribution maps in the Pincevent study contains, for example, some 600 points.

The program accepts as data the *x* and *y* coordinates of a set of points representing the positions of sites or artefacts on a map. Its goal is to find (locally) optimal clustering configurations from one to some user-specified maximum number of clusters. The *k* cluster configuration, which is also referred to as the *k*th clustering stage, is simply a division of all of the points in the analysis into one of *k* different clusters. Each cluster in a configuration is defined by its centroid or centre of gravity, which is the point with the mean *x* and mean *y* values of all of the objects belonging to the cluster. The cluster radius as measured by its RMS, root mean squared deviation, is an index of the size or compactness of a cluster. The clustering criterion that the program is designed to minimize is the SSE or sum squared error. It is calculated as the sum over all points in the analysis of the squared Euclidean distance from each point to the centroid of the cluster to which it is assigned. Graphic display of the SSE values against clustering stage can be used to identify those configurations with better clustering. Probably the best way to understand how the procedure works is by looking at the examples of the analysis of settlement distributions such as the one showing clustered patterns in our original paper (Kintigh & Ammerman 1982, fig. 2).

The first application of the *k*-means approach to spatial analysis actually involved the Magdalenian site of Pincevent near Montereau in the Paris Basin. Some of the initial results of this study were presented in our 1982 paper. More recently, a fuller account of this work has been prepared by Simek & Larick (1983). It is worth commenting on the reasoning that went into the choice of Pincevent for the initial study. Both the excavations and the report on the site are exemplary (Leroi-Gourhan & Brézillon, 1972). Careful attention has been paid to the visual inspection of artifact distributions and a coherent interpretation of the occupation of the site has been developed. The aim of our study was not to 'test' the conclusions of the excavators but rather to see whether or not the *k*-means approach could 'replicate' what they, on the basis of extensive first-hand experience, had arrived at. In other words, the logic of evaluation would be reversed and directed toward the method itself. If the *k*-means approach could not reproduce an approximation of what they found, there would be grounds for doubting the usefulness of the method.

In all, ten classes of material from Pincevent (Section 36) were individually analyzed. Almost all of them showed, according to the SSE criterion, a good configuration at the level of three clusters (Simek & Larick, 1983, fig. 6). When the centroids for this level are plotted on the site plan, they are seen to occur at or near each of the three habitation structures recognized by the excavators. Each habitation area is centred on a rock-lined hearth which provides an outside source of contextual information for evaluating the analyses of artifact distributions. Other inflection points in the SSE curves suggest that it might be productive to examine the configurations at higher cluster levels as well. When this is done for six or nine clusters, there is a clearer sense of what is meant by the recognition of multiple levels in spatial patterning. As mentioned above, much of the spatial organization of the site would seem to be connected with the three habitation units. But on a more refined spatial scale, other patterns within the larger ones can be seen as well. For example, the centroids for stone tool classes tend to occur closer to the hearth features, whereas those for the bone classes are often found on one side or have more peripheral locations.

An attempt was also made in the Pincevent study to indicate the relative abundance of points in a distribution belonging to a given cluster. This was done by plotting at the centroid position the rank order, measured in terms of the number of items or points within a class of material, that is associated with each cluster in a configuration. In the case of a three-cluster configuration, each centroid would have, for example, a rank from 1 to 3, which would provide an index of its relative abundance. It is possible in this way to display the configurations for several different classes on the same map. When this is done for all ten classes of material at the three-cluster level, it is found that centroids in the same part of the site tend to have similar ranks. Much the same pattern (that is, centroids in proximity to one another having comparable values) is observed when such a display is made at the six-cluster level (Simek & Larick, 1983, fig. 11). Even with increasing spatial refinement, there appear to be recurring relationships between relative abundance and location at Pincevent. One explanation for the patterns would be that the disposal of materials after activities had been performed was the dominant factor in shaping the distributions of material at the site (Simek & Larick, 1983). This would correspond in some respects with the excavators' interpretation of fan shaped areas of deposition around each of the three structures (Leroi-Gourhan & Brézillon, 1972, pp. 247–50). It is also of interest to look at the site in a somewhat different way and focus on those clusters that seem to stand apart from the main pattern. This is done in Fig. 25.1, where attention is paid to what might be termed the residual pattern of clusters at Pincevent. Even at the nine-cluster level, there are relatively few clusters that stand by themselves. Interestingly, only four classes of material – backed bladelets, cores, endscrapers and metacarpals – are observed among such isolated clusters, which appear in most cases to be related to lithic reduction activities. While these clusters represent only a minor component of artifact distributions in terms of the overall volumes of material occurring at the site, their patterning may provide valuable information about how space was used at the site.

The other case study in which the *k*-means approach has been used is the site of Le Flageolet I near Bézenac in the Dordogne (Rigaud, 1969, 1982). Extensive work has been done on the Aurignacian levels at the site by Simek (1984). There is a marked contextual contrast between the open air site at Pincevent and the rock shelter at Le Flageolet. In fact, the large fallen limestone blocks that occur at the latter would appear to make the actual task of spatial analysis considerably more difficult. But the complex partitioning of the site space by the blocks may offer some unexpected advantages (Simek, 1984). Again one is fortunate in being able to examine a site where excavations are of high quality and most of the rock shelter has now been exposed. In the space available here, only a few examples from the Le Flageolet study can be presented.

In Fig. 25.2, the configuration at the 11-cluster level is shown for the Aurignacian pieces occurring in couche VIII–I at Le Flageolet. The large blocks that fell at a time prior to this occupation and that divide up the site space are also shown. The centroids of the clusters fall in the open areas between the blocks and there are more clusters in the northern half of the site than in the southern half. In Fig. 25.3, the centroids and RMS circles for nine artifact classes and also cobbles used for hearths are displayed at the two-cluster level, which represents a good clustering configuration for all of the classes. The site space can be seen to be organized in terms of two main areas. It is worth noting that the centroids in the northern area all occur in much the same location and seem to be associated (on one side) with the cobble centroid (indicated by a star). In the southern area, the RMS circles tend to be larger and the centroids are also more dispersed. When a higher level of clustering is displayed as in Fig. 25.4, this difference between the two areas is still seen. There is a fairly tight overlapping of the RMS circles in several places on the north side where discrete subdivisions of space can now be recognized. The patterning on the southern half again seems to be more diffuse with only a few of the circles (representing different artifact classes) overlapping at any point. The patterns seen at the more refined scale of Fig. 25.4 reflect more closely the actual topography of the occupation surface.

As part of the Le Flageolet study, Simek (1984) also looked at the relationships between location and relative abundance for different classes of material and found that a different situation obtained than the one observed at Pincevent. While at lower cluster levels there was the kind of homogeneity among accumulations seen at Pincevent, much greater heterogeneity was encountered at Le Flageolet when levels of greater spatial refinement were examined. That is, at a given location the relative proportion of pieces belonging to one class of material differed from that of another class, when higher level configurations are displayed as in Fig. 25.4. The method used by Simek (1984) for studying this question at Le Flageolet and Pincevent was based upon star diagrams of the kind used in geology to show relative proportions. Simek (1984) suggests that the differences between the two sites may have to do with somewhat different practices with regard to refuse disposal which are in turn linked with the topographies of the sites. There are few constraints on movement throughout the site space at Pincevent and ease of movement permits greater organization of the disposal of material. At Le Flageolet, the large blocks put greater limitations on movement and, at the same time, the narrow spaces between them provide convenient places for disposing of refuse. If activities are performed on the surfaces of the blocks, as is likely, and specific activities take place in somewhat different locations, then one would expect the material swept or dropped from the various blocks to be more heterogeneous (among artifact classes) between locations. There is clearly some irony here: the constrictive topography of Le Flageolet may account for a closer mapping between where things are worked or done and where they are abandoned or disposed. If this interpretation is correct, it has wide implications for Palaeolithic studies.

Some final words should be said about current efforts with regard to the development and implementation of the *k*-means approach. Three separate areas will be mentioned here. The first concerns trying to find better ways of dealing with what might best be called the 'matching of maps'. Some of the most interesting patterns or results are those that arise when maps of different classes of objects are displayed at the same time. We have seen this in both of the case studies described above. What is needed are better ways – both more objective and also, if possible, computer assisted – of defining artifact 'zones' (Simek, 1984) and of measuring the degree of concordance between pairs or multiples of maps. In the latter case, use can perhaps be made of the centroid positions themselves and their associated RMS radii, although some correction for differences in the total number of points on respective maps may be

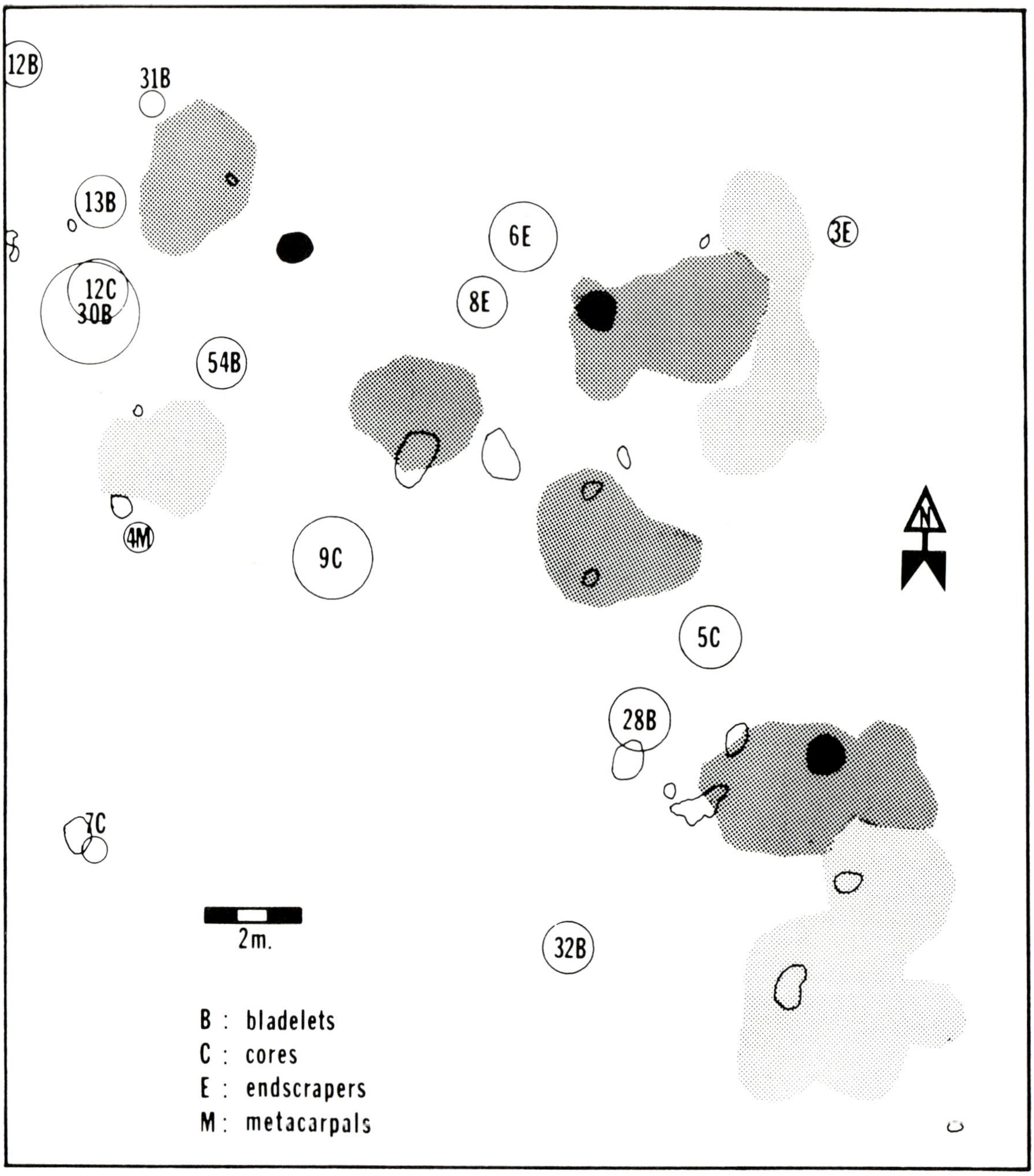

Fig. 25.1. The nine-cluster configuration for ten classes of material at Pincevent. The three rock-lined hearths are shown in black. The dark shading represents areas where the clusters of many different classes densely overlap. The lighter shading indicates areas in which there is a less dense overlapping of clusters. Those clusters which stand apart in isolation are shown individually. The numbers refer to the number of points belonging to a given cluster.

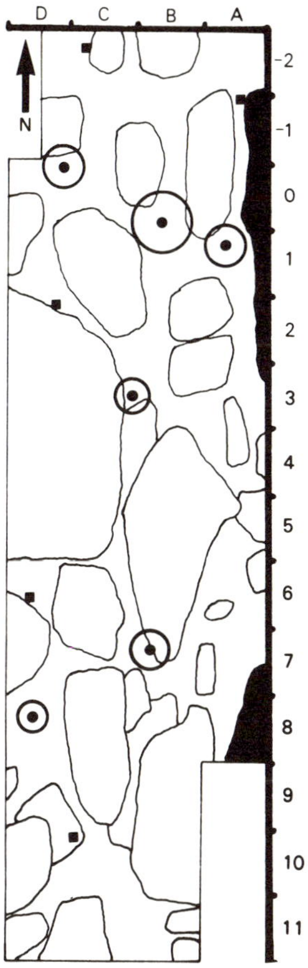

Fig. 25.2. The 11-cluster configuration for Aurignacian pieces in couche VIII–1 at Le Flageolet.

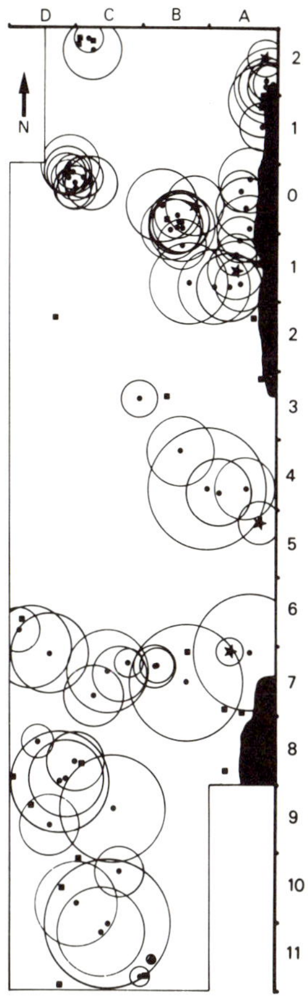

Fig. 25.4. High level clustering configuration for nine artifact classes and cobbles in couche VIII–1 at Le Flageolet. Note that different clustering levels, selected on the basis of the inflection points in the respective SSE curves, are used for different classes of material here.

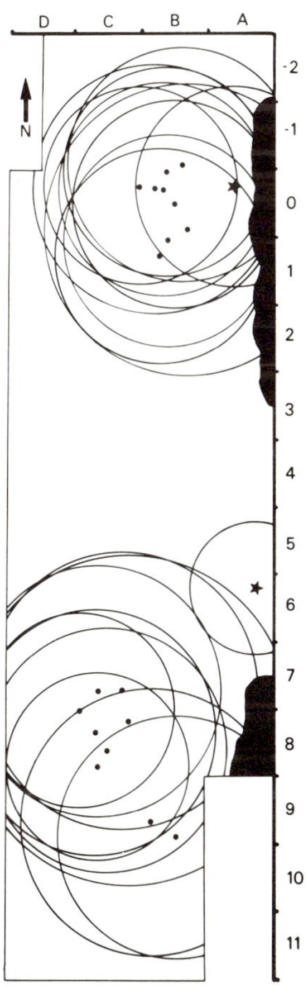

Fig. 25.3. The two-cluster configuration for nine artifact classes and cobbles (centroids indicated by stars) in couche VIII–1 at Le Flageolet.

required. In addition to an overall measure of concordance between maps, it may be of interest to characterize the form that 'matching' takes: that is, how the RMS circles overlap one another. For instance, in some cases they are centred on one another, while in others they form chains.

A second area involves trying to develop ways of linking the information obtained from refitting studies (e.g. Cahen *et al.*, 1980; see also his and other papers in this Conference volume) with the patterns recognized by the *k*-means approach. One possibility would be to treat what is known from refitting as part of the contextual information available for a site. A more analytic treatment would actually be to count the 'joins' within and between the respective clusters in a given *k*-means configuration. The idea would be to develop, in effect, a refitting index which would be printed as part of the summary statistics of a *k*-means run.

The third area concerns the overall comparison between sites or different layers at the same site. For example, there are at Le Flageolet several Perigordian levels in addition to the Aurignacian layers. It is of obvious interest to see whether or not the same kinds of patterning are expressed in the various layers and also the ways in which individual layers differ from one another. Such comparisons are admittedly ambitious undertakings. It would be helpful if some of the more routine comparative work could be computer assisted. Human analysts are likely to remain for some time, however, those best suited for making headway on such a complex problem. Even if such attempts are not all that successful, they may make a substantial contribution in terms of the development of heuristic strategies for solving spatial problems.

Bibliography

Cahen, D., Karlin, C., Keeley, L. H. & Van Noten, F. (1980). 'Mèthodes d'analyse technique, spatiale et fonctionnelle d'ensembles lithiques'. *Helinium,* **20**, 209–59.

Cliff, A. D. & Ord, J. K. (1975). 'Model building and the analysis of spatial pattern in human geography'. *Journal of the Royal Statistical Society (Series B)*, **37**, 297–348.

Doran, J. E. & Hodson, F. R. (1975). *Mathematics and Computers in Archaeology*. Harvard University Press: Cambridge, Mass.

Hodder, I. & Orton, C. (1976). *Spatial Analysis in Archaeology*. Cambridge University Press.

Hodson, F. R. (1971). 'Numerical typology and prehistoric archaeology'. In: *Mathematics in the Archaeological and Historical Sciences*, edited by F. R. Hodson, D. G. Kendall & P. Tautu. Edinburgh University Press: Edinburgh.

Johnson, A. E. & Johnson, A. S. (1975). '*k*-means and temporal variability in Kansas City Hopewell ceramics'. *American Antiquity,* **40**, 283–95.

Kintigh, K. W. & Ammerman, A. J. (1982). 'Heuristic approaches to spatial analysis in archaeology'. *American Antiquity,* **47**, 31–63.

Leroi-Gourhan, A. & Brézillon, M. (1972). 'Fouilles de Pincevent: essai d'analyse d'un habitat Magdalénien (la section 36)'. *Supplement à Gallia Préhistoire No. 7.* CNRS: Paris.

Rigaud, J. P. (1969). 'Note préliminaire de la stratigraphie du gisement du "Flageolet I" (commune de Bézenac, Dordogne)'. *Bulletin de la Société Préhistorique Française,* **66**, 73–5.

Rigaud, J. P. (1982). *Le Paleolithique en Périgord: Les Données du Sud-Ouest Sarladais et Leurs Implications*. These de Doctorat d'État es Sciences. Université de Bordeaux I: Bordeaux.

Simek, J. F. (1984). A *k*-means approach to the analysis of spatial structure in Upper Palaeolithic habitation sites: Le Flageolet I and Pincevent Section 36. British Archaeological Reports International Series S205. Oxford: BAR.

Simek, J. F. & Larick, R. R. (1983). 'The recognition of multiple spatial patterns: a case study from the French Upper Paleolithic'. *Journal of Archaeological Science,* **10**, 165–80.

Simon, H. A. (1969). *The Sciences of the Artificial*. MIT Press: Cambridge, Mass.

Simon, H. A. (1980). 'Cognitive science: the newest science of the artificial'. *Cognitive Science,* **4**, 33–46.

26

Circulation of Solutrean foliate points within the Perigord, SW France

ROY R. LARICK

Abstract

Region-wide circulation of artifact cherts, from geologic sources to habitation sites, is summarized for the late Solutrean occupation of the Perigord region of southwest France. The paper reports the spatial context for the production of foliate points; a method by which the regional movements of these items, from production to discard, may be identified; the evidence for such movements within and around the Perigord; and the possible socioeconomic significance for the pattern of movements. It is shown that foliate points travelled considerable distances, and that those made in one area were likely to have been transported to almost any other area within, but not beyond, the geographic limits of the Perigord. On this region-wide scale, foliate points must be seen to have circulated from production sites to their sites of discard, while the Solutreans, themselves, must have been quite mobile as well.

Introduction: the mobility of Paleolithic populations

One of the more interesting archaeological questions of the day is the significance of evidence for regional level social relationships during the European Upper Paleolithic. At one extreme, certain aspects of, say, the Aurignacian Industry or polychrome painting, have truly widespread material effects. The patterns suggest a very broad level of cultural relationships that stretched across numerous environmental and geographic zones and that, as such, are difficult to interpret systemically. On the other hand, more localized traditions for decorated and other complexly treated implements suggest the transmission of information that may have aided groups in maintaining social and economic relationships across regional space. Several archaeologists have worked on this aspect of archaeological reconstructions to conclude that regional relationships among prehistoric hunter–gatherers were indeed complex, and that explanations can be sought neither among the adaptations of contemporary groups nor among current techno–environmental hypotheses (Conkey, 1984; Cross, 1983; Denbow, 1984; Jones, 1983; Larick 1983*a*; Parkington, 1984; Wobst, 1978, 1983).

Among the portable evidence of the Paleolithic, Solutrean foliate points stand out as the chipped-chert implements whose symbolic functions may have surpassed the utilitarian. Foliate points are among the most complex of Paleolithic artifacts, exhibiting a diversity in production techniques, finishing retouch, shape and surface form, and raw materials, when they are viewed across the archaeological sites of southwest France. With this diversity in mind, Smith suggested the possibilities for the symbolic functions of these items (1965, 1966, 1973). Nevertheless, recent evaluation finds limitations in reconstructing the flow of information within Paleolithic society using this kind of artifact (Larick, 1983*a*, 1983*b*).

The question that faces any hypothesis for the symbolic functions of portable artifacts, but especially tacitly utilitarian implements, is where (in a regionally spatial sense) would the transmission of information have taken place in relation to where the implements lie archaeologically? In other words, theoretical indications of relationships among the regional contexts for the production, use, and discard of symbolic artifacts must be sought. Wobst, one of the first to suggest the information hypothesis (1977), defines the specific 'social context of use', within which objects with complex forms might be expected to pass information. This context is based on the social relationships that obtain between users of the objects – the amount of 'social distance' that separates users (1977, p. 324). However, neither Wobst, nor any subsequent writer, has specified if this concept of social distance, or the relationships within which properly distant actors participate, has any spatial referent that can be employed in archaeological reconstructions.

Foliate points and their makers were part of a society that was in all probability quite mobile, especially when viewed with the resolution afforded by Pleistocene evidence. Utilitarian tools may have had functional histories that spanned multiple uses at numerous places within a region, making their contexts of discard wholly unrelated, spatially, to their contexts of use. Finally, the horizon of time that may be resolved with Pleistocene evidence is of paramount importance for the functional hypothesis. Even if, theoretically, a locational context of symbolic use can be specified, we cannot assume that its depositional form would have remained spatially stable for the amount of time that constitutes the briefest resolvable stratigraphic horizon of Pleistocene time. Thus the discard of exhausted materials was a spatially extensive process that structured the distribution of artifacts on a regional scale. Before proceeding with functional interpretations of the 'style zones' for utilitarian implements, we must determine the significance of the context of discard which, only barring post-depositional modifications, forms the archaeological context.

Solutrean foliate points and their raw materials

Fragments of foliate points typically comprise 10–50% of all retouched tools within Solutrean archaeological assemblages (Fig. 26.1). Four recognized types fall readily into two classes of production technology:

those usually made on flakes and thinned with invasive percussion retouch (laurel leaves and unifacial points);[1]

those made on blades and finished primarily through pressure retouch (shouldered points and willow leaves).

Few exceptions to these rules exist. Nevertheless, a small number of laurel leaves may have been finished with pressure retouch (but this was not at all common and such determinations are debated) and laurel leaves were, very infrequently, made on blades.

Each class has its raw material prerequisites. The flake points required massive chert rocks to produce the appropriate flake blank. For laurel leaves, the chert had to be durable enough to endure the repeated shocks of Solutrean bifacial thinning. The Solutrean technique struck down directly upon the edge of the biface to remove the flakes that left the typically thin, wide, and flat invasive scars on the surfaces of these delicate tools. Only a few chert types available in the Perigord met these requirements and, among them, Senonian lepispheric nodules and Cenozoic chalcedonies stand out as the prime materials for laurel leaves.

Senonian nodules are the most widely available of all cherts within the Perigord. Nodules vary greatly in size but are frequently large and massive. Although they usually possess inclusions, faults, cavities, fossils, and other imperfections, their ubiquity compensated for their flaws – a nodule or flake proving to be mechanically improper could easily be rejected for another. This reason alone may explain why laurel leaves were made most frequently from Senonian cherts. Cenozoic chalcedonies were the next most frequently used materials, offering extreme strength as well as large raw material packages. No other Perigord chert provided an edge as tough as that of a large chalcedony flake against the heavy blows of Solutrean thinning techniques.

Blade points had different production requirements and, on the whole, were made from different raw materials. High quality blade blanks were the primary consideration. For the majority of these points, once the correct blade was struck, it was only a matter of completing details of its form with pressure retouch. Blades for shouldered points varied little in size and shape; they were short (usually less than 120 mm long), narrow (less than 20 mm wide), thin (less than 3 mm thick) and, above all, flat along their length. Willow leaves required long blades as their only outstanding technological prerequisite.

Preferred cherts for blade points lay immediately under the cortex of several Upper Cretaceous nodular cherts. Maastrichtian nodules were particularly sought, their large size, homogeneous texture, great resiliency, and general flawlessness being a blademaker's dream. As aesthetically pleasing attributes seem to have been more important for blade points, many brightly coloured and fine-textured rare Senonian cherts and Liassic jaspers were used. A price was paid for these materials as they are often too small and irregular to provide the proper blank. Few chalcedonies were used as their fibrous structure prevented both the striking of blades and secondary retouch with pressure techniques.

It was not necessarily completed foliate points that circulated throughout the Perigord. As direct evidence for movement pertains only to the cherts upon which the points were made, there is little indication of where or when the final form of the points was achieved during the movements of cherts. Unfortunately, this is a question that must be answered with debitage; material that has been most poorly treated in all but the most recent excavations of Solutrean sites. Nevertheless, a fair estimate of the spatial trajectory of completion may be made for both types of point. The sequence may have been slightly different for each.

Laurel leaves would have been at least roughed out while still close to the quarry. In that they require a large blank to produce a relatively small tool, it would have been impractical to carry blanks very far, at least before the initial shaping. Moreover, the relatively massive blanks would tend to hide imperfections that might later prevent completion. A better strategy might be to all but complete the most critical final thinning before transporting the implement very far. However, this last stage is the most critical. Breaks will most likely occur as the biface attains its thinnest cross-section. Thus the most sure strategy finished the laurel leaf where it had begun. The spatial consequences of this are obvious – laurel leaves are likely to have travelled in their completed form.

For blade points, the critical act of production comes when the blade is struck for a blank. It is likely that core preparation and blade striking took place near the quarry, as this is the procedure reconstructed for other Upper Paleolithic industries in the region (Bordes, 1967). Depending upon the degree of elaboration desired in a shouldered point, secondary retouch was either summary: to regularize the distal point and to create the proximal tang; or complex: to cover the entire surface with colateral parallel, and sometimes obliquely effected, pressure retouch. This step, however, might well have occurred at

[1] Unifacial points are not considered in this study. Their very infrequent appearance in well-provenienced late Solutrean archaeological levels suggests that their presence in poorly provenienced collections of similar age frequently results from the mixing of archaeological levels, either through natural processes or during excavation.

any time and distance after the striking of the blank. The mass removed is inconsiderable and the operation places no undue stress on the blank.

Finally, there is some evidence that foliate points were heat-treated before the finishing retouch was applied (Bordes, 1969; Collins, 1973; Robins *et al.*, 1978). While it appears that some foliate points were so treated, evidence for the extent of use of this procedure is lacking. Two extremes may be hypothesized. On one hand, the nine points that are sure candidates for treatment may have been treated fortuitously – the effects of heat on chert having been known but not employed systematically. Current evidence may indicate no more than this effect. On the other hand, that any evidence exists suggests a sophisticated theoretical and practical knowledge of heat and its effects on chert, on the part of Solutrean flintknappers. Indeed, foliate point technology was complex and sophisticated and might well be investigated in terms of a craft specialization (Cross, 1983; Larick, 1983*b*).

Systematic heat treatment would have added another intermediate step between quarrying the chert and completing the point. Nevertheless, in light of the prerequisites of economy that apply to the mechanical processes of production, foliate points probably were thermally treated close to the quarry. Thus it may be argued that, for both types of points, all but their final form was completed early in their spatial trajectories and that, for laurel leaves, completion was likely to have been very close to the site of production. While the site of completion for blade points can only be speculated, it is assumed that it too occurred early in the object's history. Consequently, although it is the movement of cherts that is measured in this study, I will speak of the circulation of the foliate points themselves.

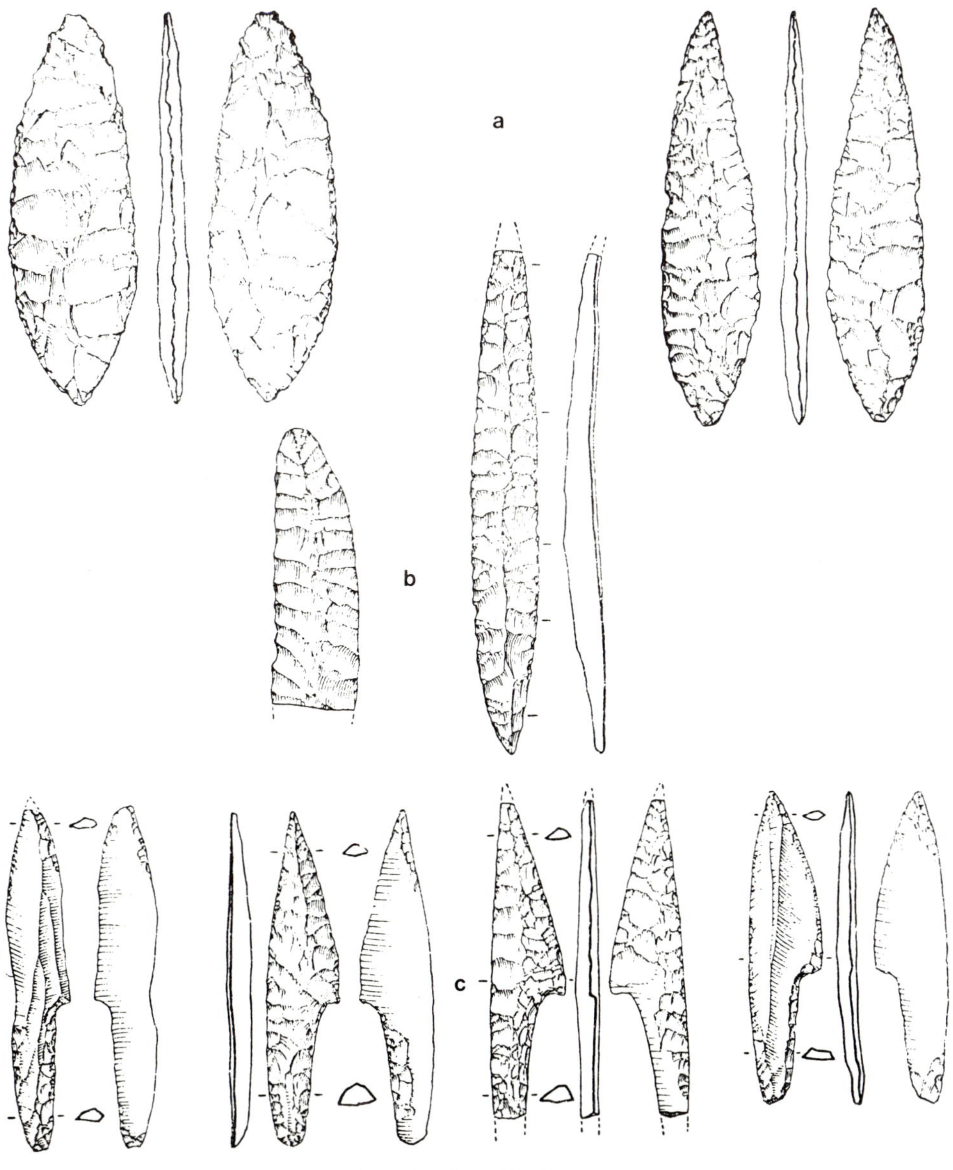

Fig. 26.1. Late Solutrean foliate points (after Smith, 1966): (*a*) laurel leaves (Jean-Blancs: collection Daleau); (*b*) willow leaves (Fourneau-du-Diable: Solutreen superieur III); (*c*) shouldered points (Fourneau-du-Diable: Solutreen superieur III).

The regional context for circulation

Like any other natural resource, foliate point cherts have a limited number of sources distributed over the earth's surface. Theoretically, any such chert will have a geological source that lies a measurable distance from the place where it was discarded as an exhausted foliate point. As such, a gauge for the circulation of foliate points is sought in their raw materials. A minimum straight line distance and direction of movement is established by linking a discarded foliate point chert with its potential geographic range of sources within and around the Perigord. This measure must be considered the absolute minimum pattern for the movement of each attributed point. Such correlations are possible for the five types of chert upon which foliate points were most frequently made. As the quantitative method for defining the range of sources has been outlined (Larick, 1986) and described in detail (Larick, 1983*a*), it will be summarily described in this presentation.

To illustrate the operation of the method, the geological processes that have structured the geographic distribution of chert-bearing carbonates across the Aquitaine Basin (Fig. 26.2) must be summarized. The frequent reappearance of a shallow sea and its associated lagoons and lakes, within this sedimentary basin, set the stage for the deposition of limestones and the development of Pleistocene topography within and around the Perigord. One major result was that from the Jurassic period of early Mesozoic sea transgressions, to just after the final Cenozoic regression, the cherts used for foliate points precipitated within several different environments. Each environment determined, somewhat differently, the size, shape, and crystalline form of the silica rocks – the attributes of physical structure that were most important to foliate point technology.

Cherts from the two eras have recognizably different structural features. Mesozoic cherts are marine in origin and have a nodular morphology and a microquartz crystalline structure. The aquatic Cenozoic cherts have, mostly, a bedded morphology and a generally chalcedoneous structure. Moreover, the structural class of each era can be subdivided into more specific types, with three types having two varieties each. All types separate according to their position within the lithostratigraphic sequence of each era and their names follow these distinctions. Thus one Cenozoic and four Mesozoic cherts were used for foliate points (from youngest to oldest):

Cenozoic (Eocene to Oligocene) chalcedoneous blocks;
Maastrichtian (Upper Cretaceous) fossiliferous (nodules and blocks);
Senonian (Upper Cretaceous) lepispheric nodules (blonde and black);
Ligerian (Turonian) lepispheric nodules;
Liassic (Lower Jurassic) jasper plaquettes.

To sum up, Solutreans produced foliate points from basically five different kinds of cherts, whose sources are known to lie within or just adjacent to the Perigord. As these materials are recognized by several attributes of physical structure that varied throughout the precipitations of two geologic eras, their sources can usually be narrowed to specific levels within this sequence. Fortunately, this is only half the story for the identification of chert sources within the Perigord. Solutrean quarryers did not have to descend through several hundred metres of limestone to gather, for example, their Liassic jaspers. Through a major turn of geologic events, they needed only access to the eastern edge of the region for this particular material.

Late Mesozoic and early Cenozoic orographic uplift folded and ruptured the vertical sequence of chert-bearing beds to orient it somewhat horizontally within the Perigord. Beginning in the northeast, towards the Massif Central, folding and faulting progressed to the southwest, leaving newly formed edges of the beds to outcrop obliquely at the surface (Fig. 26.3). This process transformed the lithostratigraphic sequence to range geographically across the entire region. As one travels northeast, from the southwest corner of the Perigord (the area that felt the fewest effects of uplift), increasingly older formations appear at the surface.

The result is that Cenozoic deposits concentrate in the west and south, Senonian limestones sit in the centre of the region, and the Jurassic carbonates constitute a band that arcs from the north, to the east, and then to the southeast, along the flanks of the Massif Central. Thus lithostratigraphic source determinations have horizontal significance and a simple geographic pattern of resources emerges within the region. The centrally located Senonian limestones form the 'central resource zone' of the Perigord. All of the region's rock shelters have weathered into these deposits, and Senonian nodules, the most abundant type of Perigord chert, can always be found within several hundred metres of these shelters. Consequently, it is no accident that all of the Paleolithic rock shelter habitation sites lie either well within or just towards the edge of this zone, that includes the traditional geographic subdivisions of the White, Central, and Black Perigords.

The Cenozoic and Mesozoic Jurassic formations that encircle the central zone contain the Cenozoic chalcedonies, Maastrichtian cherts, and Liassic jaspers. Moreover, the Ligerian nodules originate in the earliest Upper Cretaceous (Turonian) deposits that lie to the south of the Perigord, in the Haut-Agennais. While these four cherts cover a variety of structural types, they share a technological potential and a locational specificity that warrants a single class. When they are of adequate size, they offer the greatest potential for producing the most desirable foliate points – they were highly valued for these implements. Locationally, their sources all lie beyond the central zone of resources and Paleolithic occupation. Although Solutreans occupied the rock shelters of the central zone, they valued the cherts of the encircling 'peripheral resource zone' for their foliate points.

As with any archaeological approach, methodological and empirical problems are to be found with this scheme. For example, no technique has yet been able to link

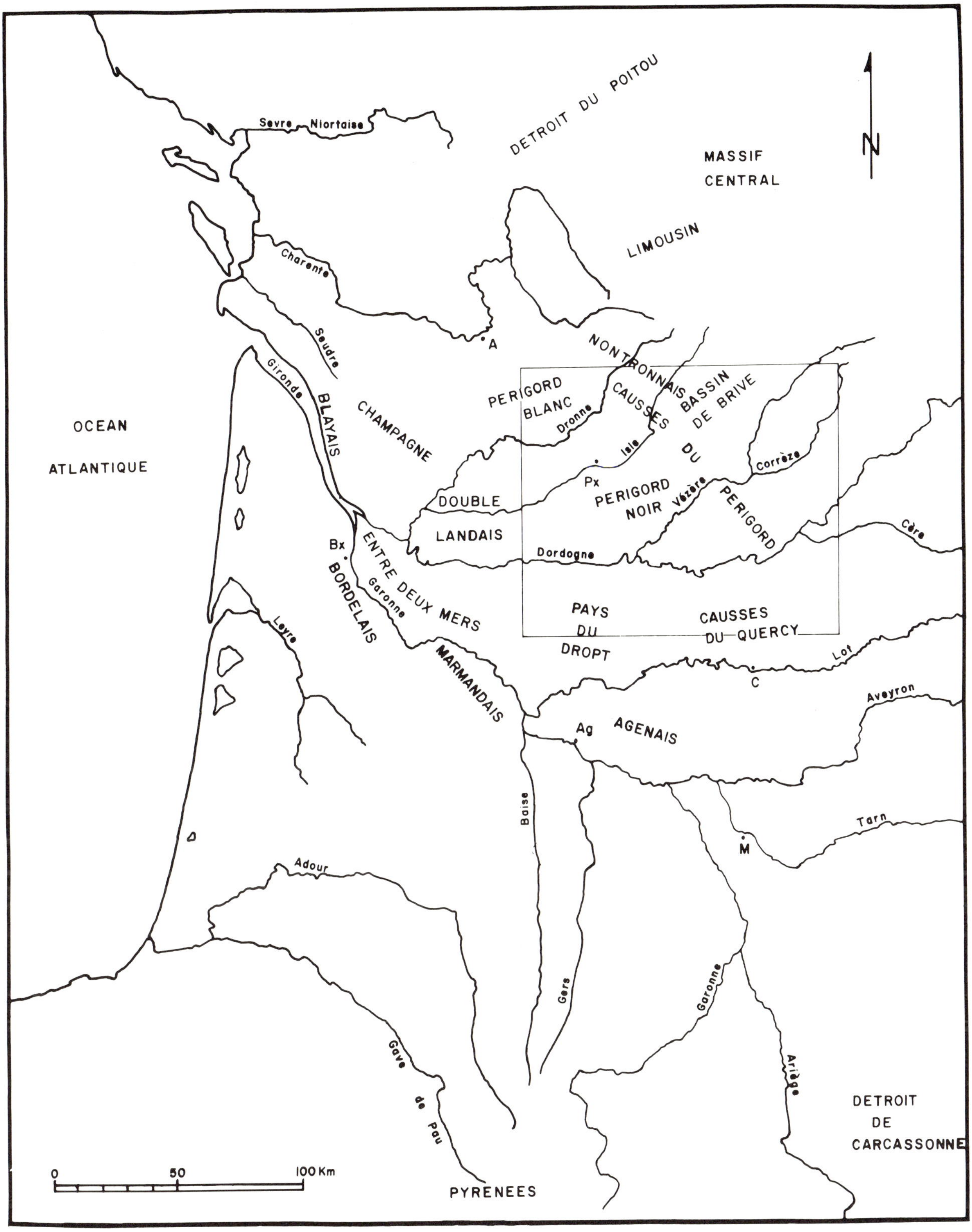

Fig. 26.2. Geographic provinces of the Aquitaine Basin and the Perigord (after Vigneax, 1975).

individual specimens to specific *in situ* sources. Whether the method is macroscopic or microscopic, chemical or physical, there remains a geographic range of sources to which any single artifact can be attributed. Moreover, archaeologists have yet to identify any specific sources within the Perigord that are known to have been quarried during the Paleolithic. It is the primary geologic sources, identified through contemporary earthmoving, that provide almost all current information for the availability of cherts in and around the Perigord.

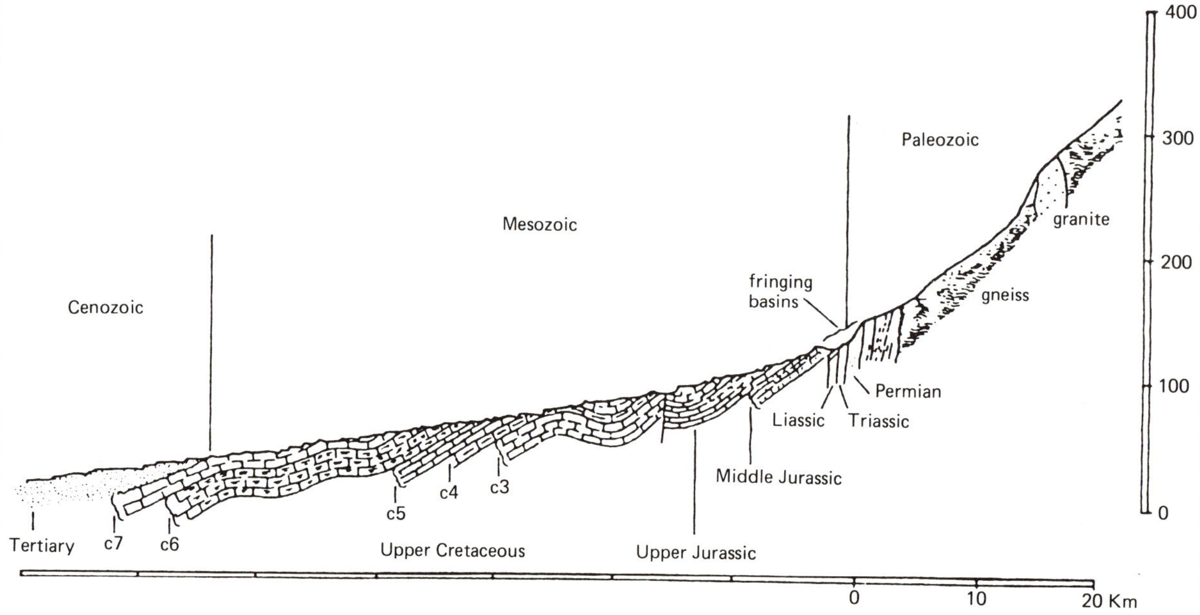

Fig. 26.3. Lithostratigraphic profile of the Perigord (NE–SW) (after Texier, 1979).

As such, these 'contemporary sources' provide the primary locational data with which the movement of chert implements may be gauged. Contemporary sources are specifically defined as places on the ground surface where good-quality chert rocks have been recovered in some abundance, usually as a fortuitous consequence of modern development. It is, however, misleading to link an artefact to any single contemporary source that has little specific archaeological significance. Paleolithic populations quarried cherts from secondary or derived geological deposits only at or near the surface, whose locations are likely to differ somewhat from the recognized contemporary sources (Larick, 1984*a*). Hence, we require a method to estimate the most probable geographic range of sources for any given artefact that has been made on a known type of chert; *k*-means cluster analysis provides the basis for such a estimate.

k-means analysis is able to define the number, size, and locations for the most dense geographic clusters of sources for each type of Perigord chert. These are termed the 'analytical sources' for the region. Figs. 26.4 and 26.5 summarize the distribution of analytical sources for central and peripheral chert types that have been sampled from approximately 750 contemporary sources throughout the Perigord.[2] *k*-means analysis provides spatial centroids for the clusters as well as a radius for the area that includes approximately 68% of the contemporary source locations held by each cluster.[3]

With this method, the distance from a site to a 'source' for a specific chert type is measured to the centroid of the analytical source for that type that lies closest to the site. In summary, *k*-means analysis confirms the hypothesized geographic organization of Perigord chert resources using only the data provided by the contemporary sources themselves. Most importantly, in the analytical sources, *k*-means analysis finds the strongest, most clear-cut geographic patterns for the availability of chert types within the complicated distribution of contemporary sources that constitutes the locational datum for gauging the movement of foliate points.

The context of discard for foliate points

Seven foliate point assemblages comprise the regional archaeological sample. All assemblages fit into the culture–

[2] The locations for most of the known contemporary sources for Perigord cherts are documented in the following publications: Bricker, 1975; Demars, 1982; Gaussen, 1980; Geneste, 1985; Larick, 1983*a*; Morala, 1980 and 1983; Rigaud, 1982; and Turq, 1984. Samples from most of these locations are maintained by the Direction des Antiquités Préhistoriques d'Aquitaine, Bordeaux.

[3] Ammerman *et al.* (this volume) summarize the applications of *k*-means analysis to spatial problems in archaeology. See Larick (1983*a*, 1986) for complete descriptions of cherts and contemporary sources used to define the analytical sources through *k*-means analysis.

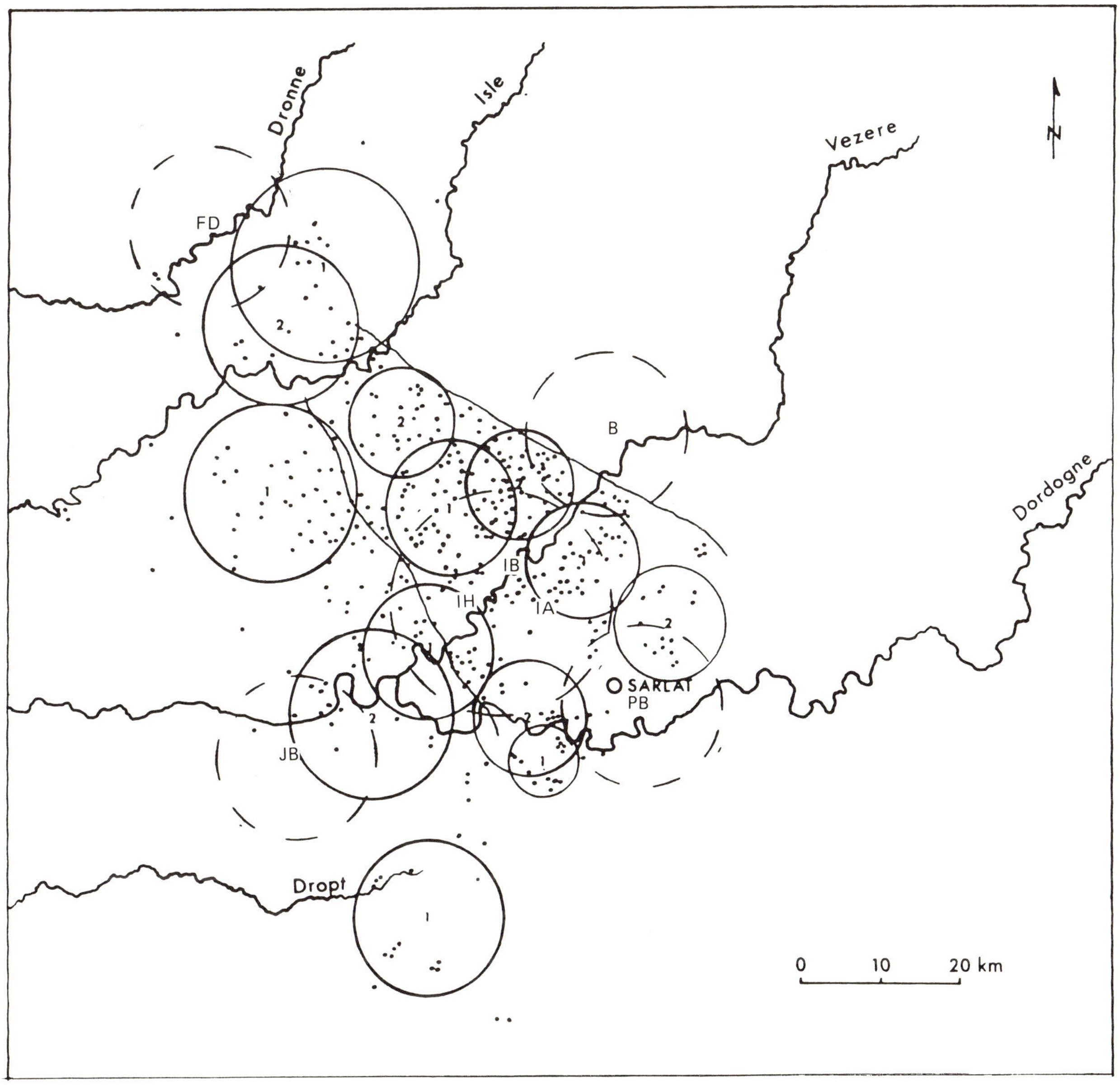

Fig. 26.4. Resource areas for central zone (Senonian) chert types: (1)=blonde Senonian nodules; (2)=black Senonian nodules. Interior sites: (LH) Laugerie-Haute; (LR) Le Ruth; (LA) Laussel. Exterior sites: (FD) Fourneau-du-Diable; (BA) Badegoule; (PB) Pech-de-la Boissière; (JB) Jean-Blancs. (L) geologic limit of peripheral resource zone.

stratigraphic categories of the Middle, Upper, and Final Solutreans defined by Smith (1966). Although minimizing the temporal range of the sample receives the highest priority for this type of study, foliate points are drawn from this typological range of assemblages under two considerations. First, limiting the population to any one of these categories, or to a smaller resolvable stratigraphic bracket, would not provide a sample of sufficient size or areal coverage. The main consideration is, however, theoretical. It is difficult to maintain Smith's distinctions in light of his and more recent evidence.

For example, the distinction between the Upper and Middle Solutreans involves, primarily, the presence of shouldered points in assemblages. Likewise, Final Solutrean assemblages separate from the Upper Solutrean primarily on the absence of backed bladelets. Very low quantities of these two types normally appear in Perigord sites, and fewer yet in the interior sites. Nevertheless, current excavations at Combe Saunière (Sarliac-sur-l'Isle, Dordogne), have located several late Solutrean archaeological levels that are rich in small fragments of both shouldered points and backed bladelets; that form, respectively, 19% and 40% of the slightly more than 200 retouched tools. These high proportions result not from the industrial predilections of the Solutrean inhabitants being so different from those at any other Solutrean site, but from fine-mesh waterscreening the entire deposit.

Indeed, had it been excavated with turn of the century methods, Combe Saunière would seem typical of Upper or Final Solutrean sites of the Perigord – a small, thin deposit that might contain a few more or less complete shouldered points and one or two small fragments of laurel leaves and

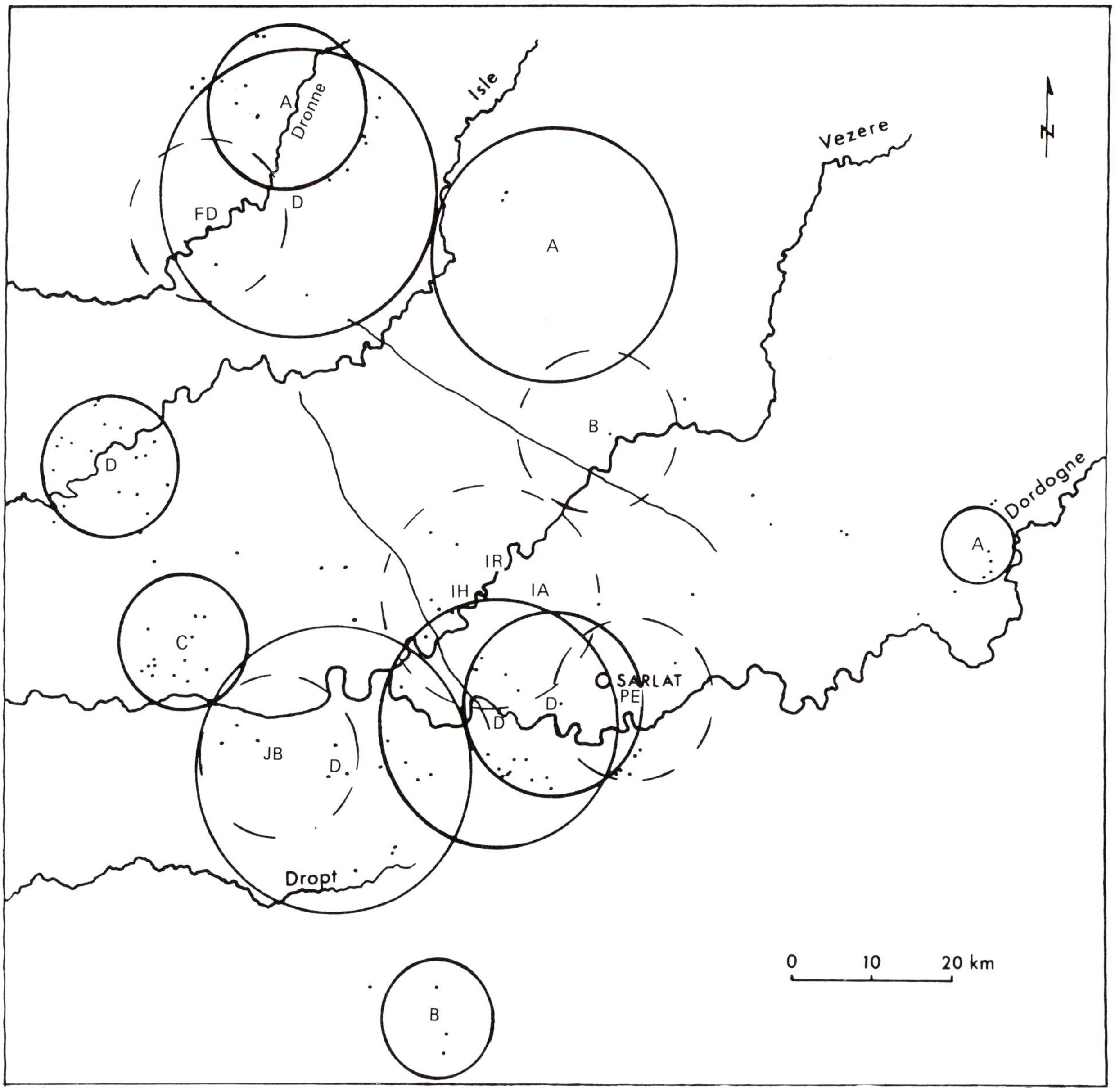

Fig. 26.5. Resource areas for peripheral zone chert types: (A)=Liassic jaspers; (B)=Ligerian nodules; (C)=Maastrichtian nodules; (D)=Cenozoic chalcedonies and argillites. Interior sites: (LH) Laugerie-Haute; (LR) Le Ruth; (LA) Laussel. Exterior sites: (FD) Fourneau-du-Diable; (BA) Badegoule; (PB) Pech-de-la Boissière; (JB) Jean-Blancs. (L) geologic limit of peripheral resource zone.

backed bladelets. In contrast, Laugerie-Haute, the site where the Middle and Final distinctions were defined, has never been excavated with the methods required to either provide an adequate areal sample of the deposit or recover small fragments of small artefacts (F. Bordes, personal communication).

In addition, new sedimentological and palynological evidence puts the Middle and Final Solutrean Levels of Laugerie-Haute within the time frame of the Upper Solutrean at other sites that have had similar analyses (Laville & Texier, 1972; Leroi-Gourhan, 1968) (Table 26.1). For this analysis, consequently, a Final Solutrean has not been distinguished from the Upper Solutrean and, at Laugerie-Haute, foliate points from Middle Solutrean collections have been combined with those attributed to the Upper Solutrean to form what I term the 'late' Solutrean.[4] All late Solutrean assemblages can be attributed to levels, areas, or entire sites, that were occupied during the final climatic subphase of Wurm III, followed by the 'Laugerie' Interstadial, and finishing after the first two subphases of Wurm IV, as they have been defined at Laugerie-Haute and other sites by Laville (1975) and Laville & Texier (1972). This period commenced around 20000 BP, and ended at approximately 17000 BP, giving a duration of somewhere between 1500 and 2500 years. While recognizing the difficulty of making such correlations on sedimentological evidence (see Hemingway, 1980, p. 56), this period does seem to bracket the late Solutrean occupation of southwest France. Moreover, it is acknowledged that the seven assem-

[4] See Larick (1983*a*) for a more thorough analysis of this problem.

Table 26.1. *Chronostratigraphy of the Late Solutrean in the Perigord*

Climatic Phases (After Laville, 1975)	Laugerie-Haute West (after Smith, 1966) Level	Culture	Laugerie-Haute East (after Smith, 1966) Level	Culture	Le Malpas (after Montet-White, 1973) Level	Culture	Abri Fritsch (Indre) (after Allain, 1979) Level	Culture
Würm IV Perigord III 'Dryas'			2 4–8	Magdalenian 3 Magdalenian 2			1–3	Magdalenian I (Badgegoulian)
Würm IV Perigord II ('Lascaux' Interstadial)			9–16	Magdalenian 1	1 (Upper)	Upper Solutrean	4–6	Magdalenian 0 (Badgegoulian)
Würm: IV Perigord I	0 1	Magdalenian 0 Final Solutrean	18–20 21	Magdalenian 0 Final Solutrean	1 (Lower)	Upper Solutrean	7	Upper Solutrean
Würm III/IV ('Laugerie' Interstadial)	2–3 5–7	Final Solutrean Upper Solutrean	22–24 25–28	Final Solutrean Upper Solutrean	2	Upper Solutrean	8–9	Upper Solutrean
Würm III Perigord XIV	8–10	Middle Solutrean	29	Middle Solutrean	3	Upper Solutrean	10	Upper Solutrean
Würm III Perigord XIII	11	Middle Solutrean	30	Middle Solutrean				
Würm III Perigord XII	12	Lower Solutrean	31	Lower Solutrean				

blages sampled from within about 2000 years allow few possibilities for contemporaneity among the represented occupations. While I wish to illustrate how these seven site locations participated in a region-wide system that circulated foliate points, the evidence must be reconstructed across a period that, while among the shorter technological sequences bracketed within the Upper Paleolithic, is rather long by absolute standards.

The sites have been chosen from all subregions of the Perigord to represent the known regional pattern of late Solutrean sites.[5] The seven lie both within the inner part and just at the edge of the central resource zone, the area that normally comes to mind when the geographic and cultural region of the Perigord is mentioned. Few Solutrean sites lie outside this central zone and few of these have yielded more than a few foliate points each – the significance of which must be weighed against their techniques of excavation. Nevertheless, the fewer than 8000 square kilometres that constitute the central zone hold about 80% of all French sites known to have been used during this brief technological manifestation.

Three 'interior' sites (Laugerie-Haute, Laussel, and Le Ruth), lie near to each other in the middle Vézère Valley. They represent the dense accumulation of Solutrean (and other Upper Paleolithic) occupations that are known from this area. White (1980, 1982, pp. 172–3) concludes that this pattern reflects an Upper Paleolithic preference for river valley settlements. If preference can fully explain this pattern, that is, if geological factors can be eliminated from having structured this distribution, then preference was clearly for the middle Vézère Valley, not just any valley. Indeed, Upper Paleolithic sites focus on this single area (White, 1982, figure 2). Nevertheless, geological factors in the middle Vézère Valley *are* perfect for preserving rock shelter occupations. Finally, all three sites lie within a few hundred metres of sources for high quality Senonian nodules.

The four 'exterior' sites (Badegoule, Pech-de-la Boissière, Jean-Blancs, and Fourneau-du-Diable) are found in geologically transitional areas between the two resource zones. They lie near to peripheral deposits, but within the Senonian limestones that normally weather to form rock shelters (Coniacian and Santonian).[6] Importantly, all four sites lie within a few kilometres of densely clustered sources for peripheral cherts. Nevertheless, as the exterior site shelters have formed in Senonian deposits, they, like those of the interior, have central zone chert sources very close at hand. To sum, the exterior sites are located in drainages that abut the edges of the Cretaceous landscape on almost all sides of the central zone. In addition, these sites lie in close proximity to geologically determined resources other than cherts. Nevertheless, while the peripheral chert resources that lie close to the exterior sites are varied and rich, most of the other geological resources of the outlying regions are not. The regions that border the Perigord have, traditionally, been thought less varied and rich, in terms of primary productivity.[7]

[5] See Larick (1983*a*) for a complete description of the seven Solutrean sites and the composition of their foliate point raw materials assemblages.

[6] The site of Badegoule sits at the base of a former low cliff face of Senonian limestone. It is impossible to determine whether this face contained a rock shelter during the Solutrean period of the late Pleistocene. Consequently, Badegoule can be attributed to neither a rock shelter nor an open-air occupation.

[7] For example, the rugged Corrèze to the northeast, the dry 'causses' to the

Table 26.2. *Central/peripheral chert ratios for resource fields and sites*

Site	Resource field	Site Assemblage
Interior sites		
Laugerie-Haute	10.0:1	2.0:1
Laussel	15.0:1	0.7:1
Le Ruth	18.0:1	3.2:1
Mean=	14.3:1	1.7:1
STD dev=	4.0	1.3
Exterior sites		
Badegoule	12.0:1	1.1:1
Pech-de-la Boissière	4.0:1	1.8:1
Jean-Blancs	2.6:1	1.5:1
Fourneau-du-Diable	2.7:1	0.6:1
Mean=	5.3:1	1.3:1
STD dev=	4.5	0.5

Archaeological patterns for circulation

Documentation of circulation proceeds with a global assessment of the scale of raw materials movements, followed by summaries of patterns found for the major types of chert. As space permits only a brief report, the presentation addresses only the scale of circulation and neither the patterns found at each site nor those for the range of point and chert types. The global analysis begins by estimating the availability of central zone cherts in relation to peripheral materials around each of the seven sites. Following the reservations expressed on the archaeological significance of contemporary sources, they are used only to compute the availability of chert types within specific areas and not to estimate the distances travelled by individual foliate points.

A circular 'resource field' is defined to lie around each site within a radius of 10 km which, while somewhat arbitrarily imposed, represents what is considered to be the maximum round-trip that can be made in eight to ten hours, within the central zone. Consequently, sources found within the resource fields are considered to have been acquired locally to the sites. The fields appear in Figs. 26.4 and 26.5 as broken circles around the sites to graphically emphasize the locations of sites in relation to localized distributions of contemporary and analytical sources.[8] Contemporary sources are tallied within each field for central and peripheral types so that the proportion of central to peripheral sources – termed the availability ratio – can be computed. The result of the operation appears in the left-hand column of Table 26.2.

A fairly clear distinction can be seen among these proportions according to the situation of the sites within the central zone: the three interior sites have similar ratios that average about 14 : 1, which is to be expected, as these sites lie within just several kilometres of each other. The small proportion of peripheral sources found within the interior site resource fields comprises scattered sources for low quality Cenozoic argillite plaquettes that sometimes sit within the unconsolidated surface deposits of the central zone plateaus. There is little evidence for any systematic use of these materials during the paleolithic, and no evidence at all for their use in making foliate points.

Exterior site availability (*c.* 5 : 1) evince the transitional nature of the exterior site locations. While central type sources still predominate, peripheral sources are three times more abundant within the exterior fields than they are within the interior fields. The exception of Badegoule results from its location, just off the inside edge of a peripheral Jurassic formation that contains very few sources for any siliceous materials. However, Badegoule does lie near a substantial source for Liassic jaspers, a very important material for the production of shouldered points. Consequently, when computed in this way, Badegoule's high availability ratio does not reflect the strategic importance of its location – near to a sparsely distributed but important resource. Thus the resemblance between the ratios for the interior sites and Badegoule, is meaningless. Badegoule is an exterior site with an interior assemblage, a fact reflected in the relative abundance of Liassic jaspers with its assemblage.

Next, an analogous ratio is computed for the foliate point assemblages of each site. Foliate points of central and peripheral chert types are tallied within each assemblage to produce an assemblage ratio. These values appear in the right-hand column of Table 26.2. Although a slight difference exists between the average of the ratios by site placement, it is probably not significant. One immediately notices the sharp contrast between the patterns for the two ratios across the two types of site placement. Whereas the availability ratio varies appreciably according to placement, the assemblage ratio is much more similar across all seven sites.

The significance of this pattern is obvious. Peripheral chert points were discarded about as frequently at interior sites as they were at exterior sites. The presence of peripheral cherts at interior sites, in quantities that almost equal those of the central cherts, shows that they were transported inwards from the peripheral zone. Thus a minimal trajectory for regional movements of foliate points can be established for the peripheral cherts. They began their movements within the peripheral ring of sources and probably passed through the ring of exterior sites, with some not ending their travels until having reached the interior sites. When discarded at interior sites, peripheral cherts had travelled about 30–50 km from their sources.

A dual character of the pattern of movements must be noted. On one hand, the central component of foliate point raw materials accumulated at each site. While peripheral cherts were valued, central-type Senonian cherts are about

east and southeast, the acidic Cenozoic plateaus to the southwest, and the sandy Landais and Double, to the west and northwest, are all much less well endowed, in terms of agricultural potential, than the central zone of the Perigord.

[8] The resource fields for the three interior sites have, on Figs. 26.4 and 26.5, been combined into one larger circle for visual simplicity. The outer limit of this field lies approximately 10 kilometres beyond the perimeter of the small triangle formed by the interior sites. Nevertheless, the availability of chert types around each of these sites has been estimated individually with the standard 10 km radius resource field.

equally represented in all assemblages. In that many Senonian materials have analytical sources within 5–20 km of any site, a large part of each assemblage might seem to have been made, used, and discarded locally. However, there is another aspect of these apparently local materials that suggests more mobility than can be shown at this level of analysis.

I have not been able to specify the sources for several Senonian nodular cherts whose knapping characteristics rival the better peripheral cherts. These were, as might be expected, the ones most commonly used for foliate points. Thus there is a good chance that many Senonian type foliate points circulated throughout the central zone. If this is the case, they could have had a range of movement that was not much less than that of the more visibly displaced peripheral cherts. Consequently, although the sources for Senonian cherts lie close to the sites where similar cherts were discarded, this does not necessarily indicate that these materials were procured locally.

On the other hand, the peripheral cherts provide clear evidence for truly region-wide movements of foliate points. Table 26.3 illustrates that some of them travelled from sources in one area of the periphery to exterior sites located on an opposite edge. For example, almost all assemblages contain significant amounts of four types that have outlying peripheral sources: Bergeracois nodules; Mussidan blocks (both with Maastrichtian sources in the extreme west and southwest), Liassic jaspers (sources in the extreme east and northeast), and Agennais chalcedonies (sources in the Cenozoic zone to the south of the Perigord). Arriving at an exterior site from an analytical source on the opposite edge of the periphery, these cherts would have travelled between 50 and 80 km. Clearly, late Solutrean populations circulated foliate points from all edges of the Perigord toward other peripheral edges, as well as to the centre itself.

While it is evident that circulation of foliate points can be traced throughout this region, it is possible that relationships existed between the Perigord and other regions with Solutrean sites. The more likely regional candidates include areas of the Charente, to the north, and of the Lot region, to the south. My only experience with foliate point cherts from outside the Perigord is with two sites located in the Charente: le Placard and Roc-de-Sers. In both cases, the range of materials used is absolutely different from the raw materials available within the Perigord, and from the cherts discarded in the Solutrean sites of the Perigord. Although foliate points from the Lot region have not been studied systematically, researchers familiar with the cherts of the area believe that foliate points from Lot sites are made on materials that can be expected from the sources known from that region.

My own research satisfies me that I have seen no more than about 20 of more than 2000 foliate points from Perigord sites whose raw materials could be termed beyond the range of variation known for Perigord cherts. Few researchers, however, would say that the range of variation is, in fact, known. Lastly, while I have not seen any Solutrean pieces from the Pyrenees, I suspect that the older attributions for some of the material are probably false, especially those that draw ties between foliate points from Pyrenees sites, and cherts from the Bergeracois or the Charente (Bahn, 1977, p. 251). Bergeracois cherts are frequently confused with other types that have somewhat similar patterns of color and texture, but whose crystalline structure is subtly different. In any event, these attributions should be verified through close scrutiny.

Table 26.3. *Chert frequencies at seven Solutrean sites*

Type	Site							Type total
	LH	LA	LR	BA	PB	JB	FD	
Senonian	165	25	42	17	95	187	203	734
Maastrichtian	12	21	2	1	28	27	157	248
Chalcedonies	47	6	9	6	21	45	99	233
Jaspers	14	1	2	8	9	12	59	105
Exotics	9	1	0	0	7	19	53	89
Site total	247	54	55	32	160	290	571	1409

Discussion: the socioeconomic context for circulation

A method for reconstructing the movements of late Solutrean foliate points has been presented. The method juxtaposes two twin-tiered locational patterns:

- the natural distribution of chert resources into central and peripheral zones;
- the cultural placement of occupation sites into interior and exterior locations.

The geographic pattern is a natural effect that can be explained in terms of geomorphological processes that have shaped this region over several hundred million years. However, the bimodal pattern of site placement becomes meaningful only when the composition of raw materials assemblages is viewed against the zonation of chert resources within the Perigord. Nevertheless, this pattern may be explained through a complex of relationships that include more than the locations of chert sources and inhabitable rock shelters, as well as factors of geologic preservation. This section attempts to outline a few of those relationships.

It is obvious, for example, that the region-wide circulation of cherts proves that Solutrean (and other Upper Paleolithic) populations used a region that was much larger than the one which contains their most visible refuse. In other words, the distribution of central zone rock shelter sites does not represent the extent of the Upper Paleolithic territory in the Perigord. Yet, to the contrary, the raw material base for the most common artefacts of this archaeological culture shows no signs of having been transported in from outside the areas that immediately surround the Perigord. Indeed, a scale of circulation has been identified to include the entire Perigord and its immediately neighbouring areas. Thus there is no evidence for linkages with distant regions such as the Pyrenees, the Massif Central, or the Paris Basin.

It is unfortunate that current knowledge of the entire peripheral region is limited. For example, recent archaeological syntheses are beginning to document the use of natural resources during the Paleolithic within relatively small areas of the central zone (e.g. Duchadeau-Kervazo, 1982; Rigaud, 1982; White, 1980). However, similar syntheses for the areas that border the Perigord address entire geographic regions (e.g. Debenath, 1974; Lenoir, 1983; Le Tensorer, 1979); they synthesize resource use at a much more general level.

It is absolutely necessary that the unique archaeological record of the central zone's rock shelters be considered before the social significance of the 'settlement' pattern is understood.

The apparent density of sites and diversity of assemblages within the interior of the central zone remains a problem. This pattern invites the application of a central place model – one that posits a centralized hierarchical interaction among specialized sites to account for location and diversity in assemblages. However, such a model would require that rather permanent settlements had organized the transport of foliate points across these relatively long distances. Moreover, central place models necessarily interpret the large interior sites to be base camps from which groups set out to procure resources from distant sources. It appears that the models for resource procurement, from a catchment to a base, do not account for these patterns. Hypotheses for centralized procurement must assume that the thick cultural deposits reflect large occupations of hunter–gatherers. Such an assumption disregards the depositional nature of these sites. Large Paleolithic deposits do not necessarily reflect large-scale occupations of long duration (Bordes, 1975; Larick, 1984*a*).

Chert sources and the central zone's rock shelters are fixed-base resources – two important factors of Paleolithic subsistence that did not move. Both types of features entertained brief visits from social groups throughout the Pleistocene. While sources had cherts taken from them, rock shelter deposits received these materials as refuse. Thus sites should hold a record of the visitors who discarded materials while moving from one resource to another. The pattern of sites should, therefore, represent temporary emplacements for a region-wide system of moving units. Viewed with the resolution that Pleistocene evidence from the Perigord permits, social units must be seen to have moved among sites and sources of all types. While the locations of their temporary camps were stationary, the hunger–gatherers themselves were not. Thus the refuse of the interior shelters holds the key to the distance and direction of foliate point movements throughout the Perigord. Simply, in moving from any set of resources on or near the periphery, to any other, either on an opposite edge or within the centre, groups were likely to pass by way of the interior cluster of sites. Such movements fostered the chance to abandon, at these sites, a few foliate points that had arrived from other areas.

It may be that the system of circulation for foliate points and the use of Perigord sites requires the theoretical inverse of a central place model. The focal points of the system may well have been within and around the exterior ring of sites and not in the dense debris of the middle Vézère Valley. For example, it has been suggested that foliate point technology was a specialized craft. The production of points may have been limited to a small group of individuals who controlled access to the appropriate knowledge, skills, and materials. Under such a system, the location of centres for specialized production should be at the exterior sites, where specialists would have had the most efficient physical access to the greatest possible range of raw materials within any one area. Confirmation for this system would find a great diversity of raw materials within the *débitage* of the exterior sites. However, the greatest diversity of completed foliate point cherts should continue to lie in the middle Vézère Valley, where discarded points accumulated during several centuries of circulation.

Conclusion

The static characteristics of a fixed-base resource can document the circulation of Paleolithic material technology across a Pleistocene landscape. It has been shown that foliate points travelled throughout the Perigord. Further, while points seem to have been unlimited by any boundaries within the region, there is little evidence that they traversed the outer edges of the peripheral resource zone. On one hand, few sites outside the periphery hold any kind of Solutrean implement and, on the other hand, few cherts in Perigord sites can be attached to sources that lie beyond the periphery. This result supports prior hypotheses that limit the extent of the Perigord 'culture zone' to within a region that is the approximate geographic equivalent of the combined resource zones (c.f. Smith, 1966).

An explanation of process for the movements of hunter–gatherers and their material culture must commence at a general level. Although it is clear that Perigord Solutreans moved regularly and circulated their points thoroughly, the social process of movement can be defined in only the most general terms. The cherts discarded at a site reflect relationships that occupants carried out with other parts of the region during a very brief period. It is evident that the residue left at a site arrived from other sites left behind. Further, some material aspects of the activities that had been current at that site would be visible only in sites occupied at a later date.

Yet, it must not be assumed that these results evince a complex network of simultaneously interacting settlements. To do so risks assuming an overly complex set of socioeconomic relationships within an overly stable social environment. It is obvious that regular, but regionally uncentralized, movements of Upper Paleolithic groups between sites could accumulate a diversity of discarded materials across an entire occupied region.

Bibliography

Allain, J. (1979). 'Les rapports de l'Homme et de son milieu d'apres les fouilles recentes de la vallée de la Creuse'. In: *Le Pleistocene supérieur en France. Illiéme* Colloque Franco-Sovietique. Editions du Laboratoire de Paléontologie Humaine et de la Préhistoire, Marseille.

Bahn, P. (1977). 'Seasonal migration in South-west France during the late glacial period'. *Journal of Archaeological Science,* **4**, 245–57.

Bordes, F. (1867). 'Considérations sur la typologie et les techniques dans le Paléolithique'. *Quartär,* **18**, 25–55.

Bordes, F. (1969). 'Traitment thermique du silex au Solutréen'. *Bulletin de la Société Préhistorique Française,* **66**, 197.

Bordes, F. (1975). 'Sur la notion de sol d'habitat en préhistoire paléolithique'. *Bulletin de la Société Préhistorique Française,* **72**, 139–44.

Bricker, H. M. (1975). 'The provenience of flint used for the manufacture of tools'. In: *Excavation of the Abri Pataud*, edited by H. L. Movius. *American School of Prehistoric Research Bulletin,* **30**, 194–201.

Collins, M. B. (1973). 'Observations on the thermal treatment of chert in the Solutrean of Laugerie-Haute, France'. *Proceedings of the Prehistoric Society,* **39**, 461–6.

Conkey, M. W. (1983). 'To find ourselves: art and social geography among prehistoric hunter–gatherers'. In: *Past and Present in Hunter Gatherer Studies*, edited by C. Schrire, pp. 258–76. Academic Press, Orlando.

Cross, J. R. (1983). 'Organizational options for non-subsistence production in the late Archaic'. Paper presented at *the 48th Annual Meeting of the Society for American Archaeology*, Pittsburgh.

Debenath, A. (1974). Recherches sur les terrains quatérnaires charentais et les industries qui leurs sont associées. Thèse de Doctorat d'Etat ès Sciences, l'Université de Bordeaux I.

Demars, P.-Y. (1982). 'L'Utilisation du silex au Paléolithique supérieur: choix, approvisionment, circulation: l'example du Bassin de Brive'. *Cahiers du Quatérnaire*, vol. 5. Editions du Centre National de la Recherche Scientifique, Paris.

Denbow, J. R. (1983). 'Prehistoric herders and foragers of the Kalahari: the evidence for 1500 years of interaction'. In: *Past and Present in Hunter Gatherer Studies*, edited by C. Schrire, pp. 175–94. Academic Press, Orlando.

Duchadeau-Kervazo, C. (1982). Recherches sur l'occupation paléolithique dans le bassin de la Dronne. Thèse de Troisième Cycle, l'Université de Bordeaux I.

Gaussen, J. (1980). 'Le Paléolithique supérieur de plein air en Périgord: secteur Mussidan-Saint Astier, moyenne vallée de l'Isle'. XIV supplément à *Gallia Préhistoire*. Editions du Centre National de la Recherche Scientifique, Paris.

Geneste, J.-M. (1985). 'Analyse lithique d'industries mousteriennes du Perigord: une approche technologique du comportment des groupes humains au Paleolithic Moyen'. Thèse de Docteur no. 2, l'Université de Bordeaux I.

Hemingway, M. F. (1980). 'The initial Magdalenian in France'. *BAR International Series 90*, 2 vols., Oxford.

Jones, R. (1983). 'Hunters and history: a case study from western Tasmania'. In: *Past and Present in Hunter Gatherer Studies*, edited by C. Schrire, pp. 27–66. Academic Press, Orlando.

Larick, R. R. (1983*a*). The circulation of Solutrean foliate point cherts: residential mobility in the Perigord. Unpublished PhD dissertation, State University of New York at Binghamton.

Larick, R. R. (1983*b*). 'Technological complexity and the use of land during the European Upper Paleolithic'. Paper presented at *the Third International Conference on Hunting and Gathering Societies*. Werner-Reimers Stiftung, Bad Homburg, West Germany.

Larick, R. R. (1984). 'Contemporary, Pleistocene, and analytical sources for Paleolithic artifact cherts in the Perigord, SW France'. Paper presented at *the 49th Annual Meeting of the Society for American Archaeology*, Portland.

Larick, R. R. (1986a). 'Perigord cherts: an analytical frame for investigating the movement of Palaeolithic hunter gatherers and their resources'. In *The Scientific Uses of Flint and Chert*, vol. I of the Proceedings of the Fourth Flint Symposium, edited by G. de G. Sieveking and M. B. Hart. Cambridge University Press.

Larick, R. R. (1986b). 'Paleolithic occupations of Perigord rock shelters viewed from the cherts found within Le Flageolet I, Couche V'. In: *Upper Paleolithic Settlement Patterns in Europe*, edited by H. Berke, J. Hahn, and J. Kind, pp. 29–38. Urgeschichtliche Materialhefte Bd 6. Archaeologica Venatoria, Tubingen.

Laville, H. (1975). 'Climatologie et chronologie du Paléolithique en Périgord: étude sédimentologique des dépôts en grottes et sous abris'. *Editions du Laboratoire de Paléontologie Humaine et de Préhistoire, Etudes Quaternaires*, Mémoire 4. L'Université de Provence.

Laville, H. & Texier, J.-P. (1972). 'De la fin du Würm III au debut du Würm IV: paléoclimatologie et implications chronostratigraphiques'. *Compte Rendu de l'Academie des Sciences de Paris,* **275(D)**, 329–32.

Lenoir, M. (1983). Le Paléolithique des basses valleés de la Dordogne et de la Garonne. Thèse de Doctorat d'Etat ès Sciences, l'Université de Bordeaux I.

Leroi-Gourhan, Arl. (1967). 'Le Badegoulien de l'Abri Fritsch: climat et chronologie'. *Bulletin de la Société Préhistorique Française,* **67**, 95–9.

Leroi-Gourhan, Arl. (1968). 'Dénomination des oscillations würmiennes'. *Bulletin de l'Association françaises de l'Etude du Quaternaire,* **4**, 281–8.

Le Tensorer, J.-M. (1979). Recherches sur le Quaternaire en Lot-et-Garonne: stratigraphie, paléoclimatologie et préhistoire paléolithique. Thèse de Doctorat d'Etat ès Sciences, l'Université de Bordeaux I.

Montet-White, A. (1973). 'Le Malpas rockshelter'. *University of Kansas Publications in Anthropology*, volume 4, Lawrence.

Morala, A. (1980). Observations sur le Périgordien, l'Aurignacien et leurs materières premières lithiques en Haut-Agennais. Mémoire du diplôme de l'Ecole des Hautes Etudes en Science Sociale, l'Université de Toulouse.

Morala, A. (1983). Précisions sur les gites des deux types du silex en Haut-Agennais'. *Bulletin de la Société Préhistorique Française,* **80**(6), 169.

Parkington, J. (1984). 'Soqua and Bushmen: hunters and robbers'. In: *Past and Present in Hunter Gatherer Studies*, edited by C. Schrire, pp. 151–74. Academic Press, Orlando.

Rigaud, J.-Ph. (1982). Le Paléolithique en Périgord: les données du sud-ouest Sarladais et leurs implications. Thèse de Doctorat d'Etat ès Sciences, l'Université de Bordeaux I.

Robins, G. V., Seeley, N. J., Symons, M. C. R. & McNeil, D. A. C. (1978). 'Identification of ancient heat treatment in flint artefacts by ESR spectroscopy'. *Nature,* **276**, 703–4.

Smith, P. E. L. (1965). 'Some Solutrean problems and suggestions for further research'. In: *Miscelena en Homenaje al Abate Henri Breuil*, volume 2, edited by E. Ripoll Perello, pp. 391–408.

Diputacion Provincial de Barcelona, Instituto de Prehistoria y Arqueologia.

Smith, P. E. L. (1966). 'Le Solutréen en France'. *Publications de l'Institute de Préhistoire de l'Université de Bordeaux,* Mémoire 5.

Smith, P. E. L. (1973). 'Some thoughts on variations among certain Solutrean artefacts'. In: 'Estudios Dedicatos al Professor Dr. Luis Pericot', edited by J. Maluquer de Motes. *Universidad de Barcelona, Instituto de Prehistoria y Arqueologia, Publications Eventuales,* **23**, 67–75.

Texier, J.-P. (1979). Recherches sur les formations supérficielles du Bassin de l'Isle. Thèse de Doctorat d'Etat ès Sciences, l'Université de Bordeaux I.

Turq, A. (1984). Observations sur les industries du Würm ancien du bassin hydrographique de la Lemance. Thèse de Troisième Cycle, l'Université de Bordeaux I.

Valensi, L. (1960). 'De l'origine des silex protomagdaléniens de l'Abri Pataud, Les Eyzies (Dordogne)'. *Bulletin de la Société Préhistorique Française,* **57**, 80–4.

Vigneaux, M. (1975). *Aquitaine Occidentale.* Masson, Paris.

White, R. K. (1980). The Upper Paleolithic of the Périgord: a topographic approach to subsistence and settlement. Unpublished PhD dissertation, University of Toronto.

White, R. K. (1982). 'Rethinking the Middle/Upper Paleolithic transition'. *Current Anthropology,* **23**, 169–92.

Wobst, H. M. (1977). 'Stylistic behavior and information exchange'. *University of Michigan Museum of Anthropology Anthropological Papers,* **61**, 317–42.

Wobst, H. M. (1978). 'The archaeo-ethnology of hunter–gatherers or the tyranny of the ethnographic record in archaeology'. *American Antiquity,* **43**, 303–9.

Wobst, H. M. (1983). 'Paleolithic archaeology – some problems with form, space, and time'. In: *Hunter–Gatherer Economies,* edited by G. N. Bailey, pp. 220–5. Cambridge University Press.

27

Recent work on sources of Italian flint

L. H. BARFIELD

While the typological study of prehistoric flint assemblages in Italy has been developing towards an advanced state of mathematical systematisation in recent years, few people have taken up research into sourcing Italian flint. In this article the author surveys recent developments in this field in Italy, including two projects connected with his own research.

Flint occurs in many of the widespread Jurassic, Cretaceous and Tertiary formations throughout Italy and Sicily (Squyres, 1975) (Fig. 27.1). Much of this flint, however, was not suitable for the manufacture of artifacts, so that the study of flint distribution based on geological publications alone does not provide an accurate picture of potential sources of supply available to prehistoric man.

Other types of rock were also used for flaked tool production, usually in areas where suitable flint was not available. For example, jasper was employed in parts of northern Tuscany, phtanite in central Emilia and quartz in the central and western Alps (Barfield, 1981; Cremaschi, 1981).

Lack of suitable rocks also led to the development of procurement of flint by direct access or exchange, over considerable distances, while more exotic rocks, such as obsidian and quartz, were traded into areas where local sources of flint were available (Williams Thorpe *et al.*, 1979; Barfield, 1981).

The identification of flint sources in Italy is greatly assisted by the fact that the colour, patterning and texture

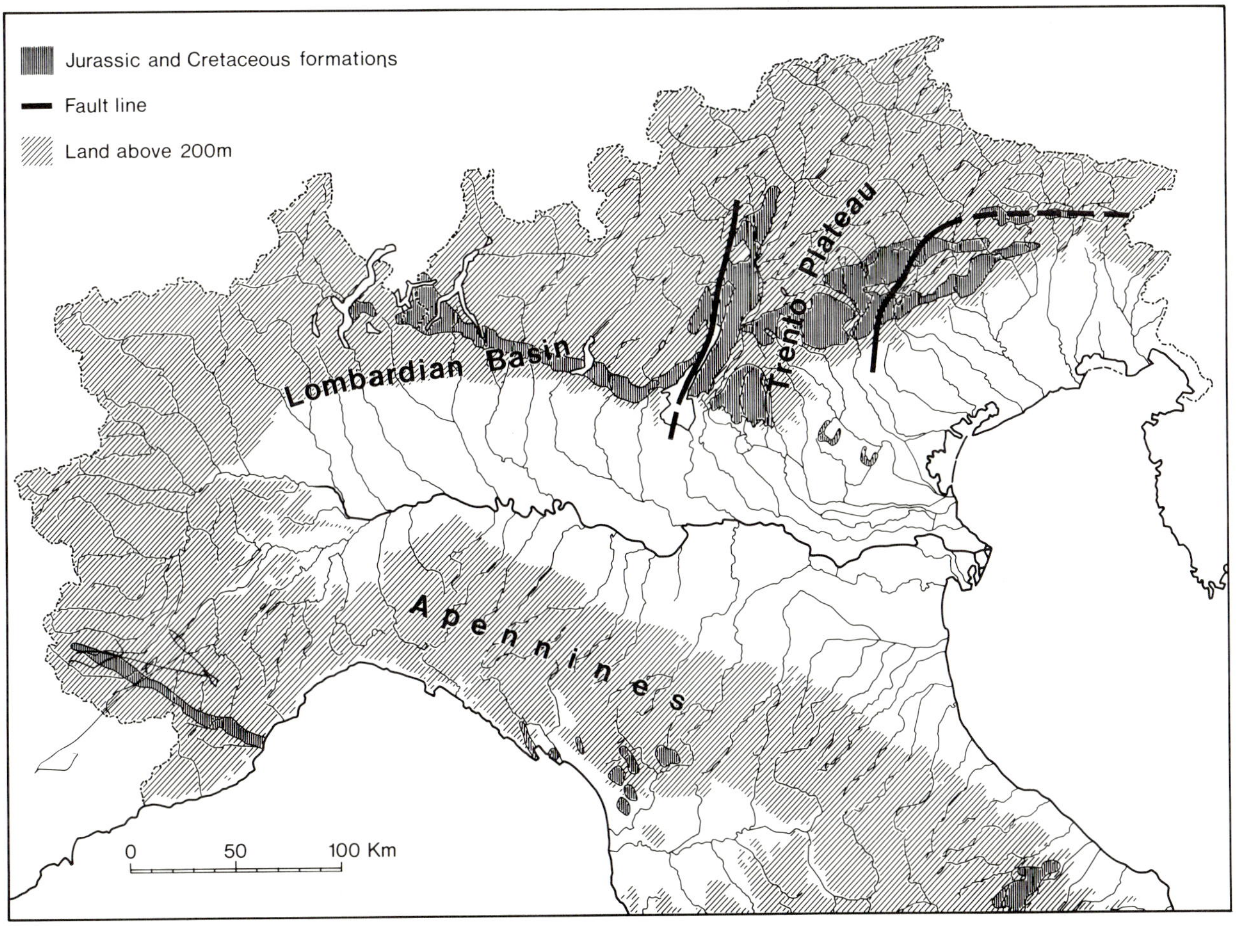

Fig. 27.1. Northern Italy – distribution of Jurassic and Cretaceous deposits.

of flint often varies between individual geological formations, so that varieties can be distinguished macroscopically and, even though one specific type may out-crop over a considerable area, an approximate geographical location can often be established on these criteria alone.

Systematic research into flint sourcing has been initiated during the past decade by M. Cremaschi, in his investigation of the sources of flint used on prehistoric sites in Emilia. Initially using colour coding and a study of the micro-organisms in the flint to determine the source, his more recent work has utilised colour coding alone. Cremaschi, in a series of studies relating to Emilia, has demonstrated that the exploitation of sources of raw material followed a clear chronological pattern, with a limited movement of flint during the Late Mesolithic (Cremaschi, *in* Biagi *et al.*, 1981), followed by a well-developed trade in flint from the southern Alps dominating assemblages throughout the Neolithic (Cremaschi, *in* Cazella *et al.*, 1976; Cremaschi, *in* Bagolini, 1981; Cremaschi, 1981). In the ensuing Chalcolithic period this transpadane trade was curtailed, as can be seen in the later assemblages at Spilamberto (Cremaschi, *in* Bagolini, 1981) and in the Bell Beaker settlement at Sant'Ilario (Cremaschi, *in* Barfield *et al.*, 1975). During this period poorer quality local flint and related rocks found in the Apennines, such as phtanite, were used for the chipped stone artifacts. Although this development has only been closely studied on a few sites, a superficial examination of other assemblages confirms this picture.

Cremaschi's examination of the Alpine flint used in Emilia revealed striations on areas of original preserved cortex, which suggests that surface collection may have been the usual method of procurement rather than direct quarrying of the flint at source. Cremaschi also suggested that the most probable source area of Alpine flint was the Monti Lessini to the north of Verona (Cremaschi, 1981).

Other work on flint sources has been carried out by Fedele and Giraudi in their examination of limited flint resources in the area of Piedmont (Fedele & Giraudi, 1978).

Flint in the province of Brescia

Complementing the research conducted by Cremaschi south of the Po, the author has been investigating the question of flint supply to Neolithic and Chalcolithic sites mainly in the eastern part of the province of Brescia.* This study fell into two parts: firstly a geological survey of the usable flint resources in the Brescian Alps and secondly an empirical examination of flint artifacts and debitage on prehistoric settlements located between the pre-Alps and the Po river.

The geological survey, carried out in 1980, used as a basis the detailed 1 : 50 000 geological map prepared by the University of Parma (Istituto di Geologia, Parma, 1972). The original survey strategy had been to walk transects across the most closely out-cropping stratified sequences of Jurassic and Cretaceous deposits in order to collect samples of flint. In practice this approach proved impossible, as, in most areas, the surface geology was obscured by soil and impenetrable vegetation, so that the survey was eventually conducted by using the nearest existing paths, and, more profitably, by following cuttings made by road construction. In this way transects covering more than 30 km were covered across the pre-Alps between Lake Iseo and Lake Garda. Samples housed in the Museo di Storia Naturale in Brescia were also studied. In addition, excursions were made into the mountains of the province of Bergamo to the west, as well as to the Monti Lessini and to Monte Baldo in the province of Verona, east of Lake Garda (Fig. 27.2).

The results of this survey, which to some extent must be regarded as provisional, point to some significant conclusions. Most important is the fact that although flint occurs abundantly in many formations in the Brescian pre-Alps in Cretaceous and Jurassic strata, very little was found that was suitable for knapping into tools, as it is either of a very poor quality in itself, or else it has been fragmented *in situ* by the folding of the geological strata. This same picture appears true also for the province of Bergamo, where poor quality, fragmented, flint is the rule. However, to the east of Lake Garda abundant deposits of extremely good quality flint are present, which are matched by extensive evidence for prehistoric and recent flint extraction.

This contrast in the quality of flint between the areas east and west of Lake Garda can be explained, to some extent, by the fact that, although similar geological phases are represented in the sequences in both areas, the two are separated by a major fault line which runs along Lake Garda and to the west of the Val di Sarca to the north. This fault line divides the Trento Plateau to the east from the westerly Lombardian Basin (Gaetani, 1975) (Fig. 27.1). The calcareous formations in the Lombardian Basin have been profoundly convoluted, which has led to a fragmentation of flint nodules and strata. In the Trento Plateau however, Jurassic and Cretaceous deposits remain still largely bedded horizontally and these contain undamaged flint.

Only three locations were identified to the west of Lake Garda where workable flint was found. Two of these related to a single out-crop of Jurassic limestone, the *Medolo* formation, lying immediately to the west of Lake Garda (Fig. 27.2). On Monte Pizzocolo, at a height of 1265 m a.s.l. and 1200 m above the level of the lake, the author had already found in 1974 an extensive scatter of debitage from quarrying of the underlying grey flint of the *Medolo*. The same formation appears near the summit of Monte Covolo, at the southern limits of the Brescian pre-Alps. Here again good quality flint and some evidence of quarrying was recovered. The prehistoric settlement at the foot of Monte Covolo, well below the level of the flint out-crop, was excavated by the author between 1971 and 1973. The continuous occupation of this site, from the Late Neolithic through to the start of the Middle Bronze Age,

* The project has been supported by grants from the British Academy and the University of Birmingham, and owes much to the advice and assistance of C. Baroni and P. Biagi.

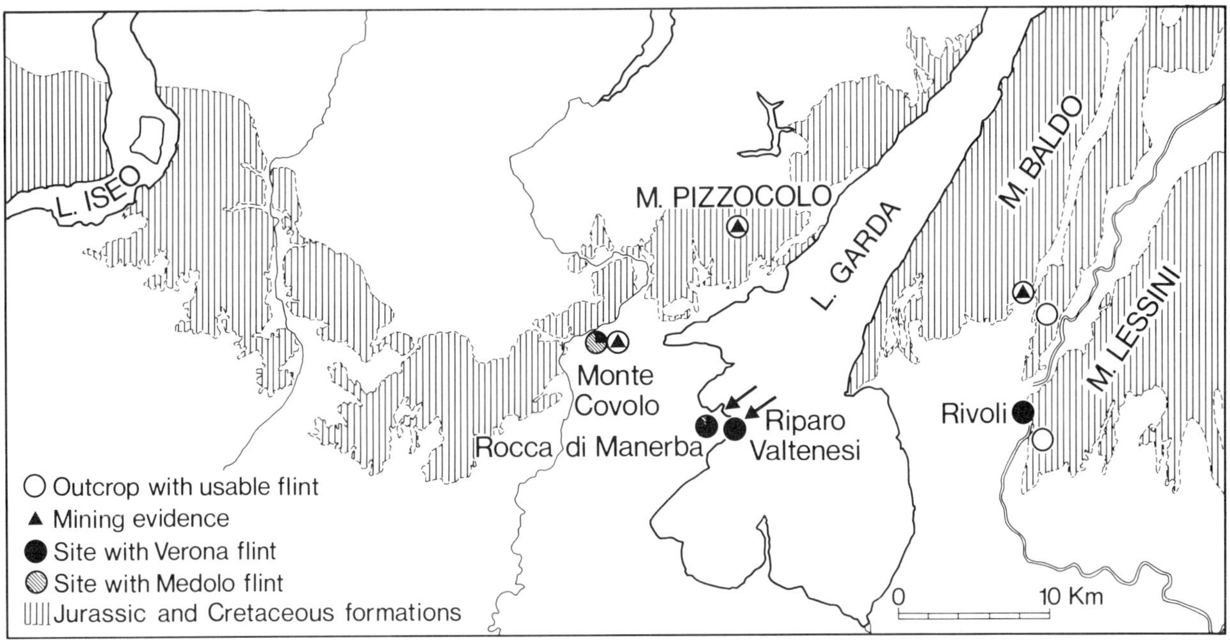

Fig. 27.2. Lake Garda with areas of Jurassic and Cretaceous formations. Flint sources and principal sites. Arrows indicate flint supply direction from source.

can be explained in relation to the exploitation of the flint deposits of the mountain (Barfield *et al.*, 1975–6, and 1977–9). At both Monte Pizzocolo and Monte Covolo the flint was of a matt-surfaced, grey variety with slight banding of lighter or darker grey.

A third source of usable flint can be postulated near Quarena di Sotto to the north-west of Brescia. Samples of red and buff Cretaceous flint from here have been collected by the Natural History Museum of Brescia. The location of these deposits was not, however, identified by our survey and the evidence from settlement assemblages suggests that this was not exploited in prehistoric times.

An excursion further west to the province of Bergamo studied flint deposits in the valley of the Brembo river. Although numerous strata of flint could be observed along the sides of this river, none was suitable for knapping.

The visits to the areas of Monte Baldo and the Monti Lessini revealed a totally different picture, with extensive deposits of good quality flint available in the higher mountain areas in the lower strata of the *Biancone* formation of limestone at the interface between Jurassic and Cretaceous deposits. The flint in these deposits was still horizontally bedded and unfractured by tectonic movement. Evidence of flint mining was also widely distributed in both the mountain ranges. Different coloured flints were closely stratified one above the other, the principal colour varieties being (1) a light speckled flint with a base colour ranging from buff, brown, greenish grey to black; (2) a pure yellow opaque flint; (3) reddish opaque flint. All this flint had very glossy surfaces.

The study of flint assemblages from settlements in the Bresciano and Mantovano formed the second part of the project. A detailed classification of the range of flint varieties from the sites was established, based on an empirical division by colour, patterning and texture. This produced a necessarily over-elaborate typology, which, for the purposes of this study, can be simplified into three main groupings: (1) *Medolo* matt, grey flint with slight banding from the Brescian sites of Lake Garda; (2) Glossy flint corresponding to the first two categories of Verona flint observed in the *Biancone*; and (3) unprovenanced varieties.

Following up observations already made by Biagi on flint from his excavations, Early and Middle Neolithic assemblages were studied on the central Po plain (Fig. 27.3). Early Neolithic settlements at Vhò (Campo Ceresole, Campo Costiere and Campo Serra Mattina) and Dugali Alti-Ostiano (Biagi, 1979), produced a remarkably standard range of flint which matched the basic Verona series with speckled grey–green, through buff, to brown flint predominating over opaque yellow flint. Middle Neolithic sites of the Middle Square-Mouthed pottery stages, in the same area, had a similar range of flint with, in addition, other flint colours. Although these other flint varieties had not been observed during the geological survey, the fact that they are present on the contemporary Square-Mouthed Pottery settlement at Rivoli Veronese would indicate that these too came from the Verona area. At Casatico and Casotte-Ostiano, both of Middle Square-Mouthed Pottery date, this wider range of flint included a very distinctive variety – one with a matt surface rich in angular microfossils, and a granular structure (Biagi *et al.*, 1983).

We can see from this study that, as on the plain to the south of the Po river (Cremaschi, 1981), while most of the flint came from the Verona area there is a difference in resources being exploited between the Early Neolithic and Middle Neolithic.

A more detailed analysis was carried out on material

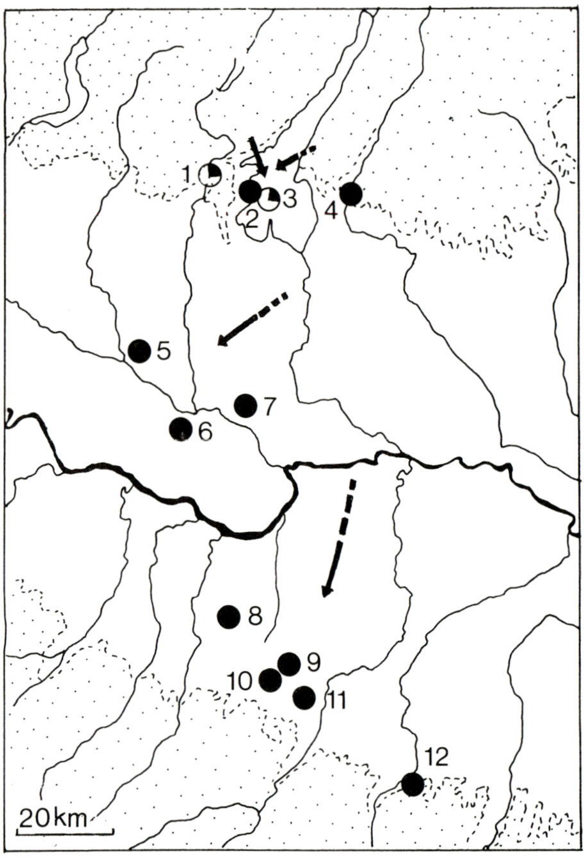

Fig. 27.3. The central Po Plain with Neolithic sites. Open circles, sites using 'Medolo' flint; filled circles, sites using 'Verona' flint; segmented circles site ratio of Medolo/Verona flint; arrows indicate supply direction: (1) Monte Covolo; (2) Rocca di Manerba; (3) Riparo Valtenesi; (4) Rivoli; (5) Ostiano; (6) Vho; (7) Casatico; (8) Campegine; (9) Rivaltella; (10) Albinea; (11) Chiozza; (12) Spilamberto.

from three sites in the area between the Brescian pre-Alps and Lake Garda: Monte Cololo, the Rocca di Manerba and the Riparo Valtenesi, Manerba (Fig. 27.2). These are all multiperiod sites, whose main occupation phases fall between the Late Neolithic (Lagozza) and the Early Bronze Age (Polada).

Monte Covolo, as we have already shown, is one of the few sources of usable flint to the west of Lake Garda and this deposit may explain the presence of the settlement. The site has produced evidence of an initial early occupation relating to a late Square-Mouthed Pottery tradition, which could not be differentiated stratigraphically from much more abundant Lagozza remains. Thereafter the site was occupied through the Chalcolithic (White Ware), Bell Beaker, Early Bronze Age (Polada Culture) into the Middle Bronze Age. The abandonment of the site would appear to relate to the latest use of flint as a raw material in this part of northern Italy (Barfield *et al.*, 1977–9).

The varieties of flint represented on this site remain constant throughout the sequence, being dominated by the local Jurassic *Medolo* grey flint (60.8%), although a significant proportion (26.4%) is comparable with basic *Biancone* range. The latter, we can provisionally suggest, was brought from the Monte Baldo–Monti Lessini area across Lake Garda; although it is difficult to explain why so much of this material should be brought to a site closely involved with the production of *Medolo* flint, unless it was considered superior material (Fig. 27.4).

Of particular interest at Monte Covolo are two fragments of a flint dagger found in the Bell Beaker levels. This object, besides being the only dagger from the site, is also of a unique deep red flint speckled with white (Barfield *et al.*, 1975–6, figs. 50, 23). It is an object traded into the site from a distance certainly further than the Veronese. Similar flint has been observed in assemblages from the Trentino and the Marche. This dagger may have reached Monte Covolo through some form of prestige exchange, in view of the fact that local flint was available in abundance; although breakage and discard by a visiting traveller cannot be discounted. Exotic material of a different kind is represented by a small fragment of Sardinian obsidian from an unstratified context, in all likelihood dating to the Lagozza period of occupation on the site (Williams Thorpe *et al.*, 1979).

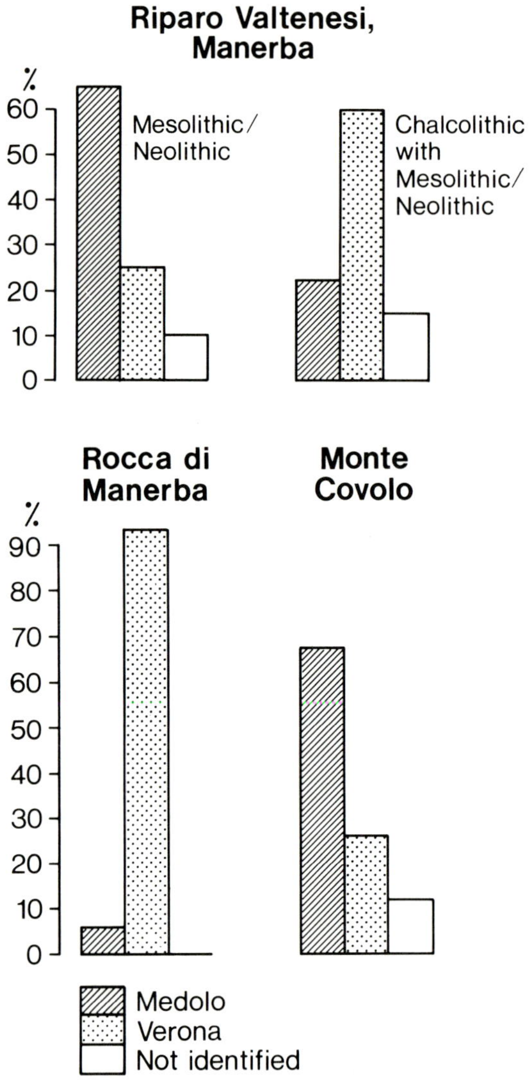

Fig. 27.4. Proportions of different types of flint used on sites at Monte Covolo and Manerba.

The Rocca di Manerba on Lake Garda is a strategically located site, occupied initially in the Square-Mouthed Pottery period. The abundant, and mainly unstratified, flint industry can, however, be in large part attributed to a more intense Lagozza occupation (Borrello, 1973). The flint industry here is dominated by the basic Verona range (94%) and only 6% of the 'local' *Medolo* flint (Fig. 27.4). From this site, the flint deposits at a lower altitude on Monte Baldo across the lake would in any case have been more accessible than those of the summits of Monte Pizzocolo and Monte Covolo on the Brescian side. Sardinian obsidian was again present on this site and is presumably also of Lagozza date (Williams Thorpe *et al.*, 1979).

The Riparo Valtenesi at Manerba had two main phases of prehistoric activity (Barfield, 1978, 1983). The first dates to the Late Mesolithic–Early Neolithic transition, in which they grey *Medolo* flint (65%) predominates over the basic Verona range (25%) (Fig. 27.4). In the Chalcolithic, the second period of use, the proportions appear reversed. In fact, Chalcolithic flint from disturbed contexts alone has been examined so far and it is likely that the *Medolo* flint is all residual from Mesolithic–Early Neolithic levels. Obsidian from this site, so far not identified as to source, is securely stratified in the Chalcolithic levels and is the latest that we have from a securely dated context in northern Italy (Barfield, 1983).

The mining activity on Monte Pizzocolo has not been closely dated, the only distinctive tool types being small core artifacts of 'Campignian' type. Such tools are sometimes associated with Chalcolithic assemblages in the Monti Lessini and are invariably restricted to areas of flint mining (Barfield, 1971). The presence of sickle blades of Early Bronze Age type in 'Verona' flint might suggest that the Pizzocolo mining activity was not of Bronze Age date.

Comment should be made here concerning the quality of flint used in the Chalcolithic cemeteries at Remedello, Volongo and Fontanella near the centre of the Po Plain, which are contemporary with the mortuary structures of the Riparo Valtenesi, Manerba. At all these sites an opaque white flint predominated, which is entirely different from that found on the other sites so far discussed. This is remarkable, since the procurement site at Monte Covolo, which has an approximately contemporary occupation phase, lies on the same River Chiese as Remedello, some 40 km to the north (Fig. 27.5).

The technology of the Remedello flint work is also different from that of the northern sites, being of a generally higher quality and more standardised production than any from sites along the edge of the Alps. This difference in quality of workmanship is indeed very apparent when the Remedello arrowheads are compared with those from the Riparo Valtenesi at Manerba. A simple scatter diagram also shows a significant size difference in flint arrowheads between the two sites, besides clear morphological and the qualitative variations; the Remedello arrowheads tend to be longer and also the leafshaped arrowhead is absent from Remedello (Fig. 27.6). The quality of workmanship of the Remedello industry is

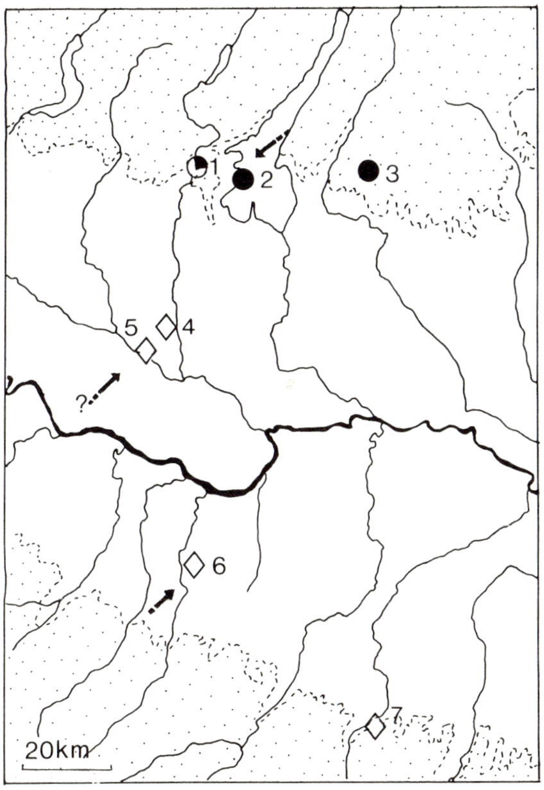

Fig. 27.5. The central Po Plain with chalcolithic sites. Open circles, sites using 'Medolo' flint; filled circles, sites using 'Verona' flint; diamonds, sites using various Apennine sources; segmented circle site ratio of Medolo/Verona flint; arrows indicate supply direction: (1) Monte Covolo; (2) Riparo Valtenesi; (3) Le Colombare; (4) Remedello; (5) Fontanella; (6) Sant'Ilario; (7) Spilamberto.

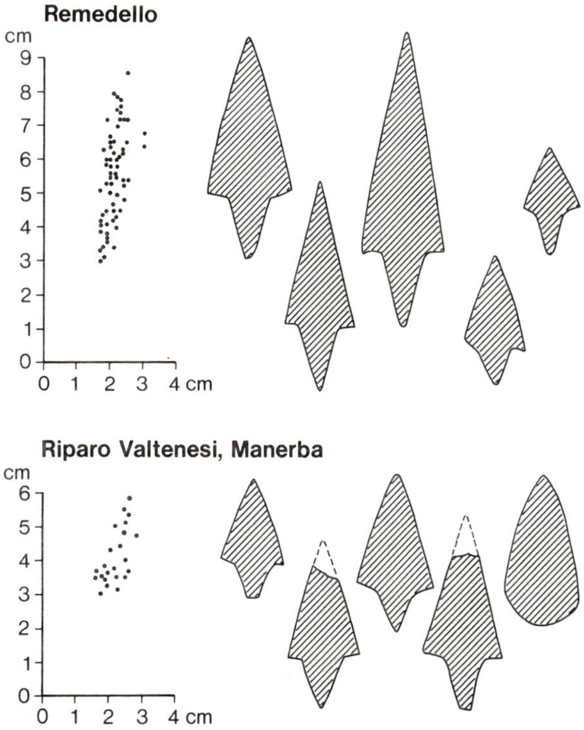

Fig. 27.6. Arrowhead types and sizes at Remedello and Riparo Valtenesi, Manerba.

perhaps best matched in the Rinaldone area of northern Tuscany, even though similar raw material has not yet been identified from that area. A north Apennine source would thus appear to be the most probable, even though hitherto it has usually been assumed that flint came to Remedello from the southern Alps (Barfield, 1981). It is important to notice here that there is an apparent correlation between typological differences in arrowheads and source of raw material; suggesting, perhaps, regional traditions of craftsmanship related to individual flint sources.

It can further be noted that the flint used for the manufacture of arrowheads and crescents from the chalcolithic burial cave of Buca di Andrea in the province of Bergamo (Poggiani Keller, 1980, and 1983), are of a variety which are not represented among the assemblages so far discussed. This would suggest that, so far unlocated, local deposits east of Lake Iseo were also in use at this time.

The overall picture that we obtain from these sites on the Lombard Plain tends to confirm that identified by Cremaschi for Emilia. The Early and Middle Neolithic sites on the Plain rely exclusively on flint from sources in, or near, the *Biancone* of Verona; with a wider range of flint in use during the Middle Square-Mouthed Pottery stage (Fig. 27.3). During the Chalcolithic at Remedello, probably Apennine sources were being used (Fig. 27.5). We cannot yet generalise about the development on the west side of Lake Garda since we do not have evidence from the whole cultural development. Both Lombard *Medolo* and Verona *Biancone* flint would seem to have been exploited through the whole period from Early Neolithic to the Chalcolithic, with the Manerba sites making almost exclusive use of the flint from across Lake Garda during Lagozza and Chalcolithic times.

In general terms, changes in flint use in Emilia and on the central plain of the Po Valley are comparable with those recognised in southern France, where 'blonde' flint from one main source area was widely distributed during the Chassey period of the Neolithic, whereas from the start of the Late Neolithic or Chalcolithic there was a switch to diverse and local flint sources (Guilaine, 1980; Mills, 1983). Hodder suggested that Late Neolithic cultural diversity reflected a breakdown of an integrated and uniform cultural system through population growth and stress (Hodder, 1979). The pattern seen in the Po Valley, while comparable, lacks the evidence of increasing population density in the Chalcolithic. If anything site density decreases (see Anghinelli, 1978, and 1981).

The picture presented here may be only a preliminary one, but it does demonstrate the potential of this approach to understanding the complexity of cultural development in northern Italy as well as showing that source areas can be clearly isolated by survey and assemblage study. The importance of the fault line along the axis of Lake Garda separating Brescian from Veronese sources is most evident as is the contrast between Apennine and south Alpine sources. Further work on both sides of Lake Garda will be needed to confirm and fill out the evidence presented here.

The dagger survey

A second flint project, currently being completed, concerns the production and distribution of flint daggers in northern Italy.

Daggers of flint appear in the archaeological record in Italy during the Chalcolithic period, contemporaneously with the first occurrence of copper ones and continue in production and use into the Early Bronze Age. Although considerable interest has been paid to these daggers in the past, their study from Chierici (1885) to Cornaggia Castiglioni (1971) has been essentially typological. Little attention has been paid to the technology of production or to the source of the raw material.

Flint daggers have been recorded from more than 174 sites in northern Italy, but, since the majority are unassociated finds, their precise dating is somewhat unclear. It can be suggested, however, from the evidence available, that the finer and larger examples are the earliest and are Chalcolithic in date.

The distribution of these objects is predominantly concentrated in the Po Plain and the pre-Alps of the Veneto, where finds in context are almost always from individual inhumation burials (Fig. 27.7). From this it can be suggested that many of the isolated finds, which are usually undamaged and found over the same area, are also derived from unrecorded burials of the same type and that the distribution of daggers itself mirrors the tradition of single inhumation burial in northern Italy in contrast to the collective burial sites to the west. The absence of daggers in the west has been explained in terms of a difference in the role of grave goods between the two burial traditions (Barfield, 1985*a*).

Leaving aside more detailed typologies, Italian flint daggers can be divided into two main groups: those made by the bifacial flaking of cores and those on long large blades with, usually, only unifacial retouch (Fig. 27.8). The former are found throughout Italy, although especially abundant in the north, while the latter are almost exclusively restricted to central and southern Italy (Colini, 1898–1902; Barfield, 1985*a, b*) (Fig. 27.9).

An examination of the blade daggers in several museum collections suggests that they are manufactured from a very limited range of light brown flint (Munsell colour 10 YR/2) and would thus appear to have been produced by a specialist tradition from only one or a few centres. Their distribution throughout central Italy would have been effected by exchange in a manner analogous to the distribution of Grand Pressigny and banded flint daggers and daggers of other flint in France. Bocquet (1974) in a similar way has contrasted the specialisation of imported flint daggers at Charavines (Savoy), France, with the associated impoverished local flint tradition.

The bifacial daggers, on the other hand, are usually produced in the same flint as the rest of the associated flint industry, and thus represent an artifact within the competence of local flint working traditions.

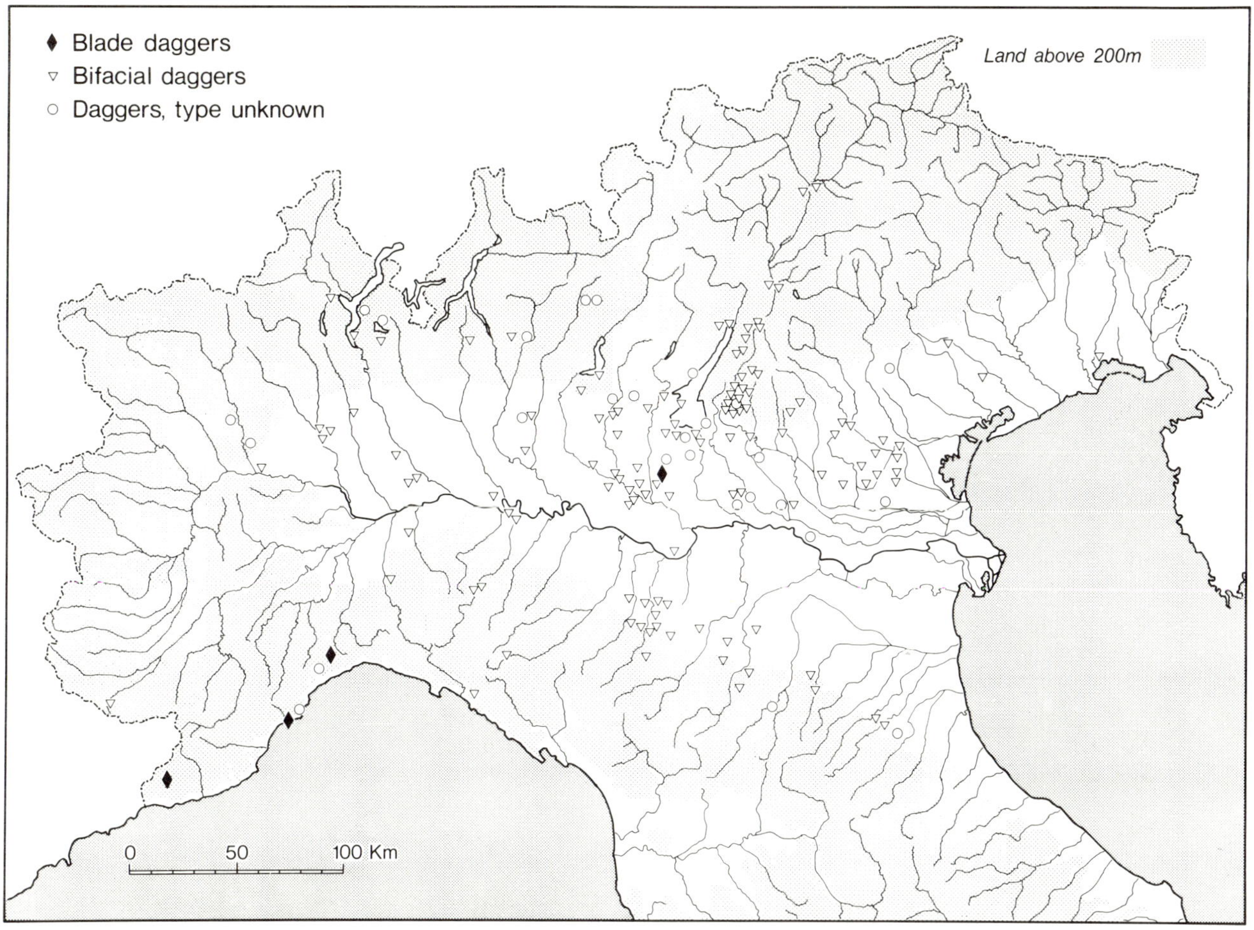

Fig. 27.7. Distribution of flint daggers in northern Italy.

The study of the sources of the flint daggers in northern Italy has not yet been completed. Several of those on the Po Plain are made of flint which corresponds well with the Verona speckled flint. At Remedello and Fontanella, as we have seen, the source is unknown but possibly Apennine, while at Spilamberto the local phtanite was used. This all points to fairly local production. This distinction between a local production of bifacial daggers and specialist blade daggers may be an over-simplification of the actual situation. Flint itself for the 'local' industry often had to be imported from some distance, while at Monte Covolo the unique bifacial dagger of red speckled flint demonstrates that even bifacial daggers could be brought to a site where two different types of more local flint were in use. The latter object may have arrived by an exchange mechanism analogous to that used by blade daggers.

That some basic difference in exchange mechanism and prestige value may distinguish blade from bifacial daggers, is suggested by the presence of daggers of southern French banded flint in northern Italy.† Blade daggers occur in Liguria, where at least two are of banded flint, and where their presence can be explained partly in terms of the lack of suitable local flint. Another example, however, was found in the central Po Valley at Asola, close to Remedello, which is more than 400 km from source (Fig. 27.9). It would be interesting to suggest that this traffic in flint daggers from France was reciprocated in copper daggers from northern Italy, as exemplified in the example from the Dolmen d'Orgon (Provence), generally considered to be of Italian origin (Sauzade, 1979).

Clearly, the evidence to substantiate the model suggested here is still inadequate and the dichotomy perhaps an over simplification of what must have been complex and unstable arrangements of production, distribution and specialisation. Exchange mechanisms must have differed according to local availability of flint, with some communities perhaps specialising in flint production in order to fit into exchange networks. Changes in these networks and the exploitation of different sources at different periods of time would add to this complexity of interpretation. Likewise the degree of specialisation envisaged for the production of either type of dagger cannot be directly assessed. However, if the production of blade daggers was on a more specialised footing then we should further consider the possibility of a direct technological or even blood relationship between the producers of Grand Pressigny, banded flint and Italian

† I am grateful to Dr J. Courtin for information relating to the probable source and distribution of banded flint in southern France.

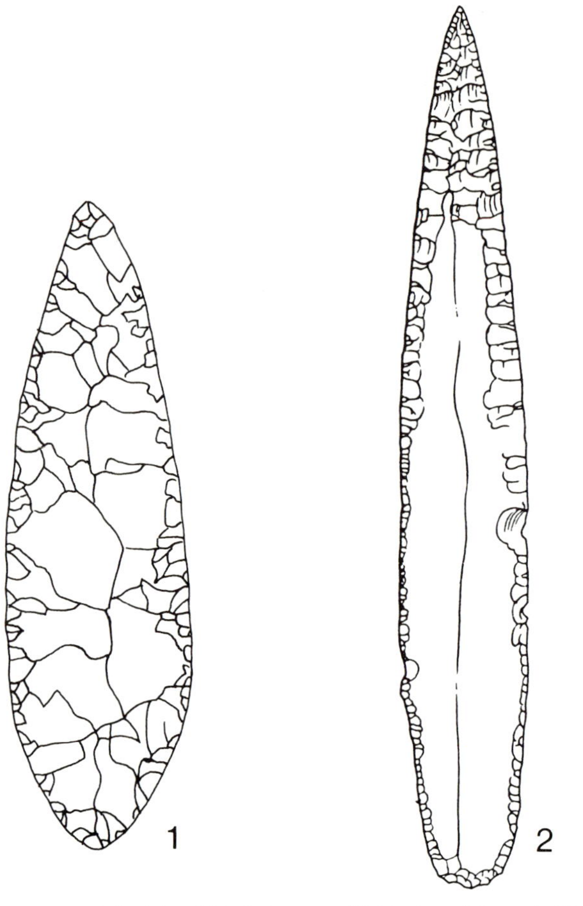

Fig. 27.8. Bifacial and blade daggers: (1) Mariana (Mantua); (2) 'near Naples'.

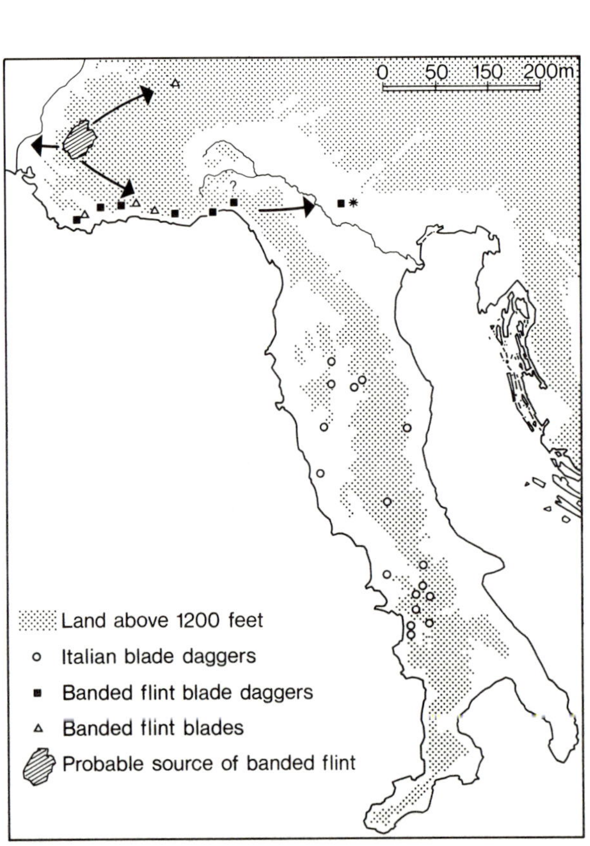

Fig. 27.9. Distribution of French, banded flint daggers in northern Italy and Italian blade daggers in central and southern Italy. The location of Asola is marked with an asterisk.

blade daggers, which all show such similar techniques of manufacture and patterns of dispersal.

Bibliography

Anghinelli, S. & A. (1978). 'Riconoscimento paletnologico (Mantova-Cremona)'. *Preistoria Alpina,* **14**, 281–94.

Anghinelli, S. & A. (1981). 'Ricerche e studi in provincia di Mantova'. *Atti del 1° convegno archeologico regionale della Lombardia*, Milano, pp. 287–90.

Bagolini, B. (1981). *Il neolitico e l'età del rame; ricerca a Spilamberto S. Cesario 1977–1980*. Tamari, Bologna.

Barfield, L. H. (1971). *Northern Italy before Rome*. Thames and Hudson, London.

Barfield, L. H. (1978). 'Riparo Val Tenesi, Manerba, scavi 1976'. *Benacus,* III, 5–24.

Barfield, L. H. (1981). 'L'eneolitico e l'antica età del bronzo in Lombardia'. *Atti del 1° convegno archeologico regionale della Lombardia*, Milano, pp. 139–56.

Barfield, L. H. (1983). 'The chalcolithic cemetery in the Riparo Val Tenesi, Manerba del Garda'. *Antiquity,* LVII, 116–23.

Barfield, L. H. (1981). 'Patterns of north Italian trade, 5000–2000 bc. In: Barker, G. & Hodges, R. (eds) *Archaeology and Italian Society. Papers in Italian Archaeology II*, BAR International Series, **102**, 27–51.

Barfield, L. H. (1985*a*). Burials and boundaries in chalcolithic Italy, *Cambridge Conference on Italian Archaeology 1984*. In press.

Barfield, L. H. (1985*b*). Sir William Hamilton's chalcolithic collection. *British Museum Italic Symposium 1983*. In press.

Barfield, L. H., Biagi, P. & Borrello, M. A. (1975–76). 'Scavi nella stazione di Monte Covolo, Parte I. *Annali del Museo di Gavardo,* **12**, 7–149.

Barfield, L. H., Barker, G., Chesterman, J., Pals, J-P. & Vooriips, A. (1977–79). 'The excavations at Monte Covolo, Part II. *Annali del Museo di Gavorda,* **13**, 5–88.

Barfield, L. H., Cremaschi, M. & Castelletti, L. (1975). 'Stanziamenti del vaso campaniforme a Sant'Ilario d'Enza (Reggio Emilia)'. *Preistoria Alpina,* **11**, 155–99.

Biagi, P. (1979). 'Stazione neolitica a Ostiano (CR) loc. Dugali Alti: scavi 1980'. *Preistoria Alpina,* **15**, 25–38.

Biagi, P., Barker, G. W. W. & Cremaschi, M. (1983). *La stazione di Casatico di Marcaria (Mantova) nel quadro paleoambientale ed archeologico dell'olocene antico della Val Padana centrale*. Istituto Universitario di Bergamo, Studi Archeologici, Vol. 2.

Biagi, P., Castelletti, L., Cremaschi, M., Sala, B. & Tozzi, C. (1981). 'Popolazione e teritorio nell'appennino tosco-emiliano e nel tratto centrale del bacino del Po, tra il IX° ed il X° millennio'. *Emilia Preromana,* **8**, 14–36.

Bocquet, A. (1974). 'Les poignards néolithiques de Charavines (Isère) dans le cadre de la civilisation Saône-Rhône'. *Études préhistoriques,* **9**, 7–17.

Borrello, M. A. (1973). 'Hallazgos de materiales de tradicion Lagozza en la Rocca di Manerba, prov. de Brescia'. *Benacus,* I, 27–33.

Cazzella, A., Cremaschi, M., Moscoloni, M. & Sala, B. (1976). 'Siti neolitici in località Razza di Campegine (Reggio Emilia)'. *Preistoria Alpina,* **12**, 76–126.

Chierici, G. (1885). 'Nuovi scavi nel sepolcreto di Remedello'. *Bullettino di Paletnologia Italiana,* XI, 138.

Colini, G. A. (1898–1902). 'Il sepolcreto di Remedello Sotto nel Bresciano ed il periodo eneolitico in Italia'. *Bullettino di Paletnologia Italiana,* XXIV–XXVIII.

Cornaggia Castiglioni, O. (1971). 'La Cultura di Remedello'.

Memoria della Societa Italiana di Scienze Naturali e del Museo di Storia Naturale, Milano, **xx**, fasc. 1, pp. 7–79.

Cremaschi, M. (1981). 'The source of the flint artefacts for the Central Po Plain and Apennine sites, between the 7th and 2nd millennium BC'. In: Engelen, F. H. G. (ed.), *Third international symposium on Flint, Staringia,* **6**, 139–42.

Fedele, F. & Giraudi, C. (1978). 'Litologia preistorica del Piemonte; i distribuzioni degli affioramenti di selce'. *Gruppo di studio del Quaternario padano, Quaderno*, Vol. 4, Parma.

Gaetani, M. (1975). 'Jurassic stratigraphy of the southern Alps: a review'. In: Squyres, C. H. (ed.), *Geology of Italy*, Tripoli, pp. 377–402.

Guilaine, J. (ed.) (1980). *Le Groupe de Véraza et la Fin des Temps Néolithiques*. Editions du CNRS. Toulouse.

Hodder, I. (1979). 'Economic and social stress and material culture patterning'. *American Antiquity,* **44**, 446–54.

Istituto di Geologia dell'Università di Pavia (1972). *Carta Geologica dell'Adamello*, Pavia.

Mills, N. (1983). 'The Neolithic in southern France'. In: Scarre, C. (ed.), *Ancient France*, pp. 91–145, Edinburgh.

Poggiani-Keller, R. (1980). *La sezione archeologica del museo della Valle*, Museo della Valle, Zogno.

Poggiani-Keller, R. (1983). 'L'età del rame'. In: *Archeologia Lombardia*, Nuovo Banco Ambrosiano, Milano, pp. 36–82.

Sauzade, G. (1979). 'Le dolmen des Gavots à Orgon (Bouches du Rhône)'. *Bulletin du Musée d'Histoire Naturelle de Marseille*, **XXXIX**, 65–80.

Squyres, C. H. (ed.) (1975). *Geology of Italy*, Tripoli.

Williams Thorpe, O., Warren, S. E. & Barfield, L. H. (1979). 'The sources and distribution of archaeological obsidian in northern Italy'. *Preistoria Alpina,* **15**, 73–92.

28

Danubian raw material distribution patterns in eastern central Europe

JACEK LECH

Summary

This paper is an interim report on the author's research on the distribution and supply of siliceous rocks to the Danubian early farming communities in the eastern part of central Europe in the second half of the fifth millennium and the first half of the fourth millennium BC (in conventional ^{14}C dates).

Introduction

Raw materials of stone are a good indicator of the range of mutual contacts among the early farming communities. The clearly differentiated siliceous rocks that occur in central Europe provide good opportunities for research in this field. But research has not taken full advantage of this situation. The distribution of siliceous rocks in the Vistula Basin is well known. However, everywhere there are still considerable gaps in our knowledge of the distribution of these raw materials to the early farming settlements. The present report covers the ground unevenly and refers mainly to the territory of Poland and Czechoslovakia. All the raw materials mentioned in the text have been described in detail in a previous paper (Lech, 1981).

One of the first people to study the supply of the siliceous raw materials to the early farming communities was Professor Tadeusz Sulimirski (born in Kobylany, in Poland, in 1898: died in London in 1983). This paper is dedicated to his memory.

Supplies to the first farming communities

In the Oder and Vistula River Basins, flint may be found at Danubian I Linearbandkeramik (LBK) first farming settlements as much as 200 km from the outcrop or flint mine source from which it was obtained, usually when alternative large-scale raw material extraction centres nearer at hand did not exist (Fig. 28.1).

Good examples of this long-distance supply are the predominance of Jurassic-Cracow flint in the settlements of Lower Silesia (Niemcza in the Wałbrzych district; Strachów and Skoroszowice in the Wrocław district) and in the Rzeszów region (Rzeszów site 16; Boguchwała and Kraczkowa) and the predominance of chocolate flint supplied from the north-east fringe of the Holy Cross Mountains (see Schild, this volume) in the sites of Kujavia (Brześć Kujawski and Przybranowo in the Włocławek district; Łojewo in the Bydgoszcz district and others).

In all these cases it is likely that expeditions were sent out specifically to collect flint. However, exchange of such raw materials apparently also took place, as is shown by the existence at some sites of small amounts supplied from remote sources, which are often only a few per cent of the total assemblage at the receiving settlement (Kaczanowska & Lech, 1977, table 1; Lech & Leligdowicz, 1980, pp. 174–7). Thus, exchange probably accounts for the presence of small quantities of obsidian on the majority of sites in southern Poland and for that of chocolate and Świeciechów flint at Cracow-Olszanica (Milisauskas, 1976, pp. 33–47) and Cracow-Mogiła 63, as well as for Jurassic-Cracow and Świeciechów flint in Kujavia (Fig. 28.1). There is often no apparent economic reason for such exchanges, which appear to be due to relations of a social character between distant communities. However, it is to be noted that nearer supplies such as the striped flint from Krzemionki mine were not exploited at this time, and that Volhynian flint had a much smaller range than in later periods.

In terms of distance we may note that small supplies of the Jurassic-Cracow flint even reached LBK settlements on the Danube River and have been discovered at Štúrovo in western Slovakia, about 260 km from source; at Vedrovice-Zábrdovice and Bohušice in southern Moravia, between 260 and 300 km from the natural deposits and even at Křimice near Pilsen in western Bohemia, nearly 450 km from the geological source. At Štúrovo radiolarite was the dominant raw material and perhaps originated in the Tata region, on the other side of the Danube, approximately 30 km away. From the quantitative point of view, as far as the number of imports is concerned, obsidian occupies second place at this site. Single specimens of the Volhynian flint type reached Štúrovo, probably from the east, at the same time as the obsidian. The nearest deposits of the Volhynian flint are between 350–400 km from Štúrovo.

Eastern Slovakia was a separate raw material region. Here, the obsidian deposits were exploited, as were a few other raw materials of local significance. Obsidian was processed in numerous workshops. From there, the main distribution route led to Little Poland. Workshops processing obsidian are also known in eastern Slovakia, at linear pottery culture sites which were synchronous with the LBK

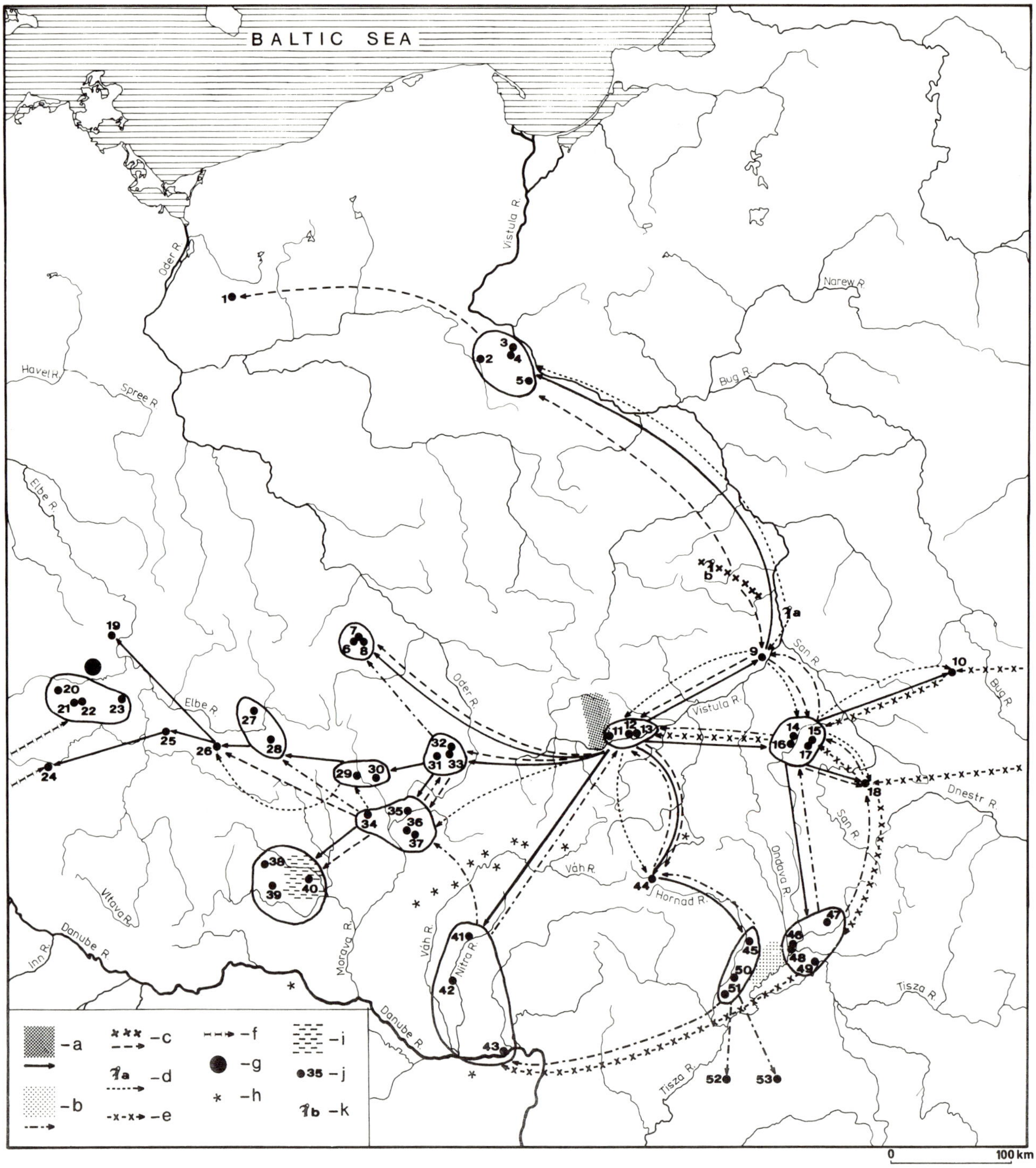

Fig. 28.1. The distribution of siliceous rocks to the first farming communities in eastern central Europe at about 4400–4000/3900 BC. (*a*) – Jurassic-Cracow flint deposits and directions of raw material distribution; (*b*) – obsidian deposits and directions of raw material distribution; (*c*) – chocolate flint deposits and directions of raw material distribution; (*d*) – grey–white spotted flint from Świeciechów-Lasek and directions of raw material distribution; (*e*) – directions of Volhynian and Dnestr flint distribution; (*f*) – directions of Bavarian striped hornstone distribution; (*g*) – Skršin quartzite deposits; (*h*) – radiolarite deposits; (*i*) – hornstones of Moravský Krumlov type deposits; (*j*) – archaeological sites: *Poland* (1) Żuków, dist. Szczecin; (2) – Łojewo, dist. Bydgoszcz; (3) – Przybranowo, dist. Włocławek; (4) – Straszewo, dist. Włocławek; (5) – Brześć Kujawski, dist. Włocławek; (6) – Niemcza, dist. Wałbrzych; (7) – Strachów, dist. Wrocław; (8) – Skoroszowice, dist. Wrocław; (9) – Samborzec, dist. Tarnobrzeg; (10) – Tarnoszyn, dist. Zamość; (11) – Cracow-Olszanica; (12) – Cracow-Mogiła 63; (13) – Cracow-Pleszów; (14) – Rzeszów, site 16; (15) – Łańcut, dist. Rzeszów; (16) – Boguchwała, dist. Rzeszów, site 1; (17) – Kraczkowa, dist. Rzeszów; (18) – Kormanice, dist. Przemyśl, site 1; *Czechoslovakia* (19) – Chabařovice, dist. Ústí nad Labem, site důl Petri; (2) – Chotěbudice, dist. Louny; (21) – Malá Černoc, dist. Louny; (22) – Velká Černoc, dist. Louny; (23) – Slaný, dist. Kladno; (24) – Křimice, dist. Pilsen-North; (25) – Říčany, dist. Praha-East; (26) – Bylany, dist. Kutná Hora; (27) – Libčany, dist. Hradec Králové; (28) – Úhřetice, dist. Chrudim; (29) – Mohelnice, dist. Šumperk; (30) – Brníčko, dist. Olomouc; (31) – Otice, dist. Opava; (32) – Neplachovice, dist. Opava; (33) – Kravaře-Kouty, dist. Opava; (34) – Mostkovice, dist. Prostějov; (35) – Lipnik, dist. Přerov; (36) – Prusy, dist. Přerov; (37) – Žopy, dist. Kroměříž; (38) – Bohušice, dist. Třebíč; (39) –

of west Slovakia and Little Poland (Fig. 28.1). One example is the workshop from feature 9 in the settlement at Kopčany, in the district of Michalovce, dated by S. Šiška to the Barca III phase.

As well as obsidian, single specimens made of the Volhynian flint, probably from the Upper Dnestr River Basin and of other kinds of flint from that area, are found in east Slovakian linear pottery culture sites. Jurassic-Cracow flint is also present at Velké Trakany and other sites in the Trebišov district of Slovakia. Among 94 pieces recovered from Malé Raškovce, also in Trebišov district, 87.2% were made of obsidian, 4.3% of Jurassic-Cracow flint and 8.5% of other undefined types of chert. Flint from the south Polish Jura, Volhynian flint and other flint from the Upper Dnestr are also found in Bükk culture settlements such as Kračava (Michalovce district) and Ždaňa (Košice district) in the same region.

At the LBK settlement of Vedrovice-Zábrdovice in southern Moravia, however, the local deposits of hornstone of Moravský Krumlov type were the most important type of raw material. Among 652 artefacts from feature 037, and from units 023B/63 and *Široká u lesa* at this site, Moravský-Krumlov hornstone amounted to 78.8%, Jurassic-Cracow flint to 4.3%, Baltic erratic flint 2.3% and radiolarite or jasper to 0.15%. Thirteen per cent of the artifacts were unidentified, and presumably from local sources (Ondruš, 1976). Similarly, obsidian and chocolate or Świeciechów flint were of no importance in the settlement of Olszanica and Mogiła 63 in Little Poland (Fig. 28.1).

Though long-distance raw material sources of flint are of little statistical significance at these sites, it is clear from an analysis of the 037 and 098 workshops at Vedrovice-Zábrdovice that these materials *were* being made into tools at the settlement (Ondruš, 1976; Lech, 1983, pp. 49–52).

At the well-known LBK settlement of Bylany in the Kutná Hora district in eastern Bohemia, Jurassic-Cracow flint was very important and isolated artefacts of chocolate and Świeciechów flint were also recovered. The Bylany settlement is situated more than 300 km from the nearest Jurassic-Cracow flint deposits, about 425 km from the chocolate flint deposits and approximately 475 km from the Świeciechów mine. Besides these raw materials from the Vistula River Basin, other sources of supply include quartzite of Skršin type, Cretaceous erratic flint, and porcellanite and hornstone of Moravský Krumlov type. Different kinds of exotic western European flint have also been found recently at the settlement of Langweiler 9 in the Aldenhovener Platte in Western Germany (Löhr, Zimmermann & Hahn, 1977, pp. 160–5).

Such data, though not complete, allows us to state that all the LBK communities of central Europe exploited raw materials obtained from various, sometimes very distant, sources. Those raw materials were, to a large extent, mined products and their distribution, as well as satisfying the raw material requirements of these communities, appears also to fulfil other purposes of ritual or social exchange.

We do not yet have percentage calculations for the raw materials from Bylany. But we can say that, apart from Cretaceous erratic Baltic flint, this was mainly flint from the southern Polish Jura, and to a lesser extent quartzite of the Skršin type. Visual differentiation of the raw material is incomplete for some phases of settlement at this site. The results so far obtained seem to indicate that, in the early phase of the settlement, Cretaceous erratic flint was dominant and Moravský Krumlov hornstone was also of some significance. It seems that quartzite of Skršin type was lacking at this period and that the Jurassic-Cracow flint, which was important later only occurs here as single specimens. One specimen of Świeciechów flint can be related to the early phase of the settlement. This raw material has also been recorded by J. K. Kozłowski among the raw materials of the older LBK phase in the Mohelnice settlement, in north-west Moravia (1970, p. 74). It probably indicates that the Bylany settlement was getting its supplies of erratic flint from this direction (Fig. 28.1).

The known distribution of the siliceous rocks among the LBK settlements in the second half of the fifth millennium BC is summarised above. Striped hornstone is first exploited in Bavaria at the end of the fifth millennium, and in the first half of the fourth millennium. The beginnings of its widespread distribution can be related to the Linearbandkeramik culture (Vencl, 1971, p. 78).

At this time, striped Bavarian hornstone reaches some settlements in western Bohemia, approximately 100–200 km from the deposits, for example at the site of Velká Černoc and Malá Černoc, in Louny district and Křimice, in the Pilsen North district. In the eastern direction, the farthest distance is reached by this raw material at the beginning of the fourth millennium, when it is found in settlements of the stroke pottery culture, and also occurs on many sites in Bohemia and, on a smaller scale, in Moravia. This shows that the vast multidirectional distribution of raw materials observed in Linearbandkeramik settlements persists in later settlements of the stroke pottery culture and even later. As will be shown below similar phenomena are recorded in the early Lengyel and Tisza settlements (Fig. 28.2).

Supplies to later Danubian communities

In the fourth stroke ornamented pottery phase at Předměřice settlement (district of Hradec Králové), investigated by Vávra, among 231 specimens from features numbers 7 and 8, 73.6% were made of erratic 'Baltic' flint, 4.8% of chocolate flint, 3.5% of Jurassic-Cracow flint, 1.3% of

Bojanovice, dist. Znojmo; (40) – Vedrovice-Zábrdovice, dist. Znojmo; (41) – Topolčany-Jacovce, site Medzivody; (42) – Nitra, site 'Přemyslova ulíca'; (43) – Štúrovo, dist. Nové Zámky; (44) – Poprad-Matejovce; (45) – Ždaňa, dist. Košice, site Baxaser; (46) – Kopčany, dist. Michalovce; (47) – Krčava, dist. Michalovce, site Stoček; (48) – Malé Raškovce, dist. Trebišov; (49) – Velké Trakany, dist. Trebišov, site Szélmalomdomb; *Hungary* (50) – Boldogkövaralja, dist. Borsod-Abaúj-Zemplin; (51) – Megyaszó Nagy Répás, dist. Borsod-Abaúj-Zemplin; (52) – Hortobágy near Debrecen, site Árkuspart; (53) – Debrecen-Tocopart; (*k*) – Tomaszów, dist. Radom-mine of chocolate flint.

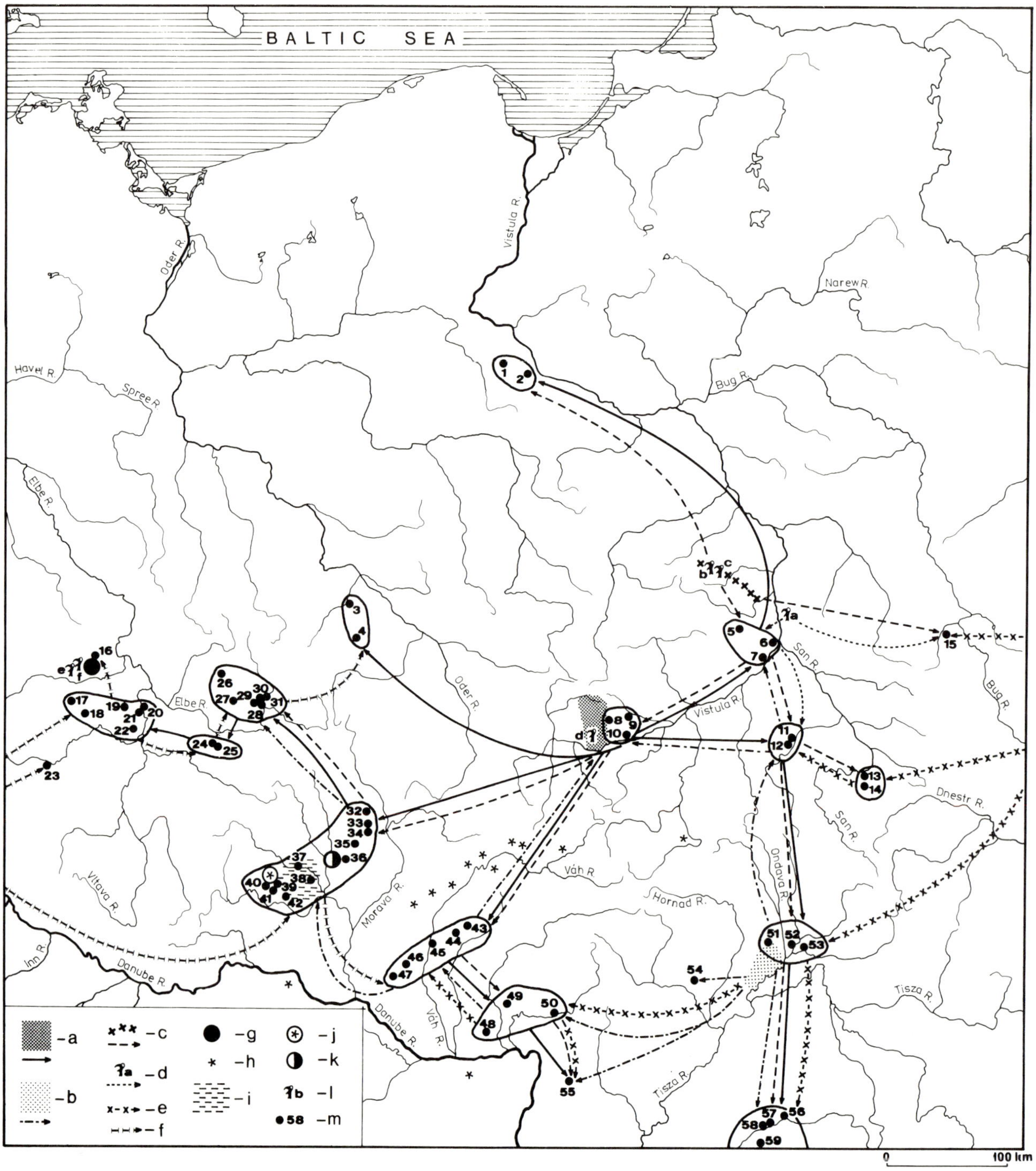

Fig. 28.2. The distribution of siliceous rocks in the later Danubian communities in eastern central Europe at about 4000/3900–3500 BC. (*a*) – Jurassic-Cracow flint deposits and directions of raw material distribution; (*b*) – obsidian deposits and directions of raw material distribution; (*c*) – chocolate flint deposits and directions of raw material distribution; (*d*) – grey-white spotted flint from Świeciechów-Lasek and directions of raw material distribution; (*e*) – directions of Volhynian and Dnestr flints distribution; (*f*) – directions of Bavarian striped hornstone distribution; (*g*) – Skršin quartzite deposits; (*h*) – radiolarite deposits; (*i*) – hornstones of Moravský Krumlov type deposits; (*j*) – 'plasma' of Boskovštejn type deposits; (*k*) – Jurassic flints and hornstones of Stránská skála type deposits; (*l*) – mines: (b) – Tomaszów, dist. Radom, mine of chocolate flint; (c) – Wierzbica, dist. Radom, site 'Zele', mine of chocolate flint; (d) – Sąspów, dist. Cracow, mine of Jurassic-Cracow flint; (e) – Tušimice, dist. Chomutov, mine of Tušimice quartzite; (f) – Bečov, dist. Most, mine of Bečov quartzite; (*m*) – archaeological sites: *Poland* (1) – Konary, dist. Bydgoszcz, site 20; (2) – Gustorzyn, dist. Włocławek (3) – Gniechowice, dist. Wrocław, site 2; (4) – Niemcza, dist. Wałbrzych; (5) – Opatów, dist. Tarnobrzeg; (6) – Malice, dist. Tarnobrzeg; (7) – Samborzec, dist. Tarnobrzeg; (8) – Iwanowice, dist. Cracow; (9) – Niedźwiedź, dist. Cracow; (10) – Cracow-Mogiła 48; (11) – Rzeszów, site 16; (12) – Rzeszów, site 18; (13) – Przemyśl-Budy, site 1; (14) – Kormanice, dist. Przemyśl; (15) – Werbkowice, dist. Zamość; *Czechoslovakia* (16) – Hrobčice, dist. Teplice; (17) – Malá Černoc, dist. Louny; (18) – Chrášťany, dist. Rakovnik; (19) – Buštěhrad, dist. Kladno; (20) – Řež, dist. Praha-East; (21) – Černý Vůl, dist. Praha-West; (22) – Řeporyje, dist. Praha-West; (23) – Křimice, dist. Pilsen North; (24) – Miskovice, dist. Kutná Hora; (25) – Bylany, dist. Kutná Hora; (26) – Bukvice,

obsidian and 0.4% of striped Bavarian hornstone and porcellanite. The Předměřice settlement is about 280 km from the nearest Jurassic-Cracow flint deposits; 300 km from the striped Bavarian hornstone deposits, 400 km from the chocolate flint deposits and 420–450 km from the obsidian deposits. It is difficult to define the source of the erratic 'Baltic' flint: the type which was dominant at this site. This could come there from the north, east or the west. In any case it must have travelled a much smaller distance than the raw materials mentioned above.

In the Hradec Králové district, Jurassic-Cracow flint occurs at almost all the sites of the period, both those of LBK and the stroke ornamented pottery culture. In addition, striped Bavarian hornstone has been found at stroke pottery sites at Nový Bydžov and Plotiště; chocolate flint has been found at Chlum and Smiřice, and obsidian at Smiřice. Quartzite from the deposits in the north-western Bohemia about 150–180 km away rarely occurs among the raw materials at stroke pottery sites in this area. Isolated discoveries of obsidian also occur in settlements of the younger phases of the stroke pottery culture in Bohemia, sometimes associated with fragments of Moravian painted pottery (Steklá, 1959, pp. 32–3).

In Moravia itself, striped Bavarian hornstone occurs in small amounts, mainly at sites of the stroke pottery culture (e.g. Určice-Záhumení in the Prostějov district) or the Moravian painted pottery culture (e.g. at Boskovštejn in the Znojmo district and Vyškov-Dědice in the loco district) sometimes with the pieces of stroke pottery (e.g. at Mošovice in the Znojmo district) (Fig. 28.2). This is the most easterly occurrence of this raw material, about 280–360 km from the nearest source.

In the Moravian painted pottery settlements local raw materials are dominant; but side by side with them are found single specimens of Polish raw materials. Jurassic-Cracow flint and chocolate flint occur quite often. Obsidian also occurs sporadically. Workshop sites of this period are often found close to deposits of the local raw materials. One example is the site at Jezeřany-Maršovice (Znojmo district) dated by Rakovský to the ɪb/ɪɪa phase of the Moravian painted pottery culture. Among 1312 artefacts from features 1–52 at this site, the local Moravský Krumlov hornstone is absolutely dominant (93.2%). The hornstone was worked on a large scale. Numerous big cores occur side by side with the nodules of this raw material, which come from the same source as those at the Vedrovice-Zábrdovice settlement. Raw materials from distant sources constitute only 1.2% of the finds, so one may believe that they were of little or no technological importance. Among raw materials of non local origin, 0.5% consists of erratic 'Baltic' flint from north-eastern Moravia, that is, from deposits relatively close by. The nearest *in situ* deposit is that of the Jurassic-Cracow flint (0.4%) about 260 km away; those of chocolate flint are about 430 km away, and those of obsidian (0.1%) (one specimen) about 340–400 km. The other 5.7% are undefined raw materials, burned specimens and the like. In the next, ɪɪb painted pottery phase the site preserved its workshop character.

In Moravia, at the settlement of Horakov, (Brno suburb district) belonging to the ɪb phase of the painted pottery culture, flint of Stránská Skála type from the nearby deposits was the basic raw material (museum collection, Brno). Moravský Krumlov hornstone, Jurassic-Cracow and chocolate flint are also found in considerable quantites at this site and radiolarite and erratic 'Baltic' flint also occurred.

In the west of Slovakia, a much bigger variety of raw materials was exploited at this time. In the ɪ Lengyel phase of the Bardoňovo settlement (in the Nové Zámky district) 554 artefacts were recovered. Local raw material, a form of grey chert, predominated. It comprised 43.9% of the total assemblage. Obsidian was also important and made up a further 18.4% of the total, and lesser amounts were found of Jurassic-Cracow flint – (9.4%), red radiolarite – (4.3%), and other undefined raw materials – (24%) (Fig. 29.2). Chocolate and Volhynian flints do not occur at Bardoňovo, but they were found at another Slovakian site, that of Koláry (Lučenec district). Here obsidian was absolutely dominant but Jurassic-Cracow flint and local white hornstone were also present; no red radiolarite was found. According to Pavúk the settlement at Koláry also belongs to the first phase of the Lengyel culture in Slovakia, but is later than the settlement at Bardoňovo. Perhaps this is the reason for the difference in the raw material supply of the two sites. The settlements at Bardoňovo and Koláry are situated 60 km away from each other, in a west–east direction. At Koláry much more obsidian is found, but deposits of this raw material were 60 km nearer to Koláry than to Bardoňovo. Volhynian flint travelled with the obsidian from the east. And radiolarite, absent at Koláry, was able to reach Bardoňovo from the nearest deposits, in the middle Váh River Basin or from the region of Vienna – to the west of the settlement or even from the south – from the region of Tata, in Hungary. Bardoňovo was about 40 km from Tata, and Koláry about 90 km. In both cases it would be necessary to cross the Danube. The settlements were

---

dist. Jičin; (27) – Novy Bydžov, dist. Hradec Králové; (28) – Plotiště, dist. Hradec Králové; (29) – Chlum, dist. Hradec Králové; (30) – Předměřice, dist. Hradec Králové; (31) – Smiřice, dist. Hradec Králové; (32) – Kostelec on the Hané R., dist. Prostějov; (33) – Určice, dist. Prostějov, site Záhumení; (34) – Otaslavice, dist. Prostějov, site Glosy; (35) – Vyškov-Dědice, site 'Cihelna'; (36) – Horakov, dist. Brno-Country; (37) – Nová Ves u Ivančic, dist. Brno-Country; (38) – Jezeřany-Maršovice, dist. Znojmo; (39) – Střelice, dist. Znojmo; (40) – Boskovštejn, dist. Znojmo; (41) – Mašovice, dist. Znojmo; (42) – Těšetice-Kyjovice, dist. Znojmo; (43) – Bánovce on the Bebrava R., dist. Topolčany; (44) – Velké Hoste, dist. Topolčany; (45) – Pečenady, dist. Trnava; (46) – Bínovce, dist. Trnava; (47) – Budmerice, dist. Trnava; (48) – Bardoňovo, dist. Nové Zámky; (49) – Santovka, dist. Levice; (50) – Koláry, dist. Lučenec; (51) – Egreš, dist. Trebišov, site Hôrka; (52) – Velké Raškovce, dist. Trebišov; (53) – Čičarovce, dist. Trebišov; *Hungary*; (54) – Szerencs-Taktaföldvár, dist. Borsod-Abaúj-Zemplin; (55) – Aszod, dist. Pest; (56) – Esztár, dist. Berettyóújfalu; (57) – Herpály, dist. Berettyóújfalu, site Földvar; (58) – Berettyóújfalu, site Szilhalom; (59) – Zsáka, dist. Berettyóújfalu, site Markó.

220–250 km from the nearest Jurassic-Cracow flint sources and 390–420 km away from the Polish chocolate flint deposits.

At Santovka (Levice district) 409 artefacts from the beginning of phase II of the Lengyel culture were studied. Unlocalised raw materials from Slovakia were dominant. Among the rest obsidian was the most common type of raw material with 12.5% of the collection, next the radiolarite with 8.8%, Jurassic-Cracow flint with 8.3%, the chocolate flint comprised 5.9% and the Volhynian flint only 1%. It may be noticed that the multi-directional contacts with the remote territories already recognised in the Linearbandkeramik culture settlement persist and even grow stronger in this period. The chocolate flint found at Santovka is derived mainly from the mine at Tomaszów but, as well as this source, chocolate flint from the Wierzbica-Zele mine has also been found here. This type of flint was processed on a large scale at the Opatów settlement in the Tarnobrzeg district (Więckowska, 1971, pp. 127–8). Santovka is situated at the same distance from the deposits as the settlements at Koláry and Bardoňovo.

The most interesting raw material inventories from settlements in the west of Slovakia are those from Bánovce on the Bebrava (Fig. 28.2). 341 specimens of flaked artefacts from the end of the second phase of the Lengyel culture were obtained at this site during the investigations of 1977–78. Jurassic-Cracow flint probably from the mine at Sąspów was the basic raw material. This amounted to 68.9% of the total finds at Bánovce. Other raw materials in use comprise chocolate flint from the mines at Tomaszów and Wierzbica (7%), radiolarite (5.6%), and Cretaceous erratic 'Baltic' or Volhynian flint (0.3%) (one specimen). The nearest deposits of Jurassic-Cracow flint are about 180 km from the settlement. The flint mine at Sąspów is about 200 km away and the mines of chocolate flint are about 350 km away. The predominance of raw material carried from a long distance at Bánovce is interesting because only 40 km from the settlement, in the Vlára River Basin, there are abundant radiolarite deposits. Bánovce thus serves as an example of a settlement of the later period systematically utilising raw materials from the remote deposits, as do the sites of Bylany, Brześć Kujawski, Niemcza, Skoroszowice and Strachów in the Linearbandkeramik culture period.

The settlement was situated between the valleys of two big rivers – the Váh and the Nitra, on the route from the Polish deposits to the west of Slovakia. Large amounts of the northern raw materials and the lack of the intermediate settlement centres suggest that it had direct contacts with the communities of the Lengyel–Tisza complex in the region of Cracow. Though expeditions from the settlement to the mine territories in the south of Polish Jura are also probable, the artefacts made of chocolate flint are a proof of contacts with the inhabitants of the Cracow region. Jurassic flint was reaching Bánovce in precore forms or as initially worked cores, which were further processed and exploited in the settlement. Workshops preparing such forms occur on the mine territories – for example workshop 1/1971 at Sąspów flint mine (Lech, 1975). Radiocarbon dates obtained at Sąspów indicate that the flint at this site was also exploited at the same period as phase II of the Lengyel culture (Lech, 1981, p. 48).

The Bánovce settlement, located in the northern part of the west Slovakian Danubian settlement region, being the nearest settlement to the flint deposits in Poland, was probably an important intermediary point in the distribution network of raw materials carried from the north to the territories of the west Slovakia. Polish flint travelled from this site along the Nitra River to more distant contemporary settlements (Fig. 28.2).

Raw material distribution patterns confirm the existence of permanent, direct and indirect, contacts between the communities living on both sides of the Carpathian Mountains, also visible in the ceramic styles. Steady supplies of raw materials from the north, observable in particular phases of the Danubian in the west of Slovakia, are at the same time a proof of settlement continuity on the loess areas in the Vistula River Basin (see Kulczycka-Leciejewiczowa, 1979, pp. 101–2).

In spite of the great variety of local flint and the abundance of the available deposits, we can also see in the settlements of Vistula River Basin the occurrence of raw materials from distant regions. The same holds true for the mining settlements. In the numerous finds from Opatów (Tarnobrzeg district), a settlement exploiting chocolate flint from the Wierzbica-Zele mine, beside 94.26% of this raw material, 5.00% of obsidian, 0.13% of Świeciechów flint and 0.17% of Jurassic-Cracow flint were found. In the Cracow-Mogiła 48 settlement exploiting nearby deposits of Jurassic-Cracow flint, this raw material comprised 98% of the raw material inventory. Besides that there was 1% of the chocolate flint and 0.9% of obsidian (Kaczanowska & Lech, 1977, table II). The Opatów settlement is situated about 50 km from the flint deposits in the region of Wierzbica, about 130 km from the deposits of Jurassic-Cracow flint and about 260 km from the nearest obsidian deposits. The Cracow-Mogiła 48 settlement is about 150 km from the nearest chocolate flint deposits and more than 200 km from the obsidian deposits. In general terms, both settlements in Little Poland must have been partly synchronous with the settlements from western Slovakia discussed above.

Very interesting discoveries of stroke ornamented pottery and materials of the Tisza Culture were made by J. Vizdal at Velké Raškovce (Trebišov district) in east Slovakia in the years 1971–2. Stroke pottery occurred in a few features at this site (Vizdal 1973). This pottery is derived from the other side of the Carpathian Mountains, from the loess areas of Little Poland. The most likely route for movement between the two regions is along the Wisłok and the San River Valleys, through the passes of the Carpathian Mountains and along the valleys of the Ondava and the Laborec. In this case it is difficult to say if we are dealing with the migration of a human group making the pottery or with the adoption of a pottery style as a result of contact between two communities. In any case, this is the furthest south-easterly find of pottery of this type (Kulczycka-Leciejewiczowa, 1979, p. 100). The flaked stone assemblage from the site consisted of 128 specimens, more than a half of which were made from local raw materials coming from an undefined source. Imported raw materials included obsidian (12.5%), Jurassic-Cracow flint (7.8%), Volhynian flint (0.9%), and flint of the Dnestr type (0.9%). At the adjoining site Čičarovce, also in Trebišov district, which is dated by Vizdal to the late Tisza culture and by Šiška to the beginning of the Tiszapolgár culture, an assemblage of 127 artefacts included 33.1% made of obsidian, 33.1% of local or unlocated raw materials and 33.9% of the Volhynian flint. The visible increase in the amount of the Volhynian flint in this horizon foreshadows its later dominance in the later Tiszapolgár culture settlements.

Towards the end of the Neolithic period in eastern Slovakia, at the beginning of the development of the Tiszapolgár culture, many different types of raw material were in use. Volhynian flint did not at first win a dominant position. A typical stone artefact inventory from this period is provided by Gašaj's excavations at Egreš-Hôrka, also in the Trebišov district. Eighty-one flaked stone artefacts were found in the settlement, which belongs to the Tiszapolgár-Csöszhlom-Oborín culture. Among these local obsidian comprises 44%, chocolate flint 36%, Jurassic-Cracow flint 4%, Volhynian flint 4%, and 12% are of undefined raw material (Fig. 28.2).

Jurassic-Cracow flint reaching the settlements in the eastern Slovakian lowland travelled about 200–250 km. Volhynian flint also arrived with Jurassic-Cracow flint, presumably as a result of indirect exchange from the settlements in the region of Przemyśl and Rzeszów, along the San and the Wisłok River route (Fig. 28.2). This exchange pattern is confirmed by a study of the flaked stone from Velké Trakany and Velké Raškovce. In both these settlements local raw materials were dominant. However, essential changes in the exchange pattern seem to take place during the existence of the latest settlements – Čičarovce and Egreš. The share of raw materials from the distant deposits shows a clear increase at this time. This indicates the existence of contacts with at least two settlement regions in the loess areas of Little Poland – one on the Upper Vistula and the other in the region of Przemyśl.

Among the latter settlements, which already belong to the central European Aeneolithic, the direction and the intensity of contacts indicated by the distribution of raw materials show a partial change. Generally, in this period blank and ready tools are, to a much greater degree, a subject of distribution than natural nodules, pre-core forms and initially worked cores. This general trend does not always hold true. In the region discussed, the late fourth millennium cemeteries of the Tiszapolgár culture in eastern Slovakia are an exception (Vizdal, 1977; Lichardus, 1980; Lichardus-Itten, 1980).

Conclusions

The Danubian I–LBK first farming communities, in the Vistula and the Oder catchment areas, came from the south. In the second half of the fifth millennium BC, these communities had lively contacts with their areas of origin in the south, on the other side of the Carpathians and the Sudeten region. In the south this is shown by the presence of Polish raw materials from the Vistula River Basin; in the north by the presence of southern obsidian, as well as by the parallel rhythm of the stylistic changes in pottery in the two regions. These facts clearly confirm the presence of long-distance direct contacts in LBK between Little Poland and Silesia, and between the region of the Holy Cross Mountains and Kujavia; contacts which can be connected with expeditions seeking for flint. Complex multi-directional systems of indirect exchange between north and south are also shown by the occurrence of obsidian at Skoroszowice, and of chocolate flint at Żuków (Szczecin district), Skoroszowice and Bylany, and that of Jurassic-Cracow flint at both Brześć Kujawski, Bylany and Křimice.

North–south contacts continued in the fourth millennium in sites of the Lengyel and Tisza culture. But, with the passing of the time the intensity of these contacts decreased, or changed in character. This is clearly visible when we compare the situation among early and late Lengyel groups on both sides of the Carpathians. This change reflects the gradual adaptation of the early farming communities to the natural environmental conditions of the region into which they migrated, whereby deposits of raw materials closer to the settlements were used to a larger extent. Another important reason for the change in the direction of contacts at the end of Neolithic and at the beginning of the Early Eneolithic periods can be found in the intensified cultural relations with the east, also accompanied by intensification of flint supplies from the Volhynian upland and the Dnestr Basin. In the eastern part of central Europe, north–south contacts decreased in importance in the second half of the fourth millennium and in the first half of the third millennium; whereas in the whole eastern part of central Europe the role of east–west relations began to grow in significance.

Bibliography

Kaczanowska, M. & Lech, J. (1977). 'The flint industry of Danubian communities north of the Carpathians'. *Acta Archaeologica Carpathica*, **17**, 5–28.

Kozłowski, J. K. (1970). 'Z badań nad wytwórczością krzemieniarską w kulturze ceramiki wstęgowej rytej'. In: *Z badań nad kulturą ceramiki wstęgowej rytej*, ed., J. K. Kozłowski, pp. 73–94. Kraków: Polskie Towarzystwo Archeologiczne. Oddział w Nowej Hucie.

Kulczycka-Leciejewiczowa, A. (1979). 'Pierwsze społeczeństwa rolnicze na ziemiach polskich. Kultury kręgu naddunajskiego'. In: *Prahistoria ziem polskich*, ed., W. Hensel, vol. 2, Neolit, ed., W. Hensel & T. Wiślański, pp. 19–164. Wrocław-Warszawa-Kraków-Gdańsk: Zakład Narodowy imienia Ossolińskich.

Lech, J. (1975). 'Neolithic flint mine and workshops at Sąspów, near Cracow'. *Tweede Internationale Symposium over Vuursteen. 8–11 Mei 1975 – Maastricht. Staringia*, vol. 3, pp. 70–1. Nederlandse Geologische Vereniging.

Lech, J. (1981). 'Flint mining among the early farming communities of central Europe'. *Przegląd Archeologiczny*, **28**, 5–55.

Lech, J. (1983). 'Flint mining among the early farming communities of central Europe. Part II – The basis of research into flint workshops'. *Przegląd Archeologiczny*, **30**, 47–80.

Lech, J. & Leligdowicz, A. (1980). 'Die Methoden der Versorgung mit Feuerstein und die lokalen Beziehungen zwischen den Siedlungen und Bergwerken im Weichselgebiet während des 5. bis 2. Jt. v. u.Z.' In: *Urgeschichtliche Besiedlung in ihrer Beziehung zur natürlichen Umwelt*, ed. F. Schlette, pp. 151–84. Wissenschaftliche Beiträge der Martin-Luther-Universität Halle-Wittenberg 1980 - 6 (L15). Halle (Saale).

Lichardus, J. (1980). 'Zur Bedeutung der Feuersteingewinnung in der jüngeren Steinzeit Mitteleuropas'. In: *5000 Jahre Feuersteinbergbau. Die Suche nach dem Stahl der Steinzeit.*, eds G. Weisgerber, R. Slotta & J. Weiner, pp. 265–70. Bochum: Deutschen Bergbau-Museum.

Lichardus-Itten, M. (1980). 'Silexknollen als Beigabe in Gräbern der frühkupferzeitlichen Tiszapolgár-Kultur'. In: *5000 Jahre Feuersteinbergbau. Die Suche nach dem Stahl der Steinzeit.*, eds G. Weisgerber, R. Slotta & J. Weiner, pp. 279–83.

Löhr, H., Zimmermann, A. & Hahn, J. (1977). 'Feuersteinartefakte'. In: Der Bandkeramische Siedlungsplatz Langweiler 9, eds R. Kuper, H. Löhr, J. Lüning, P. Stehli & A. Zimmermann, *Rheinische Ausgrabungen*, **18**, pp. 131–266. Bonn: Rheinland-Verlag GMBH.

Milisauskas, S. (1976). *Archaeological Investigations on the Linear Culture Village of Olszanica*. Wrocław-Warszawa-Kraków-Gdańsk: Zakład Narodowy imienia Ossolińskich.

Ondruš, V. (1976). 'Neolitické dílny z Vedrovic-Zábrdovic'. In: *Sborník prací filozofické fakulty Brněnské univerzity*, 24–5, Řada archeologicko-klasická (E), 20–1, pp. 133–9. Brno: Univerzita J. E. Purkyně.

Steklá, M. (1959). 'Vztahy mezi keramikou vypíchanou a malovanou'. *Acta Universitatis Carolinae – Philosophica et Historica*, **3**, 31–8.

Sulimirski, T. (1957–59). *Polska przedhistoryczna. Część II. Drugie tysiąclecie przed Chr.*, Londyn: Polski Uniwersytet na Obczyźnie.

Sulimirski, T. (1960). 'Remarks concerning the distribution of some varieties of flint in Poland', *Światowit*, **23**, 281–307.

Vencl, S. (1971). 'Současny stav pozàní postmesolitických štipanych industrii w Československu'. In: *Z badań nad krzemieniarstwem neolitycznym i eneolitycznym*, ed., J. K. Kozłowski, pp. 74–99. Kraków: Polskie Towarzystwo Archeologiczne & Muzeum Archeologiczne w Krakowie.

Vizdal, J. (1973). *Zemplín v mladšej dobe kamennej*. Košice: Východoslovenské vydavatelstvo, NP pre Zemplínske múzeum v Michalovciach.

Vizdal, J. (1977). *Tiszapolgárske pohrebisko vo Velkých Raškovciach*, Košice: Východoslovenské vydavatelstvo, NP pre Zemplínske múzeum v Michalovciach.

Więckowska, H. (1971). 'Materiały krzemienne i kamienne z osad kultury ceramiki wstęgowej i trzcinieckiej w Opatowie'. In: *Z polskich badań nad epoką kamienia*, ed. W. Chmielewski, pp. 103–83. Wrocław–Warszawa–Kraków–Gdańsk: Zakład Narodowy imienia Ossolińskich.

29

Cretaceous chert sourcing in north east Ireland: preliminary results

D. GRIFFITHS AND P. C. WOODMAN

At the last international flint conference in Maastricht, a paper by one of the authors dealt with the problems of transportation of flint in eastern Ireland (Woodman, 1979). It was noted that Cretaceous flint (now usually referred to as chert) was being brought to sites in the centre of the Bann Valley but that due to the nature of the geological deposits in north east Ireland, it was impossible to decide whether chert was being brought into this area from the East Antrim coastal area or from sources further to the west. This short contribution reports on the progress towards sourcing chert in north east Ireland.

Archaeological and geological background

The initial problem is that we have in north east Ireland a heavy bladed Later Mesolithic lithic industry based on the use of the Larnian uniplane cores (Woodman, 1978). Retouched tools are produced through the simple secondary working of these large blades. While sites with a normal range of retouched tools and industrial waste are known in both Leinster, e.g. Dalkey Island (Liversage, 1968), or Strangford Lough, e.g. Rouch Island (Movius, 1940), the distribution of waste and retouched tools in Co. Antrim can be rather different.

On the Antrim coast, recent excavations at Bay Farm I (lower) have revealed an extensive area of chert working which may represent only a tiny remnant of the original working floor which dates to the Later Mesolithic. This area has been extensively disturbed by both raised beaches and mud flows which have left very little *in situ*. As a contrast, excavations at Newferry, Co. Antrim (Woodman, 1977) have revealed an area where Later Mesolithic occupation took place over a period of 2000 years from 5500 bc. Aside from the basal layers, there is virtually no evidence of chert working being carried out on this site. The differences between the coastal and inland sites are clearly demonstrated in Fig. 29.1.

It is therefore tempting to suggest that the products of the coastal industrial sites may have been taken to areas such as Newferry. There would appear to be good geological reasons and archaeological evidence for this conclusion. As can be seen from Fig. 29.2, Cretaceous chert has a restricted availability as Cretaceous deposits outcrop in the form of virtually the edge of a rectangle. Other exposures are precluded due to the extensive overlying basalts and it has been shown that during the last glaciation (in Ireland, the Midlandian) Scottish or Irish sea ice did not over-ride the Basaltic Antrim plateau and conveniently leave erratics in the Bann Valley. Instead it would appear that an Ice Dome developed in Lough Neagh and pushed out in several directions. Creighton (1974) noted that 'flint' was only found as an erratic on the shores of Lough Neagh and not in the southern end of the Lower Bann Valley. Obviously, therefore, there may have been a need to import raw materials into the Lower Bann Valley.

There are good archaeological reasons why chert may have been brought from the Antrim coast. Firstly, as was

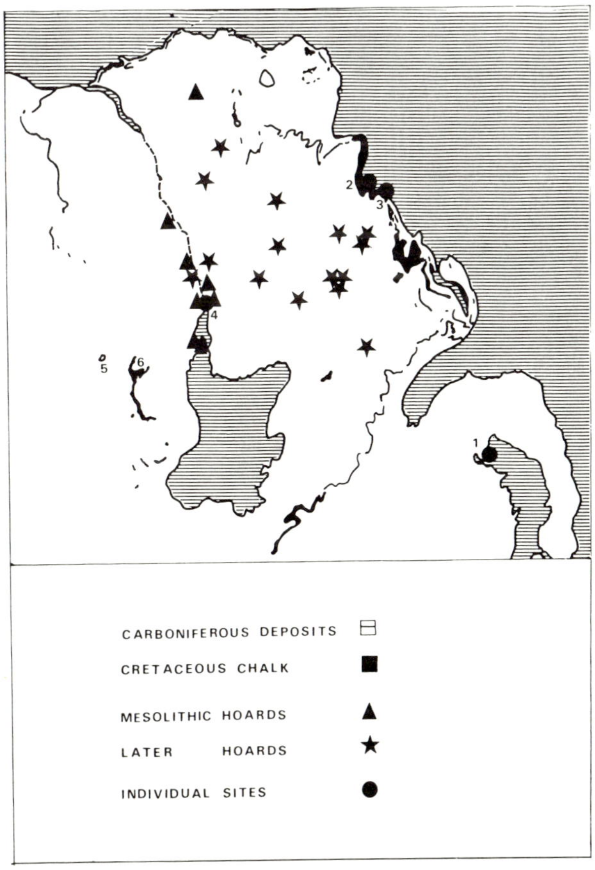

Fig. 29.1. Location of Cretaceous deposits in north east Ireland and locations referred to in text: (1) Rough Island; (2) Bay Farm, Carnlough; (3) Mad Mans Window; (4) Newferry; (5) Slieve Gallion; (6) Carmaen.

shown by Woodman (1979), the blade size found on the Bann Valley sites is identical to that found on the coast in an area where chert is freely available. Therefore, it is unlikely that the blades at Newferry are produced from local erratic flint, e.g. from the shores of Lough Neagh. There is also no evidence so far that there are anything but industrial sites on the east Antrim coast during the Mesolithic and at Bay Farm phosphate analysis has not revealed any areas of intense settlement. Therefore, combining the low potential of this area during the Mesolithic (Woodman forthcoming) with large quantities of industrial waste, it must be assumed that the products of these industrial sites were being transferred elsewhere.

While it is tempting to assume that this shows that Newferry and Bay Farm were part of the same cycle of production and consumption, they could easily be parts of two different cycles. It was therefore felt to be essential that some method be found to characterise Antrim coastal flint as distinct from that found to the west of the Bann Valley. Due to the fact that chert was, unlike mined chert, available on beaches from a range of layers in adjacent cliffs of Cretaceous white limestone, it was necessary to find a method which would quickly characterise a broad range of chert. The only hope lay in the fact that there is a stratigraphically longer series of Cretaceous deposits in the east than in the west and that it might be possible to show that there was either a narrower or different range of chert types available in the west.

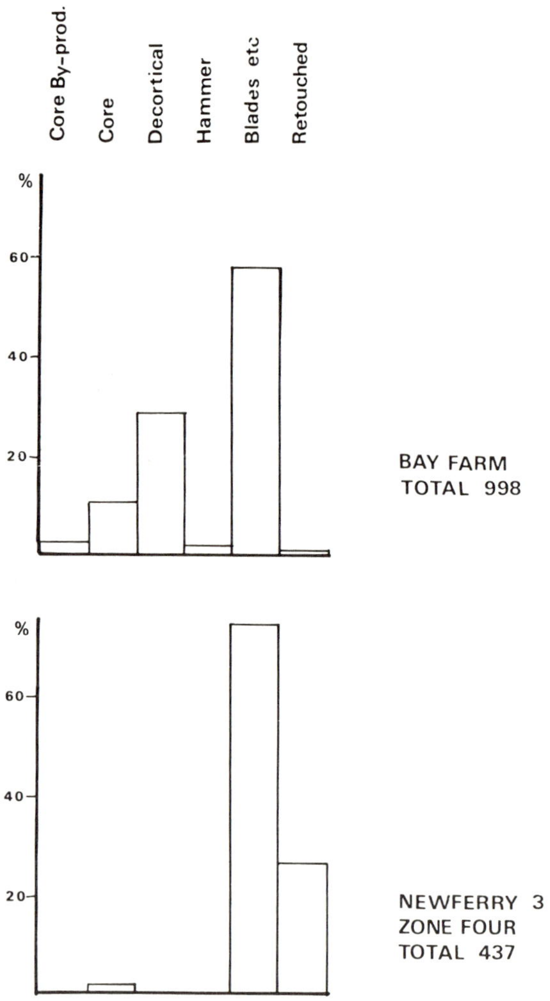

Fig. 29.2. Distribution of industrial products and retouched tools from Bay Farm, Carnlough and Newferry, Co. Antrim.

It was suggested that electron spin resonance spectroscopy might provide a relatively simple method of characterising flint as occasionally distinctive signals had been noted to be associated with particular types of chert. There was at this stage no guarantee that any distinctive characteristics would be found in any particular band of Irish flint. However, if a distinctive range of signals associated only with either the eastern or western sources was found also in the Newferry flint, then this would suggest that the flint artefacts at Newferry were derived from that particular source. Similarly a narrow range of signals from Newferry might be associated with the western sequence while a broader range would be associated with the longer eastern sequence.

As a preliminary step, one of the authors selected flint from the following locations: (1) the beach at Carnlough, east Antrim; (2) the chipping floor at Bay Farm I, lower Carnlough, chosen to ensure that the same range of material was *used* by Mesolithic flint workers as was *available* on the beaches; (3) the Later Mesolithic site at Newferry, Co. Antrim; (4) Slieve Gallion, Co. Derry; (5) the Carmaen outcrop near Moneymore, Co. Derry. In the last instance, a quarry sunk below the surface rather than a natural exposure was used. These specimens were then taken to London for preliminary analysis with electron spin resonance.

Electron spin resonance analysis

The ESR (electron spin resonance) spectra of chert samples from the known geological sources in Ireland were compared in an attempt to find some feature that is characteristic of the provenance of the sample. ESR spectra of chert artefacts were also recorded. All the spectra (except that of an artefact believed to have been quite strongly heated) showed signals due to E′ and oxygen-associated hole centres (Griffiths *et al.*, 1983*a*). In addition some spectra showed features due to manganese (Robins *et al.*, 1981), $\cdot CH_3$, $R\dot{C}H_2$ (Griffiths *et al.*, 1982) and other centres. Intercomparison of the spectra yielded a parameter that served to distinguish the two major geological sources considered. This parameter is the ratio of the amplitudes of the two lines marked A and B in Fig. 29.3. This is an empirical observation and the geochemical implications and origins of the lines and their associated features will not be discussed, save to mention that they are believed to arise from oxygen-associated hole centres in the silica lattice (Griffiths *et al.*, 1983*a* and Stapelbroek *et al.*, 1978).

Samples were taken by removing small cylindrical cores of chert with a water-cooled diamond-impregnated drill. In some cases granular samples were also used, but fine particles were avoided (Griffiths *et al.*, 1983*a*). In most cases more than one sample was taken from each nodule. The ratio of the amplitudes of the lines A and B was calculated for each sample using ESR spectra recorded at

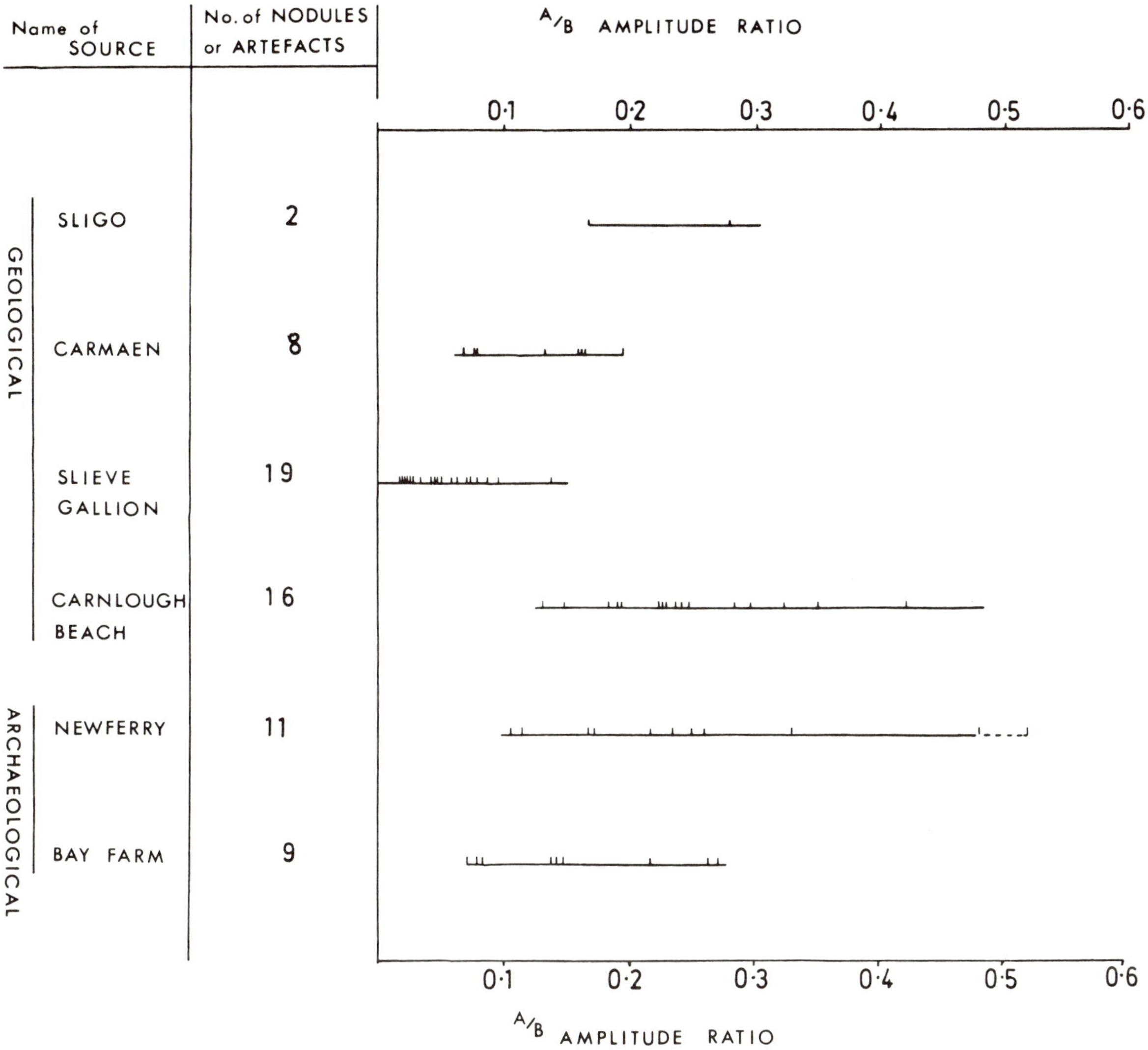

Fig. 29.3. Form of ESR curve from which the A/B ratio is derived.

identical spectrometer settings. The microwave power used in this work was 4 mW. Varying the power may change the value of the A/B ratio as well as the measurement errors.

The range of A/B ratios found in chert from each source is shown in Fig. 29.4 by the extent of the horizontal lines. Considerable variation in the A/B ratio may be found within one nodule. The ratio ranged from 0.13 to 0.25 within a single nodule from Carnlough Beach. The small vertical marks on the horizontal lines in Fig. 29.4 indicate the average value of A/B for each nodule or artefact sampled.

Experiments have shown that heating can raise the A/B ratio. The dotted portion of the Newferry range is due to a sample which may have been heated. This is suggested by the ESR spectrum of the sample (Griffiths *et al.*, 1983*b*). Another artefact from Newferry (not plotted) was almost certainly heated.

The two most thoroughly sampled geological sources have almost mutually exclusive ranges of A/B. It is interesting to note that in both the Slieve Gallion and the Carnlough beach ranges, a single nodule was sampled having a distinctly higher A/B ratio than the others, and in both cases this single nodule accounts for just over one-third of the whole range. It is of course possible that any of the ranges may be broadened or the distribution of ratios within the ranges changed by the analysis of further samples. The difference between the Slieve Gallion and Carnlough beach ranges seems nevertheless to be significant.

On the other hand, the range found at Carmaen quarry shows a partial overlap with that found on the east Antrim coast. As the Carmaen rock could be a lower part of the geological sequence, not represented on the higher ground of Slieve Gallion, this might be expected.

Comparing the archaeological with the geological ranges shows that the Newferry range (solid portion) very closely corresponds to the Carnlough beach range, not only in extent but also in the distribution within that range: the Newferry range has one artefact of markedly higher A/B ratio that accounts for just over a third of the range.

The Bay Farm samples do show a much narrower range though the explanation here could be a much smaller sample size. It could also reflect in this instance a tendency on this site to choose flint of distinctive character rather than randomly selecting nodules from the beach – in particular the rather darker glassy flint that is found towards the top of the sequence on the east Antrim coast; however, even here, it is possible to see that Bay Farm material was of distinctly different character from the more available western source at Slieve Gallion. The overlap with Carmaen is of lesser importance as this source is usually only available through below surface quarrying and so its longer (?) sequence would not have been as available to prehistoric man.

Obviously it is impossible to be certain that all the flint found in Bann Valley sites such as Newferry came from the east Antrim coast as nodules of poor quality erratic flint are available on the shores of Lough Neagh, but if these preliminary results are confirmed through the analysis of larger batches, then the A/B amplitude ratio indicates that the transportation of flint from the Antrim coastal area was an important component in the supply of raw materials to the Bann Valley.

While sourcing of flint has been an obvious problem, the question of the mechanisms by which this material was transported presents even greater difficulties. The fact that there is no significant reduction in the size of industry or tendency to rework the blades into smaller tools exhibits a certain confidence about supply which suggests that flint was not obtained through an exchange system. Therefore, the major problem is whether the Antrim coast was exploited by a low density highly mobile population or through sedentary communities based on the Bann Valley which used specialised task groups to obtain raw materials.

This analysis also suggests that it may be possible to distinguish between north eastern Irish and English flint which could also be of importance. Some Mesolithic sites in north west England are using a distinctive multi-coloured flint of unknown origin. This may be flint that has been brought from considerably further east and therefore represents evidence of human movement over a considerable distance. However, it is more likely that this flint is made from glacial erratic Irish Sea flint.

It is also becoming apparent that flint is rather more freely available round the south and south west coasts of Ireland than has been hitherto expected. This flint which has probably been brought in, by ice, off the continental shelf could well either have its own distinctive characteristics or be similar to southern English flint. Analysis of two nodules from beaches in County Sligo shows that characteristics other than the A/B ratio may need to be used. Even visual inspection of some of the large nodules of flint found on the coast of County Cork shows that certain potentially distinctive types exist. Again it may be possible in the south and south west of Ireland to distinguish local flint from material brought in from outside, i.e. to the north and east.

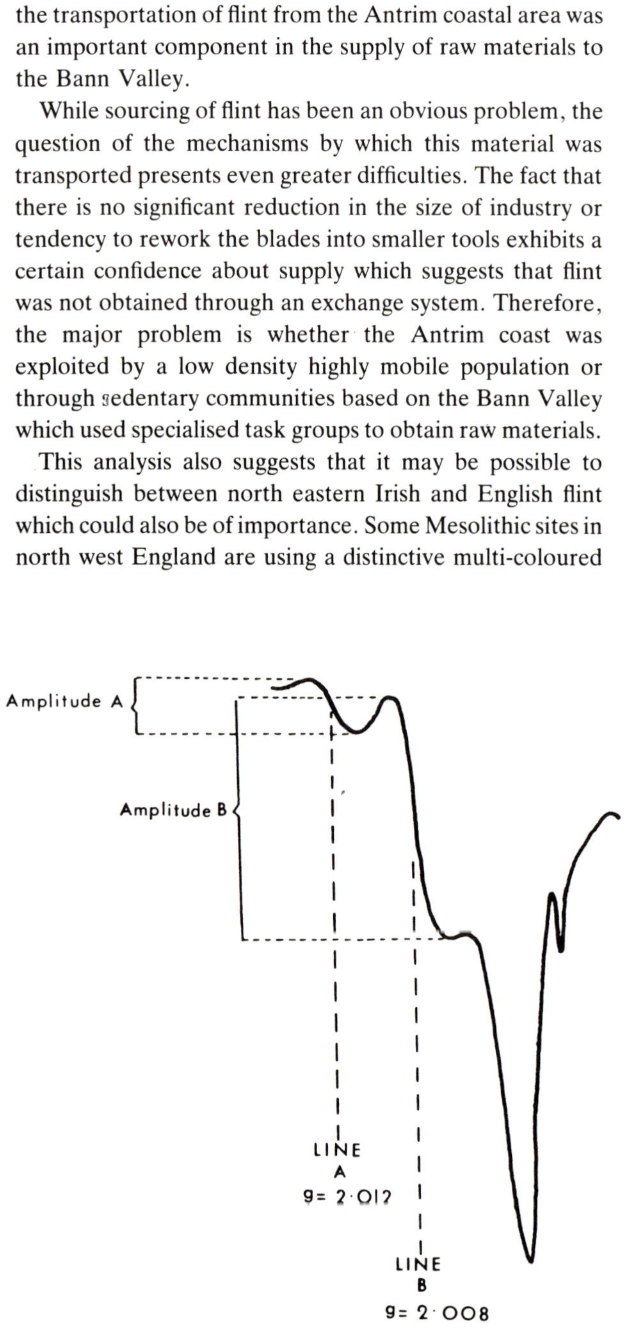

Fig. 29.4. Means of A/B ratios of samples from selected sites.

Acknowledgement

The authors gratefully acknowledge the provision of spectrometer time by Prof. M. C. R. Symons of Leicester University, and the financial support of the SERC.

Bibliography

Creighton, J. R. (1974). A study of the Later Pleistocene geomorphology of north central Ulster. (Unpublished PhD thesis, Queen's University, Belfast.)

Griffiths, D. R., Robins, G. V., Seeley, N. J., Chandra, H., McNeil, D. A. C. & Symons, M. C. R. (1982). 'Trapped methyl radicals in chert'. *Nature,* **300**, 435–6.

Griffiths, D. R., Seeley, N. J., Chandra, H. & Symons, M. C. R. (1983*a*). 'ESR dating of heated chert'. *PACT,* **9**, 399–409.

Griffiths, D. R., Seeley, N. J. & Symons, M. C. R. (1983*b*). 'Determination of chert heating conditions using ESR spectroscopy'. *4th International Flint Symposium*, Brighton, England, in press.

Liversage, G. D. (1968). 'Excavations at Dalkey Island, Co. Dublin 1956–59'. *Proceedings of the Royal Irish Academy,* **66C**, 52–233.

Movius, H. L. (1940). 'Report on a Stone Age excavation at Rough Island, Strangford Lough, Co. Down'. *Journal of the Royal Society of Antiquaries of Ireland,* **70**, 111–42.

Robins, G. V., Seeley, N. J., Symons, M. C. R. & McNeil, D. A. C. (1981). 'Manganese (II) as an indicator of ancient heat treatment in flint'. *Archaeometry,* **23**, 103–7.

Stapelbroek, M. & Griscom, D. L. (1978). 'Oxygen-associated trapped hole centers in high purity fused silica'. In: Pantelides, S. T. (editor), *The Physics of SiO_2 and its Interfaces*, Proceedings of the International Topical Conference, Yorktown, NY, USA, p. 263–7.

Woodman, P. C. (1977). 'Recent excavations at Newferry Co. Antrim', *Proceedings of the Prehistoric Society,* **43**, 155–200

Woodman, P. C. (1978). *The Mesolithic in Ireland.* BAR 58. Oxford.

Woodman, P. C. (1979). 'Problems of flint utilization within eastern Ireland'. *Proceedings of the 3rd International Symposium on Flint, Staringia No. 6.* Ed. F. H. G. Engelen, pp. 113–15. Maastricht.

30

Microwear polishes, blind tests, and texture analysis[1]

M. H. NEWCOMER, R. GRACE AND R. UNGER-HAMILTON

A cornerstone of the high-magnification approach to microwear analysis of flint tools is the identification on them of microwear polishes which, according to the leading exponent of the method, Laurence Keeley, may be linked to different contact materials: 'The most important discovery made in the course of these experiments is that the microwear polishes formed by various materials have distinctive appearances and are, indeed, distinguishable from one another' (Keeley 1980, p. 83). It is claimed that the distinctiveness of these polishes has been 'partially confirmed' by blind tests (Keeley and Newcomer 1977, p. 61) and validated by subsequent work by Moss (1983), Anderson-Gerfaud (1981) and Vaughan (1981), among others. There is, however, a reluctance among many archaeologists to accept the functional interpretations derived from such studies, and we believe this is centered on the use of microwear polishes to identify different types of materials on which the flint tools were used. On the face of it, it seems very unlikely that contact with materials as similar chemically and physically as bone and antler should produce distinctive polishes and claims that they do are backed up only with a small amount of tested data, some unclear photographs and no theory. Workers with the high-magnification method are not always agreed about the distinctiveness of polishes and recent work suggests that polishes produced by different materials can overlap, and that there are other variables which can affect the appearance of polishes, such as flint type, motion in use, duration of use and physical and chemical post-depositional effects (Holmes this volume; Unger-Hamilton 1984, pp. 112–31).

There is also dissatisfaction with other aspects of microwear methodology, especially about the frequent failure to separate observation from interpretation. Microwear analysis is far from reaching the status of a discipline like pollen analysis, and it is unacceptable for its practitioners to present us with elaborate functional reconstruction, such as Buller's (1983, p. 110) claim that retouched tools were used at El Wad to cut tendons, with no primary evidence in the form of photographs, drawings or even descriptions of an experiment showing that a replica of the tool could have been used in this way. Some investigators, including Keeley himself (in Van Noten 1978; Cahen, Keeley and Van Noten 1979; in Audouze *et al.* 1981), clearly believe that the method is so well established that little or no primary data need be presented. Others, like Moss (1983), are more careful to publish such evidence but, unfortunately, due to the problems of taking photographs with sufficient depth of field, it is difficult for other workers to identify polishes from published photographs and impossible to formulate their own interpretation of the evidence.

[1] A previous draft of this paper was presented to a symposium in Tubingen in 1985 and will be published shortly.

The blind tests

In order to investigate some of the problems of microwear polish identification, to develop an improved methodology and to monitor the progress of beginners in microwear research, we decided to run a series of blind tests. Since the original Keeley–Newcomer blind test of 1977, such tests have become a generally accepted way of checking the ability of microwear analysts to reconstruct tool function (Odell and Odell–Vereeken 1981; Gendel and Pirnay 1982); other methods, such as Vaughan's 'reliability test' (1981), in which he re-examined some of his own experimental tools, trusting that he would forget how he had used them, are clearly less satisfactory.

The present series of blind tests were carried out at the microwear laboratory of the London Institute of Archaeology. Six post-graduates in the laboratory agreed to take part in the tests. All of these were actively engaged in microwear projects using the high-magnification approach, studying Mousterian and later lithic artefacts from sites in Britain, Europe and the Near East with the aid of two Olympus Vanox microscopes and one Wild M-20 microscope in the laboratory. Most of the post-graduates had first degrees in archaeology and all had studied stone artefacts for some time before their microwear training. They also satisfied the necessary criteria for the blind test, having reasonable experience in identifying polishes thought to be characteristic of different contact materials (cf. Table 30.2) (three among them have had papers on microwear accepted for publication in reputable journals with peer review).

The tests described below were carried out in 1984–85, using the microscopes, and other equipment, with which the participants were familiar. Tools for the blind tests were made and used by Newcomer, using the same basic rules as for the Keeley–Newcomer test (19778, p. 29). In this exercise, 30 tools have been studied. These were presented

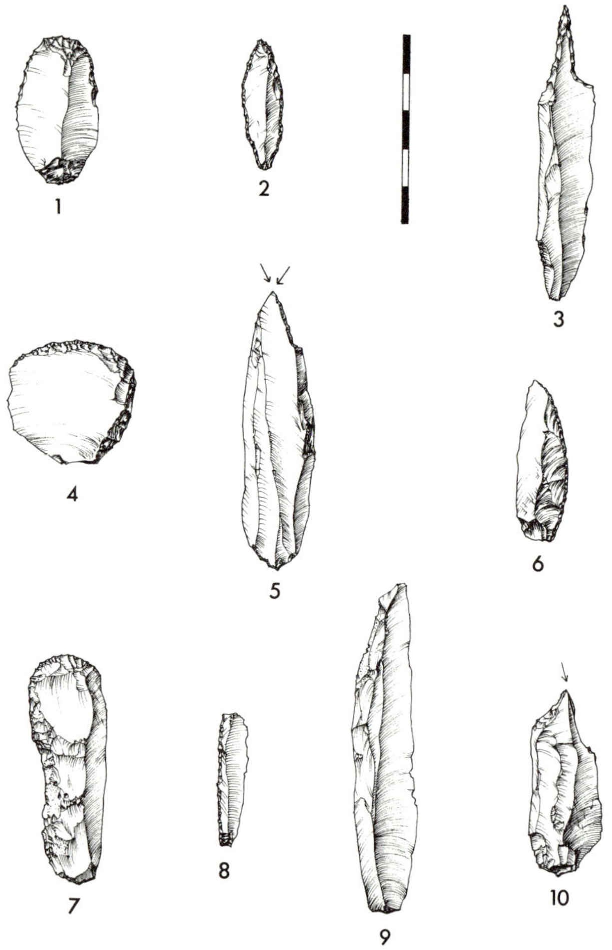

Fig. 30.1. Test tools 1–10: 1, end-scraper hafted in wooden handle with resin and beeswax and used for 15 min to scrape fresh ash wood, 2, drill-bit with distal end hafted in wooden shaft with resin and beeswax and used in bow drill for about 2 min to drill 10 holes in thick shell; 3, piercer used to cut grooves in soaked reindeer antler for 12 min; 4, end-scraper used to scrape soft limestone for 6 min; 5, dihedral burin used on both burin facets to scrape fresh ox bone for 11 min; 6, backed blade fitted to wooden arrowshaft with resin and beeswax and shot from 34 lb yew bow fives times into garden earth; 7, end-scraper made by pressure flaking with antler tine, held on hide pad, unused; 8, backed bladelet hafted with resin and beeswax on surface of wooden haft and used for 29 min to slice fresh meat; 9, unretouched blade used on proximal end of right edge to cut off fins and tail, gut and scale 750 g fresh zander or pike-perch; 10, truncation burin with burin facet used for 9 min to scrape fresh ash wood. Scale in centimetres.

to the participants in groups of ten to enable the tests to be varied as necessary.

The first ten tools were replicas of Upper Palaeolithic tools, made from a single block of flint from the Seine near the site of Pincevent (Fig. 30.1). These tools are generally easier to study functionally than the more generalised tools of the Keeley–Newcomer test since the morphology of tools like end-scrapers and piercers gives strong clues as to the part of the tool used and the motion likely to be in use. The tools were used in simulated Upper Palaeolithic tasks, for periods of time ranging from fractions of a second (the arrowtip) to 29 minutes (the backed bladelet used to cut meat) and then cleaned with detergent and handed over for analysis to the six volunteers. Further cleaning was done in an ultrasonic tank, and it was agreed that the one observer who wanted to clean the tools in a solution of NaOH should study them last. Newcomer also supplied outline drawings of the dorsal and ventral surfaces of the tools, on which could be indicated the area used and the place where a photograph to document the inferred should be taken. To discourage hedged observations, a worksheet was also supplied in which *small* spaces were provided for observations on the tool type, the part of the tool used, macroscopic wear, motion of use, evidence of hafting and other comments. The volunteers were asked to be specific about the worked material, and if possible to submit only one answer.

The results from the examination of these ten tools by five microwear analysts, and by one technologist who attempted to infer their function with a hand lens only, are not very impressive (Table 30.1). As might be expected, the part of the tool used and the motion in use were more often correctly identified than the worked material. Tool number 7, an *unused* end-scraper made by holding the tool's dorsal surface on a hide palm pad and pressure retouched with an antler tine, is particularly depressing. Only one observer suspected that the scraper edge was unused, and of the three attempts to identify a worked material, all said 'hide', though the ventral edge of the tool had not touched the hide pad. One observer said that the tool had been used for about an hour.

Tool number 9, used to de-fin and scale a fish (pike-perch or zander), also yielded poor results; one person suggesting antler, another wood and another 'dryish plants'. It is worth remarking here that, when working on archaeological material, a wrong identification of worked material is much more misleading than a mere 'don't know'.

In Table 30.2 another way of scoring the results of test tools 1–10 is given, to see whether there is a correlation between scores on the test tools and amount of experience with microwear analysis. In this table one point has been awarded for a correct identification of the part used, another for the motion in use and a third for the worked material, giving a possible total of 30 points for each volunteer. (It is obviously difficult to score such tests objectively, and some conventions for part credits had to be established. If the worked material was antler and the answer was 'bone or antler' a partial credit was given. No points were substracted for wrong guesses.)

There is a rough correlation between experience and score in Table 30.2, but the high marks achieved by the technologist working with only a hand lens gives pause for thought. This person attempted to identify worked material in only one case, but nevertheless did very well in identifying both the area of the tool used and the movement in use. Most participants in this test admitted that their identification of worked material was influenced to some degree by the tool's morphology as well as by what Holley and Del Bene (1981, p. 346) call 'a suite of other attributes of wear'. For a tool such as the backed bladelet used to cut

Table 30.1. *Scores for test tools* 1–10. *Five observers, all using microwear analysis (technologist with hand lens excluded). Part credit given if, for example, answer submitted was 'bone/antler' and correct answer was 'antler'*

	Tool	Part used	Motion	Worked material
1	End-scraper on wood	4 (and 1 partial)	3 (and 2 partial)	1
2	Drill on shell	1	1	0
3	Piercer grooving antler	5	0	(2 partial)
4	End-scraper on stone	4 (and 1 partial)	4 (and 1 partial)	0
5	Burin scraping bone	4 (and 1 partial)	(1 partial)	(3 partial)
6	Backed blade shot into earth	4 (and 1 partial)	3	1
7	Unused end-scraper	0	0	(1 partial)
8	Backed bladelet cutting meat	3 (and 1 partial)	3	1 (and 3 partial)
9	Blade cutting and scaling fish	3 (and 1 partial)	1 (and 3 partial)	0
10	Burin scraping wood	4	3	(1 partial)

Table 30.2. *Scores for test tools* 1–10. *Observer awarded one point each for used area, use motion and worked material (possible total* = 30 *points)*

	Score		
Observer	Correct	Partially correct	Experience (years)
I	12	4	(Technologist with hand lens)
II	3	10	1
III	8	3	1.5
IV	12	3	2
V	15	2	1
VI	15	5	7

meat (test tool 8), this has important implications, since the identification of meat as a possible worked material, by four out of five analysts, was based more on a lack of edge damage and the tool's overall delicate design than on the presence of meat polish.

For tools 1–10, one of us (Grace) made an independent investigation of the polish characteristics, using the image processing form of texture analysis described in our previous paper (Grace, Grahame and Newcomer this volume). Flint texture analysis can be used to identify the texture (grey level) microwear polishes and to compare textures produced by different contact materials. For purposes of comparison the texture analysis data is expressed on a scatter diagram where unused flint textures are placed close to the origin, as in the previous paper (above p. 68 and Fig. 8.9).

In the earlier paper a small sample of well-developed polishes said to be characteristic of various contact materials were shown to be well separated in texture. The larger and more comprehensive sample available from the blind test experiments described above simulates more closely the range of microwear polishes found on archaeological material. The texture analysis was carried out in the same manner as in the previous investigation,[2] a more considerable number of unused flint textures were plotted to provide a fair comparison with the experimental samples (Fig. 30.2).

The scatter diagram clearly illustrates that all samples of unused flint textures cluster around the origin, indicating the distinctiveness of the unused flint textures from polish textures. It is equally obvious that polishes produced from contact with different materials are spread throughout the diagram and do not cluster into distinctive groups according to contact material. For example, point a is one of the polished areas most similar and point b is the polish most different from unused flint; but both of these polishes are from the same tool, used for sawing wood for 15 minutes. Both polishes are present on the mid-point of the used edge, the difference is that point a represents polish from the ventral surface and point b from the dorsal surface (Fig. 30.3). These polishes are the result of sawing, a use motion that would not be expected to produce different polishes on different surfaces as both were touching the wood throughout the task.

The texture analysis technique used for the scatter diagram would seem to be capable of clustering unused flint textures and distinguishing them from polished textures, even though there is considerable variation between these unused flint textures.

Among the samples of unused flint plotted in Fig. 30.2 are examples from three different sources, including English chalk flint from Potters Bar (Hertfordshire),

[2] In this paper the D statistic has been replaced by the ASM statistic which takes into account the problems of differential sensitivity of film emulsions discussed in the previous communication (see Grace, Graham and Newcomer 1986 for details).

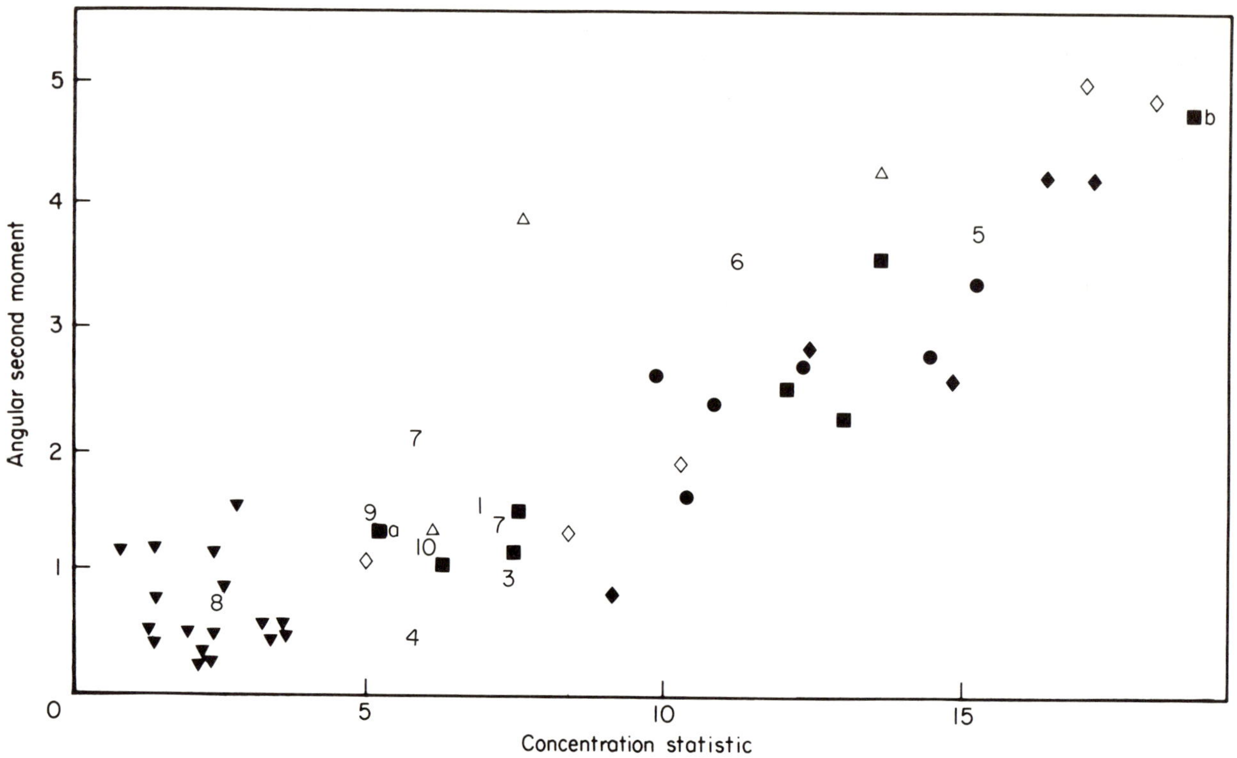

Fig. 30.2. Scatter diagram whose origin represents unused flint so that the positions of points represent the differences of polished areas from unused flint: ▼ unused; ■ wood; ● antler; ◆ bone; ◇ plant; △ hide; a and b, wood sawing. The numbers represent the positions of the test tools.

Newton and Brandon (Norfolk), as well as the Seine flint of the test tools. Therefore, if the technique is capable of clustering unused flint textures which would be expected to have significant variation among themselves, then, for example, antler polishes similar to each other and distinct from other polishes should cluster in the same way.

To test this assumption the difference values of various polishes from a model antler polish have been calculated in the same way as before, producing a scatter diagram in which the origin represents antler polish (Fig. 30.4). It can be seen that antler polishes do cluster to some extent; but are not mutually exclusive of other microwear polishes in the same way that unused flint textures are. The two polishes nearest to the origin and therefore most similar to model antler polish are in fact wood; and within the cluster of antler polishes there are examples of wood, plant and some bone polishes.

Using the two scatter diagrams (Figs. 30.3 and 30.4) we may examine the positions of the blind test tools Nos. 1–10, as distinguished by the texture analysis technique, and compare them with the results of the tests carried out by standard optical microwear analysis. Firstly, we compare the test identification of each peace with its relationship to unused flint textures as shown in Fig. 30.3.

Test tool number 1 was used for scraping wood for 15 minutes and clusters with some other samples of wood polish; even though visually it differs from them.

Tool 2, the drill bit, retained no polish, which perhaps never developed or was lost through edge damage; in any case, the tool could not be analysed.

Tool 3 was used for grooving antler and had only isolated areas of polish. The analysis of one of these small areas places the polish near to the same group that contains tool 1, rather than with antler polishes.

Tool 4 was used for scraping stone for 6 minutes. It does not cluster with other polishes and shows the least difference from unused flint of all the test tools having polish. This indicates a lack of polish development, which would make it difficult to identify, and no analyst could identify this polish.

Tool 5 was used for scraping bone for 11 minutes and clusters with the main group of antler/bone polishes. All the analysts identified this as bone or antler polish, though the nearest point on the diagram represents a polish produced by whittling wood for 20 minutes.

Tool 6 was used as a projectile point and has polish similar in texture to that on tool 5. Although the contact material is unknown the polish produced is of the same nature and intensity as bone/antler polish, but no conclusions can be drawn from this.

Tool 7 was unused and is the most problematical of the test tools. In most cases the microwear analysts stated that this tool was used, and three out of four analysts identified it as hide polish. The texture analysis found that certain areas of the flint along the retouched edge produced a surface that was significantly different from unused flint. An area of the

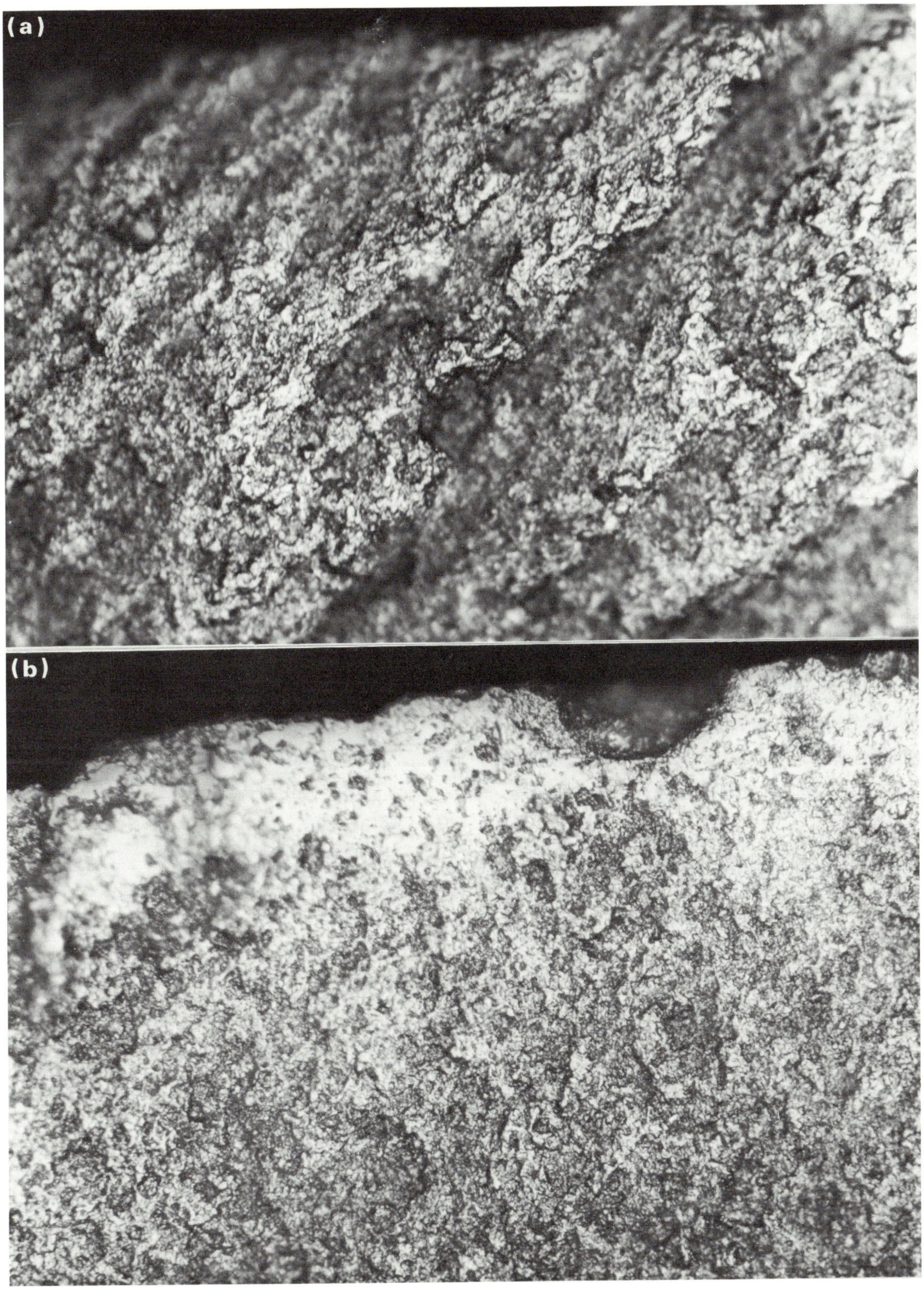

Fig. 30. 3. Polishes produced by sawing wood for 15 min.
(*a*) Polished area represented by point a in Fig. 30.2.
(*b*) Polished area represented by point b in Fig. 30.2.

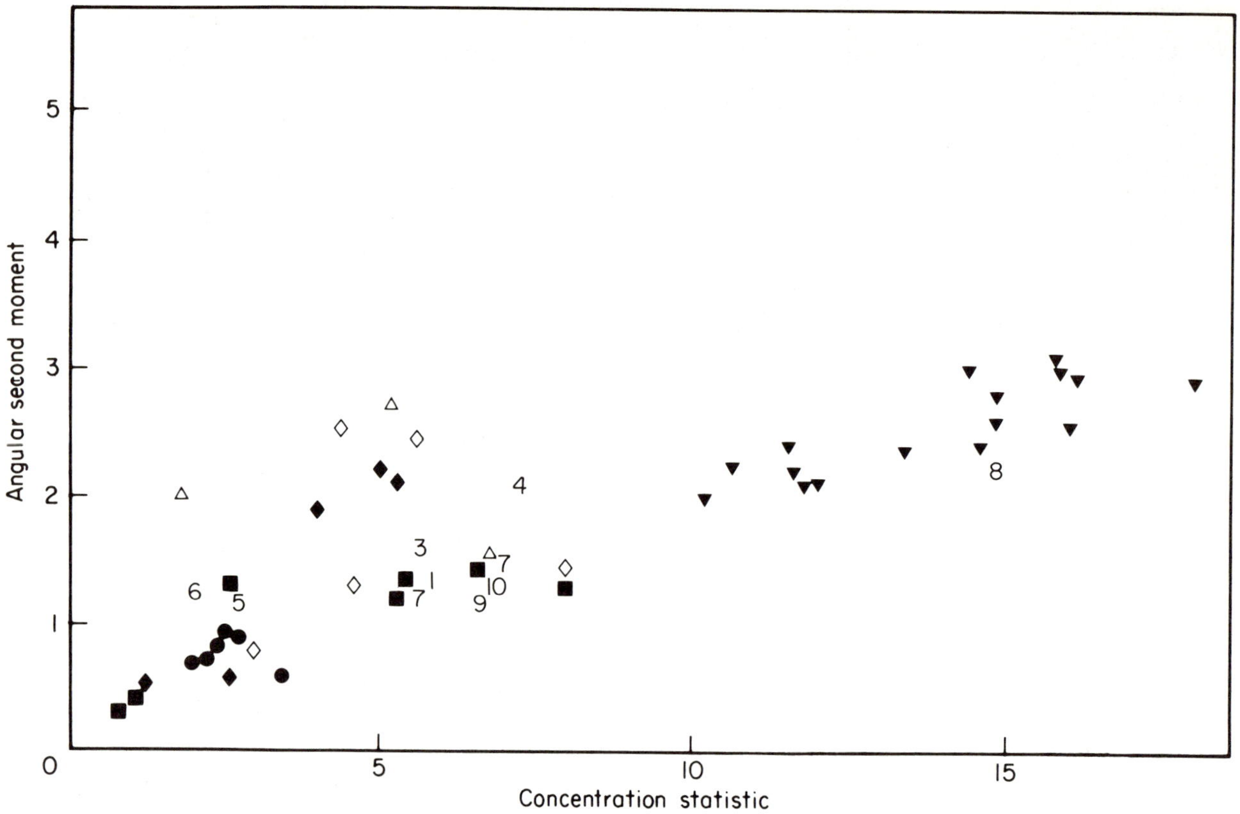

Fig. 30.4. Scatter diagram whose origin represents antler polish so that the position of points represent the difference of polished areas from model antler polish. Key as in Fig. 30.2. The numbers represent the positions of the test tools.

central portion of the ventral surface also had evidence of surface alteration from the texture analysis, though no polish was visible through the microscope.

Tool 8 was used for cutting meat for 29 minutes. This produced no significant surface alteration and so the piece clustered with unused flint. No characteristic polish was identified by the microwear analysts, but some did interpret the function of the tool correctly. This identification, however, was based on the morphology of the backed bladelet. The absence of a developed polish on a cutting tool suggests use on a contact material such as meat, that produces little or no polish (Fig. 30.5). The correct identification of this tool could not have been based on polish characteristics, as the texture analysis demonstrates that no significant surface alteration took place during use.

Tool 9 was used for cutting and scraping fish for 20 minutes, and clusters with wood polishes. This means, in terms of the texture analysis, that it had the same nature and intensity as wood polishes. In one case it was identified as wood polish in the test.

Tool 10 was used for scraping wood for 9 minutes and occurs in the same group as test tools 1 and 9, associated mainly with wood.

As we have shown above, polish textures do not cluster simply according to contact material, so the test tools would not be expected to cluster with polishes of the same contact material. The significance of their positions on the scatter diagram is that these demonstrate why incorrect identifications of polish can be made. For example, if a polish has little separation from unused flint this means the polish is insufficiently developed to exhibit characteristics that indicate the contact material. Tool 4, used for scraping stone, is such a case (Fig. 30.6). Furthermore, if two polish textures produced by contact with different materials are similar to each other, as in the case of tools 9 (fish) and 10 (wood), then the probability is that they will be identified optically as the same (Fig. 30.7). Of course, with prehistoric tools the inaccurate identification would go unnoticed. Only with experimental tools whose function is known is it possible to detect such errors!

These results of the texture analysis of the test tools are based on the texture of the polishes alone, and therefore the similarities and differences are due to polish features such as brightness, smoothness and the spatial arrangement of texture features within the scanned area. The fact that the scanned area is small but of very high resolution (see Grace, Graham and Newcomer 1986) means that sub-visual texture features are being measured, and it could be argued that features such as comet-shaped pits, edge rounding and reticular patterning would not be characterised by this method. However, these features are not sufficiently consistent in their association with a specific contact material to be of value for identification purposes. They are indicators of a possible function rather than being charac-

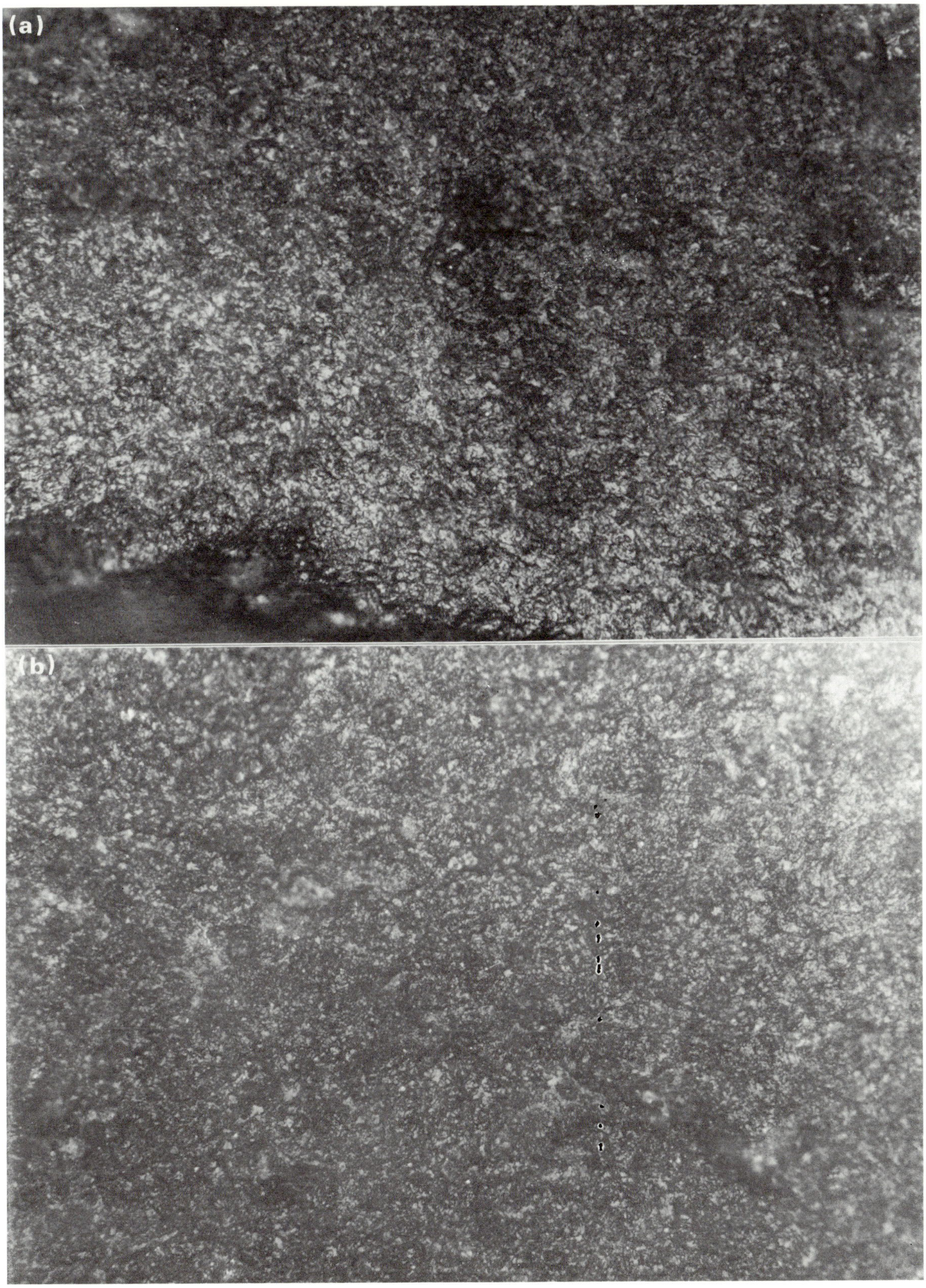

Fig. 30.5. Test tool 8 used for cutting meat for 29 min.
(*a*) Used area. (*b*) Unused area.

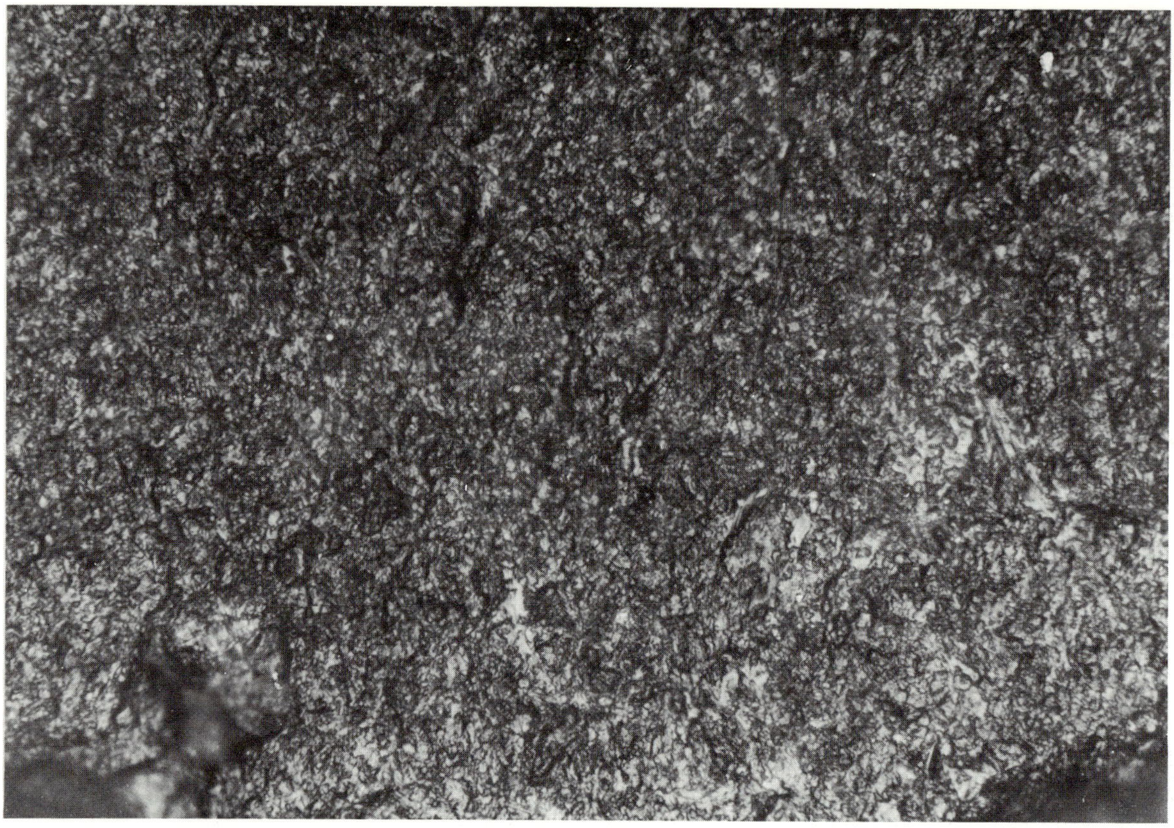

Fig. 30.6. Test tool 4 used for scraping limestone for 6 min.

teristic of a specific function. For example, the reticular patterning often associated with wood polish is caused by the original topography of the flint surface rather than being a product of contact with one material. A reticular pattern can be created on an uneven flint surface by a variety of materials. Conversely, with a smooth topography the same function will not produce a reticular pattern (Keeley 1980, p. 35). Edge rounding is specified as a characteristic of dry hide working (Keeley 1980, p. 50), but again it is possible to produce edge rounding with any material of sufficient hardness, as the rounded working edge of test tool 4, used to scrape limestone, clearly indicates. All the analysts noted that edge rounding was present and five out of six identified the tool as being used on hide, mainly on this supposedly diagnostic feature, as the polish on the tool was insufficiently developed to be identified. Therefore, we believe that these feature cannot be used as exclusive indicators of a specific material.

If features of attrition and intra-polish distributions are a function of the combination of the original flint topography and the hardness of the contact material, the problem becomes one of how to differentiate between materials that have a similar hardness. To use an example from the blind test: what is the difference in hardness between fresh wood and fish, as revealed by their relative effects on similar flint surfaces? The difference in hardness would only be detectable if the duration of use was held constant. If, as in the case of the blind test, a tool is used for scraping wood for 9 minutes, and for scraping fish for 20 minutes, then the relationship between hardness of the contact material and tool use duration could combine together to produce an equivalent abrasion factor, resulting in similar polishes being produced by contact with different materials.

One way in which tools having similar microwear textures can be separated according to contact material is by examining the characteristics of the used edges of tools for their suitability for use on materials of different hardness. When the group of textures that cluster around the origin of Fig. 30.4 are examined, they are separable according to contact material by their edge characteristics. For example, the tools used on bone and antler are burins, a piercer and the tip of a blade; as opposed to the backed blade used on plant material and the concave unretouched edges of flakes that were used on wood. This is not simply a correlation of tool type with function, as within the bone/antler tool group different tool-types (burins, piercers and blades) are represented. The tools are typologically different, but the worked edges on these tools have similar characteristics whether they occur on burin bits or on tips of blades. But a burin bit is simply not efficient for whittling wood, and a concave lateral edge of a flake cannot be used for grooving antler. Therefore the edge characteristics of tools can determine their functional capabilities; so that the function of tools having similar microwear polishes can be identified by their suitability for a particular task, in terms of their edge characteristics.

The importance of these edge characteristics and the morphology of the tools is demonstrated by the results

Fig. 30.7. (*a*) Test tool 10 used for scraping wood for 9 min.
(*b*) Test tool 9 used for cutting and scraping fish for 20 min.

Table 30.3. *Identification of worked materials rubbed for* 10 *min on ventral surface of flint flake*

Tool	Actual worked materials	Observer III	Observer IV	Observer V	Observer VI*
11	Hide	Hide	Dry hide?	Hide	—
12	Antler	Bone/antler	Soaked antler?	Bone	—
13	Wood	—	Fresh hide?	Plant	—
14	Antler	Bone	Antler/bone	Bone	—
15	Ferns	Unused	Wood?/fish?	Wood	—
16	Hide	Unused	Meat?/unused	Hide	—
17	Wood	Post-depositional damage/meat	Wood?/reeds	Plant	—
18	Bone	Bone/antler/wood	Bone?	Antler	—
19	Ferns	—	Reed/wood?	Wood	—
20	Bone	Antler	Bone/antler	Antler	—

* No attempts made as polishes too poorly developed.

Table 30.4. *Identifications of worked materials cut for* 10 *min with an unretouched edge*

Tool	Actual worked materials	Observer III	Observer IV	Observer V	Observer VI
21	Antler	Ferns	Bone	Ferns	Plant
22	Wood	Wood	Hide	Wood	—
23	Ferns	Antler	Antler	Bone	Antler
24	Bone	Bone/antler	Ferns	Antler	Bone
25	Hide	Hide	Hide	Hide	Hide
26	Ferns	Bone/antler	Bone	Antler	Bone
27	Antler	Antler/bone	Wood	Bone	Bone/antler
28	Hide	Wood	Ferns	Wood	Wood
29	Bone	Hide?	Antler	Hide	Bone
30	Wood	Wood?	Wood	Ferns/wood	Wood

obtained with test tool 8, whose function was determined as meat cutting, even though no polish was present.

In order to confirm our interpretation of the blind test data, two further sets of blind tests were undertaken by the same individuals. With the second set of test tools (Nos. 11–20), we attempted to concentrate on polish identification, eliminating the clues to worked material provided by tool shape and edge damage. The test tools were a series of unretouched flakes of Brandon flint whose bulbar surfaces had been rubbed against several types of contact materials. The volunteers were not told what materials the experimenter (Newcomer) would use, but did know that the ventral surface of each flake would be rubbed against a contact material for 10 minutes. The experimenter paired the flakes, so that each of the five worked materials (hide, antler, wood, ferns and bone) was rubbed against two flakes. The microwear analysts did not know this had been done, and only Observer 5 managed to pair the flakes successfully, although the materials inferred were mostly misidentified (Table 30.3). Observer 4's relative success with these tools was achieved by ignoring the conventionally defined polish characteristics and trying to assess the hardness or softness of the worked materials.

The results from test tools 11–20 raised further doubts about the distinctiveness of microwear polishes, but as some of the observers suggested that perhaps the nature of the contact (rubbing) was producing uncharacteristic polishes, so another ten test tools were produced for examination (Nos. 21–30). These were used in a more normal fashion, all ten being made from a block of Brandon flint and having one of their unretouched cutting edges used to slice the same paired materials as in tools 11–20. This time the volunteers were told what materials would be used, and even that they would be in pairs. Despite this, and despite the fact that the polishes seemed better developed than in test tools 11–20, no one was able to pair the five worked materials correctly (Table 30.4). (Due to pressure of time, none of the test tools 11–30 were subjected to textural analysis, nor did we ask the technologist to infer worked materials with the hand lens.)

The results of the work on test tools 21–30 are in some ways the most troubling of all. Despite knowing the extent of edge damage, the range of contact materials and the fact that they were paired, and despite the possibility of making and using their own copies of the test tools in similar flint, most admitted difficulty in matching their experimental polishes to the test pieces.

The result of the series of blind tests described above would seem to give less cause for optimism than did the Keeley–Newcomer test of 1977. In the earlier test the identification of worked materials also provided the most difficult task for the microwear analyst, but the conclusions then that 'in many cases' the material could be identified and that the test was 'partial confirmation of the distinctiveness of the various microwear polishes' (Keeley and Newcomer 1977, p. 61) seem rather optimistic to Newcomer today. In the present series of tests, there has been no convincing demonstration that anyone can consistently identify worked material by polish type alone; if someone could, then a well-developed wood polish, for example, would never be mistaken for anything else.

While the results of the experiments described above sugest that the accepted criteria for identification of specific contact materials are ambiguous, we do not mean to suggest that study of microwear polishes should be abandoned: broader distinctions between polishes characteristic of soft and hard contact materials may well be identifiable. We believe, however, that the identification of stone tool function should be based on a wide range of data, including evidence of stone tool morphology and edge damage as well as on the usual range of microwear features including polish. Where microwear polishes are concerned, we feel that more needs to be done, both to enable these to be characterised objectively by techniques such as texture analysis (Grace, Grahame and Newcomer this volume) and also to determine the nature of polish formation which is still open to considerable doubt.

Acknowledgements

While respecting their anonymity, we would like to thank all the volunteers who took part in the blind tests. The opinions expressed are of course our own. We owe a special debt to Roy Hodson for his continuing interest in this project and for much helpful advice.

Bibliography

Anderson-Gerfaud, P. (1981). *Contribution methodologique à l'analyse des microtraces d'utilisation sur les outils préhistoriques*. Thèse de 3me cycle, Université de Bordeaux.

Audouze, P., Cahen, D., Keeley, L. H. and Schmider, B. (1981). 'Le site Magdalénian du Buisson Campin à Verberie (Oise)'. *Gallia Préhistoire*, **24**(1), 99–143.

Büller, H. (1983). 'Methodological problems in the microwear analysis of tools selected from the Natufian sites of El Wad and Ain Mallaha. In Cauvin, M-C. (ed.) *Traces d'Utilisation sur les Outils Neolithiques du Proche Orient*, pp. 107–26. Lyon: Travaux de la Maison de l'Orient 5.

Cahen, D., Keeley, L. H. and Van Noten, F. L. (1979). 'Stone tools, tool kits and Human behaviour in prehistory'. *Current Anthropology*, **20**(4), 661–3.

Gendel, P. A., and Pirany, L. (1982). 'Microwear analysis of experimental stone tools: further test results'. *Studia Prehistorica Belgica*, **2**, 251–65.

Grace, R., Graham, I. D. G. and Newcomer, M. H. (1985). 'The Quantification of microwear polishes'. *World Archaeology*, **17**(1), 112–20.

Grace, R., Graham, I. D. G. and Newcomer, M. H. 1986. 'Preliminary investigations into the mathematical characterisation of wear traces on flint tools' (this volume).

Holley, G. A. and Del Bene, T. A. (1981). 'An evaluation of Keeley's microwear approach'. *Journal of Archaeological Science*, **8**, 337–52.

Keeley, L. H. (1980). *Experimental Determination of Stone Tool Uses: a microwear analysis*. Chicago: University of Chicago Press.

Keeley, L. H. and Newcomer, M. H. (1977). 'Microwear analysis of experimental flint tools: a test case'. *Journal of Archaeological Science*, **4**(1), 29–62.

Moss, E. H. (1983). *The Functional Analysis of Flint Implements: Pincevent and Pont d'Ambon: two case studies from the French final Palaeolithic*. British Archaeological Reports, International Series, no. 177. Oxford.

Odell, G. and Odell-Vereecken, F. (1981). 'Verifying the reliability of lithic use-wear assessment by "blind tests": the low-power approach'. *Journal of Field Archaeology*, **7**(1), 87–120.

Unger-Hamilton, R. (1984). *Method in Microwear Analysis: Sickles and Other Tools from Arjourne, Syria*. Ph.D. thesis. London University.

Van Noten, F. (ed.) (1978). *Les Chasseurs de Meer*. Dissertationes Archaeologicae Gandenses, XVIII. Ghent Belgium.

Vaughan, P. (1981). *Lithic Microwear Experimentation and the Functional Analysis of a Lower Magdalenian Stone Assemblage*. Ph.D. dissertation, University of Pennsylvania.

UNIVERSITY
OF BRISTOL
LIBRARY

87 01340